Routledge Handbook of Latin America and the Environment

The *Routledge Handbook of Latin America and the Environment* provides an in-depth and accessible analysis and theorization of environmental issues in the region. It will help readers make connections between Latin American and other regions' perspectives, experiences, and environmental concerns.

Latin America has seen an acceleration of environmental degradation due to the expansion of resource extraction and urban areas. This Handbook addresses Latin America not only as an object of study, but also as a region with a long and profound history of critical thinking on these themes. Furthermore, the Handbook departs from most treatments on the topic by studying the environment as a social issue inextricably linked to politics, economy, and culture. The Handbook will be an invaluable resource for those wanting not only to understand the issues, but also to engage with ideas about environmental politics and social-ecological transformation. The Handbook covers a broad range of topics organized according to three areas: physical geography, ecology, and crucial environmental problems of the region. These are key theoretical and methodological issues used to understand Latin America's ecosocial contexts, and institutional and grassroots practices related to more just and ecologically sustainable worlds.

The Handbook will set a research agenda for the near future and provide comprehensive research on most subregions relative to environmental transformations, challenges, struggles and political processes. It stands as a fresh and much needed state of the art introduction for researchers, scholars, post-graduates and academic audiences on Latin American contributions to theorization, empirical research and environmental practices.

Beatriz Bustos is an associate professor in the Department of Geography at the University of Chile. Her research focuses on resources geography and the sociopolitical transformations that exploitation of natural resources produces in rural communities. Her work ranges from examining the geography of commodities such as salmon, copper, wine, agro-industries, coal, lithium and green hydrogen, to rural livelihoods under neoliberal extractive economies. More recently she is researching ideas of rural citizenship.

Salvatore Engel-Di Mauro is Professor at the Department of Geography and Environmental Studies of SUNY New Paltz (US), teaching courses on physical and people–environment geography as well as on socialism. He is Senior Editor for *Capitalism Nature Socialism* and Reviews Editor for *Human Geography*. He has recently published *Socialist States and Environment* and, with George Martin, *Urban Food Production for Ecosocialism*. His research areas include soil contamination and acidification, urban food production, and socialism and the environment.

Gustavo García-López is an engaged researcher, educator, and apprentice organizer, from the islands of Puerto Rico. He has experience in transdisciplinary social-environmental studies. His work is situated at the intersection of ecology and the political, postcolonial/decolonial, and Latin American and Caribbean studies. He engages with commons and commoning, autogestion, and environmental and climate justice issues. He is part of the JunteGente collective in Puerto Rico, the Post-Extractive Futures initiative, the Climate Justice Network and the Undisciplined Environments blog. He lives uprooted from his lands but finding home and guiding stars in his daughter Maia. He is held in life by broad networks of care and nourishment, of people, spirits, memories, and ecologies.

Felipe Milanez is Assistant Professor at the Institute for Humanities, Arts and Sciences Professor Milton Santos, at the Federal University of Bahia, Brazil. Author of *Memórias Sertanistas: Cem Anos de Indigenismo no Brasil* and *Guerras da Conquista*, with Fabrício Lyro, his work and activism focus on the violence against environmental defenders, the genocide of indigenous peoples and ecocide. More recently, his research dedicates to learn with indigenous art, anti-colonial epistemologies and political ecologies from Abya Yala.

Diana Ojeda is Associate Professor at Cider (Center for Interdisciplinary Development Studies) at Universidad de los Andes in Bogotá, Colombia. Her work analyses processes of "green grabbing," dispossession, and state formation from a perspective that combines feminist political ecology and critical agrarian studies. More recently, her research has focused on pesticide use in Colombia.

Routledge Handbook of Latin America and the Environment

Edited by Beatriz Bustos,
Salvatore Engel-Di Mauro,
Gustavo García-López, Felipe Milanez,
and Diana Ojeda

Routledge
Taylor & Francis Group
LONDON AND NEW YORK

Cover image: Pau-Brasil. Arissana Pataxó, 2020

First published 2023
by Routledge
4 Park Square, Milton Park, Abingdon, Oxon OX14 4RN

and by Routledge
605 Third Avenue, New York, NY 10158

Routledge is an imprint of the Taylor & Francis Group, an informa business

British Library Cataloguing-in-Publication Data
A catalogue record for this book is available from the British Library

Library of Congress Cataloging-in-Publication Data
Names: Bustos, Beatriz (Bustos Gallardo), editor. | Engel-Di Mauro,
Salvatore, editor. | García López, Gustavo A., editor. | Ojeda, Diana, editor. |
Milanez, Felipe, editor.
Title: Routledge handbook of Latin America and the environment /
edited by Beatriz Bustos-Gallardo, Salvatore Engel-di Mauro, Gustavo García-López,
Felipe Milanez, Diana Ojeda.
Other titles: Handbook of Latin America and the environment
Description: Abingdon, Oxon ; New York, NY : Routledge, 2023. |
Series: Routledge international handbooks | Includes bibliographical references and index.
Identifiers: LCCN 2022053844 (print) | LCCN 2022053845 (ebook) |
ISBN 9780367361860 (hardback) | ISBN 9781032478364 (paperback) |
ISBN 9780429344428 (ebook)
Subjects: LCSH: Latin America--Environmental conditions. |
Environmental degradation--Latin America. | Environmental sociology--Latin America. |
Environmental economics--Latin America. | Environmental justice--Latin America. |
Environmental policy--Latin America.
Classification: LCC GE160.L29 R68 2023 (print) | LCC GE160.L29 (ebook) |
DDC 363.70098--dc23/eng20230328
LC record available at https://lccn.loc.gov/2022053844
LC ebook record available at https://lccn.loc.gov/2022053845

ISBN: 978-0-367-36186-0 (hbk)
ISBN: 978-1-032-47836-4 (pbk)
ISBN: 978-0-429-34442-8 (ebk)

DOI: 10.4324/9780429344428

Typeset in Bembo
by SPi Technologies India Pvt Ltd (Straive)

To our chosen families who supported us through these times.

For all the environmental defenders and community leaders who were killed defending Pachamama, and to those who every day share their labor, care, and commitment to support the network of life in our Abya-Yala, whose struggles and experiences inspired and mobilized the ideas in this book.

For a green future.

Por justicia, dignidade y kidunguenewn nguen a todes.

Contents

Figures

Tables

Contributors

Alberto Acosta is an Ecuadorian economist and university professor. He was a Member of the International Tribunal for the Rights of Nature, Minister of Energy and Mines (2007). President of the Constituent Assembly (2007–2008) and candidate for the Presidency of the Republic (2012–2013). Partner of struggles of social movements inside and outside his country. Author of several books

Iñigo Arrazola Aranzabal is a Doctoral student of geography at the Federal University of Bahia (UFBA), Brazil. His work and interests orbit around the neoliberal re-structuration processes of agribusiness, the socio-territorial conflicts it generates, its consequences for the least privileged actors along the agri-food production networks, and the type of responses communities develop against them. Member of the Collective of Critical Geography of Ecuador and currently working with the Comissão Pastoral da Terra in the state of Bahia, Brazil, in land and water struggles, with a special focus on how beliefs and popular religiosity frames the way people and local actors orientate through these conflicts.

Andrés León Araya holds a PhD in Anthropology from the City University of New York (CUNY) and is the current director of the Center for Political Research (CIEP) of the University of Costa Rica, where he also teaches in the Political Science and Anthropology departments. His research focuses on the intersections between land and political power in the process of state formation in Central America. Also, he is starting a new research project on the transnational technologies of the coup d'état in Latina America.

Tatiana Roa Avendaño is an environmentalist, educator, researcher and gardener, and a member of CENSAT Agua Viva/Friends of the Earth Colombia. She is an engineer, holds a Master's degree in Latin American Studies from the Universidad Andina Simón Bolivar (Quito, Ecuador), and is currently a PhD candidate in Political Ecology at the Center for Latin American Research and Documentation (CEDLA), University of Amsterdam. She has written numerous articles and books on extractivism, water justice, and energy sovereignty. As an activist she is a member of Oilwatch, the Alianza Colombia Libre de Fracking, and the Mesa Social Minero Energética y Ambiental por la Paz. She has also collaborated with diverse research groups, such as Alianza por la Justicia Hídrica, and the Latin American Social Sciences Council (CLACSO) Working Groups on Political Ecology of Abya-Yala and Territorialities in Dispute and Re-existence. She has been a member of the international committee of *Ecología Política* magazine for more than a decade.

Katia R. Avilés-Vázquez is a researcher, partner in many struggles, and manager of projects related to agroecology and climate justice. She holds a PhD in Geography from the University of Texas at Austin, where she studied Cultural Ecology and the Politics of Small Farmers in Puerto Rico. Her research highlights community-based adaptations, and she approaches the topic from a grassroots activism perspective. She has intentionally only co-authored articles, book chapters, and white papers, as part of her commitment to knowledge as a collective creation effort. She managed the Caño Martín Peña Ecosystemic Restoration Project for five years, which led to the approval of the public disclosure of the project's environmental documents. She directed the Model Forest Office. After María, she has focused her work on developing local capacities and distributing resources for local entities. Her work and activism have been highlighted in local and international media outlets, including *The Intercept*, *Democracy Now* and *The Guardian*, and she was the recipient of the EPA Environmental Champion Award and the ESF Graduate of Distinction Award. Currently, she directs the Institute of Agroecology, which works to promote research and secure resources for the protectors and managers of the land, particularly agroecological producers.

Denilson Baniwa, was born in Mariuá, Rio Negro river, Amazonas. He is an indigenous artist, he is indigenous and artist, and his being indigenous leads him to invent another way of making art. He is a visual artist and communicator with artistic and social processes based on the Amazonian Indigenous Movement and his transit through the non-Indigenous universe. As an activist for the rights of indigenous peoples, he gives lectures, workshops, and courses. He is also an advertiser, creator of digital culture and hacker, contributing to the construction of indigenous imagery in different media such as magazines, films and TV series. @denilsonbaniwa

Gerónimo Barrera de la Torre is a PhD student in the Institute of Latin American Studies at the University of Texas at Austin. His research focuses on the political ecology of forest conservation, historical geography, post-statist geographies, and critical cartographies. He has worked with Indigenous Chatino people and campesino communities in Oaxaca, México, regarding local knowledges about their environment and Chatino understanding of their landscape through their language. He is working with the same communities looking into the conservation and management programs of their communal forests and codirecting a collaborative documentary on carbon offsetting projects in the region.

Diogo de Carvalho Cabral is an Assistant Professor in Environmental History at Trinity College Dublin. Before that, he was a British Academy-funded Newton International Fellow based at the Institute of Latin American Studies/School of Advanced Study, University of London. His academic awards include the Journal of Historical Geography Best Paper Prize (2016) and an honourable mention in the Milton Santos Prize (2017). He is the author of *Na Presença da Floresta: Mata Atlântica e História Colonial* (2014) and *Metamorfoses Florestais: Culturas, Ecologias e as Transformações Históricas da Mata Atlântica*, co-edited with Ana Bustamante (2016).

Alejandro Camargo is an Assistant Professor in the Department of History and Social Sciences at Universidad del Norte in Barranquilla, Colombia. He is interested in nature–society relations in riverine environments, agrarian change, the history of development, climate change adaptation, and environmental conflicts.

Juan Antonio Cardoso is a cultural anthropologist currently working in the development sector. His areas of interest include environmental history, political ecology, and the anthropology of development. He has conducted research on social elites and the production of nature in the Colombian Caribbean region.

María Soledad Castro-Vargas is a PhD Candidate at Institut de Ciència i Tecnologia Ambientals, Universitat Autònoma de Barcelona (ICTA-UAB). Her research is focused on analysing the relationship between chemicals, socio-natures and uneven development.

Combining approaches from political ecology and ecotoxicology, her doctoral thesis is grounded in the case of the pesticide-contaminated wetland Térraba-Sierpe in Costa Rica. The objective of her research is to understand why and how pesticide-contaminated water-scapes are being produced in Costa Rica in relation with the pesticide complex. She holds a BSc on Natural Resources Management and an MSc in Water and Coastal Management.

Elga Vanessa Uriarte-Centeno is the Executive Director of Amigxs del M.A.R., an environmental organization that promotes environmental and climate justice from the radical transformation of the economic system and the eradication of the colonial policies of the countries of the global North, with which she has worked for ten years. She holds an interdisciplinary Bachelor's degree in Social Sciences with an emphasis on Latin American and Caribbean Studies, and a Master's degree in Cultural Management and Administration from the University of Puerto Rico. She is currently pursuing a Juris Doctor degree at the Interamerican University of Puerto Rico School of Law. She has worked in multiple social projects and non-profit organizations as an artist and/or coordinator. She has excelled in environmental movements, as a popular educator in human rights and in the rescue of urban spaces through the creation of cultural spaces.

The Colectivo de Geografía Crítica del Ecuador was created in 2012 in response to the urgent need to produce geographical knowledge and resources from a critical perspective. The collective brings together geographers, social scientists, and activists concerned with radical geography practice and discourse of understanding the production and appropriation of space from the perspective of different actors and logics, making visible the ways in which territories and territoriality are marginalized by capitalism, patriarchy and colonialism. From this perspective, the group collaborates with ecologist, indigenous and feminist organizations that resists capitalist dispossession in its various forms.

Paul Cooney is a professor in the Economics Department at the Catholic University (PUCE) in Quito, Ecuador.

Ximena Insunza Corvalán is a Lawyer at the Universidad de Chile, with an LLM from McGill University. She holds a Master's in Public Policy from the Faculty of Economics and Business, University of Chile, is an Assistant Professor in the Department of Economic Law and a researcher at the Center for Environmental Law, Faculty of Law, University of Chile. Alternate Minister of the Second Court of Santiago. She has been a PhD student in Law, Universidad de Chile, from 2017 to date. Her latest publications are (co-authored with Cordero Luis), "Citizen participation in air decontamination plans. An evaluation without romance," *Journal of Environmental Law*; and (co-authored with Bustos, Beatriz and Yasna Contreras) *Policy Brief: The National Rural Policy 2020: Contributions to the Constituent Process* (2021).

Raquel Echenique Llovet is a Chilean illustrator. She was born in Catalonia in 1977 and after spending her childhood and adolescence in France, she moved to Chile, where she graduated as a designer at the Catholic University. She is the daughter and granddaughter of artists, with her powerful colors and strokes she has illustrated more than thirty books in different publishers. She was part of the Siete Rayas collective, a pioneer in the dissemination and appreciation of illustration in Chile. She has received various recognitions including the 2012 Hummingbird Medal, Fundación Cuatrogatos 2017 award, Banco del Libro de Venezuela award, the Chilean Amster-Coré award, and the IBBY 2019 Honor List. She has participated in various group shows and in 2015 she exhibited at PLOP! Gallery with her exhibition "Bandada". She continues to be dedicated to book illustration, perfecting her trade day by day.

Vanessa Empinotti is Assistant Professor in Rural Policy and Planning in the Spatial Planning Program at Federal University of ABC-UFABC, in Brazil. An agronomist by training, she completed her PhD in Geography at Colorado University in Boulder. She studied participatory practices and water governance at the São Francisco River Basin in the northeastern region of Brazil. Her research employs Critical Political Ecology to analyze institutional arrangements, environmental governance, and power relations. She has been studying drought and its influence over water governance in megacities, water security in marginalized settings, and rural/urban relations in territorial planning. She is also the director of eco.t—Political Ecology, Planning and Territory Research Group and co-coordinates the Water Insecurity Network in Brazil as part of a regional HWISE (Household Water Insecurity Experiences) initiative.

Fernando Estenssoro Saavedra is an academic in the Institute for Advanced Studies of the University of Santiago de Chile. Doctor in American Studies; Magister in Political Science; Degree in History; Geographer. He is a specialist in contemporary world politics and geopolitics. In recent years, he has researched the geopolitical dimension of the environmental debate. He is the author of numerous articles and books. Among his main works are: *Global Environmental Geopolitics of the XXI Century. The Challenges of Latin America* (2019); *History of the Environmental Debate in World Politics, 1945–1992: The Latin American Perspective* (2014); *Environment and Ideology: Background for a History of Political Ideas at the Beginning of the XXI Century* (2009).

Georges F. Félix is an Assistant Professor within the Centre for Agroecology, Water and Resilience (CAWR), based at Coventry University (England, UK). At CAWR, he co-chairs the Stabilisation Agriculture Programme, teaching and conducting research on the reconstruction of farming systems in the aftermath of disasters, with a geographical scope that includes extensive fieldwork in regions of Europe, Africa, the Americas, and the Caribbean. He is also co-founder of Collective Cultivate!, is linked to SOCLA, and has been a lecturer at the University of Puerto Rico at Utuado.

Pabel Camilo López-Flores has a PhD in Sociology (from the Scuola Normale Superiore & University of Milan 'Bicocca', Italy). Currently, he is an Associate Researcher in the Postgraduate in Development Sciences, Universidad Mayor de San Andrés CIDES-UMSA (Bolivia) and a Postdoctoral Researcher at the Institute of Latin American and Caribbean Studies IEALC-UBA, University of Buenos Aires (Argentina). His research topics are mostly situated in the fields of political sociology and the sociology of territory and of social movements, in an interdisciplinary perspective. He is a member of the Working Group: "Territorialities in dispute and re-existence" of the Latin American Council of Social Sciences (CLACSO).

María Juliana Flórez Flórez is Associate Professor at the Instituto de Estudios Sociales y Culturales Pensar (Pontificia Universidad Javeriana, Bogotá). She studied psychology (Universidad Católica Andrés Bello, Caracas), a Cooperation and Development specialization (Universidad de Barcelona) and a Critical Social Psychology Master's and PhD (Universitat Autònoma de Barcelona). From a feminist and decolonial perspectives, her research interests are around social movements, communitarian economy and post-developmental alternatives. She is the author of *Lecturas emergentes. El giro decolonial en los movimientos sociales* (Vol. I) and *Subjetividad, poder y deseo en los movimientos sociales* (Vol. II) (2015).

Maria Fragkou is Associate Professor in the Department of Geography at the University of Chile. She is one of the professors in charge of the Political Ecology and Socioenvironmental conflicts laboratory of her department, and one of the founding members of the Latin

American Conference on Political Ecology. Her research focuses on the construction of water scarcity through a multi-scale analysis of national water policies, local and regional development dynamics, and domestic water uses and habits. She studies cities and urban phenomena from an urban metabolic perspective, combining quantitative and qualitative takes on urban socio-environmental problems. She is interested in issues of urban sustainability, water management, energy policies, and water and environmental justice.

Rodrigo Fuster is an Associate Professor in the Department of Environmental Sciences and Renewable Natural Resources at the University of Chile. He teaches Integrated Water Resources Management, and researches on regional aspects of water management, using a complex systems approach. He is an agronomist, with a Master's and a PhD in Environmental Science from the Autonomous University of Barcelona. As a teacher, he has directed several undergraduate and graduate theses. His work has focused interest in understanding the territorial conflicts associated with water and the development of tools to support decision-making under the institutional framework of the Chilean model of water management, conducting research in different basins throughout Chile.

Sandra Rátiva Gaona is a Colombian environmentalist and activist, a working partner within the Cooperative of Renewable Energies (Onergia) in Mexico, and a doctoral student in the Graduate Program of Sociology at the Institute of Social Sciences and Humanities of the BUAP, Mexico.

Paola Bolados García has a degree in Philosophy from the University of Salvador, Argentina, and a PhD and a Master's degree in Anthropology from the Catholic University of the North, Chile. She works as an Academic at the Faculty of Social Sciences and as a Researcher at the Center of Advanced Studies at the University of Playa Ancha. Her research interests address the subjects of extractivism, socio-environmental conflicts, feminist political ecology, and ecofeminism. Her recent research and publications focus on water problems in Chile, as well as in the so-called Sacrifice Zones, where the dimensions of gender and feminism have acquired particular relevance.

Carol E. Ramos Gerena studies topics related to agroecology, critical food systems education, participatory planning, land use policies, and food sovereignty. Carol herself has worked in governmental and non-governmental organizations that support community development projects in her country. For nearly a decade, she has promoted agroecology and collaborated on environmental restoration in Puerto Rico. Ella Carol is currently pursuing a PhD in urban and regional planning at the University at Buffalo-SUNY with a focus on food systems planning. Carol completed her master's degree in Environmental Planning at the University of Puerto Rico, focused on the sustainable planning of agroecological initiatives in K-12 public schools. His bachelor's degree in Biology was completed at the Mayagüez Campus of the University of Puerto Rico. Currently, she works at the UB Food Lab as coordinator of a two-city action research initiative to promote urban agriculture policies designed by and for communities of color in Buffalo and Minneapolis.

Chachi González (she) is muralist, co-director and one of the original 8 founders of Colectivo Moriviví. Moriviví is a collective of women artists who produce public and community art from Puerto Rico and together with the Puerto Rican diaspora in the US. Their work intends to democratize art and bring the narratives of Puerto Rican communities into the public sphere to create spaces in which they are validated. Independently, Chachi is an interdisciplinary artist under the pseudonym s.alguien, with a strong background in Engraving, Sculpture, Drawing and Painting. @arte.s.alguien

Liza Grandia has collaborated with environmental, social, and agrarian justice movements in the Maya lowlands and lived for seven years in remote Q'eqchi' communities of northern Guatemala and Belize. Her first two books document land grabs in this region. Since 2012, she has been an Associate Professor of Native American Studies at UC Davis, where she directs the Indigenous Research Center of the Americas. In 2017–19 she was awarded a Mellon Foundation "New Directions" fellowship to pursue studies in toxicology and environmental epidemiology for her new work on pesticides and other environmental hazards.

Edwin A. Hernández-Delgado is a Marine Biologist, with a M.Sc. in Environmental Microbiology of Tropical Waters, and a Ph.D. in Tropical Biology with a specialization in Coral Reef Ecology. He is a Professor and Affiliate Researcher at the Interdisciplinary Program and at the Department, Environmental Sciences at the University of Puerto Rico, Río Piedras Campus. He is a Research Fellow of the Center for Applied Tropical Ecology and Conservation at UPRRP and a Senior Scientist at Sociedad Ambiente Marino. His research interests include ecological restoration, and human and climate change impacts on coral reef, seagrass and fish communities.

Felipe Irarrazaval is postdoctoral researcher and adjunct professor at the Instituto de Estudios Urbanos y Territoriales at the Pontificia Universidad Católica de Chile and at the Center for the Study of Conflicts and Social Cohesion (COES). He received a PhD in Human Geography from The University of Manchester. His research examines resource governance, particularly extractive industries in Latin America, through the lens of global production networks, political geography, and urban studies. He has researched in Peru, Bolivia, and Chile, and has published his research in journals, including *Economic Geography*, *Political Geography*, *Annals of AAG*, *Capitalism Nature and Socialism*, and *EURE*.

Ana Isla is a Professor Emerita, Brock University, Canada. Her doctoral dissertation examined the structure and the functioning of the complex Canada–Costa Rica debt-for-nature investment relationship and the projects developed by two non-governmental organizations. She conducts research on two areas: subsistence economies in the Amazonia rainforest of Peru, and mining in Latin America.

Manuel Bayón Jiménez is Master in Urban Studies from FLACSO-Ecuador and in Human Rights from UNED, Spain. He is currently coordinating the Contested Territories Amazonía research project at FLACSO and PhD Student at University of Leipzig. He worked at the National Strategy Centre for the Right to Territory (Cenedet), and in Acción Ecológica as a researcher of the Route for the Truth and Justice for Nature and Peoples. He has published papers in different indexed journals (*Antipode*, *City*, *Geoforum*, *Iconos*, *Political Ecology*), and co-authored the book *The Jungle of White Elephants* (2017). He is also a member of the Collective of Critical Geography of Ecuador.

Ailton Krenak is a prominent indigenous thinker and philosopher, and the author of five books, among them *Ideas to Postpone the End of the World*, professor honoris causa at the Federal University of Juiz de Fora.

Amalia Leguizamón is Associate Professor of Sociology and core faculty at the Stone Center for Latin American Studies at Tulane University. Her research examines the political economy of the environment in Latin America, particularly Argentina's swift agrarian transformation based on the early adoption and intensive implementation of genetically modified soybeans. Her work has been published in *The Journal of Peasant Studies*, *Latin American Perspectives*, and *Geoforum*, among other outlets. She is also the author of *Seeds of Power: Environmental Injustice and Genetically Modified Soybeans in Argentina* (2020).

Nashieli Rangel Loera is an anthropologist. She was born in Guadalajara, Mexico, where she graduated in Sociology with a specialization in Latin America. She obtained her master's degree (2004) and her doctorate (2009) in Social Anthropology from the State University of Campinas-Unicamp, Brazil. Currently, she is a professor and researcher at the Department of Anthropology of the Philosophy and Humanities Institute of Unicamp. Her research interests center around issues related to the way of life of peasant populations, social movements, territorialities, the state, and the social effects of contemporary collective claims.

Angus Lyall is a political and economic geographer located in Quito, Ecuador. His ethnographic research explores uneven development in Latin America through the lens of resource governance and rural livelihoods and labor. In particular, his work has focused on oil governance in the Amazon and labor regulation in Andean agroindustrial enclaves. He is a member of the Collective of Critical Geography of Ecuador and teaches critical social and spatial theory at the Universidad San Francisco de Quito (USFQ).

Rocío Silva-Santisteban PhD, is a Principal Professor at Antonio Ruiz de Montoya Univerity, Peru. She is also a writer, journalist, and human rights activist. Rocío is the former Executive Director of the National Human Rights Coordinator and is a prominent public commentator on issues of human rights, culture, and the environment.

Daniela Manuschevich is Assistant Professor in the Department of Geography at the Universidad de Chile in Santiago. Her background is in biology and ecosystem ecology conservation. She received her PhD from the College of Environmental Science and Forestry of the State University of New York, where she held a Fulbright Fellowship. She is conducting research on land use change and tree-farm expansion. Recently, she has been working on how peasants live at the margins of expanding tree farms in the Araucanía region, focusing on land use modeling and forest policy.

Givânia Maria da Silva is a Substitute Professor at the University of Brasília (UnB). She holds a Master's degree in Public Policy and Education Management (2012) from the UnB and is currently a doctoral student in Sociology at the same university. She is Associate Researcher of the Association of Black and Black Researchers (ABPN), the Center for Afro-Brazilian Studies (Ceam/UnB/Brazil) and Geppherg/UnB, and a researcher of the research group Cauim/UnB. She is also a Co-founder of the National Coordinator of Articulation of Quilombos (Coordenação Nacional de Articulação de Quilombos—CONAQ).

Diana Carolina Murillo Martín is a Doctoral candidate in Sociology of the social and political policies at the University of Zaragoza (Spain) and a member of the Collective of Critical Geography of Ecuador. Her research interests focus on political ecology and critical geography. She has worked in socio-environmental issues in Ecuador and Colombia and she is currently studying the territorial tensions in relation to the páramos' delimitation policy (Andean high mountain ecosystem) in Colombia.

Finn Mempel is a PhD researcher at the Institute of Environmental Science and Technology (ICTA) at the Autonomous University of Barcelona. His research focuses on the historical transformations in the geography and social metabolism of the global soybean complex, as well as the discourses shaping governance interventions in soybean production and trade.

María Gabriela Merlinsky has PhDs in Social Sciences (from UBA) and in Geography (from Paris 8). She is a Senior Researcher at CONICET, based at the Gino Germani Research Institute (UBA), and coordinator of the Environmental Studies Group of the Urban Studies

Area (GEA-AEU-IIGG). She is a Regular Professor at the Faculty of Social Sciences of the University of Buenos Aires. She has published books, book chapters, and articles in specialized journals on vulnerability, risk and the city; the political ecology of water; and collective action, environmental conflicts and public policies.

Iliana Monterroso is a Scientist and Coordinator of Gender and Social Inclusion Research with the Equal Opportunities, Gender, Justice, and Tenure team at the Center for International Forestry Research (CIFOR), Guatemala. She is a member of the gender coordination team of the CGIAR Research Program on Forests, Trees, and Agroforestry.

Talita Furtado Montezuma is a Professor of Law at the Universidade Federal Rural do Semi-árido in Brazil.

Mariana Mora is Associate Professor-Researcher at the Center for Research and Advanced Studies in Social Anthropology (CIESAS) in Mexico City. She holds a PhD in Anthropology from the University of Texas at Austin. Her scholarship is situated within critical race theories, gender studies, decoloniality, and the political. She is author of the book, *Kuxlejal Politics: Indigenous Autonomy, Race and Decolonial Research in Zapatista Communities* (2018). She is also part of the continental Anti-Racist Action Research Network (Red Investigación Acción Anti-Racista, RAIAR), the Collective to Eliminate Racism in Mexico (Copera) and the Decolonial Feminist Network in Mexico.

Mina Lorena Navarro is a research professor at the Graduate School of Sociology of the Institute of Social Sciences and Humanities of the BUAP, Mexico.

Grettel Navas is a Postdoctoral Researcher at the Institute of Environmental Science and Technology, Universitat Autònoma de Barcelona (ICTA-UAB). Her current research interests are political ecology, environmental justice, pollution, and public health. Grettel is part of the Direction and Coordination Group of the global Environmental Justice Atlas (EJAtlas) that documents environmental conflicts and resistance movements worldwide and a member of the Latin-American working group 'Ecología(s) política(s) desde el Sur/Abya-Yala' from CLACSO.

Vlocke Negro is an engraver and urban artist from Mexico City. His recurring themes have to do with social and socio-environmental movements unleashed by neocolonialism in Latin America. He is currently developing a series of graphics and collages where he breaks down the official images of the banknotes and creates a new discourse where ideological and material practices such as colonization and extractivism are revealed. He has exhibited his work in cities such as Bogotá, Quito, Mexico City and Berlin. @vlocke_negro

Jesús J. Vázquez-Negrón is a Puerto Rican organizer, advocate, popular educator, and activist that works in the intersections of environmental justice, agroecology, food sovereignty, and climate justice at the national and international levels. He has been working collectively for the past 12 years with rural, urban, and coastal communities organizing mutual support efforts, political education workshops, dialogues, capacity training, and just recovery initiatives with family farms where people work and live. He is the National Coordinator of Organización Boricuá of Ecological Agriculture of Puerto Rico, a 30-year-old national platform composed by farmers, peasants, farm workers, and is also a food sovereignty activist who promotes and practices agroecology as a tool to achieve food sovereignty and social justice on the archipelago. He works and collaborates internationally in the Latin American Coordinator of Rural Organizations, the Food Sovereignty Alliance, La Vía Campesina and the Climate Justice Alliance.

Raquel Neyra is an economist and doctor in sociology. She is an activist, advising civil institutions in Cajamarca in environmental defense, as well as being a researcher in socio-environmental conflicts, violence, and coloniality in Peru, with a focus on political ecology and ecological economics. She is also a visiting professor of the doctorate in Economics at the Universidad Nacional Agraria la Molina, a collaborator with the EnvJustice project of the ICTA-UAB with the cases of Peru and a member of the STAND-UGR project and of the GT Political Ecologies from the South/Abya Yala of Clacso. She is also the author of the book *Socio-environmental Conflicts in Peru: Violence and Extractivism* and several articles and book chapters on the subject of environmental justice.

Chris O'Connell, PhD, is a postdoctoral research fellow at the School of Law and Government in Dublin City University, Ireland. His current research project examines the relationship between climate change, environmental destruction and contemporary forms of slavery in Bolivia and Peru.

María Carolina Olarte-Olarte is an associate professor at the Faculty of Law of Universidad de los Andes (Bogotá, Colombia). She is a critical feminist lawyer who holds a PhD in Law from Birkbeck College, University of London. Her research interests include the legal geographies at the interface of constitutional law and property law; social protest and public space; debt and feminism; and the socio-economic dimensions of transitional justice. Her current research and projects focus on the legal geographies of subsoil property and the links between debt and dispossession by use.

Maria Teresa Oré is a Graduate and Master of Sociology from the Pontificia Universidad Católica del Perú (PUCP). She worked as assistant professor and member of the Board of Directors in the Water Resources Master Program Department CCSS (PUCP). She was co- coordinator of the Environmental and Society Studies Group (GEAS) and coordinator of the interdisciplinary project "State and Water Scarcity in the Ica River Basin". She has been a visiting professor at universities in Bogotá, Cali, Quito and La Paz. Her publications include various books on water scarcity in Latin America, in addition to various articles in Peruvian and international books and journals. She is currently a researcher for the project "The sustainability of groundwater", which involves research in seven countries: India, Algeria, Morocco, Tanzania, USA (California), Peru and Chile. The project is led by Amsterdam University in cooperation with the CCSS Department (PUCP) and is sponsored by the European Union.

Evelyn Moreno Ortiz is a member of Citizens of the Karst, dedicated to education for the conservation of the Puerto Rican Karst region, and co-founder of the Institute for Permaculture of Puerto Rico, where she writes about health and teaches about healthy and vegetarian cooking. She is co-author (with Luz E. Cuadrado Pitterson, Lucilla Fuller Marvel, Mariecel Maldonado LaFontaine, and Mari A. Villariny Marrero) of the book *Planificación para un Puerto Rico Sostenible: Fundamentos del Proceso*.

Laura Ortiz Hernández, known as Soma Difusa, is a Colombian illustrator and muralist. She studied graphic design at the National UniversityLater, she became involved in independent publishing events and fanzines, and had the opportunity to work with a group of muralist women through which she approached street art. She is interested in talking about collective and personal feelings expressed through customs. Within her compositions she draws a lot of hands and flowers because she thinks that they can express many different things without the need for faces. @somadifusa

Arissana Pataxó belongs to the Pataxó people in Bahia, Brazil. She is a visual artist and art and Patxôhã language teacher at the State Indigenous School Coroa Vermelha. As an artist, she uses different media to provide an interlocution about her people and issues that cross their lives. She has a degree in Arts from the School of Fine Arts of Bahia (UFBA), Master in Ethnic and African Studies (CEAO- UFBA) and is a doctoral researcher in visual arts at UFBA. @arissanapataxoportfolio

Jorge Perez-Quezada graduated as an Agronomist from the University of Chile, and obtained his MS and PhD degrees at the University of California, Davis, specializing in ecosystem ecology. Currently, Dr. Perez-Quezada is a professor in the Department of Environmental Science and Renewable Natural Resources at the Faculty of Agricultural Sciences of the University of Chile. His research is focused on estimating carbon stocks and greenhouse gas fluxes in natural and managed ecosystems, as well as on sustainable management planning. He is the Academic Director of the Master's program in Territorial Management of Natural Resources at the University of Chile.

Marco Pfeiffer holds a Bachelor in Agronomy and a Master's degree in Geology from the University of Chile, along with a PhD in Environmental Sciences from the University of California, Berkeley. He is currently an Assistant Professor at the Department of Soil Science and Engineering at the University of Chile. His research interest is in understanding the environmental factors that explain the spatial and temporal distribution of soils, as well as how soil variability gives rise to an array of different interactions with the other spheres of planet Earth.

Patricio Pliscoff is an Associate Professor in the Institute of Geography and Department of Ecology at the Universidad Católica de Chile. The focus of his research is biogeography. Their research lines are the description and mapping of biodiversity patterns at different spatial and temporal scales, applying spatial analysis techniques. He was working in the characterization of the spatial distribution of ecosystems and species in Chile and Latin America and relating these analyses of the spatial distribution of biota with climate change. He has also been part as a fellow of the Intergovernmental Science-Policy Platform on Ecosystem Services and Biodiversity (IPBES).

Roberto Thomas Ramírez has been formed in the heat of diverse collective and community experiences that have been fundamental to his life, and to how he relates and connects, in terms of his dedication and character. First from the culture accompanying his mother in community presentations of popular theater, then in the university, and for eleven years he has accompanied and facilitated processes of organization and community development with the regional organization Ecodevelopment Initiative of Jobos Bay (IDEBAJO), which is a community-based organization that serves as an umbrella for various and diverse organizational structures in the southeast-central region of Puerto Rico. IDEBAJO seeks to facilitate grassroots organizing processes for the transformation of excluded communities through the defense of the natural and cultural heritage, and endogenous sustainable community development. Roberto currently serves as General Coordinator of this organization contributing to the organizational growth, the emergence of various socio-productive projects, and the birth of a space for training and popular education called the School of Formation and Community Training, from where he works with the recognition and development of community knowledge, offering training in various technical and organizational areas in order to strengthen community autonomy and the values of solidarity, environmental, economic and social justice.

Angélica M. Reyes Díaz from Juncos, Puerto Rico, works at the Institute for Research and Action in Agroecology. She has a bachelor's degree in Social Sciences in Research-Social Action from the UPR in Humacao. She has worked on projects about the relationship of residents with their natural environment and how they can work to improve it. These projects were mostly aimed at educating and strengthening the relationship of communities with their environment, emphasizing the importance of improving the quality of life of residents through their connection with nature.

Marissa Reyes Díaz is a farmer and organizer, co-founder of the Güakiá agroecological collective and member of the Boricuá Organization of Ecological Agriculture. A native of Juncos, she has a bachelor's degree in Wildlife Management from the University of Puerto Rico in Humacao. Since high school, she has developed an interest in biology, science, and society. She took an agroecology course with the El Josco Bravo project in 2014. In 2015 she channeled her knowledge with social interests and started a conversation with a group of people to develop an agroecological project. Since then there is Güakiá, Agroecological Collective. She currently also works at the non-profit organization El Puente- Enlace Latino de Acción Climática coordinating the Walk "Puerto Rico against Climate Change" and the mobilization for just transitions and climate justice. She believes in collective work, community power, food sovereignty, solidarity economy, alliances and collaboration.

Thea Riofrancos is an Andrew Carnegie Fellow (2020–22) and an Associate Professor of Political Science at Providence College. She researches resource extraction, renewable energy, climate change, green technology, social movements, and the left in Latin America. She is the author of *Resource Radicals: From Petro-Nationalism to Post-Extractivism in Ecuador* (2020), co-author of *A Planet to Win: Why We Need a Green New Deal* (2019), and is currently writing *Extraction: The Frontiers of Green Capitalism*. Her peer-reviewed research has appeared in *World Politics*, *Perspectives on Politics*, *Global Environmental Politics* (forthcoming), and *Cultural Studies*. Her writing has appeared in *The New York Times*, *The Washington Post*, *Foreign Policy*, and *The Guardian*, among others.

Diana Jiménez Thomas Rodríguez is a PhD candidate in International Development at the University of East Anglia and the University of Copenhagen. Her primary research interests are development studies, (feminist) political ecology, environmental justice, Latin American politics, and feminist studies.

Facundo Rojas has a PhD in Geography. He is Professor of Epistemology of Geography and Geographic Information Systems II at the Faculty of Philosophy and Letters, National University of Cuyo (Mendoza, Argentina). He is a member of the editorial committee of the Boletín de Estudios Geográficos. He is an Adjunct Researcher of the Environmental History Group of the Argentine Institute of Nivology, Glaciology and Environmental Sciences (IANIGLA) of National Council for Scientific and Technical Research (CONICET) and specializes in historical geography, environmental history, and political ecology. He is currently working on the following projects: "Climate History in San Juan, La Rioja and Catamarca. Climate-hydrological variability and socio-territorial processes during 18th to 21st centuries."; "Highlands" and "From land to table: an environmental history of vitiviniculture in the Americas."

Nadia Romero Salgado holds a Master's in Cultural Studies from Universidad Andina Simón Bolívar-Quito and a Diploma in Political Ecology from CLACSO, Argentina. She is a member of the Collective of Critical Geography of Ecuador and an activist for the social memory of the impacts of neoliberalism and the financial crisis in Ecuador. Her academic interests are placed at the interface of political ecology, critical geography, and social memory, with an emphasis in the social and environmental impacts of neoliberalism, the extractivism of the shrimp industry in the mangrove forests, and economic injustice resulting from financial crime. She currently works as an independent researcher and in educational processes for social organizations.

Arnim Scheidel is a Ramón y Cajal research fellow at the Institute of Environmental Science and Technology, Universitat Autònoma de Barcelona (ICTA-UAB). His research focuses on the political ecology of development, environmental justice, ecological economics, and agrarian and environmental change. Arnim is part of the Direction and Coordination Group of the global Environmental Justice Atlas (EJAtlas) that documents environmental conflicts and resistance movements worldwide.

Lise Sedrez is Associate Professor in History of the Americas at the Universidade Federal do Rio de Janeiro and a CNPq researcher. She is co-editor of the book series Latin American Landscapes, and chief-editor of the Online Bibliography on Latin America Environmental History. Lise is also a founding member of the Sociedad Latino-Americana y Caribeña de Historia Ambiental. A Rachel Carson Center Fellow (2015–16), Lise's recent publications include *The Great Convergence: Environmental Histories of BRICS*, with S. Ravi Rajan, and *A History of Environmentalism: Local Struggles, Global Histories*, with Marco Armiero.

Eduardo Silva is Professor and Friezo Family Foundation Chair in Political Science at Tulane University. His publications include *Challenging Neoliberalism in Latin America and Reshaping the Political Arena, from Resisting Neoliberalism to the Second Wave of Incorporation* (co-editor). His recent articles have appeared in *Politics and Society*, *Extractive Industries and Society*, the *European Review of Latin American and Caribbean Studies*, and *Latin American Politics and Society*.

Bárbara Oliveira Souza has a PhD in Anthropology from UnB, and is a researcher associated with the Center for Afro-Brazilian Studies (UnB) and with Tterra/UnB, and a volunteer professor at the University of Brasilia linked to NEAB/CEAM/UnB, where she teaches courses on racial issues, traditional peoples and communities, and sustainability.

Ezra Spira-Cohen is a PhD Candidate in the Department of Political Science at Tulane University.

Natalia Dias Tadeu is an Environmental Manager (School of Arts, Sciences and Humanities, University of São Paulo, Brazil), Master and PhD in Environmental Sciences (Institute of Energy and Environment, University of São Paulo, Brazil) and Postdoctoral Researcher (Faculty of Sciences, University of the Republic, Uruguay). She is a member of Uruguay's National System of Researchers (SNI), Associate Editor of the Journal *Ambiente & Sociedade* (Brazil) and currently researches topics related to water policy, water governance, and hydrosocial territories in Uruguay. She is a professor in the Multidisciplinary Seminar in the Faculty of Social Sciences at the University of the Republic (UDELAR, Uruguay) and a researcher at the SARAS Institute (Uruguay).

Melina Ayelén Tobías holds a PhD in Social Sciences (UBA) and Geography (Paris III). She is an Assistant Researcher at CONICET, based in the Environmental Studies Group within the Urban Studies Area, Gino Germani Research Institute. Her research interests are urban water policy in Buenos Aires, the development of water and sanitation infrastructure networks, and the tension for sustainability in metropolitan watersheds. She is an Assistant Professor of Methodology in Social Work (National University of José C. Paz) and Head of Practical Works in Environmental Sociology in Sociology, UBA.

Anahí Urquiza is Associate Professor at the University of Chile, with an experience in teaching and doing research of over 15 years. She is a Social Anthropology and Master in Anthropology and Development from the University of Chile. She has a PhD in Sociology from the University of Munich, Germany (Ludwig Maximilian Universität München). Anahí teaches Methodology, Research Processes, Theory of Social Systems and Socio-Environmental Issues. Her field of research focuses on the environment and society relationship, particularly in water vulnerability, poverty and energy transitions, participation, governance and resilience when facing climate change.

José Santos Valderrama currently serves as Manager of Sustainable Agriculture and Environment at Hispanic Federation Puerto Rico, where he manages programs and leads advocacy and public policy campaigns for the benefit of the agricultural sector and rural communities in Puerto Rico. From 2016 to 2021, he worked at the Puerto Rico Model Forest where he held roles as a consultant, community organizer, and general coordinator. He is a founding partner and president of Cooperativa de Trabajo Cabachuelas (CABACOOP) in Morovis where he serves as a community social worker in charge of the Community Outreach Program. He is also a founder of Finca Graciana in the municipality of Hatillo in Puerto Rico and was part of the board of directors of the Institute for Research and Action in Agroecology. He has been involved with environmental and community organizing issues since 2012, including participation in the Coalition of Anti-Incineration Organizations in the fight against the waste incineration project in Arecibo. He holds a master's degree in community social work, certifications in school social work and development of solidarity agricultural microenterprises; and he is also an agroecological promoter.

Karolien van Teijlingen is a Human Geographer with a PhD in social sciences. Her research addresses the transformations generated by the extractive industries, and the related discussions about development and its alternatives. With theoretical roots in political ecology and critical geography, she is particularly interested in the multi-scalar power relations that such transformations involve. Currently she is a member of the Collective of Critical Geography of Ecuador and a postdoctoral researcher at Radboud University, The Netherlands.

Diana Vela-Almeida is a Postdoctoral researcher at the Norwegian University of Science and Technology and a member of the Collective of Critical Geography of Ecuador. Her academic interests are placed at the interface of political ecology, ecological economics, and critical geography and her research can be grouped into three core themes. First, extractivism and environmental governance, particularly related to multi-scalar relations of power and political participation in complex natural resources-based economies. Second, the analysis of socio-territorial conflicts motivated by actions against green neoliberal schemes. Finally, feminist and decolonial theorization for the study of historical and emerging proposals for economic transformation.

Melissa Moreano Venegas is a Professor at the Department of Environment and Sustainability of the Simón Bolívar Andean University in Quito, Ecuador. She is a member of the Collective of Critical Geography of Ecuador, the CLACSO's working group on Political Ecologies of Abya-Yala, and the Latin American and the Caribbean Platform for Climate Justice. She currently investigates the transition to green capitalism amidst the construction of post-extractive futures; the role of the financialization of nature in ecosystem conservation strategies; and the criminalization of the so-called environmental defenders.

Juan Pablo Soler Villamizar is a researcher in the area of energy and climate justice at CENSAT Agua Viva, and a member of the Comunidad de Sembradoras de Territorios Aguas y Autonomías (Sowers Communities of Territories, Waters and Autonomies) — SETAA Communities, which is a community organization that suffers the effects generated by the Hidroituango dam in Colombia. He is a co-founder of the Colombian Movement in Defense of Territories and People Affected by Dams—Movimiento Ríos Vivos, which articulates communities affected by hydroelectric dams in six departments of Colombia. He is also a member of the Movimiento de Afectador por Represas—MAR (Movement of People Affected by Dams) in Latin America and a member of the Group of Work Energy and Equity. In his personal and professional development, he has promoted the dissemination of knowledge and the implementation of community initiatives for self-management and autonomy.

Lucrecia Wagner has a Degree in Environmental Management (National University of the Center of Buenos Aires Province) and a PhD in Social and Human Sciences (National University of Quilmes, Buenos Aires). She is a researcher at the National Scientific and Technical Research Council (CONICET, Argentina), and a member of the Environmental History Group at the Argentinean Institute of Snow, Glaciology and Environmental Sciences (IANIGLA). Her main research topics are environmental conflicts, specially related to extractive activities, with emphasis on environmental legislation and social participation. She is a professor in the PhD program in Social Sciences at the National University of Cuyo (UNCuyo), and in the Masters in Environmental and Territorial Policies at the University of Buenos Aires (UBA).

Part I

Introduction

1 Suturing the open veins of Latin America, building epistemic bridges[*]

Latin American environmentalism for the 21st century

Beatriz Bustos, Salvatore Engel-Di Mauro, Gustavo García López, Felipe Milanez, and Diana Ojeda

This book is an attempt to provide an in-depth analysis and theorization of environmental issues in Latin America. Latin American political processes as well as natural resources have been of worldwide significance (e.g., the 1992 Rio Conference; the 1st World Social Forum). Yet few books address this topic, and none do in a multidisciplinary and socio-ecologically encompassing way. The environment remains a contentious ground for states, firms, social movements, and wider society. As an editorial collective, we propose that Latin America is not only an object of study, but also a region replete with critical thinking on these themes. We hope this Handbook will be a useful resource for those wanting not only to understand the issues, but also to engage with ideas about environmental politics and social transformation. The Handbook is an attempt to set a research agenda and provide illustrations of environmental transformations, challenges, struggles and political processes.

As any other humanly inhabited space or as a general politically and culturally demarcated area, Latin America exceeds what generates it as an ideological construct, with deep colonial roots. Formalized boundaries or notions may shift according to the outcomes of social struggles and attendant imaginaries (for instance, the reduction of Indigenous Peoples to an overarching, rather extraneous "Latinness," or the arbitrary inclusion or exclusion of Caribbean places), but what constitutes Latin America has nevertheless a material basis in the many and varying physical, ecological, and social processes that make for broadly shared characteristics, both tragic and disastrous and inspiring and healing. In this volume, it is the latter, the social dynamics, that form the focus and entry point in recounting and explaining the many facets of the lands whose veins remain terribly open, to borrow a regrettably still apt phrase from the late Eduardo Galeano. But at least a few of the veins may be closing and undergoing repair, thanks to the indefatigable efforts of numerous communities, who include many unrecognised scholars. The suturing of the open veins is therefore not only an economic or environmental challenge, but also an epistemic one.

The unique physical environments and ecosystems traversing Latin America, redolent with the effects of thousands of years of human interaction, are not just made up, nor can they be. Insisting on taking the environment into consideration is important, regardless of how arbitrary regional constructs ultimately are. As organisms, we are always enmeshed in and interdependent with other organisms and physical forces. The mighty rivers incising the Amazonian plains, the El Niño Southern Oscillation, and the Zika virus are just some of the myriad possible examples

---

[*] Authors listed in alphabetical order.

DOI: 10.4324/9780429344428-2

of the multiple socio-environmental relations that constitute Latin America. We pay greater attention to people's relations with the rest of nature, in their political, economic, and cultural dimensions, without thereby reinforcing a still predominant humanity–nature dichotomy, where fictions of "pristine nature" reign and have been imposed through colonialism. In this sense, our attempt is to contribute to dismantling the intellectual and ideological trappings that persist after formal decolonization and to forging an intellectual culture that values, recovers, and promotes ways of knowing among the colonized that have long been suppressed. This is the reasoning behind the notion of environment herein taken, instead of nature, which is a physically more encompassing level of abstraction that includes processes and beings that exist without us. To politicize the biophysical acknowledges that the web of life is far greater than what the social struggle can capture. Viewing the biophysical environment in expressly political terms does not reduce the rest of nature to a human plaything. The matter is to bring into focus the social causes that inflict deprivations, vulnerabilities, harm, and deadly horrors on many people, like the environmental destruction wrought through extractivism and the increasingly frequent and more intense disasters magnified by climate change. These have been and remain political problems in the context of Latin America. At the time of writing, COVID-19 mortality in the region has surpassed a million, with a disproportionate death toll among Indigenous and Black populations due to structural racism and continued colonial violence. To understand this staggering and heart-wrenching figure as the inevitability of viral diffusion would exonerate the deadly politics underlying the ongoing public health catastrophe. The capitalist destruction of entire ecosystems, the enormously unequal health care provisions, the sustained destructive interference of foreign powers (mainly the US) are all consummately political decisions, at multiple scales involving highly unequal global inter-relations, that lead to the decimation primarily of communities who have been undermined by ruling-class policies and violent repression. Recourse to the latter strategy has been on the rise again in multiple countries over the past couple of years, more recently in Colombia. Deadly force is also being applied passively, and, in the case of the Bolsonaro government (Brazil), by wilful negligence, as, among other aims, a genocidal policy against Indigenous Peoples. Mortal pathogens, like all forms of calamity and environmental destruction, are always drenched in social relations of power. There is no neutrality to the environment, as technical experts and mass media would typically portray it. In this analytical framing, we tread the path of many political ecologists, including those autochthonous to the region, in an open and wide epistemic dialogues around critical perspectives to nature and the environment.

Understanding the environment in this critical and relational way is even more important now because, over the past decades, environmental degradation has intensified with the expansion of extractive economies (mining, logging, plantation agriculture, industrial farming, and fishing, among others) and urban areas, even as the boom cycle of extractivism has recently run its course, for now. The environmental politics associated with these major shifts are traversed by the recrudescence of a rightward (including fascistic) turn, even in countries where socialism-leaning governments had previously held sway a decade earlier. These recent political economic developments have intensified the already highly unequal and often deadly effects of climate change and now also of the COVID-19 pandemic. Heightened repression and increasing right-wing influence are also shaping environmental policies, such that so-called "green" solutions are becoming yet another instrument of deprivation for the many, along with violent dispossession, especially for Black and Indigenous Peoples. One illustration is the tendency to kick out local communities to conserve or plant forests with the aim of offsetting carbon emissions from the world's wealthy. Another example is the dirty war (e.g., the ousting of Evo Morales in Bolivia) waged to secure rare earths for renewable energy technologies, like solar panels and hydrogen fuel cells.

As should be expected, there is another side to this story, one that often goes untold but provides the basis of brighter prospects and futures for the region and the rest of the world. Aside from the everyday grassroots struggles that produce leftist governments and resist extractivism and gendered and racialized dispossession, many communities throughout Latin America—from amongst others, peasant, Afro-descendent, Indigenous, and feminist collectives—have developed a panoply of practical examples of alternative ways of living and understanding that help build or have already achieved egalitarian and ecologically sustainable worlds. These struggles are global and continuous with centuries-long efforts to end colonialism and rebuild not only a new society, but a new world. Such struggles and lived alternatives are behind the rich array of contributions to environmental politics from Latin American thinkers. However, they remain largely unknown or under-appreciated in the Anglophone academic world. Hence, a main point of this collection is to build bridges among scholars who share these concerns but are separated by linguistic and other social barriers. Doing so can simultaneously lift up and make more widely known the insightful and critical scholarship that endows Latin America as a place where many kinds of conversations and practices have been emerging that can help rebuild a constructive human interaction with the rest of nature.

Rationale and outline of the volume

We hope that the chapters and diverse approaches included in this collection will push forward debates as open and collaborative as the collective process of bringing this book to life has been. Along the four years this book took to be in your hands or on your screen, we worked to think about how its contents could best reflect the epistemic alternatives we aimed to share. Each of us lives in different countries and works in different institutions and learning to coordinate north–south academic practices in a pandemic world has been a process of solidarity and caring. Further, as politically committed people, we lived the social revolts that sprouted in each of our respective countries. These have fed into our debates and editorial decisions. How could we not discuss ecological fascism when the Bolsonaro regime was implementing genocidal policies and the Trump administration was targeting Latin American migrants and undermining climate change policies? How could we not consider extractivism in an environmentally oriented handbook when Chile was revolting against inequality? How could we not include a chapter on violence against environmental defenders when Colombia is the country with the highest number of activists killed ever recorded? How could we not discuss disaster politics when Puerto Rico still endures the aftermath of Hurricane Maria? We live and breathe Latin America's tensions, and this compels us to devote ourselves to building a better future and expanding livable presents. We have the privilege of being able to speak multiple languages, of being heard more than most people are, but we also struggle to build academic bridges, to prevent losing our arguments in translation. We know what it is like to be edited for clarity, but also understand the underlying imposition of northern academic hegemony and the richness that Anglophone readers miss by not being able to read in Spanish or Portuguese (just as we miss by not being conversant in one or another of the 448 or so Indigenous languages spoken in Latin America, in Abya Yala). Thus, our editorial guidelines aimed to represent how we dream academia could be, looking for authors, to the best of our collective abilities, from the whole region, from the Rio Grande to Tierra del Fuego, from the Pacific to the Atlantic.

It was important to us to have a balance in geographical diversity of landscapes, cultures, experiences from the Lacandon Forest to the Patagonian fjords, from the Amazonian basin to the driest deserts of Atacama, from the Caribbean Islands to the Chaco, and in the social composition of authorship, aiming to promote the voices of women, people from racialized groups, and insisting on having Latin American authors in chapter references. We invited Spanish-speaking scholars and helped with the translation of their work. In this, we wanted to avoid imposing a form of academic writing typical of the Anglophone academic worlds, so that readers can get a sense of the complexity that Spanish and Portuguese academic work offers. We believe that in conveying ideas across such contexts the mode of expression matters as much as content when building epistemic communities. It is not just about writing in your language but also about understanding each other's ways of thinking. Finally, we also invited younger scholars to provide a platform for new voices. Along the way, some of those had to withdraw from the project, we also struggled to find authors from all countries, autonomous territories and nations. We are deeply thankful to the authors who stuck with us in this project, through COVID-19, increased care responsibilities, growing precarity, job changes, political mobilization, and state repression. When it seemed that the world was falling to pieces, our meetings helped us see a common purpose, a light under conditions of intense uncertainty, and the power of collective thought and action. Reading the chapters showed us that there is such promising scholarly work in the making, that, as messy as this moment was, it was also a moment of deep commitment to building a new kind of academia, one that stands in close relation with political and environmental struggles.

Scholars in Latin America have long reflected on environmental conflicts, movements and grassroots resource-management practices, their relation to the violence of conquest, colonial, classed, racialized, and gendered power relations, and the possibilities for alternative socio-ecological arrangements. This book features general discussions and appraisals of the state of the art of key approaches—concepts, theories and frameworks, and methodologies—to understanding the socio-environmental context of Latin America, in its regional and global dimensions. We cover theoretical and conceptual contributions that are also relevant for a broader audience, including comparative approaches and challenges of knowledge production about the region and its localities. We emphasize theoretical approaches developed from within Latin America, as knowledge that is traditionally marginalized in hegemonic academia, but which has made significant contributions to understanding environmental issues often beyond Latin America itself. We also include frameworks developed by Black, Indigenous and other scholars and activists from the region.

Following this introductory chapter (Part I), Part II provides a discussion of the diversity of environmental and ecological systems and problems of the region, from an interdisciplinary perspective. The chapters include analysis of Latin America's environmental history, particularly that of the agricultural boom of the 19th century (de Carvhalo and Sedrez), land use and degradation patterns in relation to global commodity markets, government policies and geopolitical forces (Manuschevich et al.); the vulnerabilities of different ecosystems to climate change (Pliscoff), including coastal ecosystems, linked to other social dynamics such as deforestation and overfishing (Hernández-Delgado); neocolonial extractivism in relation to the "open veins of Latin America" thesis (Grandia); current forms of environmental (neo)colonialism linked to extractivist dynamics (O'Connell and Silva Santisteban); and water scarcity problems, linked to large-scale mining, urban expansion, and neoliberal policies (Fragkou et al.). By the end of this part, readers will get a clear sense of the trajectories, turning points, actors, and current issues confronting environments in Latin America.

Part III offers a political economy entry in Latin American critical environmental thought: critical theories of development, including the political economy of resource extraction across history and up to and including present-day neo-extractivism (Leguizamón); the ecological debts accrued from environmental degradation (Roa); the trajectories of climate adaptation policies and their lack of attention to social justice (Merlinsky and Tobías); the politics of environmental disasters and risks, including the social unrest and mobilizations that emerge in these contexts (Camargo and Cardoso); toxic pollution from pesticides in agro-industrial production, related to global capitalism (Castro and Mempel);contested ideas about resource extraction, between radical resource nationalism and anti-extractivism (Riofrancos); political ecology of labor that links communities and plantations through the social production of nature and value (León Araya); dependency and ecologically unequal exchange theories (Cooney); the distribution of resource rents and its implications for dependency and inequality, also with regard to the "open veins" thesis (Irarrázaval). Together, the chapters in this part provide a closer look at the challenges that Latin American environmental economies and policies face over the past century and into the future.

Part IV focuses on how social movements and grassroots community organizations are facing environmental degradation and the violence embedded in struggle over resources. The cases across the chapters are challenging the ideas of development as inherently good, and of humans as separate from nature, and looking beyond, to the deep interconnections that create and sustain life, understood as good living, collectively, with nature (*buen vivir, vivir bien, vivir sabroso*). The essays cover a diversity of struggles and actually existing alternatives against extractivism, for territoriality (political recognition and re-appropriation), and for communal governance of the environment, as well as associated institutional reforms. The chapters more specifically discuss the creation of sacrifice zones in energy/mining, agro-export, forestry and the gender dimension in the ensuing conflicts (Bolados); environmental conflicts and violence, linked to expanding social metabolism and marked by colonial-racial inequalities (Neyra); structural racism and injustices against Quilombo communities in the context of the COVID-19 pandemic (Da Silva and Oliveira), re/territorialization as a historically-grounded, dynamic daily practice of indigenous resistance (Mora); Indigenous autonomies and their relation to self-determination, "the political" and societal transformation (López Flores); the new forms of "green" biopiracy in forests, linked to sustainable development and climate policies (Isla); landless peasants' and workers' movements engaging in land occupations and encampments and demanding land reform (Loera); and violence against environmental defenders (Navas, Thomas and Scheidel). By the end of this part, readers will understand why Latin America is a key region to study environmental activism and alternative futures.

Part V delves into the institutional and political contributions from Latin America to think about environment–society relations. It begins with a brief history of environmental thought in Latin America, from limits to growth to eco-development and sustainable development (Estenssoro). Then, a group of chapters addresses salient topics in environmental thought and action, ranging across a wide range of topics: *buen vivir* and degrowth as transformative alternatives to development (Acosta); social cartographies as an alternative methodology for production, sharing, and representation of knowledge about territories in conflict (Barrera); the rights of nature (Insunza); tenure reforms for collective governance of protected areas and forestlands as commons (Monterroso); and the historical environmental policy innovations resulting from social mobilizations (Spira-Cohen and Silva).

Finally, the chapters in Part VI propose future paths and emerging debates within the region to build opression-free futures. Here, authors share the contributions from feminist political ecology and ecofeminism, centered on the critique of hegemonic productions of nature, care of the body-territory (cuerpo-territorio), and the defense of the commons (Ojeda); decolonization and communalization, vis-à-vis subalternization, relationality of territories, reproduction of life, and use value (Olarte and Flórez); the emancipatory contributions of Indigenous, ecologist, and feminist movements (Colectivo de Geografía Crítica del Ecuador); agroecology and food sovereignty initiatives as forms of resistance contributing to self-reliance, climate justice, and disaster resilience (Félix); social movements centered on reproduction of life, ecological and interdependent sensibility, and affective relations of care (Navarro, Rátiva and Furtado);

Indigenous ontologies, epistemologies and knowledges emerging from the relationship with plants and nature (Krenak and Milanez); autogestion and decolonization against the colonial politics of death (García-López, Avilés-Vázquez, Moreno, Ramírez, Vázquez Negrón, Valderrama, Reyes, Ramos and Uriarte); and community initiatives for just energy transitions, based on harvesting sun and water, but also communal relations and labours (Soler).

To accompany the chapters, we have also included five works of art that attest to the multiple ways in which theory and action are entangled in the region. Environmental defense takes place in the streets and classrooms. Denilson Baniwa has interpreted the powerful history of the creation of the universe by the Amazonian entity Yebá Buro, a powerful mythological grandmother of his nation Baniwa and other indigenous societies in the Upper rio Negro. Raquel Echenique (https://www.instagram.com/raquelechenique/?hl=es), from Chile, shares with us the hopes and paths that led to the Chilean new constitution, hopefully an ecological one. Sharon González Colón, co-founder and co-director of Colectivo Moriviví, an all-women artistic collective from Puerto Rico (www.colectivomorivivi.com/), created the work "Abrazo entre fronteras" ("Hug between borders"), visualizing the interconnections between bodies and territories across Latin America and Africa. Soma Difusa, a muralist and graphic artist from Colombia, chose to represent the region through her piece "Construcción interior" ("Internal Construction") through the collective work of ants, planting over the over-exploited landscape of the region. Vlocke, a Mexican artist and activist, portrays the violence of mining through his piece "Contamina," which alludes to both the destruction that the mine brings and to the strength that the resistance against it engenders. The book cover's painting from Arissana Pataxó, artist from the Pataxó people, in the state of Bahia, Brazil, represents planting trees, planting love, against the history of the extraction of Brazilwood (redwood).

In sum, we propose in this book to understand the environment as political. We do so collectively by drawing on a diversity of theoretical frameworks and practical experiences, using as a strength the diversity of backgrounds, including our varied geographical coordinates and networks. The chapters herein underscore the importance of grasping that how research is performed matters in framing and providing an understanding of environmental concerns. Accordingly, we wish to underline the importance of participatory-action research methods and the ethics of doing research in and from Latin America, even if not all chapters necessarily reflect this directly. Finally, the content of the Handbook shows how Latin American scholarship is replete with ideas, studies, and research practices of global significance and must no longer be overlooked in traditional treatments of regional issues, as it often is. After all, we are all part of a bewilderingly differentiated unity, part of the collective of existences who is Earth, with different experiences of life but sharing the experience of co-existing, building liveable presents and imagining viable futures.

Part II

Biophysical processes and environmental histories

2 Latin American ecosystems vulnerability

Patricio Pliscoff

Biodiversity and climate change

Biodiversity refers to the expression of the different levels of biological organization on the planet, which includes the genetic level, species, populations, and ecosystems. The expression of these levels of organization in space, which would be characterized as the landscape, is also considered an expression of biodiversity (Tukiainen et al. 2019). Each of these levels expresses its diversity in different ways, but the most visible for the spatial and temporal scale of humans are those variations that are observable at the level of ecosystems and species. The infinite variability that we can observe among species can be seen in the features (characteristics) that are the product of the Earth's evolutionary history, which is expressed in their different sizes, shapes, colors, etc. Ecosystems in turn represent their diversity with different functions and structures, which are defined by the control of the physical environment and evolutionary history at more local scales. The diversity of both species and ecosystems is modulated by the climate in different ways. In the case of species, evolution is influenced by climate through natural selection, which favors or disadvantages some traits over others, maintaining over time those that best reflect the climatic optimum over which the species persist (Peterson et al. 2019).

One of the best-known and studied climatic controls is the regulation of temperature, with physiological basal divisions between groups of species based on their ability to regulate temperature (Newbold et al. 2020); for example, the differentiation in fauna between ectotherms and endotherms. At the ecosystem level, climate exerts control on a regional scale, determining the dominant structure through the relationship between climate and vegetation. This is one of the most studied relationships and is one of the main themes on which the discipline of biogeography has focused (Morueta-Holme et al. 2015). It is possible to establish evident correlations between temperature and precipitation ranges and the expression of vegetation on the Earth's surface (Box 2016). These correlations become much more evident at continental scales, where it is possible to represent the variability of the main forms of vegetation organization (Biomes), using a gradient of precipitation and temperature (Mucina 2019). These relationships can be made more accurate and representative at regional scales, by adding other environmental variables such as moisture and radiation. Climate also helps to explain some of the most important patterns on which biodiversity is organized; for example, gradients of species diversity (richness). Diversity is distributed in space in a non-heterogeneous way, in both terrestrial and marine environments throughout the planet (Mora et al. 2011; Worm and Tittensor 2018). The best-known pattern of diversity is the increase of diversity toward the equator and the decrease of diversity toward the poles (latitudinal pattern of richness). It could be established from this pattern that species diversity is controlled in a linear way by temperature, which can be corroborated for some biological groups (Saupe et al. 2019). But this is not a consistent pattern, since

DOI: 10.4324/9780429344428-4

there are several groups that do not respond to this gradient, presenting their diversity concentrations at high latitudes, closer to the poles than to the equator (Saupe et al. 2019). Another pattern of general diversity in nature, which has historically attracted the interest of biogeographers, is the altitudinal pattern of diversity. Diversity tends to decrease as one goes up in elevation on an altitudinal gradient. This has been consistently found in different parts of the planet (Rahbek et al. 2019). This pattern may be inversely related to temperature variability, as it also decreases with elevation. But like the latitudinal pattern, it is not a rule for all biological groups and can also be strongly determined by human influence on natural landscapes.

Climate determines where plants and animals can live. All species on the planet have combinations of climatic factors (e.g. radiation, moisture, temperature, etc.) in which they can survive and reproduce (Whittaker 1970). The set of combinations that define the presence of a species is known as its ecological niche. The niche of a species can be characterized by the combination of the different environmental variables that allow a species to persist in a given place (Chase and Leibold 2003). Among these environmental variables, climatic variables are of great importance because they define the range of distribution of species. Range is defined as the area in which a species is present, where two elements can be identified; the extension of occurrence, which is defined by a unit of distance (e.g., latitude) or area (e.g., square kilometers), and the area of occupation, which are the locations within the extension of occurrence that is currently occupied by the species (Mace et al. 2008). Climate change can affect the range of species by modifying the current conditions of their ecological niches, resulting in range contractions and expansions (Guisan and Thuiller 2005). The concept of ecological niche has its origin in the work of ecologists like Grinnell and Elton, who sought to establish the role of species and identify how they make use of the resources available in a habitat. The concept of ecological niche developed by Hutchinson (1957), refers to the niche as a hyper volume of n dimensions where the environmental conditions (factors) in which the species can persist are present. This definition is also key in the conceptualization of the species distribution models developed below, which relate directly to geographic space through the notion of multidimensional n-variable space, which a species inhabits or could potentially inhabit. It is here that the distinction is made between the fundamental niche and the realized niche, where the former indicates the total space of the variables and the latter specifies the geographical area the species in fact inhabits (Guisan and Zimmermann 2000).

Modeling species and ecosystems distributions

For the analysis of the effects of climate change on species and ecosystems, the application of environmental niche modeling or species distribution modeling techniques are the most appropriate tools for assessing the impacts of climate change (Bonebrake et al. 2018). These models are generated using data on the presence or absence of species and variables that define the environmental space, which is then projected onto the geographical space. The modeling technique allows a relationship to be established between geographic positions and the presence or absence of the entire range of variables where the points are located. Some modeling techniques also allow for the incorporation of absences, which allows for better discrimination of the descriptive variables and, therefore, a greater degree of adjustment of the model. The most commonly used variables are climatic and topographic variables. The former is the most used, mainly due to the ease of collection, and the global high-resolution databases (~ 1 km) with free access and use. The most widely-used strategy for selecting variables for modeling comprises a set of predefined climate variables, such as the 19 variables in the global database Worldclim (Fick and Hijmans 2017).

The ability provided by distribution models to project the probability of a species' presence in a geographical space, allows for the generation of projections for the range of distribution of species according to the available future climate scenarios. At present, there are a large number of climate models available—the latest report of the Intergovernmental Panel on Climate Change (2014) reports 37 global circulation models (GCM)—representing different scenarios of climate change on a global scale, which can be used to project the current distribution range of species. One of the most relevant points for future projections is the management of uncertainties related to the variability of global models (Beaumont et al. 2008). New species distribution modeling approaches allow the analysis of different types of uncertainties; for example, those related to the combination of future climates that do not exist at present (non-analogous climates), and it may be possible to identify these areas in projections, and thus generate more reliable predictions (Uribe-Rivera et al. 2017).

Climate change impacts on biodiversity

A species' range responds to climate change, expanding as new populations are added to its peripheral boundaries and contracting when populations at these boundaries disappear (Lenoir and Svenning 2015; Pecl et al. 2017). There are two patterns of species response that are most recognized on a global scale, the first being the pattern of latitudinal variation, where species tend to seek colder conditions than they currently do, moving toward the poles as they compensate for global warming conditions (Sturm et al. 2001; Chapin et al. 2004). The second pattern is one of altitudinal variation in distribution ranges, in which species move toward higher areas, where colder conditions currently exist, and, in addition, there are conditions of greater heterogeneity of relief that allow the effects of warming in lower areas to be cushioned (Grabherr et al. 1994; Hickling et al. 2006; Lenoir et al. 2008). The response of species to climate change is individual and difficult to predict (Hughes 2000), although it may be that in those species that share similar traits it is possible to identify a similar response (Kullman 2002). Projections can also be made according to the extent of distribution, as it is known that species with restricted distribution ranges are more likely to be sensitive to climate changes than those with wider distributions (Johnson 1998), and species that are annual, shrubby, and have short life cycles react more quickly to climate changes (Lenoir et al. 2008). Despite the existence of patterns of latitudinal and altitudinal variation, the response of each species will be influenced by how the group of species with which it interacts reacts and the degree of intensity of that interaction.

Climate change ecosystem vulnerability

Vulnerability to climate change is understood as the degree to which species or ecosystems are threatened by climate change (Gonzalez et al. 2010; Ordonez 2020). The identification of the most vulnerable areas has become a relevant tool for developing adaptation strategies (Comer et al. 2019; Kling et al. 2020). However, the preference for focusing only on average changes in external climate conditions, which is defined as exposure, has been recognized as a weakness in vulnerability analyses. In fact, exposure is only one of the elements that define vulnerability. The other elements are sensitivity and adaptive capacity: Sensitivity is the degree of change in an ecosystem after being subjected to disturbances, while adaptive capacity, or resilience, is the ability of ecosystems to return to their original state after being subjected to disturbances. These last two elements are key to defining vulnerability, especially at regional or local spatial scales, in which their level of vulnerability is not only defined by the magnitude of climatic variations (Li et al. 2018). Analysis of vulnerability to climate change in ecosystems are taking advantage of the increasing availability of spatial information that accounts for different types of natural or anthropogenic disturbances (e.g., mapping of fires or land-use change). This information is now available for any point on the planet at detailed spatial scales. Therefore, the identification of the most sensitive areas is now much more accurate and can be combined with all available climate scenario information (Kling et al. 2020).

Working with this type of information allows us to identify the vulnerability to climate change of ecosystems that are not included in global assessments due to their very restricted spatial scale. For example, in the case of Latin America, one of the areas that is most sensitive to climate change is the tropical Andes, an area that corresponds to an altitudinal band of different amplitude depending on the country. Where a direct relationship has developed between Indigenous communities and the territory, the impacts identified for the ecosystems have a direct impact on those communities, reducing the availability of water sources due to the retreat of glaciers or the changing phenological behavior of crops. This type of impact can currently be mapped with great spatial accuracy, allowing for the linking of the socio-ecological impacts of climate change with those related to natural ecosystems.

Latin American and Caribbean ecosystems

Latin America and the Caribbean make up no more than 15% of the Earth's surface (around 2 billion hectares), yet it is one of the areas of the planet with the greatest diversity of species and ecosystems in the world. Some 40% of the planet's biodiversity and 25% of its forest area is found in Latin America and the Caribbean (UNEP-WCMC and IUCN 2016), in addition to 6 of the world's 10 megadiverse countries (Brazil, Colombia, Ecuador, Mexico, Peru, and Venezuela). The region's ecosystems have been recognized as one of the most valuable on the planet for their ability to contribute to the quality of life of their inhabitants, with nature's contribution estimated as at least US$24.3 trillion per year, which is equivalent to the total gross product of the countries in the region, and 40% of the world's capacity to produce nature-based materials (IPBES 2018).

At the same time, the most recent assessment of the state of terrestrial ecosystems in Latin America and the Caribbean, developed within the framework of the International Union for Conservation of Nature (IUCN) Red List of Ecosystems assessment tool, identified 85% of the forest areas of the Americas as potentially threatened (Ferrer-Paris et al. 2019), related to changes in the coverage of natural ecosystems at different time scales. Habitat conversion, fragmentation, and over-exploitation are the major direct drivers of biodiversity loss, loss of ecosystem functions, and diminishing nature's contributions to human wellbeing, from local to regional scales, in all biomes. Habitat degradation due to land conversion and agricultural intensification; drainage and wetland conversion; urbanization and other new infrastructure; and resource extraction are the greatest direct threats to nature's contributions to people and biodiversity in the Americas. Human-induced climate change is also becoming an increasingly important direct driver of biodiversity loss, amplifying the effects of other drivers (habitat degradation, pollution, invasive species, and overexploitation) through changes in temperature, rainfall, and the nature of some extreme events. The main drivers of adverse trends in biodiversity and nature are expected to intensify in the future, increasing the need for improved policy and effective governance if biodiversity, and nature's contributions to people, are to be maintained (Díaz et al. 2019).

Climate change vulnerability assessments of Latin American and Caribbean ecosystems

To assess the climate-change vulnerability of Latin American and Caribbean ecosystems, two hierarchical levels of the classification system of the WWF world's ecosystems were defined as units of analysis: the level of biomes and the level of ecoregions. In Latin America and the

Caribbean, eleven biomes and 189 ecoregions are identified (Dinerstein et al. 2017). Two criteria were defined to carry out the vulnerability analysis; exposure to climate change, which for this analysis is understood as the change in the future with respect to the current climate, as measured by the variables of mean annual temperature and annual precipitation. The Worldclim 2.1 database (Fick and Hijmans 2017) was used as a source for the definition of current climate, which was defined as the climate for the period of 1970–2000. The future scenario was based on the MIROC6 global circulation model from the CMIP6 global simulation set (Tatebe et al. 2019). This model shows a high degree of correlation with the other GCMs that are part of the global simulation set to be used in the next IPCC assessment (Fasullo 2020). Two socio-economic pathways (SSP) were used, one reflecting a global situation of strong adaptation to climate change (SSP2-4.5), and the other representing a business-as-usual situation which is considered to be the worst possible scenario, with global temperature increases above 4 degrees (SSP5-8.5). The future scenarios were defined for the furthest simulation date, the period 2070–2100.

For the definition of exposure, the difference between the future temperature or precipitation model and the current climate was calculated. Once the difference was obtained, the values were normalized using a linear rescaling from 0 to 1 (from minor to major change). In the case of precipitation, the rescaling was done for both negative and positive differences, which were added to obtain the final change values. In the case of sensitivity, a landscape modification global cover was used (Gosling et al. 2020). This identifies a land cover impact gradient, ranging from areas with less human impact (areas dominated by natural vegetation) to those that have suffered a greater degree of perturbation (urban, agricultural, forest areas, etc.). Like exposure, the landscape modification values were rescaled to obtain a scale of values from 0 to 1, from lesser to greater degree of modification. Finally, the exposure (X-axis) and sensitivity (Y-axis) axes were plotted in a scatterplot graph. The results for temperature and precipitation were compared, changing the Y-axis sensitivity with the optimistic (SSP2–4.5) and business-as-usual (SSP5–8.5) scenarios.

The results show heterogeneity in the spatial patterns of vulnerability to climate change at the biomes level. The biomes with the highest degree of vulnerability and exposure are tropical dry forests and Mediterranean forests and shrublands. These are mainly found in lowland areas of Mesoamerica and the Caribbean, on the northern coast of Brazil (in the case of tropical dry forests), and in central-northern and central Chile (in the case of the Mediterranean forests and scrublands). These ecosystems will suffer the greatest changes due to the increase in average annual temperature and the decrease in precipitation. The combination of changes in both climatic variables will generate major impacts in these ecosystems, which can mainly be related to the movement of species to wetter and colder areas, changes in the structure of the ecosystems, and extinctions of species that are more sensitive to climatic variations and that have a lower range of distribution.

It is relevant to relate these results to the current level of protection, as many of the ecosystems present in these biomes are poorly represented within the national systems of protected areas. For example, the central area of Chile, where the Mediterranean forests are located, is the least-protected area in the country and is where the largest number of people and the greatest amount of agricultural and forestry activities are concentrated (Fuentes-Castillo et al. 2019). In the case of tropical dry forests, a relevant ecosystem is the Caatinga, which has already been recognized as critically endangered using the IUCN Red List Ecosystem Assessment Framework (Ferrer-Paris et al. 2019).

When comparing the differences between the SSP scenarios, no major differences in precipitation patterns are observed between the optimistic and the (more pessimistic) business-as-usual scenarios. If there are changes in the case of temperatures in the business-as-usual scenario,

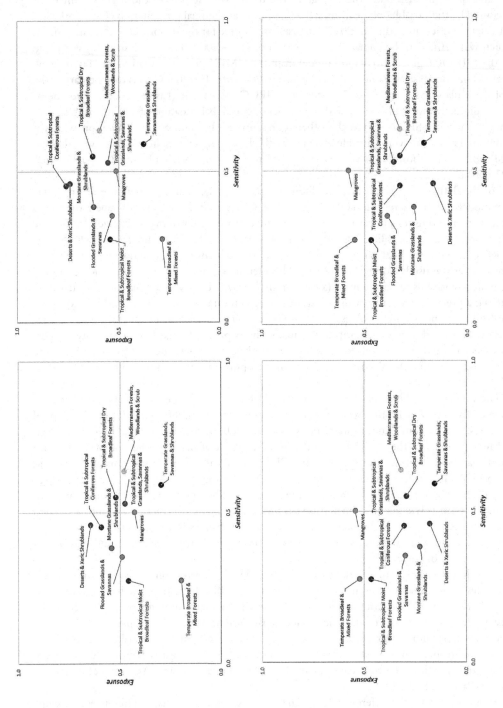

Figure 2.1 Climate change vulnerability scatter plots for temperature (upper figures; left SSP2–4.5, right SSP5–8.5) and precipitation (lower figures; left SSP2–4.5, right SSP5–8.5).

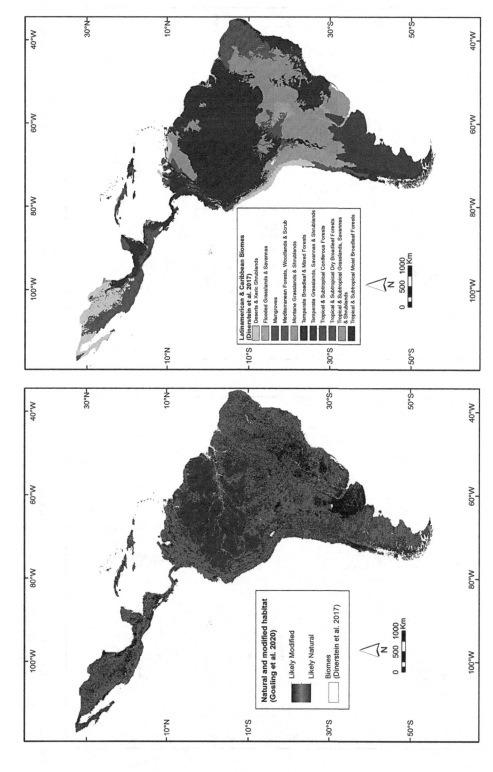

Figure 2.2 Distribution of Biomes (top/right), and Natural and modified habitat (bottom/left) in Latin American and Caribbean, following Gosling et al. 2020.

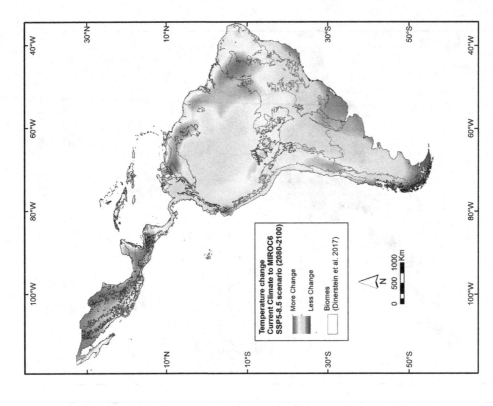

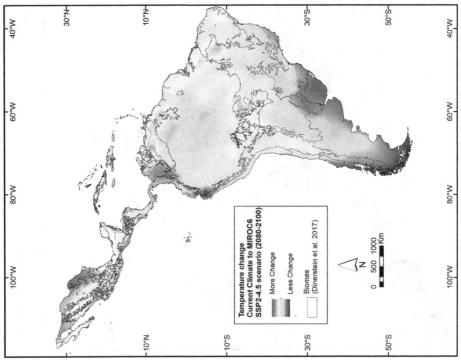

Figure 2.3 Temperature changes (this page) and precipitation changes (next page) cfrom current climate to future scenario in Latin American and Caribbean using MIROC6 GCM model.

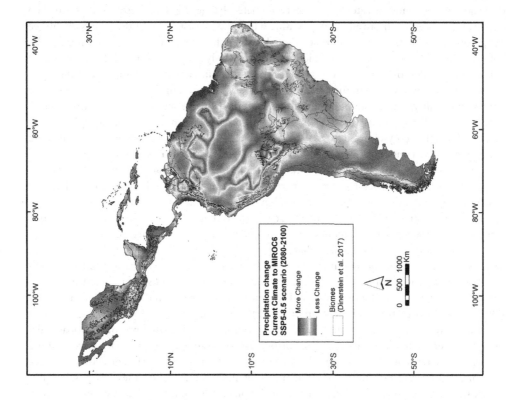

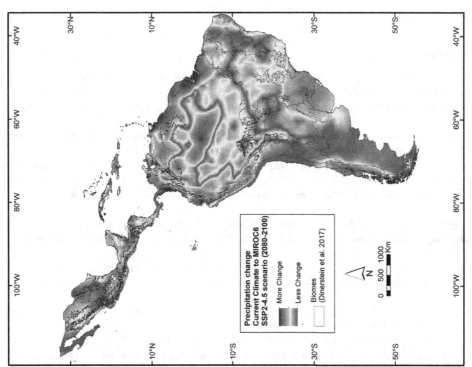

Figure 2.3 (Continued)

the major biomes show greater exposure to climate change. Finally, these results should be understood only as one more antecedent within a framework of analysis of the effects of climate change on biodiversity. The incorporation of threats such as fires can modify or reinforce the results obtained. It should also be considered that the analysis of exposure has been generated with climate models that reflect average values of temperature and precipitation. The incorporation of the frequency of extreme weather events is also important for a more complete view of the vulnerability of ecosystems under different climate change scenarios. The spatial scale of representation of ecosystems (biomes) is extremely broad, so it must also be considered that highly vulnerable ecosystems such as wetlands are not reflected in these analyses. All these elements must be considered within a conservation planning framework for Latin American and Caribbean ecosystems to advance toward better adaptation to ongoing climate change.

Latin American countries have taken part in the main multilateral agreements related to climate change (e.g., the Paris Agreement). They have taken an active part in the definition of global adaptation strategies. Most countries have defined their nationally determined contributions (NDCs), which is the most concrete way to generate efforts to reduce emissions and adapt to the impacts associated with climate change (Bárcena et al. 2020). One tangible way to relate the ecosystem impacts identified in this work to national adaptation efforts would be through the incorporation of the most vulnerable ecosystems into new networks of protected areas. The ongoing framework of the Convention on Biological Diversity, which defines global targets for 2030 (a post-2020 global biodiversity framework), promotes increasing the coverage of protected areas from 15% (the current figure) to 30% in terrestrial ecosystems on a global scale. There are several proposals that identify priority areas to advance to this 30% target, combining not only ecosystems' importance but also the representation of other relevant ecological processes like carbon capture and storage (Dinerstein et al. 2020). Based on these proposals, Latin American countries could define new areas for the protection of biodiversity and better represent the most vulnerable ecosystems, making progress in adapting to climate change.

References

Bárcena, Alicia, Joseluis Samaniego, Wilson Perez, and José Eduardo Alatorre. 2020. The climate emergency in Latin America and the Caribbean: the path ahead – resignation or action? ECLAC Books, No. 160 (LC/PUB.2019/23-P), Santiago: Economic Commission for Latin America and the Caribbean (ECLAC).

Beaumont, Linda J., Lesley Hughes, and A. J. Pitman. 2008. "Why is the choice of future climate scenarios for species distribution modeling important?" *Ecology Letters* 11(11): 1135–1146.

Bonebrake, Timothy C., Christopher J. Brown, Johann D. Bell, Julia L. Blanchard, Alienor Chauvenet, Curtis Champion, … Greta T. Pecl. 2018. "Managing consequences of climate-driven species redistribution requires integration of ecology, conservation and social science." *Biological Reviews* 93(1): 284–305.

Box, Elgene Owen. 2016. "World bioclimatic zonation." In *Vegetation structure and function at multiple spatial, temporal and conceptual scales*, edited by Elgene Owen Box, 3–52. Cham, Switzerland: Springer International.

Chapin, III F. Stuart, Terry V. Callaghan, Yves Bergeron, M. Fukuda, J. F. Johnstone, G. Juday, and S. A. Zimov. 2004. "Global change and the Boreal Forest: Thresholds, shifting states or gradual change?" *Ambio* 33: 361–365.

Chase, Jonathan M., and Leibold, Mathew A. 2003. *Ecological niches: linking classical and contemporary approaches*. Chicago: University of Chicago Press.

Comer, Patrick J., Jon C. Hak, Marion S. Reid, Stephanie L. Auer, Keith A. Schulz, Healy Hamilton, Regan L. Smyth, and Matthew M. Kling. 2019. "Habitat climate change vulnerability index applied to major vegetation types of the western interior United States." *Land* 8 (108): 1–27.

Díaz, Sandra, Josef Settele, Eduardo S. Brondízio, Hien T. Ngo, John Agard, Almut Arneth, Patricia Balvanera, et al. 2019. "Pervasive human-driven decline of life on Earth points to the need for transformative change." *Science* 366 (6471): 1–10.

Dinerstein, Eric, Anup Joshi, Carly Vynne, Andy T. L. Lee, Félix Pharand-Deschênes, M. França, Sanjiv Fernando, et al. 2020. "A "global safety net" to reverse biodiversity loss and stabilize earth's climate." *Science Advances* 6 (36): 1–14.

Dinerstein, Eric, David Olson, Anup Joshi, Carly Vynne, Neil D. Burgess, Eric Wikramanayake, Nathan Hahn, et al. 2017. "An ecoregion-based approach to protecting half the terrestrial realm." *BioScience* 67 (6) (June): 534–545.

Fasullo, John T. 2020. "Evaluating simulated climate patterns from the CMIP archives using satellite and reanalysis datasets." *Geoscientific Model Development Discussions* 13 (February): 1–26.

Ferrer-Paris, José Rafael, Irene Zager, David A. Keith, María A. Oliveira-Miranda, Jon Paul Rodríguez, Carmen Josse, Mario González-Gil, Rebecca M. Miller, Carlos Zambrana-Torrelio, and Edmund Barrow. 2019. "An ecosystem risk assessment of temperate and tropical forests of the Americas with an outlook on future conservation strategies." *Conservation Letters* 12 (2): 1–10.

Fick, Stephen E., and Robert J. Hijmans. 2017. "WorldClim 2: New 1-km spatial resolution climate surfaces for global land areas." *International Journal of Climatology* 37 (12): 4302–4315.

Fuentes-Castillo, Taryn, Rosa A. Scherson, Pablo A. Marquet, Javier Fajardo, Derek Corcoran, María José Román, and Patricio Pliscoff. 2019. "Modelling the current and future biodiversity distribution in the Chilean Mediterranean hotspot. The role of protected areas network in a warmer future." *Diversity and Distributions* 25 (12): 1897–1909.

Gonzalez, Patrick, Ronald P. Neilson, James M. Lenihan, and Raymond J. Drapek. 2010. "Global patterns in the vulnerability of ecosystems to vegetation shifts due to climate change." *Global Ecology and Biogeography* 19 (6): 755–768. https://doi.org/10.1111/j.1466-8238.2010.00558.x.

Gosling, Joe, Matt I. Jones, Andy Arnell, James E.M. Watson, Oscar Venter, Andrea C. Baquero, and Neil D. Burgess. 2020. "A global mapping template for natural and modified habitat across terrestrial Earth." *Biological Conservation* 250 (March): 108674. https://doi.org/10.1016/j.biocon.2020.108674.

Grabherr, Georg, Michael Gottfried, and Harald Pauli. 1994. "Climate effects on mountain plants." *Nature* 369: 448.

Guisan, Antoine, and Niklaus E. Zimmermann. 2000. "Predictive habitat distribution models in ecology." *Ecological Modeling* 135: 147–186.

Guisan, Antoine, and Wilfried Thuiller. 2005. "Predicting species distribution: Offering more than simple habitat models." *Ecology Letters* 8 (9): 993–1009. https://doi.org/10.1111/j.1461-0248.2005.00792.x.

Hickling, Rachael, David B. Roy, Jane K. Hill, Richard Fox, and Chris D. Thomas. 2006. "The distributions of a wide range of taxonomic groups are expanding polewards." *Global Change Biology* 12 (3): 450–455.

Hughes, Lesley. 2000. "Biological consequences of global warming: Is the signal already apparent?" *Trends in Ecology & Evolution* 15: 56–61.

Hutchinson, G. Evelyn. 1957. "Concluding remarks." *Cold Spring Harbour Symposium on Quantitative Biology* 22: 415–427.

Intergovernmental Panel on Climate Change. 2014. *Climate change 2013 – The physical science basis: Working group I contribution to the fifth assessment report of the intergovernmental panel on climate change.* Cambridge: Cambridge University Press.

IPBES. 2018. *The IPBES regional assessment report on biodiversity and ecosystem services for the Americas*, edited by Jake Rice, Cristiana Simão Seixas, María Elena Zaccagnini, Mauricio Bedoya-Gaitan, Natalia Valderrama, Christopher B. Anderson, Mary T. K. Arroyo, et. al. Bonn, Germany: IPBES.

Johnson, Chris N. 1998. Species extinction and the relationship between distribution and abundance. *Nature* 394(6690): 272–274.

Kling, Matthew M., Stephanie L. Auer, Patrick J. Comer, David D. Ackerly, and Healy Hamilton. 2020. "Multiple axes of ecological vulnerability to climate change." *Global Change Biology* 26 (5): 2798–2813.

Kullman, Leif. 2002. "Rapid recent range margin rise of tree and shrub species in the swedish scandes." *Journal of Ecology* 90: 68–77.

Lenoir, Jonathan, Jean-Claude Gégout, Pablo Marquet, Patrice de Ruffray, and Henri Brisse. 2008. "A significant upward shift in plant species optimum elevation during the 20th century." *Science* 320 (5884): 1768–1771.

Lenoir, Jonathan, and Jens-Christian Svenning. 2015. "Climate-related range shifts - a global multidimensional synthesis and new research directions." *Ecography* 38 (1): 15–28.

Li, Delong, Shuyao Wu, Laibao Liu, Yatong Zhang, and Shuangcheng Li. 2018. "Vulnerability of the global terrestrial ecosystems to climate change." *Global Change Biology* 24 (9): 4095–4106.

Mace, Georgina M., Nigel J. Collar, Kevin J. Gaston, Craig Hilton-Taylor, H. Resit Akçakaya, Nigel Leader-Williams, E.J. Milner-Gulland, and Simon N. Stuart. 2008. "Quantification of extinction risk: IUCN's system for classifying threatened species." *Conservation Biology* 22 (6): 1424–1442.

Mora, Camilo, Derek P. Tittensor, Sina Adl, Alastair G. B. Simpson, and Boris Worm. 2011. "How many species are there on earth and in the ocean?" *PLoS Biology* 9 (8): 1–8.

Morueta-Holme, Naia, Kristine Engemann, Pablo Sandoval-Acuña, Jeremy D. Jonas, R. Max Segnitz, and Jens-Christian Svenning. 2015. "Strong upslope shifts in Chimborazo's vegetation over two centuries since Humboldt." *Proceedings of the National Academy of Sciences* 112 (41): 12741–12745.

Mucina, Ladislav. 2019. "Biome: evolution of a crucial ecological and biogeographical concept." *New Phytologist* 222 (1): 97–114.

Newbold, Tim, Laura F. Bentley, Samantha L. L. Hill, Melanie J. Edgar, Matthew Horton, Geoffrey Su, Çağan H. Şekercioğlu, Ben Collen, and Andy Purvis. 2020. "Global effects of land use on biodiversity differ among functional groups." *Functional Ecology* 34 (3): 684–693.

Ordonez, Alejandro. 2020. "Points of view matter when assessing biodiversity vulnerability to environmental changes." *Global Change Biology* 26 (5): 2734–2736.

Pecl, Gretta T., Miguel B. Araújo, Johann D. Bell, Julia Blanchard, Timothy C. Bonebrake, I-Ching Chen, Timothy D. Clark, et al. 2017. "Biodiversity redistribution under climate change: Impacts on ecosystems and human well-being." *Science* 355 (6332): 1–9.

Peterson, Megan L., Daniel F. Doak, and William F. Morris. 2019. "Incorporating local adaptation into forecasts of species' distribution and abundance under climate change." *Global Change Biology*, 25 (December): 775–793.

Rahbek, Carsten, Michael K. Borregaard, Alexandre Antonelli, Robert K. Colwell, Ben G. Holt, David Nogues-Bravo, Christian M. Ø. Rasmussen, et al. (2019). "Building mountain biodiversity: Geological and evolutionary processes." *Science* 365 (6458): 1114–1119.

Saupe, Erin E., Corinne E. Myers, A. Townsend Peterson, Jorge Soberón, Joy Singarayer, Paul Valdes, and Huijie Qiao. 2019. "Spatio-temporal climate change contributes to latitudinal diversity gradients." *Nature Ecology & Evolution* 3 (10): 1419–1429.

Sturm, Matthew, Charles Racine, and Kenneth Tape. 2001. "Increasing shrub abundance in the Arctic." *Nature* 411: 546–547.

Tatebe, Hiroaki, Tomoo Ogura, Tomoko Nitta, Yoshiki Komuro, Koji Ogochi, Toshihiko Takemura, Kengo Sudo, et al. 2019. "Description and basic evaluation of simulated mean state, internal variability, and climate sensitivity in MIROC6." *Geoscientific Model Development* 12 (7): 2727–2765.

Tukiainen, Helena, Mikko Kiuttu, Risto Kalliola, Janne Alahuhta, and Jan Hjort. 2019. "Landforms contribute to plant biodiversity at alpha, beta and gamma levels." *Journal of Biogeography* 46 (8): 1699–1710.

UNEP-WCMC United Nations Environment Programme World Conservation Monitoring Centre, IUCN International Union for Conservation of Nature. 2016. Protected Planet Report 2016. UNEP-WCMC, IUCN.

Uribe-Rivera, David E., Claudio Soto-Azat, Andrés Valenzuela-Sánchez, Gustavo Bizama, Javier A. Simonetti, and Patricio Pliscoff. 2017. "Dispersal and extrapolation on the accuracy of temporal predictions from distribution models for the Darwin's frog." *Ecological Applications* 27 (5): 1633–1645.

Whittaker, Robert H. 1970. *Communities and ecosystems*. New York: Macmillan.

Worm, Boris, and Derek P. Tittensor. 2018. *A theory of global biodiversity*. Princeton: Princeton University Press.

3 Soil degradation and land cover change in Latin America

Daniela Manuschevich, Marco Pfeiffer, and Jorge Perez-Quezada

Introduction

In the seminal work *Land Degradation and Society* by Blaikie and Brookfield (1986), the issue of land degradation was defined as a social problem: soil degradation, such as that caused by erosion or salinization, is not a problem until it affects society. Historically, land has had a pivotal role in the conceptualization of production and productivity. Human labor interacts with nature (land) to produce food, fiber, and fodder; if land is degraded, more labor is required to produce the same quantities. That essential idea of productivity has evolved with the development of biotechnology and the use of fertilizers and machinery; however, much of the idea of productivity is based on fossil fuels and not on human labor. All of these new technologies seem to have increased the productivity of the land at an ever-increasing pace—yet the world has witnessed a dramatic increase in the Human Appropriation of Net Primary Production (HANPP), defined as the overall biomass produced on the Earth's surface through photosynthesis (Rojstaczer, Sterling, y Moore 2001; Vitousek et al. 1986). HANPP has doubled in the last century, and will probably reach 78% if biofuels become mainstream by 2050 (Krausmann et al. 2013). The demand for land, whether for carbon sequestration, timber, or biofuels, is already competing with the demand for food production worldwide.

Simultaneously, in the name of progress, governments have facilitated the creation of resource frontiers, using agriculture sustained by fossil-fuel inputs, such as fertilizers and pesticides. This reflects a narrow-minded framework oriented to agribusiness and global exports, in which deforestation and the agrarian frontier keep advancing over naturally-forested areas (Armesto et al. 2010). As demands on global land area increase and climate change intensifies, it is more important than ever to look at the factors that drive soil and land degradation. This is particularly true for Latin America and the Caribbean (LAC), a region considered fundamental for the biosphere and human existence. The region has also witnessed high rural emigration and an overall tendency toward commercially-oriented, agro-industrial intensification. For example, Brazil and Argentina make up 9–10% of the harvested area embodied in global trade (MacDonald et al. 2015).

In this chapter, we first present an overview of soils in LAC, focusing on their most significant features. Second, we review the most recent tendencies in land-use change (LUC) and their principal drivers. Third, we offer an overview of a possible future for land and people, based on an example from collectively-managed land in Chile.

Soils in Latin America

As previously noted by Gardi et al. (2014), the LAC region has a great diversity of soils as a result of its tectonic context, geological history, contrasting topography (Figure 3.1), climatic gradient, and biome distribution (Figure 3.2). This diversity of soil-forming factors implies

DOI: 10.4324/9780429344428-5

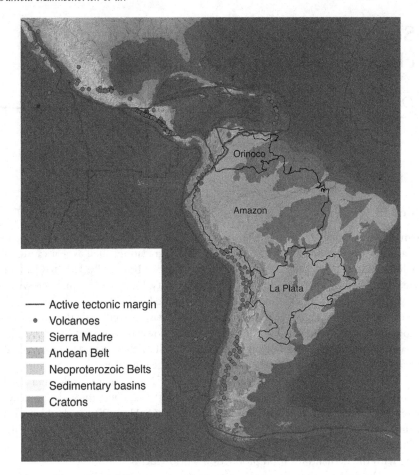

Figure 3.1 Tectonic setting and main physiographic features in LAC (based on Cordani et al. 2016).

that each of the twelve soil orders described in the worldwide soil taxonomy can be found in the region.

At a continental scale—without any intention of covering all aspects of soil diversity in the region–soil-type trends in LAC can be described based on the distribution of the main physiographic units and biomes. As this first level of analysis, we can organize the region by considering three main physiogeographical features.

The first feature is defined by tectonic processes: the active tectonic margins of the Pacific Rim of South and Mesoamerica, and the landmasses created at the margins of the Caribbean plate (Figure 3.1). The western margin of the continent is part of the Pacific Ring of Fire, consisting of subduction zones that result in a relatively high frequency of earthquakes and volcanic eruptions. This tectonic process created the most prominent mountain belts of the region the Sierra Madre system in Mesoamerica and the Andes in South America. Its northern section (31–20°N), the Sierra Madre system, lies mostly in an arid climate, with the presence of Aridisols and Entisols soil orders associated with desert biomes, where erosion and overgrazing are the main problems. In the moister highlands of Mesoamerica—in the coniferous biome—Alfisols soil orders dominate (Figure 3.2). Mollisols soil orders are present in the alluvial flats of the internal basins existing in the mountain belts, which can be made to be very

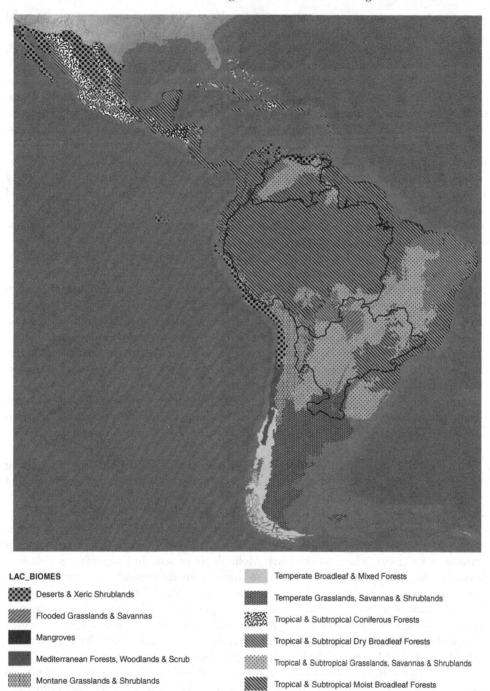

Figure 3.2 Main biomes in Latin American and the Caribbean based in the work of Olson et al. (2001).

LAC_BIOMES

Deserts & Xeric Shrublands

Flooded Grasslands & Savannas

Mangroves

Mediterranean Forests, Woodlands & Scrub

Montane Grasslands & Shrublands

Temperate Broadleaf & Mixed Forests

Temperate Grasslands, Savannas & Shrublands

Tropical & Subtropical Coniferous Forests

Tropical & Subtropical Dry Broadleaf Forests

Tropical & Subtropical Grasslands, Savannas & Shrublands

Tropical & Subtropical Moist Broadleaf Forests

productive by irrigation; however, salinization is a threat when such soils are cultivated in arid zones. Vertisols, which are clay-rich, are the dominant soils in the moister section of the continent, near the Gulf of Mexico. Vertisols are difficult to till, but if salinization is prevented, these soils can produce good yields (Krasilnikov et al. 2013).

The Sierra Madre, the Andes, and other mountain ranges traverse a wide range of climate zones, biomes, and geological formations. Toward the tropics (20°N to 18°S), the mountain ranges along the Pacific margin as well as the Caribbean islands combine active tectonics with stretches of volcanism in a more humid climate. These processes are reflected in its lush forests, with erosion-prone Inceptisols in mountainous areas and acidic Ultisols with low fertility toward moister and hotter areas, as in the Andean foothills and eastern Sierras. These soils are susceptible to erosion in the absence of human-made terraces, as has occurred in vast areas of Ecuador after reaching the highest rates of deforestation rates in South America, which affected approximately half of the area located in foothills (Moreira, Siqueira, and Brussaard 2006; Mendonça-Santos et al. 2015). In the humid volcanic areas, fertile Andisols dominate. If properly managed, these Andisols can be very productive (Ochoa-Cueva et al. 2015).

Mollisols and Vertisols are intensively used in irrigated agriculture, due to their potential for high productivity. They have been mostly used for export-oriented agriculture since the beginning of the 1900s (Bulmer-Thomas 1987). Urban sprawl over fertile land is an issue of concern throughout Latin America (García-Ayllón 2016; Thebo, Drechsel, and Lambin 2014). Despite not covering large areas, organically-rich Histosols in the high Andean wetlands, such as the *Páramos* and *Bofedales* systems, play a substantial role in regulating the water flow used for agricultural and urban consumption. *Páramos* and *Bofedales* wetlands systems are particularly vulnerable to impacts like overgrazing and the lowering of the water table due to water extraction (Buytaert et al. 2006).

Some portions of the Caribbean islands, which still harbor tropical forests, are covered with heavily-leached low nutrient Oxisols and Ultisols, being mostly unsuitable for mass-production agriculture. Oxisols and Ultisols have been overexploited due to factors detailed in the land use section.

A contiguous zone of arid and semiarid climate traverses the Andes from west to east along its southern portion (18–52°S) (Figure 3.3), a region known as the Arid Diagonal. Entisols and Aridisols dominate the landscape west of the Andes toward the north at the Atacama Desert, and east of the Andes toward the south at the Patagonian Steppes. Soils in this vast area are characterized by their salinity, many of them with indurated subsoils which severely hinder vegetation growth. The most productive soils along the Arid Diagonal are in the southern extreme in Patagonia, where nutrient-rich Mollisols are present. In Patagonia, agriculture is limited by the extreme climatic conditions sustaining a mostly livestock economy. The main burden in these arid regions is wind erosion and salinization where irrigation is employed (Rubio, Lavado, and Pereyra 2019). The western margin of the Andes south of 30°S comprises a narrow fringe of Mediterranean and Temperate climates, with Alfisols and Andisols used extensively for both traditional agriculture and agribusiness, with a significant habitat loss in the only Mediterranean hotspot of the LAC region (Armesto et al. 2010; Echeverria et al. 2006).

A second physiographic macro-unit is composed of the old geological shields (cratons) and associated belts of the Precambrian age, which comprise a large portion of the South American continent toward the northeast, and mostly in Brazil (Figure 3.1). In the tectonically stable areas, much of South America falls into geological shield regions, comprising very old rocks that could be deformed (Precambrian belts) or stable (cratons). The most extensive pattern in the region combines old crystalline rocks and tectonic stability with tropical climate, generating highly-developed soils (Ultisols and Oxisols) (Figure 3.3). These soils are heavily leached of nutrients, with high levels of iron and aluminum oxides, low pH, and low fertility. At the

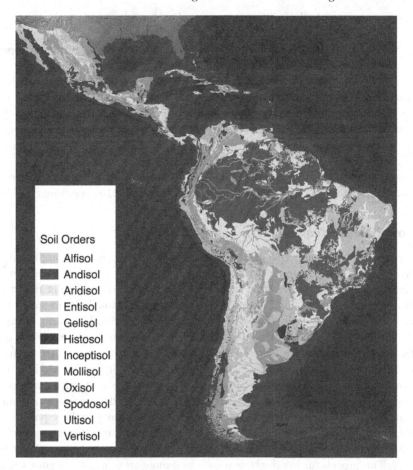

Figure 3.3 Soil Orders in LAC according to Soil Taxonomy (Obtained from USDA-NRCS, 2005, www. soils.usda.gov/use/worldsoils/mapindex/order.html).

Caatinga and Cerrado, low-fertility Oxisols remain dominant in vast areas, while Alfisols also cover a large portion, and the smaller intra-basins are mostly cultivated, containing young deposits dominated by low-natural-fertility Entisols. Additionally, more fertile soils from the orders Aridisol, Inceptisol, and Mollisol can be found in certain specific areas, which have mostly been converted to agriculture (Oliveira and Marquis 2002).

The third macro-unit, standing out for its size and importance, is composed of three main sedimentary basins: the Orinoco, the Amazon, and La Plata (Figure 3.1). Toward the north of the continent, between the Andes and the Amazon craton, lies the Orinoco River, the region's third biggest watershed (Figure 3.1), dominated by highly acidic and compacted Oxisols and Ultisols. To the south lies the Amazon basin, a vast flat area of lower altitude, dominated by extremely weathered Oxisols with very low natural fertility, which paradoxically sustain the richest plant and animal diversity on the planet—the Amazon tropical rainforest. This basin has a long history of human-induced disturbance dating back to the pre-Columbian era (Levis et al. 2017). In the lower areas along the Amazon's main tributaries, waterlogged Inceptisols are found in young, coarse sediments. Toward the west, where the Amazon encounters the Andes, Ultisols, and to a lesser extent Alfisols, are present. These Ultisols and Alfisols are characterized by being rich in clay and slightly more fertile than low-basin Oxisols (Figure 3.2); however, due to their location on steeper slopes, they are more vulnerable to pressures such as deforestation and the conversion of land use to livestock.

Finally, the Rio de La Plata basin hosts well-weathered soils (Oxisols), while the upper basin extends into Brazil's Cerrado Biome, with waterlogged Alfisols at the Pantanal Biome, and humus-rich Mollisols dominate toward the Chaco and Pampas region—the latter region hosting the most fertile and productive lands of the entire LAC (Figure 3.3). Here, loess and volcanic deposits rich in minerals and nutrients sustained extensive natural grasslands in the past, which today are almost completely converted to agriculture (Rubio, Lavado, and Pereyra 2019). Since the 1990s, no-tillage farming has spread across the Pampas, which has improved soil quality and reduced soil degradation (Taboada et al. 1998).

Keeping in mind soil variability and the diversity of processes that intermingle across LAC, in the following section we summarize the main tendencies in land-cover change.

Land-cover change tendencies in Latin America

The most significant land-use changes are related to the extension of anthropic landscapes. Between 1992 and 2015, urban areas in LAC increased by 46%, while anthropic landscapes, such as mosaic landscapes, increased nearly 15.5%. In contrast, the main net reductions are observed in natural cover. In just 23 years, the area covered by closed, broadleaved forest has decreased by 24% (Figure 3.4). In Chile, Uruguay, and Southern Brazil, the main changes have been from shrubs to evergreen trees, which might seem like forest recovery, but in fact is mostly the replacement of native shrubs by tree farms. Therefore, forestry activities are an important threat to the remaining native covers at higher latitudes of LAC (Figures 3.4 and 3.5).

Mainstream discourse portrays small farmers and forest dwellers as the main drivers of land use change and degradation. However, several authors argue that the expansion of commodity frontiers are driven by political and economic conditions that allow very large profits—largely the result of a process of land grabbing from previous dwellers, access to cheap labor, good technology, lack of regulations, and access to global markets (Kröger y Nygren 2020; Waroux et al. 2018). This process further displaces people to remote areas, usually of poor soil quality, or to the main cities and towns. In the following paragraphs, we present examples of these dynamics by focusing on the major land-cover changes in the Pacific Rim, the ancient cratons, and the three main basins.

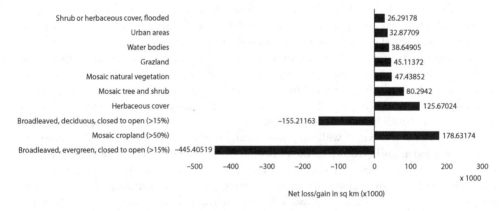

Figure 3.4 Main overview of land cover changes in Latin America and the Caribbean (1992–2015). Based on data from ESA (2017).

Figure 3.5 Main changes in terms of area in land cover between 1992 and 2015 in Latin America and the Caribbean, with net increases and reductions in the area in 1000 km². Based on ESA (2017).

Land-use change in the Pacific Rim is mostly dominated by the loss of evergreen and deciduous forest, driven by agriculture in the south and central Andes (Figure 3.4). The southeastern section of the Sierra Madre stands out as a cluster of forest loss. Contrary to mainstream discourses, deforestation in Guatemala and Honduras is driven by narco-ranchers and not by small farmers (Devine et al. 2020; Tellman et al. 2020). Cattle-ranching is used to launder money from drug and human trafficking, as cartels offer "*pago o plomo*" (a bribe or a bullet) to acquire land from locals (Devine et al. 2020). Then, narco-ranchers acquire up-to-date technology to produce commodities, such as oil palm, sugar cane, and cattle, and gain control of an area marked by migrations and drug flows toward the United States, despite the supply-side war on drugs (Tellman et al. 2020; Wrathall et al. 2020).

Known as the *arc of deforestation*, Brazil and Bolivia concentrate most of the land-use change in the central western areas of South America, which are located on cratons. Although these soils cannot maintain their productivity if not covered by native vegetation, they have been extensively converted to cattle ranching or soybean production (Davidson et al. 2007; Malhi et al. 2008; Sparovek et al. 2012). A significant portion of the southern and eastern sections, where the climate is drier, are dominated by the Caatinga and Cerrado savanna forests, which have lost almost half of their original plant cover, mainly for corn and soybean cultivation, being the second-most active frontiers of agricultural expansion in LAC (Beuchle et al. 2015; Fearnside 2001).

Finally, the main land-use changes are occurring in the three large basins. In the Orinoco Basin, natural savannas experienced rapid conversion into commercial agriculture (Etter et al. 2006). The Rio de la Plata Basin is also experiencing extensive land-use change, mostly driven by cattle-raising and soy cultivation over productive soils, encouraged by the abnormal profits generated by new agricultural technologies, infrastructure, and other factors (Waroux et al. 2018). These processes have been facilitated by governments through regulations, monetary policies, and infrastructural investments (Waroux et al. 2018). In Chile, for example, state subsidies combined with trade liberalization produced large profits, while encouraging deforestation and producing soil degradation (Heilmayr et al. 2020).

In summary, land-use change and soil degradation are not natural processes, but rather constructed through an interplay of soils, policies, and geopolitics. Land-use changes almost never occur in the most fertile and appropriate soils. While soils are essential for meeting the needs of a growing human population, it is also essential to keep in mind that demand and production are socially produced. Global demands for food and fodder are constructed through a global market that permeates national and local economies. In many cases, land-use change is driven by the northern demand for meat and drugs, through illegal activities with devastating social and ecological effects (e.g., Guatemala), while in others it is directly influenced by state policies (e.g., Brazil and Chile).

A future for land and people

In the following we describe two cases where local communities in central Chile have taken a path of conservation of both their land and related cultural heritage. These cases are remarkable, considering the intensity of Chile's neoliberal project, where common land tenure has been intentionally unrecognized (Manuschevich 2016). Aside from the Chilean case, Latin America has a rich tradition of communal and indigenous tenure and management that has adapted and survived many socio-economic transformations (Monterroso, Chap. 30 in this volume). One example is the well-known case of *terra preta de índio* (Indian black earth), where carbon and nutrient-rich soils were built up by *caboclo* and Indigenous communities in naturally poor Oxisols, and has recently gained much attention due to the global race to sequester carbon (Soentgen et al. 2017).

In Chile, communal land tenure originated in abandoned lands allocated during the Spanish colony (Castro and Bahamondes 1986). These lands were first used to harvest wood for mining activities, but when this became unprofitable, livestock were raised to produce tallow for export. In areas of low productivity, as in the foothills and mountains of the Aconcagua valley, some lands were abandoned and later occupied by impoverished people that had previously worked in farms and mines (Razeto and Sunckel 2016). These migrant workers organized forms of communal land tenure called *comuneros*. The *comuneros* form of tenure distributed land into several distinct categories: (1) *hijuelas*, small areas for private use of each community member that were used to grow irrigated crops; (2) *lluvias*, collective lands that were temporarily assigned for individual use to grow rainfed cereal crops until productivity decreased, often because of rainfall variability, and; (3) *campo común*, collective, rainfed grasslands that were used by all members for small-livestock production (usually goats), many times exceeding the land's carrying capacity, which produced severe soil degradation. This form of land tenure avoided the familiar succession (division) of land among poor peasants, while allowing the maintenance of food production. The legal recognition of these agricultural communities began in the 1940's, extended through the agrarian reform (1962–1973), and even persisted during the military-led dictatorship (1973–1989) (Solis de Ovando Segovia 2004).

The Aconcagua valley is located in central Chile, an area recognized as a global biodiversity hotspot (Alaniz, Galleguillos, and Perez-Quezada 2016), less than 1% of which is protected by the National System of State Protected Areas. The agricultural community Serranía El Asiento is composed of 106 members that own 1,700 hectares, located in the foothills of the Andes. After a subdivision process, the community decided to protect 1,000 hectares, obtaining in 2006 legal recognition as the Nature Sanctuary Serranía del Ciprés (CIEM 2016a). The community aimed at: (1) protecting a relict vegetation area that contains ancient specimens (>1,800 years) of *Austrocedrus chilensis*, a long-lived tree species whose conservation status is vulnerable in Chile; (2) establishing the bases for tourism and research, as a way to financially sustain the sanctuary; and (3) involving community members and other relevant actors in conservation (CIEM 2016a). A second example is the agricultural community Campo de Jahuel, where the 126 members decided to protect approximately 6,700 hectares out of the total of 8,500 hectares

Figure 3.6 People from the Serranía El Asiento community are proud of obtaining the legal protection of their land by the creation of the Nature Sanctuary Serranía del Ciprés. Photo: Jorge Razeto.

owned by the community (CIEM 2016b). After 10 years of fighting for the protection of this area, in February 2020 the Nature Sanctuary El Zaino-Laguna El Copín was finally established. This form of collective land tenure and legal recognition allows protecting not only the environment, but also archaeological sites present in this area and the cultural heritage of the *arrieros*—traditional producers who move their livestock up to the mountains during the summer months, when grasslands are dry in the lowlands. In these two initiatives, the long-term inhabitants of the area have decided to undertake a path of conservation of both their lands and their traditions (Figure 3.6).

Conclusions

Since land degradation is a social problem, addressing it requires good science and planning, as well as effective, science-based governance and concrete actions at the local level that can initiate the path toward conservation and restoration. At national and regional scales, the use of tools for land-use planning such as landscape management plans can help reduce or reverse land degradation. Such planning tools can be combined with bottom-up management activities from a landscape perspective at local and farm scales. Anthropic use of natural ecosystems can avoid causing land-cover change, provided that land management follows the principle of sustainable harvest, which defines the number of products that can be collected without compromising the yield of the land in the future. There are existing examples of this type of management that can be applied to both timber and non-timber products (Donoso Daille and Pérez Quezada 2018). In grasslands, for example, the sustainable harvest is known as carrying capacity, which is the maximum number of animals that an ecosystem can feed each year, allowing the regeneration of plants for the next season. One key point that this approach must consider, however, is environmental variability. This implies that conditions might change from year to year, particularly in ecosystems with high interannual variability in precipitation, as well as the fact that the climate is changing. In practice, however, LAC governments have instead facilitated the creation of resource frontiers, and conventional agricultural techniques based on fossil fuel-based inputs, such as fertilizers and pesticides. This reflects a narrow-minded framework oriented to agribusiness and global exports. In contrast, agroecological practices have a long history of application in LAC, which are based on the ecological rationales that small farmers traditionally use (Altieri and Toledo 2011; Félix, Chap. 35 in this volume).

Acknowledgments

Grant ANID ACE210006.

References

Armesto, Juan J., Daniela Manuschevich, Alejandra Mora, Cecilia Smith-Ramirez, Ricardo Rozzi, Ana M. Abarzúa, and Pablo A. Marquet. 2010. "From the Holocene to the Anthropocene: A historical framework for land cover change in southwestern South America in the past 15,000 years." *Land Use Policy* 27: 148–60.

Alaniz, Alberto J., Mauricio Galleguillos, and Jorge F. Perez-Quezada. 2016. "Assessment of quality of input data used to classify ecosystems according to the IUCN red list methodology: The case of the Central Chile Hotspot." *Biological Conservation* 204 (December): 378–85. https://doi.org/10.1016/j.biocon.2016.10.038.

Altieri, Miguel A., and Victor Manuel Toledo. 2011. «The agroecological revolution in Latin America: Rescuing nature, ensuring food sovereignty and empowering peasants». *Journal of Peasant Studies* 38 (3): 587–612.

Beuchle, René, Rosana Cristina Grecchi, Yosio Edemir Shimabukuro, Roman Seliger, Hugh Douglas Eva, Edson Sano, and Frédéric Achard. 2015. "Land cover changes in the Brazilian Cerrado and Caatinga Biomes from 1990 to 2010 based on a systematic remote sensing sampling approach." *Applied Geography* 58 (March): 116–27. https://doi.org/10.1016/j.apgeog.2015.01.017.

Blaikie, P., and H. Brookfield. 1986. "Land degradation and society." http://www.scopus.com/inward/record.url?eid=2-s2.0-0022826823&partnerID=40&md5=2367cff90ea99e2b451e0b1145cf8a8c.

Bulmer-Thomas, Victor. 1987. The political economy of Central America since 1920. Cambridge Latin American Studies. Cambridge: Cambridge University Press. https://doi.org/10.1017/CBO9780511572029.

Buytaert, Wouter, Rolando Célleri, Bert De Bièvre, Felipe Cisneros, Guido Wyseure, Jozef Deckers, and Robert Hofstede. 2006. "Human impact on the hydrology of the Andean Páramos." *Earth-Science Reviews* 79 (1): 53–72. https://doi.org/10.1016/j.earscirev.2006.06.002.

Castro, Milka, and Miguel Bahamondes. 1986. "Las comunidades agrícolas, IV Región, Chile." Ambiente y DesarrolloII (1): 111–26.

CIEM. 2016a. Plan de Manejo Santuario de la Naturaleza Serranía del Ciprés.

———. 2016b. Plan de Manejo y conservación comunitaria Complejo El Zaino - Laguna El Copín.

Cordani, Umberto G., Victor A. Ramos, Lêda Maria Fraga, Marcelo Cegarra, Inácio Delgado, Kaiser G. de Souza, Francisco Edson M. Gomes, and Carlos Schobbenhaus. 2016. "Tectonic map of South America=Mapa Tectônico Da América Do Sul." Map. CGMW-CPRM-SEGEMAR. http://rigeo.cprm.gov.br/jspui/handle/doc/16750.

Davidson, Eric A., Cláudio J. Reis de Carvalho, Adelaine Michela Figueira, Françoise Yoko Ishida, Jean Pierre H. B. Ometto, Gabriela B. Nardoto, Renata Tuma Sabá, et al. 2007. "Recuperation of nitrogen cycling in Amazonian forests following agricultural abandonment." *Nature* 447 (7147): 995–98. https://doi.org/10.1038/nature05900.

Devine, Jennifer A., Nathan Currit, Yunuen Reygadas, Louise I. Liller, and Gabrielle Allen. 2020. "Drug trafficking, cattle ranching and land use and land cover change in Guatemala's maya biosphere reserve." *Land Use Policy* 95 (June): 104578. https://doi.org/10.1016/j.landusepol.2020.104578.

Donoso Daille, María Regina, and Jorge Pérez Quezada. 2018. *Productos del bosque del sur de Chile: uso y recolección*. Editorial Universitaria, Santiago, Chile. http://repositorio.uchile.cl/handle/2250/153157.

Echeverria, Cristian, David Coomes, Javier Salas, José María Rey-Benayas, Antonio Lara, and Adrian Newton. 2006. "Rapid deforestation and fragmentation of Chilean Temperate Forests." *Biological Conservation* 4, 130: 481–94.

ESA. 2017. "Land Cover CCI Product User Guide Version 2.Tech. Rep." maps.elie.ucl.ac.be/CCI/viewer/download/ESACCI-LC-Ph2-PUGv2_2.0.pdf.

Etter, Andres, Clive McAlpine, David Pullar, and Hugh Possingham. 2006. "Modelling the conversion of colombian lowland ecosystems since 1940: Drivers, patterns and rates." *Journal of Environmental Management* 79 (1): 74–87. https://doi.org/10.1016/j.jenvman.2005.05.017.

Fearnside, Philip M. 2001. "Soybean cultivation as a Threat to the environment in Brazil." *Environmental Conservation* 28 (1): 23–38. https://doi.org/10.1017/S0376892901000030.

García-Ayllón, Salvador. 2016. "Rapid development as a factor of imbalance in urban growth of cities in Latin America: A perspective based on territorial indicators." *Habitat International* 58 (noviembre): 127–42. https://doi.org/10.1016/j.habitatint.2016.10.005.

Gardi, C, M Angelini, S Barceló, J Comerma, C Cruz Gaistardo, A Encina Rojas, A Jones, et al. 2014. *Atlas de Suelos de América Latina y el Caribe*. Luxembourg: Comisión Europea-Oficina de Publicaciones de la Unión Europea.

Heilmayr, Robert, Cristian Echeverría, and Eric F. Lambin. 2020. "Impacts of Chilean forest subsidies on forest cover, carbon and biodiversity." *Nature Sustainability* 3: 701–9. https://doi.org/10.1038/s41893-020-0547-0.

Kröger, Markus, Anja Nygren. 2020. "Shifting frontier dynamics in Latin America." *Journal of Agrarian Change* 20 (3): 364–86. https://doi.org/10.1111/joac.12354.

Krasilnikov, Pavel, Ma del Carmen Gutiérrez-Castorena, Robert J. Ahrens, Carlos Omar Cruz-Gaistardo, Sergey Sedov, and Elizabeth Solleiro-Rebolledo. 2013. The soils of Mexico. World Soils Book Series. Springer Netherlands. https://doi.org/10.1007/978-94-007-5660-1.

Krausmann, Fridolin, Karl-Heinz Erb, Simone Gingrich, and Timothy D. Searchinger. 2013. "Global human appropriation of net primary production doubled in the 20th century." *Proceedings of the National Academy of Sciences* 110 (25): 10324–29. https://doi.org/10.1073/pnas.1211349110.

Levis, C., F. R. C. Costa, F. Bongers, M. Peña-Claros, C. R. Clement, A. B. Junqueira, E. G. Neves, et al. 2017. "Persistent effects of pre-columbian plant domestication on Amazonian forest composition." *Science* 355 (6328): 925–31. https://doi.org/10.1126/science.aal0157.

MacDonald, Graham K., Kate A. Brauman, Shipeng Sun, Kimberly M. Carlson, Emily S. Cassidy, James S. Gerber, and Paul C. West. 2015. "Rethinking agricultural trade relationships in an era of globalization." *BioScience* 65 (3): 275–89. https://doi.org/10.1093/biosci/biu225.

Manuschevich, Daniela. 2016. «Neoliberalization of forestry discourses in Chile». *Forest Policy and Economics* 69: 21–30. https://doi.org/10.1016/j.forpol.2016.03.006.

Malhi, Yadvinder, J. Timmons Roberts, Richard A. Betts, Timothy J. Killeen, Wenhong Li, and Carlos A. Nobre. 2008. "Climate change, deforestation, and the fate of the Amazon." *Science* 319 (5860): 169–72. https://doi.org/10.1126/science.1146961.

Mendonça-Santos, Maria de Lourdes, Juan Comerma, Julio Alegre, Ildefonso Pla Sentis, Carlos Cruz Gaistardo, Rodrigo Vargas, Diego Tassinari, Moacir de Souza Dias Junior, Sebastián Santanayans Vela, María Laura and Corso. 2015. Regional assessment of soil changes in Latin America and the Caribbean: Food and agriculture organization. Food and Agriculture Organization of the United Nations, Roma, Italia.

Moreira, Fatima, Oswaldo Siqueira, and Lijbert Brussaard. 2006. "Soil and land use in the Brazilian Amazon." In *Soil biodiversity in Amazonian and other Brazilian ecosystems*, edited by M. Mendonça-Santos, H. G. Dos Santos, M. R. Cohelo, A. C. C. Bernardi, P. L. O. A. Macahdo, C. V. Manzanatto, y E. C. C. Fidalgo. Oxfordshire, UK: CABI Publishing.

Ochoa-Cueva, Pablo, Andreas Fries, Pilar Montesinos, Juan A. Rodríguez-Díaz, and Jan Boll. 2015. "Spatial estimation of soil erosion risk by land-cover change in the Andes oF Southern Ecuador." *Land Degradation & Development* 26 (6): 565–73. https://doi.org/10.1002/ldr.2219.

Oliveira, Paulo S., and Robert J. Marquis, eds. 2002. *The cerrados of Brazil: Ecology and natural history of a neotropical savanna*. Columbia University Press. New York.

Olson, David, Eric Dinerstein, Eric D. Wikramanayake, Neil D. Burgess, George V. N. Powell, Emma C. Underwood, Jennifer A. D'Amico, Illanga Itoua, Holly E. Strand, John C. Morrison, Colby J. Loucks, Thomas F. Allnutt, Taylor H. Ricketts, Yumiko Kura, John F. Lamoreux, Wesley W. Wettengel, Prashant Hedao, and Kenneth R. Kassem. 2001. "Terrestrial ecoregions of the world: A new map of life on earth: A new global map of teerrestrial ecoregions provides an innovative tool for conserving biodiversity". *BioScience* 51 (11): 933–8. https://doi.org/10.1641/0006-3568(2001)051[0933:TEOTWA]2.0.CO;2.

Razeto, Jorge, and Hanny Sunckel. 2016. "Trayectoria Agraria de la Comarca de Aconcagua." In *Aconcagua, ia comarca*, edited by Alejandro Canales, Manuel Canales, and Jorge Razeto, Corporacion CIEM, 93–108. El Almendral. San Felipe, Chile.

Rojstaczer, Stuart, Shannon M. Sterling, and Nathan J. Moore. 2001. "Human appropriation of photosynthesis products." *Science* 294 (5551): 2549–52. https://doi.org/10.1126/science.1064375.

Rubio, Gerardo, Raul S. Lavado, and Fernando X. Pereyra, eds. 2019. *The soils of Argentina*. World Soils Book Series. Springer International Publishing. https://doi.org/10.1007/978-3-319-76853-3.

Soentgen, Jens, Klaus Hilbert, Carolin Von Groote-Bidlingmaier, Gabriele Herzog-Schröder, Eije Pabst, and Sabine Timpf. 2017. "Terra preta de índio: Commodification and Mythification of the Amazonian dark earths." *GAIA - Ecological Perspectives for Science and Society* 26 (January): 136–43. https://doi.org/10.14512/gaia.26.2.18.

Solis de Ovando Segovia, Juan G. 2004. "Normativa legal de las comunidades agrícolas: análisis crítico del D.F.L. No. 5 de 1968 del Ministerio de Agricultura." GIA. Santiago, Chile.

Sparovek, Gerd, Göran Berndes, Alberto Giaroli de Oliveira Pereira Barretto, and Israel Leoname Fröhlich Klug. 2012. "The revision of the Brazilian Forest act: Increased deforestation or a historic step towards balancing agricultural development and nature conservation?" *Environmental Science & Policy* 16 (February): 65–72. https://doi.org/10.1016/j.envsci.2011.10.008.

Taboada, M. A., F. G. Micucci, D. J. Cosentino, and R. S. Lavado. 1998. "Comparison of compaction induced by conventional and zero tillage in two soils of the rolling pampa of Argentina." *Soil and Tillage Research* 49 (1): 57–63. https://doi.org/10.1016/S0167-1987(98)00132-9.

Tellman, Beth, Steven E. Sesnie, Nicholas R. Magliocca, Erik A. Nielsen, Jennifer A. Devine, Kendra McSweeney, Meha Jain, et al. 2020. "Illicit Drivers of Land Use Change: Narcotrafficking and forest loss in Central America." *Global Environmental Change* 63 (July): 102092. https://doi.org/10.1016/j.gloenvcha.2020.102092.

Thebo, Anne, P Drechsel, and E Lambin. 2014. "Global assessment of urban and peri-urban agriculture: Irrigated and rainfed croplands." *Environmental Research Letters* 9 (November): 114002. https://doi.org/10.1088/1748-9326/9/11/114002.

Vitousek, Peter M., Paul R. Ehrlich, Anne H. Ehrlich, and Pamela A. Matson. 1986. "Human appropriation of the products of photosynthesis." *BioScience* 36: 368–73.

Waroux, Yann le Polain de, Matthias Baumann, Nestor Ignacio Gasparri, Gregorio Gavier-Pizarro, Javier Godar, Tobias Kuemmerle, Robert Müller, Fabricio Vázquez, José Norberto Volante, and Patrick Meyfroidt. 2018. "Rents, actors, and the expansion of commodity frontiers in the Gran Chaco." *Annals of the American Association of Geographers* 108 (1): 204–25. https://doi.org/10.1080/24694452.2017.1360761.

Wrathall, David J., Jennifer Devine, Bernardo Aguilar-González, Karina Benessaiah, Elizabeth Tellman, Steve Sesnie, Erik Nielsen, et al. 2020. "The impacts of Cocaine-trafficking on conservation governance in Central America." *Global Environmental Change* 63 (July): 102098. https://doi.org/10.1016/j.gloenvcha.2020.102098.

4 Climate change impacts on Caribbean coastal ecosystems

Emergent ecological and environmental geography challenges

Edwin A. Hernández-Delgado

Climate change impacts in the Anthropocene

The wider Caribbean region is one of the most important global biodiversity hotspots threatened by human-driven climate change. Climatic changes have been one of the most important drivers of biodiversity changes over geological time scales. However, current trends in increasing temperature and rapidly increasing concentrations of atmospheric carbon dioxide and other greenhouse gases (GHG) such as nitrous oxide and methane, have no parallel in the geological record, at least over the Plio-Pleistocene, when compared to those observed during the spread of human civilizations (Petit et al. 1999). Agricultural and industrial development have become major drivers of GHG emissions, which have in turn had profound changes in biome composition and spatio-temporal distribution, leading to the emergence of novel ecosystems. The human, as a single species, has had such a permanent negative impact on its environment in a way that is radically unmatched by any other species. This has led many authors to name a new geological epoch after ourselves: the Anthropocene (Crutzen and Stoermer 2000). The transition of Anthropocene ecosystems that differ in species composition and/or ecological function from present and past systems is recognized as an almost inevitable consequence of ecological alterations resulting from land-use change and from other human-driven sources, including climate change, and has been regarded as novel ecosystems (Hobbs et al. 2006). As future novel climates will be warmer than the present and recent climates, with spatially variable shifts in precipitation and an increased risk of species reshuffling into future no-analog communities and other ecological surprises, special attention must be paid in order to model and understand such trends (Williams and Jackson 2007), and to implement appropriate management strategies.

Biodiversity has always responded dynamically to environmental disturbances across geological time scales, altering the cross-scale distribution of genes, species, composition of the biological, and the spatial extent and distribution of ecosystems, of landscapes and seascapes. But humans have accelerated the pace of change. The entire planet has been altered by humans, including major modifications of forest landscapes, accelerating species extinction due to habitat alteration, hunting or overfishing, and through changes in GHG, climate change, and ocean acidification (Thomas 2020). Therefore, recent changes in biodiversity should reflect direct and indirect responses to multiple drivers of change. Using tropical coastal and marine ecosystems as models, major drivers of change include a myriad of local human-driven factors (e.g., pollution, sedimentation, eutrophication, habitat modification, overfishing), Anthropogenic climate change (e.g., sea surface warming, hypoxia, ocean acidification, coral bleaching, mass coral mortality events), and the lack of sustainable governance and resources (e.g., education, public engagement, technology, funding, management capacity, legal framework, enforcement). This combination of factors often results in a major lack of equity and in the evolution of poverty

DOI: 10.4324/9780429344428-6

traps, largely affecting marginalized communities, and small island development states (SIDS) (Hernández-Delgado 2015), with significant colonial policies and legacies, and where socio-economic impacts of increasing temperature and hurricanes can be more severe (Hsiang 2010). Here, the term *colonial legacy* is used in the context of changes that affect island nations and regions governed by invasive colonial rulers. Such colonialism can perpetuate change in many facets of society, including political practices, government structure, governance, and economic development, which often result in non-sustainable practices, strong environmental degradation, and situations of social and environmental injustice.

The long-term consequences of local Anthropogenic impacts to Caribbean coastal ecosystems and climate-related changes are often magnified on SIDS due to their higher vulnerability to multiple climate-related impacts, their lack of access to technology, their still strong colonial policies and legacies, environmental injustice, slow ability to recover from natural disasters, the unprecedented socio-economic impacts of the COVID-19 pandemic, and their limited socio-economic resources to either mitigate or adapt to future climate change-related impacts. Also, ecological surprises, such as coral disease outbreak events, may lead to species and functional group removal from habitats (e.g. Caribbean Acroporid corals), further altering ecological functions and services and in the long term leading to climate-related local species extirpations (Hernández-Pacheco et al. 2011; Mercado-Molina et al. 2020). In the long term, this may lead to the loss of functional redundancy, to major habitat homogenization, to net food web shrinkage, and to the loss of many ecological functions and services, thereby affecting SIDS' ecological services and benefits.

There is evidence that coral reef social-ecological systems across Caribbean territories are significantly more vulnerable to mass bleaching impacts than sovereign states with stronger legal framework (Siegel et al. 2019). The colonial legacy is a critical game-changing factor influencing the ecological state of coral reefs. Further, projected sea level rise (SLR) may affect a significant proportion of coastal wetlands, increasing the vulnerability of coastal human settlements to future storm events (Nicholls 2004). This would further result in the net erosion of ecosystem resilience, thus leading to increased ecological and socio-economic vulnerability to future climate-related extreme events (e.g. hurricanes and winter swells) and SLR.

Projected climate changes across the wider Caribbean may also impose major challenges to ecology and environmental geography that remain poorly understood. Coastal human populations of Latin America and the Caribbean are already highly vulnerable as a result of environmental deterioration, non-sustainable development (Hernández-Delgado et al. 2012), stronger hurricanes (Morales Vélez and Hughes 2018), and other impacts. The region's wide variety of coastal natural resources has long supported subsistence livelihoods, fishing communities, and tourism, which have become increasingly vulnerable to extreme climatic events (Smith and Rhiney 2016). Further, over the last century Caribbean-wide coastal resources have significantly deteriorated in many areas as a result of multiple factors, including Anthropogenic pressure, poverty, and destructive human activities such as overfishing, marine pollution, and unrestricted, unsustainable development of the coast (Jackson et al. 2014).

Many Caribbean SIDS face parallel socio-economic and political challenges that could be exacerbated by projected changes in the climate and aggravated by current neo-colonial educational, political and socio-economic barriers (Machado-Aráoz 2010; Fusté 2017), which are exacerbated by a pressing colonial legacy and environmental injustice. Other concerning challenges include the declining regional economy (Córdova and Seligson 2010), increasing poverty (Leipziger 2001), income inequality (Córdova and Seligson 2010), non-sustainable tourism (Hampton and Jeyacheya 2020), declining fisheries (Jackson et al. 2014), lack of food security and sovereignty (Beckford 2012), and migration (Pellegrino 2000), especially considering the increasing frequency and strength of hurricanes (Acosta et al., 2020; Emanuel 2005), and high vulnerability to disasters (Ferdinand et al. 2012; Lloréns 2018). Further, there are increasing geopolitical challenges associated with globalized economies, colonial relations, unequal distribution of and access to resources, and justice associated with the burst of 'blue economy' (Ertör and Hadjimichael 2020). The World Bank has defined *blue economy* as the sustainable use of ocean resources for economic growth, improved livelihoods and jobs, while preserving the health, ecological functions, and benefits of coastal and marine ecosystems. It encompasses many activities, including the management of tourism, renewable energy, maritime transport, waste management, fisheries, resource conservation, and climate change.

Threats of climate change to Caribbean coastal and marine ecosystems

Seagrasses

Seagrasses are flowering plants adapted to grow submerged in shallow marine grounds, capable of forming extensive meadows, and which provide significant ecological and socio-economic services to humankind (Duarte 2002). However, seagrasses are undergoing significant chronic environmental changes as a result of coastal human population stress, including Anthropogenic climate change. Seagrasses provide key ecological services, including:

1) organic carbon production and export (Odum et al. 1959)
2) carbon storage in sediments (Russell et al. 2013)
3) nutrient cycling (Odum et al. 1974)
4) modification of water chemistry, supporting higher pH values and buffering ocean acidification trends (Unsworth et al. 2012)
5) supporting biodiversity (Glynn 1964)
6) sustaining important fisheries (Nagelkerken et al. 2000)
7) dampening wave action and stabilizing sediments and shorelines (Fonseca and Calahan 1992)
8) sustaining important trophic transfers and connectivity to adjacent habitats (Unsworth et al. 2008)

However, a combination of multiple stochastic and chronic stressors, including sediment-laden and nutrient-loaded turbid runoff, physical disturbances, invasive species, disease, commercial fishing practices, aquaculture, overgrazing, algal blooms, and global warming have resulted in extensive seagrass declines at scales of hundreds of square kilometers across the globe (Orth et al. 2006) and in the rapid evolution of novel seagrass ecosystems (Figure 4.1).

Seagrass ecosystems are vulnerable to a wide range of stochastic natural (e.g., hurricanes) and human-driven factors (e.g., runoff pulses, mechanical turbulence from propellers, turbines etc.). There are also multiple chronic local (e.g. pollution, eutrophication, sedimentation, turbidity) and regional/global, climate change-driven factors (e.g., sea surface warming, ocean acidification, SLR) that may lead to rapid loss in biodiversity, habitat fragmentation, anoxia, increasing H_2S concentration, rapid colonization by invasive species and opportunist taxa (epiphytes, cyanobacteria), and altered demersal/benthic (bottom) species composition (Hernández-Delgado et al. 2020). In the long term, these combined factors may affect the natural productivity of deeper seagrass assemblages of all species, eventually affecting ecosystem functions, services and benefits, ecosystem resilience and their socio-economic value, including its fish nursery ground role, their water filtering role, and their ability to dampen wave action and stabilize soft sediments. Therefore, the natural connectivity role of seagrasses might also be affected, potentially degrading ecological functions of its associated ecosystems.

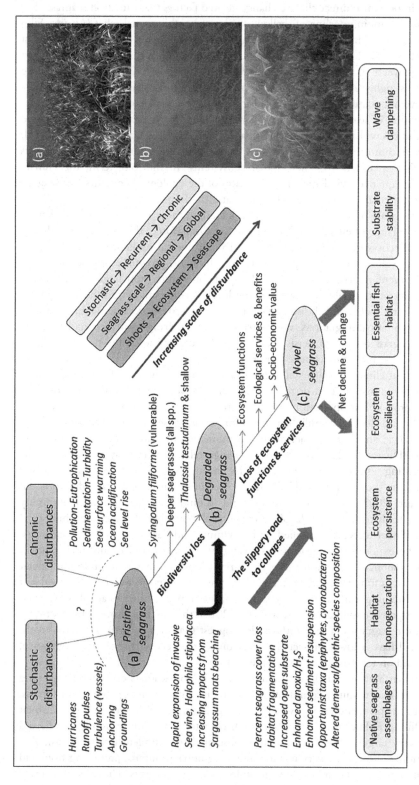

Figure 4.1 Conceptual dynamics of combined impacts from local human stressors and Anthropogenic climate change on seagrass ecosystems. Major drivers of change are illustrated by the gray boxes on the top left and on the list of disturbances on the top and left side of the image. Ovals illustrate the stages of progressive ecological change. Arrows represent trajectories of change and the outcomes of combined stochastic and chronic disturbances. Rectangles show the increasing scales of disturbances. Images correspond to ovals of progressive change: (a) Pristine native turtle grass, Thalassia testudinum; (b) Example of T. testudinum fragmented habitat under declining water quality; and (c) Example of the invasive sea vine, Halophila stipulacea, displacing T. testudinum.

There are also important indirect climate change-related factors that can affect seagrass eco-systems. With the continuing decline of shallow coral reef benthic spatial relief (Alvarez-Filip et al. 2009), the projected increase in record-breaking marine heatwave events, by mid-21st century (Arias-Ortiz et al. 2018) and on ocean deoxygenation trends (Limburg et al. 2020), it is expected that between 70 and 90% of all coral reefs will be lost by 2050 (Hoegh-Guldberg et al., 2018). This is highly concerning for the conservation of most shallow seagrass communities. Losing wave attenuation from shallow adjacent coral reefs, in combination with SLR, will enhance exposure of seagrasses to increasing wave action in the near future. By this time many seagrass meadows will also be negatively impacted by a changing climate (Unsworth et al. 2018), largely by declining protection from wave energy and wave runup, and by losing its natural ecological connectivity to other coastal ecosystems. In the long term, the climate change-related consequences will largely be a function of the ecological conditions of seagrass ecosystems, environmental history of each location, and the frequency and severity of disturbances. Further, the scale of any disturbance event, including those related to climate change, will generate a wide range of effects at a different pace as a function of the number of resilience features of seagrass ecosystems (Unsworth et al. 2015). Biological resilience features include genetic diversity, species diversity, biological traits (e.g., biomass, growth, reproduction, competition ability, energetic reserves), and the availability of continuous (non-fragmented) habitat. Also, the complexity of its trophic interactions, the level of connectivity to other ecosystems (e.g., mangroves, coral reefs), and the sustainability of bio-physical processes (e.g., sustaining mobile links through faunal migrations, water quality, water exchange, surface circulation) are fundamental for seagrass ecosystem persistence and resilience, and its ability to recover from broad-scale disturbance, including climate change.

Broader impacts of climate change will synergistically interact with multiple local human effects, most notably eutrophication or the excesses of dissolved nutrient concentration (i.e., nitrogen, phosphorous). Eutrophication may lead to enhanced phytoplankton growth, to increased turbidity and declining sunlight penetration, to increased overgrowth by epiphytic and opportunistic algae, shading by algae, and to a combination of shading and anoxia by resus-pended sediments (Duarte 2002). In turn, climate change-related effects may include a combination of contrasting effects. On one hand, increased CO_2 should lead to increased photosynthesis, but on the other it should also lead to declining calcification rates due to ocean acidification (Duarte 2002). Seawater warming trends should lead to increasing community respiration, which leads to increased dissolved oxygen solubility but also increased biological oxygen demand. Under SLR, if combined with declining and flattening shallow coral reefs, seagrasses might be exposed to long-term increasing wave energy and runup during storm events, as well as to submarine erosion and deep regression, which may result in losing habitat and in losing ecosystem functions and services. Under present and projected trends of coral reef degradation and flattening with increasing climate change impacts, fisheries are also expected to move toward seagrass ecosystems (Unsworth et al. 2018), thereby increasing potential future fishing impacts over seagrasses. These authors also suggest that, in comparison to coral reefs, climate change will have a lower projected impact on seagrass ecosystem services, resilience, global coverage, and on the cost-benefit of conservation.

According to Duarte et al. (2020), rebuilding marine life on seagrass ecosystems following chronic disturbances may require a combination of activities with different potential for success. These include: species protection (low), wise harvesting (high), space protection (high), habitat restoration (medium), pollution reduction (critical), climate change mitigation (critical), and establishing recovery targets by 2050 (assuming substantial to complete interventions). Key management actions to foster recovery include: reducing nutrient inputs, protecting from and avoiding physical impacts, and conducting large-scale restoration projects. These should engage multiple societal actors. Following Unsworth et al. (2018), specific potential actions to increase the long-term resilience of seagrasses may include:

1) increasing genetic diversity through seed banks and habitat restoration;
2) reducing physical impacts through a combination of management strategies that should include restoring adjacent coral reefs to reduce wave energy (Hernández-Delgado et al. 2018a, 2018b);
3) reducing algal/epiphytic overgrowth by restoring water quality and herbivory;
4) increasing photosynthetic productivity by improving water quality;
5) reconnecting isolated and fragmented meadows through habitat restoration;
6) restoring balanced herbivory (e.g. through marine protected areas or MPAs);
7) reducing chemical toxicity through watershed-scale management of pollutant discharges;
8) providing early warning signals of issues of concern through permanent monitoring programs;
9) increasing compliance with environmental regulations associated with seagrasses;
10) restoring and maintaining connectivity to other ecosystems.

However, the net success of ecological restoration will depend on the projected influences by recurrent marine heatwaves in the future (Aoki et al. 2020). Nevertheless, it would be paramount to enhance, by all means, efforts to build public awareness and education about the critical importance of conserving and restoring seagrass habitats (Orth et al. 2006).

Coral reefs

Coral reefs provide a habitat to well over a million species and multiple essential ecosystem services (e.g., food protein, natural pharmacological products, carbon sequestration, coastal protection, tourism-based revenue) to hundreds of millions of people throughout the tropics, particularly, on small islands (Cinner et al. 2012). Despite their importance, coral reefs are in rapid decline (Hoegh-Guldberg et al. 2017, 2018), with the rate accelerating for many coral reefs over the past decade (Hughes et al. 2017). The wider Caribbean region has lost entire monospecific biotopes of Elkhorn coral (Acropora palmata) across its range even before recurrent massive coral bleaching and coral disease outbreaks, mostly as a result of a combination of local human stressors (Cramer et al. 2020). Local factors such as fishing (Jackson et al. 2014), eutrophication (Cloern 2001), sewage pollution (Bonkosky et al. 2009), turbidity (Loiola et al. 2019), sedimentation (Rogers 1990), and poor land-use management along with altered terrestrial-marine connectivity (Ramos-Scharrón et al. 2015) are detrimental to coral reef conservation. These factors have affected the coral reef's long-term ability to recover from disturbances (Gardner et al. 2005). As a consequence of a combination of multiple disturbances, including increasing hurricanes and climate change-related stress, coral reef decline has also resulted in significant structural flattening and in an overall biological community homogenization (Alvarez-Filip et al. 2009), severely affecting reef accretion and losing its wave energy attenuation role. The persistence of reef fish assemblages and the maintenance of many ecological services are also related to the persistence and dominance of significant reef-building species (Alvarez-Filip et al. 2013). Therefore, combined human-driven stressors may cause irreversible loss of physical structure, ecological persistence, services, and resilience, which can be compounded by climate change-related factors.

The most significant climate change-related threats to coral reefs are: sea surface warming, ocean acidification, recurrent hurricanes and winter storms, and sea level rise (Figure 4.2). Sea surface warming causes recurrent massive coral bleaching events, which have increased in

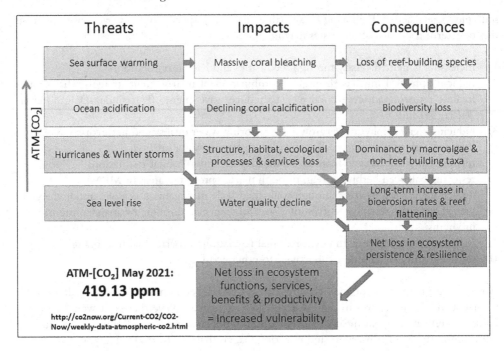

Figure 4.2 Conceptual model of threats, impacts, and consequences from Anthropogenic climate change on coral reef ecosystems. Major drivers of change are illustrated by arrows. Some can produce multiple impacts and impacts can result in multiple cumulative and synergic consequences.

frequency and intensity on regional and global scales, and are predicted to occur annually in the coming decades (Hoegh-Guldberg et al. 2007). Recent mass coral bleaching events have resulted in significant coral mortality episodes, mostly affecting reef-building taxa (Rogers and Miller 2006). Bleaching can, in turn, result in massive coral mortalities and in the loss of critical reef-building coral species (Hernández-Pacheco et al. 2011). Coral disease outbreaks have also been linked to increasing SST and mass bleaching events (van Woesik and Randall 2017). More recently, stony coral tissue loss disease (SCTLD) epizootic events have caused extensive coral mortalities across the wider Caribbean (Aeby et al. 2019). Bleaching and coral disease event trends have resulted in major long-term loss of resilience (Scheffer et al. 2001) and in ecological phase shifts in benthic community structure (Aronson et al. 2004), which often lead to macroalgal dominance over corals, and drive long-term community recovery trajectories into very different pathways following disturbances (Hernández-Delgado et al. 2014). Coral bleaching and overall reef decline can also have negative impacts on reef fish community structure, particularly on small reef dwellers. Declining reefs may also lose over 35% of their fishery productivity in the long term (Rogers et al. 2018). Therefore, long-term coral reef decline from the combined impact of local human stressors and climate change may threaten the role of coral reefs in providing food protein, risking food security and sovereignty, particularly on SIDS (Hernández-Delgado 2015).

Ocean acidification, in combination with increasing SST, can also result in declining coral calcification (Albright et al. 2008). This can, in turn, lead to a long-term coral biodiversity and reef resilience loss (Anthony et al. 2011). Further, climate change has been implicated in increasing frequency of stronger hurricanes across the Atlantic (Emanuel 2005) which could have important physical impacts on coral reefs. This may result in the loss of benthic physical structure, habitat, ecological processes, and of ecological services and benefits. In the long term, this may contribute to the increasing dominance by macroalgae and non-reef-building taxa. Finally, SLR may further contribute to increased hydrodynamic impacts by increasing water level, particularly under extreme events, increasing shoreline erosion and coastal flooding risks. Increasing atmospheric CO_2 concentration trends suggest that impacts from such threats are expected to increase in the near future. This combination of threats may lead to a net loss in ecosystem persistence and resilience and, in the long term, a loss in the ecological services, benefits, and productivity of the coral reef ecosystem, leading to enhanced risks to the persistence and resilience of other associated ecosystems such as seagrass and mangroves and to coastal human communities.

It remains poorly understood how quickly Caribbean corals can recover from bleaching and disease-related mortality, and from any of such compounding disturbances. There is evidence showing that Indo-Pacific coral reefs have a higher potential for recovery from massive coral bleaching (Adjeroud et al. 2009) than their Caribbean counterparts (Hernández-Pacheco et al. 2011; Edmunds 2013). Part of this contrasting pattern has to do with the significantly lower fish and coral functional redundancy of the wider Caribbean region in comparison to that of the Great Barrier Reef (Bellwood et al. 2004). The higher the functional redundancy, the higher the recovery potential from disturbance.

Conclusions and regional strategies for climate change adaptation

The Wider Caribbean region is one of the most important biodiversity hotspots on the planet. However, most coastal and marine ecosystems have undergone significant environmental degradation. Much of the historical impacts are associated with non-sustainable development practices and policies, largely resulting from the widespread colonial policies and legacies of SIDS. Neoliberal capitalism, massive non-sustainable tourism practices, overexploitation of natural resources, environmental injustice, lack of food security and sovereignty, increasing water scarcity, and increasingly recurrent natural disasters have resulted in declining resilience to future natural disasters and in an increased vulnerability to projected climate change impacts and SLR. Climate change has already caused major unprecedented impacts across many coastal tropical ecosystems, on public health, on local economies, and on community livelihoods. The magnitude of such impacts has become more pronounced for SIDS, and has been further exacerbated by the public health and socio-economic crises from the COVID-19 pandemics. Preparing and adapting for climate change has become a growing challenge, particularly for Caribbean SIDS. The nature and severity of impacts are likely to vary from place to place, across economic sectors, and also across social organizational scales, from individual and household to regional and trans-national scales. There are, however, multiple important actions that are needed to improve the ability of wider Caribbean nations to adapt to projected climate change impacts, while at the same time cope with pressing local priorities such as the post-pandemics and post-natural disaster socio-economic recovery.

Capacity building and technology transfer

Multiple Caribbean SIDS are still under the shadow of historical colonial policies and legacies, with limited capacity and resources to adapt and mitigate climate change. This is a fundamental component to empower Caribbean nations to address, adapt, and mitigate climate change-related problems, and to implement sustainable solutions to socio-economic and environmental problems. There is an imperative need to elaborate and implement multidisciplinary local, national, and regional capacity-building efforts (e.g., technology transfer, research training, grant writing, ecological restoration, protected marine areas management, watershed rehabilitation and management, enhanced governance, and implementation of sustainability goals). There is also a need to develop capacity building and to integrate through hands-on experiences multiple societal sectors, including traditionally underrepresented groups in the ocean science (e.g. a participatory, inclusive, and just blue economy), and to enhance science infrastructure in SIDS. This will require enhanced collaboration, cooperation, and participation of multiple sectors (e.g. government, academia, private sector, NGOs, and base communities) from multiple nations. There is also a need to foster and enhance citizen science programs for promoting public involvement in the management and conservation of natural resources. It would also be critical to train the next generation of scientists, policy-makers and communicators in science through the mobilization and integration of younger generations such as students and early career scientists.

Partnerships and financing

A successful sustainable development agenda for Caribbean nations requires partnerships between governments, the private sector and civil society, including community-based actors. Inclusive partnerships built upon principles and values, a shared vision, common goals, transparency and trust, that place people and nature as the principal resources, are needed at multiple scales: local, national, regional, and global. There is an urgent need to mobilize, redirect and unlock the transformative power of private resources (e.g. tourism, transportation, energy, and pharmaceutical sectors) to deliver on sustainable development objectives. Most Caribbean nations are in crisis (socio-economic, disaster-related, and COVID-19 pandemic-related), compounded by colonial policies and legacies, and require significant foreign investment to support recovery in multiple areas. These include public health, education, sustainable energy, infrastructure, transport, water supplies, agro-ecology, natural resource rehabilitation, as well as information and communications technologies.

Access to information, data, and knowledge

There is still an important need to incorporate local knowledge from Caribbean SIDS institutions and local traditional knowledge into science and policy-making. That would require significant opportunities for participation, integration, collaboration, and cooperation among multiple local, regional, and international stakeholders. It is important to translate technological knowledge into local languages in order to reach a popular audience. In addition, there is a need to foster enhanced funding opportunities for Caribbean SIDS and for colonial states (e.g., Puerto Rico, which lacks access to international funding sources as a US colony). Also, there is a need to foster the participation of SIDS scientists on cooperative/collaborative research projects, minimizing colonial or parachute science (*sensu* Stefanoudis et al. 2021), and improving investments on scientific training, technological infrastructure, fostering peer-reviewed publications of local scientists and students from SIDS in multiple disciplines.

Communications and awareness raising

There is a major need to bridge the gap between science, policy-makers, decision-makers, the private sector, and base communities. This will require enhanced multi-sectorial two-way communication, collaboration, cooperation, and participation, and enhanced use of social media. Further, academics still need to incorporate improved communication (e.g., education and outreach) skills in all of their projects to raise awareness regarding climate change impacts to the public. There is also an imperative need to foster enhanced social-ecological governance and management capacity building on SIDS. This should include fostering the development of region-wide programs on coral and seagrass propagation and ecological restoration, fishery recovery and enhancement, and on coastal resiliency rehabilitation. Similarly, there is a growing need to develop local and Caribbean-wide strategies to foster the implementation of sustainable development goals identified by the United Nations Environment Program.

Public health, post-natural disaster, and post-COVID-19 recovery

A key element is the need to implement multiple strategies to foster the recovery of a myriad of climate change-related public health threats. According to Hernández-Delgado (2015), these may include the development of early warning systems for the identification of heat-wave threats, architectural and engineering modifications, implementation of community response plans, massive reforestation campaigns, coastal habitats restoration, a combination of green or gray infrastructure development for mitigating SLR, improved methods for water conservation, water treatment, improved coastal agriculture practices, and prevention of coastal water pollution to reduce threats to public health and foster food security and sovereignty in SIDS. Further, Caribbean SIDS have become increasingly vulnerable to projected climate change impacts due to their already high vulnerability to natural disasters and to the COVID-19 pandemic. Therefore, there is a pressing need on many island nations to complete recovery from recent major hurricanes (e.g. need to restore access to drinking water and electricity, replace damaged infrastructure, restore public health services, education, government services, and tourism). This has further aggravated the socio-economic impacts of the COVID-19 pandemic and the need to rehabilitate declining economies and livelihoods. Thus, this combination of factors has created unprecedented vulnerability threats in SIDS to social-ecological systems already historically affected by colonial policies, legacies and environmental injustice.

The community-based management alternative

Many Caribbean SIDS have limited access to human, technological, and economic resources, which can limit governance and natural resource management capacity. An alternative to that is the implementation of community-based management. There have been many communities and organizations managing their coastal resources at the local level for a long time. However, increasing influences of external interests (e.g., investors, developers, massive tourism companies) and other new actors (e.g., foreign bit coin and crypto-investors, money launderers) do not consider or value the efforts and work carried out at the local level and often result in hoarding local natural resources. This may systematically affect water availability, reduce access to agricultural lands, and eliminate critical coastal defense lines against storm surge and SLR (e.g. coral reefs, mangroves, coastal wetlands, and sand dunes). To generate local jobs, decent and stable working conditions, and to restore local livelihoods in the context of natural disaster and post-COVID-19 pandemic recovery, at least for Caribbean SIDS, it would first be necessary to adequately temper government institutions to the current reality of increasing projected climate change threats, limited climate change adaptation and mitigation capacity, declining ability to recover from natural disaster, and socio-economic, health, institutional, and governance crisis. This will require a need to strengthen the local legal framework regarding coastal zone management and investing on local-regional marine spatial planning, which will also generate jobs for local professional planners, environmental scientists, biologists, among other key emerging professionals. This means we must also redefine and refocus blue economy strategies to prevent injustice, exclusion, and inequity at the local level.

Climate change has already had important consequences on Caribbean-wide SIDS and poor nations. It has affected terrestrial and marine ecosystems, already having adverse consequences on biodiversity, in ecosystems' ecological processes, persistence, productivity, services, benefits, and resilience. Projected climate change impacts may further exacerbate such effects due to its compounded impacts with other local human stressors, increasing vulnerability of both ecosystems and human communities, which are already affected by limited governance and management capacity, limited access to resources, and are affected by colonial policies, legacies and environmental injustice. Ecological restoration is a critical step in the process of empowering base communities to adapt to climate change impacts. However, it would not be enough if meaningful collaborative networking and functional partnerships are not immediately achieved. These conditions perpetuate poverty and inequity and it would be paramount for all societal sectors across Caribbean-wide SIDS and poor nations to join efforts and seek rapid solutions. International collaboration and cooperation is paramount to support Caribbean SIDS and poor nations to overcome these compounded conditions in order to successfully be able to cope, adapt, and mitigate projected climate change. This is the most significant geographical and social-ecological challenge against a changing climate that brings multiple obstacles but also multiple opportunities.

Acknowledgments

This collaboration was possible thanks to the partial support provided by Sociedad Ambiente Marino (https://sampr.org/). This work is dedicated to the memory of my beloved mother, Sonia, and grandmother, Sara, who passed away early in 2022.

References

Acosta, Rolando J., Nishant Kishore, Rafael A. Irizarry, and Caroline O. Buckee. 2020. "Quantifying the dynamics of migration after Hurricane Maria in Puerto Rico." *Proceedings of the National Academy of Sciences* 117(51): 32772–32778.

Adjeroud, Mehdi, Francois Michonneau, Peter J. Edmunds, Yannick Chancerelle, Thierry Lison De Loma, Lucie Penin, Loic Thibaut, Jeremie Vidal-Dupiol, Bernard Salvat, and R. Galzin. 2009. "Recurrent disturbances, recovery trajectories, and resilience of coral assemblages on a South Central Pacific reef." *Coral Reefs* 28(3): 775–780.

Aeby, Greta, Blake Ushijima, Justin E. Campbell, Scott Jones, Gareth Williams, Julie L. Meyer, Claudia Hase, and Valerie Paul. 2019. "Pathogenesis of a tissue loss disease affecting multiple species of corals along the Florida Reef Tract." *Frontiers in Marine Science* 6: 678.

Albright, Rebecca, Benjamin Mason, Chris and Langdon. 2008. "Effect of aragonite saturation state on settlement and post-settlement growth of Porites astreoides larvae." *Coral Reefs* 27(3): 485–490.

Alvarez-Filip, Lorenzo, Juan P. Carricart-Ganivet, Guillermo Horta-Puga, and Roberto Iglesias-Prieto. 2013. "Shifts in coral-assemblage composition do not ensure persistence of reef functionality." *Scientific Reports* 3: 3486.

Alvarez-Filip, Lorenzo, Nicholas K. Dulvy, Jennifer A. Gill, Isabelle M. Côté, and Andrew R. Watkinson. 2009. "Flattening of Caribbean coral reefs: region-wide declines in architectural complexity." *Proceedings of the Royal Society B: Biological Sciences* 276(1669): 3019–3025.

Anthony, Kenneth R., Jeffrey Maynard, Guillermo Díaz-Pulido, Peter J. Mumby, Paul A. Marshall, Long Cao, and Ove Hoegh-Guldberg. 2011. "Ocean acidification and warming will lower coral reef resilience." *Global Change Biology* 17(5): 1798–1808.

Aoki, Lillian R., Karen J. McGlathery, Patricia L. Wiberg, and Alia Al-Haj. 2020. "Depth affects seagrass restoration success and resilience to marine heat wave disturbance." *Estuaries and Coasts* 43(2): 316–328.

Arias-Ortiz, Ariane, Oscar Serrano, Pere Masqué, P. S. Lavery, U. Mueller, G.A. Kendrick, M. Rozaimi, A. Esteban, J.W. Fourqurean, N.J.N.C.C. Marbà, M.A. Mateo, Murray, K., Rule, M.J., and Duarte, C.M. 2018. "A marine heatwave drives massive losses from the world's largest seagrass carbon stocks." *Nature Climate Change* 8(4): 338–344.

Aronson, R.B., I.G. Macintyre, C.M. Wapnick, and M.W. O'Neill 2004. "Phase shifts, alternative states, and the unprecedented convergence of two reef systems." *Ecology* 85(7): 1876–1891.

Beckford, Clinton. 2012. "Issues in Caribbean food security: building capacity in local food production systems." In *Food Production Approaches, Challenges and Tasks*, edited by D. Krstic, I. Djalovj, D. Nikezic, and D. Bjelj, 25–40. Tech, Europe.

Bellwood, D.R., T.P. Hughes, C. Folke, and M. Nyström 2004. "Confronting the coral reef crisis." *Nature*, 429(6994): 827–833.

Bonkosky, M., E.A. Hernández-Delgado, B. Sandoz, I.E. Robledo, J. Norat-Ramírez, & H. Mattei. 2009. "Detection of spatial fluctuations of non-point source fecal pollution in coral reef surrounding waters in southwestern Puerto Rico using PCR-based assays." *Marine Pollution Bulletin* 58(1): 45–54.

Cinner, J.E., T.R. McClanahan, N.A. Graham, T.M. Daw, J. Maina, S.M. Stead, A. Wamukota, K. Brown, and Ö. Bodin 2012. "Vulnerability of coastal communities to key impacts of climate change on coral reef fisheries." *Global Environmental Change* 22(1): 12–20.

Cloern, James E. 2001. "Our evolving conceptual model of the coastal eutrophication problem." *Marine Ecology Progress Series* 210: 223–253.

Córdova, Abby and Mitchell A. Seligson. 2010. "Economic shocks and democratic vulnerabilities in Latin America and the Caribbean." *Latin American Politics and Society* 52(2): 1–35.

Cramer, Katie L., Jeremy B. Jackson, Mary K. Donovan, Benjamin J. Greenstein, Chelsea A. Korpanty, Georfrey M. Cook, and John M. Pandolfi. 2020. "Widespread loss of Caribbean acroporid corals was underway before coral bleaching and disease outbreaks." *Science Advances*, 6(17): 1–10. eaax9395.

Crutzen, P., and E. Stoermer 2000. The "Anthropocene." *Global Change Newsletter* 41: 17–18.

Duarte, Carlos M. 2002. "The future of seagrass meadows." *Environmental Conservation* 29(2): 192–206.

Duarte, C.M., S. Agusti, E. Barbier, G.L. Britten, J.C. Castilla, J.P. Gattuso, R.W. Fulweiler, T.P. Hughes, N. Knowlton, C.E. Lovelock, and H.K. Lotze 2020. "Rebuilding marine life." *Nature* 580(7801): 39–51.

Edmunds, Peter J. 2013. "Decadal-scale changes in the community structure of coral reefs of St. John, US Virgin Islands." *Marine Ecology Progress Series* 489: 107–123.

Emanuel, Kerry. 2005. "Increasing destructiveness of tropical cyclones over the past 30 years." *Nature* 436(7051): 686–688.

Ertör, Irmak and M. Hadjimichael. 2020. "Blue degrowth and the politics of the sea: rethinking the blue economy." *Sustainability Science* 15(1): 1–10.

Ferdinand, Idelia, Geoff O'Brien, Phil O'Keefe, and Janaka Jayawickrama. 2012. "The double bind of poverty and community disaster risk reduction: a case study from the Caribbean." *International Journal of Disaster Risk Reduction* 2: 84–94.

Fonseca, M.S., and J.A. Calahan 1992. A preliminary evaluation of wave attenuation by four species of seagrass. *Estuarine, Coastal and Shelf Science* 35: 565–576.

Fusté, José I., 2017. "Repeating islands of debt: Historicizing the transcolonial relationality of Puerto Rico's economic crisis." *Radical History Review* 2017(128): 91–119.

Gardner, Toby A., Isabelle Côté, Jennifer A. Gill, Alastair Grant, and Andrew R. Watkinson. 2005. "Hurricanes and Caribbean coral reefs: impacts, recovery patterns, and role in long-term decline." *Ecology* 86(1):174–184.

Glynn, P.W., 1964. "Effects of Hurricane Edith on marine life in La Parguera, Puerto Rico." *Caribbean Journal of Science* 4: 335–345.

Hampton, Mark P. and Julia Jeyacheya. 2020. "Tourism-dependent small islands, inclusive growth, and the blue economy." *One Earth* 2(1): 8–10.

Hernández-Delgado, Edwin A. 2015. "The emerging threats of climate change on tropical coastal eco-system services, public health, local economies and livelihood sustainability of small islands: cumulative impacts and synergies." *Marine Pollution Bulletin* 101(1): 5–28.

Hernández-Delgado Edwin A., C.E. Ramos-Scharrón, C. Guerrero, M.A. Lucking, R. Laureano, P.A. Méndez-Lázaro, & J.O. Meléndez-Díaz. 2012. "Long-term impacts of tourism and urban development in tropical coastal habitats in a changing climate: Lessons learned from Puerto Rico." In *Visions from Global Tourism Industry-Creating and Sustaining Competitive Strategies*, edited by M. Kasimoglu, 357–398. Rikeja, Croatia: Intech Publications. ISBN 979-953-307-532-6.

Hernández-Delgado, Edwin A., C.M. González-Ramos, & P.J. Alejandro-Camis. 2014. "Large-scale coral recruitment patterns in Mona Island, Puerto Rico: evidence of shifting coral community trajectory after massive bleaching and mortality." *Revista de Biología Tropical* 62 (Suppl. 3): 49–64.

Hernández-Delgado, Edwin A., Alex E. Mercado-Molina, and Samuel E. Suleimán-Ramos. 2018a. "Multi-disciplinary lessons learned from low-tech coral farming and reef rehabilitation practices. I. Best management practices." In *Corals in a Changing World*, edited by Duque-Beltrán, C., Tello-Camacho, E, 213–243. InTech Publ. ISBN 978-953-51-3910-0.

Hernández-Delgado, Edwin A., Alex E. Mercado-Molina, Samuel E. Suleimán-Ramos, and Lucking, M.A. 2018b. "Multi-disciplinary lessons learned from low-tech coral farming and reef rehabilitation practices. II. Coral demography and social-ecological benefits." In *Corals in a Changing World*, edited by Duque-Beltrán, C., Tello-Camacho, E., 245–268. InTech Publ. ISBN 978-953-51-3910-0.

Hernández-Delgado, Edwin A., C. Toledo-Hernández, C.P. Díaz-Ruíz, N. Gómez-Andújar, J.L. Medina-Muñiz, S.E. Suleimán-Ramos, M.F. Canals-Silander. 2020. "Hurricane impacts and the resilience of invasive sea vine, Halophila stipulacea: a case study from Puerto Rico." *Estuaries and Coasts* https://doi.org/10.1007/s12237-019-00673-4,1-21.

Hernández-Pacheco, R., E.A. Hernández-Delgado, & A.M. Sabat. 2011. "Demographics of bleaching in the Caribbean reef-building coral Montastraea annularis." *Ecosphere* 2(1): 1–13. art9.

Hobbs, R.J., S. Arico, J. Aronson, J.S. Baron, P. Bridgewater, V.A. Cramer, P.R. Epstein, J.J. Ewel, C.A. Klink, A.E. Lugo, and D. Norton. 2006. "Novel ecosystems: theoretical and management aspects of the new ecological world order." *Global Ecology and Biogeography* 15(1): 1–7.

Hoegh-Guldberg, Ove, Emma V. Kennedy, Hawthorne L. Beyer, Caleb McClennen, and Hugh P. Possingham. 2018. "Securing a long-term future for coral reefs." *Trends in Ecology & Evolution* 33(12): 936–944.

Hoegh-Guldberg, Ove, P.J. Mumby, A.J. Hooten, R.S. Steneck, P. Greenfield, E. Gomez, C.D. Harvell, P.F. Sale, A.J. Edwards, K. Caldeira, and N. Knowlton 2007. "Coral reefs under rapid climate change and ocean acidification." *Science* 318(5857): 1737–1742.

Hoegh-Guldberg, O., E.S. Poloczanska, W. Skirving, and S. Dove 2017. "Coral reef ecosystems under climate change and ocean acidification." *Frontiers in Marine Science* 4: 158.

Hsiang, S.M. 2010. Temperatures and cyclones strongly associated with economic production in the Caribbean and Central America. *Proceedings of the National Academy of Sciences*, 107(35): 15367–15372.

Hughes, T.P., J.T. Kerry, M. Álvarez-Noriega, J.G. Álvarez-Romero, K.D. Anderson, A.H. Baird, R.C. Babcock, M. Beger, D.R. Bellwood, R. Berkelmans, and T.C. Bridge 2017. "Global warming and recurrent mass bleaching of corals." *Nature* 543(7645): 373–377.

Jackson, Jeremy, Mary Donovan, Katie Cramer, and Vivian Lam. 2014. "Status and Trends of Caribbean Coral Reefs: 1970–2012." *Global Coral Reef Monitoring Network, IUCN*, Gland, Switzerland, 1–304.

Leipziger, Danny M. 2001. "The unfinished poverty agenda: why Latin America and the Caribbean lag behind." *Finance & Development* 38(001): 1–8.

Limburg, Karin E., Denise Breitburg, Dennis P. Swaney, and Gil Jacinto. 2020. "Ocean deoxygenation: a primer." *One Earth* 2(1): 24–29.

Lloréns, Hilda. 2018. "Ruin Nation: in Puerto Rico, Hurricane Maria laid bare the results of a long-term crisis created by dispossession, migration, and economic predation." *NACLA Report on the Americas* 50(2): 154–159.

Loiola, Miguel, Igor C. Cruz, Danilo S. Lisboa, Eduardo Mariano-Neto, Zellinda Leao, Marilia Oliveira, and Ruy Kikuchi. 2019. "Structure of marginal coral reef assemblages under different turbidity regime." *Marine Environmental Research* 147: 138–148.

Machado-Aráoz, Horacio. 2010. "La 'naturaleza' como objeto colonial. Una mirada desde la condición eco-bio-política del colonialismo contemporáneo." *Boletín Onteaiken* 10: 35–47.

Mercado-Molina, Alex E., Alex M. Sabat, & Edwin A. Hernández-Delgado. 2020. "Population dynamics of diseased corals: effects of a shut-down reaction outbreak in Puerto Rican Acropora cervicornis." *Advances in Marine Biology* 87(1): 61–82.

Morales Vélez, A., and K.S. Hughes 2018. Geotechnical Damages in Puerto Rico After the Passage of Hurricanes Irma and María, Módulo 4, San Juan, P.R.: Colegio de Ingenieros y Agrimensores de Puerto Rico, April 5, 2018. https://drive.google.com/file/d/1fDW5v1NrMOn60GzTVaFYzSRhXu 5ACSLi/view?usp=sharing.

Nagelkerken, I., G. Van der Velde, M.W. Gorissen, G.J. Meijer, T. Van't Hof, and C. Den Hartog. 2000. "Importance of mangroves, seagrass beds and the shallow coral reef as a nursery for important coral reef fishes, using a visual census technique." *Estuarine, Coastal and Shelf Science* 51: 31–44.

Nicholls, Robert J. 2004. "Coastal flooding and wetland loss in the 21st century: changes under the SRES climate and socio-economic scenarios." *Global Environmental Change*, 14(1): 69–86.

Odum, Howard T., P. Burkholder, and J. Rivero. 1959. "Measurements of productivity of turtle grass flats, reefs and the Bahia Fosforescente of Southern Puerto Rico." *Publications of the Institute of Marine Sciences of Texas* 6: 159–170.

Odum, Howard T., B.J. Copeland, and E.A. McMahan. 1974. *Coastal ecological systems of the United States*, 1–453. Washington, DC: The Conservation Foundation.

Orth, Robert J., Tim J. Carruthers, William C. Dennison, Carlos M. Duarte, James W. Fourqurean, Kenneth L. Heck, Anne Randall Hughes, Gary A. Kendrick, W. Judson Kenworthy, Suzanne Olyarnik, and Frederick T. Short. 2006. "A global crisis for seagrass ecosystems." *Bioscience* 56: 987–996.

Pellegrino, A. 2000. "Trends in international migration in Latin America and the Caribbean." *International Social Science Journal* 52(165): 395–408.

Petit, J.R., J. Jouzel, D. Raynaud, N.I. Barkov, J.M. Barnola, I. Basile, M. Bender, J. Chappellaz, M. Davis, G. Delaygue, and M. Delmotte. 1999. "Climate and atmospheric history of the past 420,000 years from the Vostok ice core, Antartica." *Nature* 399(6735): 429–436.

Ramos-Scharrón, Carlos, Damaris Torres-Pulliza, & Edwin A. Hernández-Delgado. 2015. "Watershed- and island-scale land cover changes in Puerto Rico (1930s-2004) and their potential effects on coral reef ecosystems." *Science of the Total Environment* 506–507: 241–251.

Rogers, Alice, Julia L. Blanchard, and Peter J. Mumby. 2018. "Fisheries productivity under progressive coral reef degradation." *Journal of Applied Ecology* 55(3): 1041–1049.

Rogers, Caroline S. 1990. "Responses of coral reefs and reef organisms to sedimentation." *Marine Ecology Progress Series* 62(1): 185–202.

Rogers, Caroline S. and Jeff Miller. 2006. "Permanent 'phase shifts' or reversible declines in coral cover? Lack of recovery of two coral reefs in St. John, US Virgin Islands." *Marine Ecology Progress Series* 306: 103–114.

Russell, Bayden D., Sean D. Connell, Sven Uthicke, Nancy Muehllehner, Katharina Fabricius, and Jason Hall-Spencer. 2013. "Future seagrass beds: can increased productivity lead to increased carbon storage?" *Marine Pollution Bulletin* 73(2): 463–469.

Scheffer, Marten, Steve Carpenter, Jonathan A. Foley, Carl Folke, and Brian Walker. 2001. "Catastrophic shifts in ecosystems." *Nature* 413(6856): 591–596.

Siegel, Katherine J., Reniel B. Cabral, Jennifer McHenry, Elena Ojea, Brandon Owashi, and Sarah Lester. 2019. "Sovereign states in the Caribbean have lower social-ecological vulnerability to coral bleaching than overseas territories." *Proceedings of the Royal Society B* 286(1897): 20182365.

Smith, Rose-Anne J. and Kevon Rhiney. 2016. "Climate (in) justice, vulnerability and livelihoods in the Caribbean: the case of the indigenous Caribs in northeastern St. Vincent." *Geoforum*, 73: 22–31.

Stefanoudis, Paris V., Wilfredo Y. Licuanan, Tiffany H. Morrison, Sheena Talma, Joeli Veitayaki, and Lucy C. Woodall. 2021. "Turning the tide of parachute science." *Current Biology* 31(4): R184–R185.

Thomas, Chris D. 2020. "The development of Anthropocene biotas." *Philosophical Transactions of the Royal Society B* 375(1794): 20190113.

Unsworth, Richard K., Pelayo Salinas-De León, Samantha Laird Garrard, Jamaluddin Jompa, David J. Smith, and James Bell. 2008. "High connectivity of Indo-Pacific seagrass fish assemblages with mangrove and coral reef habitats." *Marine Ecology Progress Series* 353: 213–224.

Unsworth, Richard K., Catherine J. Collier, Gideon M. Henderson, and Len J. McKenzie. 2012. "Tropical seagrass meadows modify seawater carbon chemistry: Implications for coral reefs impacted by ocean acidification." *Environmental Research Letters* 7(2): 024026.

Unsworth, Richard K.F., Catherine J. Collier, Michelle Waycott, Len J. McKenzie, and Leanne C. Cullen-Unsworth. 2015. "A framework for the resilience of seagrass ecosystems." *Marine Pollution Bulletin* 100: 34–46.

Unsworth, Richard K., Len J. McKenzie, Lina M. Nordlund, and Leanne C. Cullen-Unsworth. 2018. "A changing climate for seagrass conservation?" *Current Biology* 28(21): R1229–R1232.

Van Woesik, R. and C. J. Randall. 2017. "Coral disease hotspots in the Caribbean." *Ecosphere* 8(5): e01814.

Williams, John W. and Stephen T. Jackson. 2007. "Novel climates, no-analog communities, and ecological surprises." *Frontiers in Ecology and the Environment* 5(9): 475–482.

5 An environmental history of the "second conquest"

Agricultural export boom and landscape-making in Latin America, ca.1850–1930

Diogo de Carvalho Cabral and Lise Sedrez

Latin American environmental history has an old tradition, and a younger one too. Latin American intellectuals in the 19th century looked to their surroundings and wondered how nature and society were connected. In Brazil, some went as far as to suggest that deforestation and slavery were connected, and that both were stains on the young nation (Pádua 2002). In the early 20th century, scholars like Euclides da Cunha, Gilberto Freyre, and Sérgio Buarque de Holanda reflected about how societies had changed nature and how nature had shaped society. And outside the region, Berkeley School geographers like Carl O. Sauer and Robert West analyzed these relations with much attention and painstaking care (Sedrez 2009; Leal 2000). However, this was not yet what we today call environmental history. This would emerge later—the younger tradition—in the last quarter of the 20th century. It was strongly influenced by debate in US historiography, by the inspiring work of Latin Americanists such as Warren Dean and Al Crosby, and by the contribution of Latin American scholars increasingly aware of the many environmental crises experienced in the region (Soluri, Pádua, and Leal 2018; Castro Herrera 2002). In the last 40 years, the field has exploded in diversity and quantity.[1]

Latin American historians have written about sewage and mining, glaciers and extractive reserves, deforestation and historical water management, cities and animals, agricultural expansion, and plant diseases, as they see the world change around them and seek to understand the historical roots of the Latin American landscapes. In the late 1990s and the early 21st century, Latin American scholars claimed their heritage from both the old and the new environmental historiography. Some had obtained their degrees in North American academia and, as they returned to the region, sought to craft new disciplinary boundaries that were unique to the Latin American experience. In fact, Guillermo Castro Herrera (2002) argued that there was a tension between environmental history written *about* Latin America according to North Atlantic academic parameters and Latin America's own academic tradition and concerns. He lobbied on behalf of environmental history *made in* Latin America, which should incorporate, on the one hand, a tradition of denunciation and criticism of the plundering of the region's natural resources by corporations from the North Atlantic world, and, on the other hand, a strong presence in Latin American academic institutions of historical geography and social and cultural history.

Indeed, one critical characteristic of environmental history *made in* Latin America has been its embrace of interdisciplinarity. Castro Herrera highlights the pioneering work, in the 1980s, of the Chilean geographer Pedro Cunill, the sociologist Nicollo Gligo, and the economist Osvaldo Sunkel. Another critical characteristic has been the understanding of Latin American in the larger context of world trade and expropriation. Thus, Castro Herrera also includes the Argentinian historian and political activist Luis Vitale among the pioneers. In 1983, Vitale wrote *Hacia una Historia del Ambiente en América Latina*. Likewise, in 1997, the Mexican authors

DOI: 10.4324/9780429344428-7

Isabel Fernández, Alicia Castillo, Alfonso Bulle Goyri, and José and Fernando Ortiz Monasterio published *Tierra Profanada: Historia Ambiental de México*.

These two features have shaped the discipline since then and, ironically, have influenced works of environmental history written *about* Latin America. In 2006, scholars from Latin and North American countries, as well as from Spain, founded the Sociedad Latinoamericana de Historia Ambiental (SOLCHA), which became a mobile focal point for dialogue on history, the environment, and society in Latin America. Guillermo Castro Herrera was its first president. Brazil, Mexico, and Colombia had already emerged as important hubs for academic training and publication, but the forum provided by SOLCHA encouraged scholars to cross national and disciplinary boundaries. While national narratives are necessarily present, transnational studies of river and basins, forests and fauna, and pollution and biodiversity loss demanded larger, more ambitious frames and a solid education in environmental sciences and environmental humanities. Thus, SOLCHA membership included in their ranks anthropologists, geographers, political scientists, and literary critics, as well as historians. New centers for environmental history also appeared in Argentina, Ecuador, and Costa Rica, with interdisciplinary aspirations from the very beginning.

Environmental historians in Latin America, therefore, share some important premises derived from this very distinctive trajectory. First, that landscape changes are visible and they are political; they are rooted in environmental dynamics, but also in social and historical dynamics. And relationships between environment and society rarely can be explained in isolation, as they are connected to larger international contexts of appropriation and unequal power relations. Perhaps these were also the triggers for the earlier work of Cunha, Freyre, and Buarque de Holanda. They were at the heart of one of the most dramatic landscape-changing processes in the history of the region. The period roughly between the 1850s and 1930s is often compared to the 16th century in terms of in-depth transformation in land and in society. It is often also claimed that it laid the groundwork for the modern relationship between Latin American societies and their environment.

Indeed, early-modern European overseas expansion elsewhere promoted the first wave of massive tropical deforestation, among other destructive effects. In the 16th and 17th centuries, wars of conquest, enslavement, and contagious diseases led to a dramatic demographic collapse in the Americas' indigenous populations. Sheep and cattle economies trampled over the carefully-managed agro-ecosystems that had been shaped by indigenous peoples, leaving behind mere hints of that previous occupation (Melville 1994; Gade 1999). This early-colonial period is likely to have seen tropical forests expanding and becoming denser, recolonizing the lands that were previously cultivated by natives (Denevan 1992; Dean 1995; Etter, McAlpine, and Possingham 2008). However, from the second half of the 18th century, with the replacement or transformation of local human populations already concluded or nearly so, tropical forests started to be ruthlessly converted into cropland, pastures, mines, cities, and roads. Environmental changes followed population changes. From 355 million inhabitants in 1700, the world's tropical regions jumped to 775 million inhabitants in 1850, which led to an expansion in cultivated land from 128 million to 180 million hectares. Williams (2003, 334–5) estimates that this agricultural expansion caused the loss of 70 million hectares of forest.

However, the truly tectonic change would not begin until the mid-19th century. As European colonialism carved up Africa and Asia, growing world market integration fuelled the pace of tropical forest conversion. It almost quintupled, with 2.2 million hectares being cleared annually between 1850 and 1920. By the end of World War I, 152 million hectares of tropical forest had been lost (Williams 2003, 334–5). As industrialization consolidated and spread throughout western and central Europe as well as the United States and Canada—shattering age-old limits of human productivity, the margins of the world economy were called upon to contribute their lands and natural resources, spatially extending export-oriented extractive and agricultural systems. The supply of farming and forestry commodities, such as sugar, coffee, cocoa, latex, and timber, connected tropical areas to the North Atlantic economic boom that forever changed much of humanity's relationship with its ecological systems (Tucker 2000; De Carvalho Cabral and Bustamante 2016; Ross 2017).

Considering only commercial crops, this process replaced an estimated area of between 15 and 25 million hectares of Latin American forestland in the 1860–1920 period, a five- to eight-fold increase compared to the previous sixty years (Houghton, Lefkowitz, and Skole 1991). Environmental historians agree that colonial deforestation in Latin America was relatively light, with the exception of the smallest Caribbean islands like Barbados, whose sugar industry's tremendous hunger for wood and lumber quickly depleted local forest stocks (Watts 1987; Miller 2007). Dean (1995) and De Carvalho Cabral (2014) estimated that less than 10% of Brazil's Atlantic Forest was converted throughout more than three centuries of Portuguese control in the Americas. But deforestation accelerated greatly in the post-emancipation period, especially beginning in the mid-19th century. São Paulo State, for instance, had 65% of its pre-colonial rainforest destroyed between 1854 and 1935 (Victor et al., 2005), while in neighboring Paraná—whose agricultural expansion derived in part from the *paulista* frontier—30% of the state's forestland was converted to cropland between 1890 and 1937 (Gubert Filho 2010). In other regions and countries, the acceleration of human impact was less pronounced. In fact, it has been estimated that in Colombia the conversion of tropical forests (including dry forests) decreased only slightly between 1850 and 1920, as compared with the previous fifty years (Etter, McAlpine, and Possingham 2008).

The different rhythms and geographies of human impact were largely shaped by specific histories of integration into the world economy, which grew at unprecedented rates from the mid-19th century until the eve of World War I. The world GDP more than doubled between 1850 and 1913, and the volume of global trade grew almost fourfold (Maddison 2001). Arguably, more than any other region, Latin America was a creation of such world-economic expansion (Topik and Wells 1998). This is in line with Maddison's (2001) regional estimates. Between 1870 and 1913, Latin American GDP grew 336%, whereas Africa and Asia—the other two capitalist peripheries—grew at the considerably slower rates of 81 and 57%, respectively. Latin American exports grew 3.9% annually between 1850 and 1912, while the region's population grew by 1.5% annually (Bulmer-Thomas 2003).

Recently emancipated from European rule, the Latin American countries had legal, economic, and cultural systems similar to the Europeans, and their elites—in contrast to their Asian counterparts—were eager to emulate European consumerism (Topik and Wells 1998). There was already a strong demand for foodstuffs and raw materials—as well as capital available for investment—earlier in the 19th century, yet the political instability of the recently formed nation-states forestalled the quest for economic growth until the 1850s and, in most cases, the 1870s (Eakin 2008; but see also Federico and Tena-Junguito 2018). By that point, a number of liberal-minded governments had embarked on reforms that made it easier for large agricultural entrepreneurs to expropriate public and communal lands, in addition to opening up the domestic economies to European and American investments in the agriculture and infrastructure sectors. Topik and Wells (1998) called this dynamic period the "second conquest of Latin America," thus comparing it to the 16th-century work of the Iberian conquistadors that subdued native empires and other less complex polities and subjected local ecosystems and labor to European mercantile economies.

From an environmental history perspective, however, one key difference stands out between the two "conquests." As previously stated, the 16th-century demographic debacle reduced the human strain on ecosystems. Native flora was permitted to re-grow on areas that had long been cultivated by the indigenous inhabitants, 80% of whom are thought to have been forest dwellers at the time of European arrival (Dull et al. 2010). According to some authors, this explains the nadir in atmospheric CO_2 in the early 17th century, which might be used as a landmark for the Anthropocene's onset (Dull et al. 2010; Lewis and Maslin 2018). In contrast, the "second conquest" induced a substantial acceleration of land clearing from the mid-19th century, which was intimately connected with other, intense carbon-emitting processes taking place in the North Atlantic: industrialization and urbanization.

Latin America's major function in this scenario was to supply cheap calories and stimulants to the North Atlantic's burgeoning urban proletariats. This was not quite new, of course; the 17th- and 18th-century Caribbean sugar economies had served as a historical model for this sort of economic-environmental configuration (Mintz 1986; De Carvalho Cabral 2014). The "second conquest," on the other hand, took place within a new food regime, one that was primarily centered on Britain and enabled by fossil-fuelled transportation technologies like steamships and railroads. From the Latin American commodities frontiers, Britain imported tropical goods like coffee and temperate cereals (particularly wheat), as well as meat later on (McMichael 2013). Moreover, during the second half of the 19th century, the Industrial Revolution had formed three additional great economic powers (France, Germany, and the United States), each of which also wielded immense influence over Latin American economies. For instance, by 1850 over half of all Brazilian coffee exports were transported to the United States, and the country remained the primary importer of that product through the 20th century. Even so, British firms controlled most of the export business in Brazil, whether in coffee or other products (Graham 1972).

As we consider the impact of this "second conquest" on the history of the region, the conventional periodization becomes less relevant for environmental historians. The traditional narratives which rely on political events as landmarks, such as the Spanish Conquest, the Wars of Independence, or the Mexican Revolution, are based on national political accounts, and they are unable to account for vast transformations in the region. At the same time, the concept of a "second conquest" should not be limited to economic analysis. Arguably, it brought about extensive transformations of land, societies, and biodiversity across the entire region.

Unlike what happened in the 16th and 17th centuries, most of the plant species or varieties that spurred the 19th-century export boom were the result of applied science, often combined with illegal networks that supplied European research centers with genotypes (McCook 2011). Coffee, which originated on the Ethiopian plateau, made its way to Brazil via imperial networks of botanical gardens, as well as smugglers. After introducing coffee to their Javanese colony from the Malabar coast and then transferring it to the Amsterdam Botanical Gardens from the turn of the 17th century to the turn of the 18th century, the Dutch sent a young tree as a present to King Louis XIV of France in 1714. An offshoot of that plant developed at the *Jardin des Plantes* de Paris and was transported to Martinique nine years later (McCook 2011). *Coffea arabica* soon spread to French Guiana, where, in 1727, a Luso-Brazilian military secretly collected plants and seeds and carried them to Belém, the capital of the Portuguese colony of Grão-Pará and Maranhão (Dean 1995; Papavero, Teixeira, and Overal 2001).

Originally domesticated in South Asia, sugarcane was first introduced to the Caribbean in the wake of Christopher Columbus' second voyage (1493), eventually becoming one of the most successful biological invasions in the "first conquest". However, plantation-driven environmental degradation and the emergence of new plant diseases made it necessary to introduce or develop new varieties that could withstand harsher environments. A number of botanical gardens and experimental stations were built in the 1880s to produce new cane hybrids that could sustain a new cycle of agricultural growth in the region, often financed and/or manned by North Atlantic institutions, such as Harvard University (Galloway 1996; Fernandez-Prieto

2018). Supported by state-sponsored scientific institutions and international smuggling, coffee and sugarcane covered extensive areas from Costa Rica to Ecuador, from the Guyanas to southern Brazil, penetrating new ecosystems with greater tolerance to variations of climate, temperature, and humidity.

In spite of this, most plantation regimes responsible for the export boom remained reliant on what French agricultural economist François Ruf dubbed "differential forest rent." Regarding permanent crops such as coffee and cocoa, this reflected the simple economic reality that it was seldom profitable to replace aged trees with new ones on the same plot—or to plant them on lands previously used for other crops—as long as there was mature forest available elsewhere (Clarence-Smith and Ruf 1996). Even semi-permanent crops like sugarcane and banana used this method. By cutting down trees and burning the debris, neo-European colonists solved two big difficulties at once: opening space for crops and fertilizing the earth. Slash-and-burn methods (shifting agriculture) were widely known to the indigenous inhabitants, but they were not practiced to the extent that foreign market pressures would cause them to be. By slashing and burning the forest, the planters had access to those gigantic nutrient reserves that would otherwise be released into the soil only very slowly through spontaneous clearings and decomposition (De Carvalho Cabral 2014). Clearing hillsides was favored in Brazil for several reasons, including the ease of operation. The trees on high ground slid down the slopes, bringing down other trees as they went. Traveling in the Brazilian province of Espírito Santo in 1888, the Bavarian princess Therese Charlotte reported her experience with such a practice:

> As we rode along a forest slope, we suddenly heard a huge crash and crackles just above us to the left. We didn't know what that meant and we were listening for the noise. It seemed that a part of the forest above us had broken down and slipped and now the gigantic trees were falling, dragging everything with it into the valley with thunder-like crashes. A huge leafless, twig-free trunk was still in upright position when it slid, then staggered, toppled, and was thrown into the depths upside down, dragging with it all the tangle of plants in its path. It was a particularly spectacular and powerful scene. A few minutes earlier, we had passed that stretch that was now under the rubble of vegetation. Later on, we were told that the vision we had was a consequence of deforestation. Several rows of trees had been partly cut down, but only the top row had been cut down completely and it was the one that caused the rest to fall.[2]
>
> (Baviera 2013, 58–9)

Coffee is the most important agricultural product to have left its mark on the landscapes of Latin America. A perennial crop, it required no wood fuel for processing (unlike sugar, for example), yet its impact on forests was greater than that of any other agricultural item in preindustrial Brazil. Between 1788 and 1888, coffee plantations replaced 7,200 square kilometers of primary forest, which is more than three times the amount cleared for sugar until 1700, according to Dean (1995, 188). The Paraiba do Sul valley, a watershed of 60,000 square kilometers sandwiched between two mountain ranges on the southeast border of the Brazilian plateau, was the most dramatic case of deforestation. The valley's subtropical, seasonally-rainy climate and proximity to Rio de Janeiro—the empire's capital and greatest slave port in the Americas—fostered the establishment of vast slave-worked estates (Stein 1957; Marquese and Tomich 2020). In a memoir written in the 1840s to instruct his son on the establishment of a coffee estate in the valley, a *fazendeiro* described the largest clearing operations in the valley as "*infernos*" that "in less than an hour turn into ashes what nature took centuries to create" (Paty do Alferes 1878, 13–14). Planters rarely bothered to restore old, failing coffee groves; instead, they just slashed and burned new forest tracts. In 1863, an agronomic commentator warned the dangers of such practices and reported to have found "many devastated estates with very little forest left for [future] coffee planting, despite the fact that [only] about 25 years have passed since this crop was most actively developed" (Fonseca 1863, 13).

Wholesale deforestation quickly triggered systemic environmental degradation in the valley (Dean 1995). In a matter of decades, hillside soils had their most fertile layers washed away, often with the help of leaf-cutting ants, whose tunnels funneled the infiltrating rainwater, thus producing water-saturated zones capable of detonating highly-concentrated erosion along the slopes (Dantas and Netto 1996). While the local planters claimed that rainfall had declined and become more erratic, the coeval Swiss-Brazilian naturalist Emílio Augusto Goeldi analyzed meteorological reports for the city of Rio and concluded that, in reality, precipitation had become more concentrated: "With regard to quantity it does not rain less than heretofore. The total number of days of rainfall has shrunk; the relative quantity of rainfall during one day has increased" (Apud Stein 1957, 218). Although hard data are lacking, anecdotal evidence suggests that by the end of the century the dry season had expanded from just two or three months to eight or nine. In September 1872, the Belgian politician Walthère de Selys Longchamps (1875, 44) was visiting Rio Preto, in the Middle Paraiba Valley, and noted that the level of the local tributary river was very low, as the locals had told him "it hasn't rained a single drop for over 8 months!" The combination of accelerated forest conversion, local climate change, and soil erosion drove the valley's landscape transition from thriving coffee plantations to low-productivity dairy farms, as the only forest remaining stood at mountain escarpments that were not readily accessible, even for cattle. In 1926, these degraded landscapes were depicted melancholically by a novelist with considerable first-hand experience:

> The old coffee grove withered on a hillside of dry, exhausted, ant-excavated land. The bare trees, with their twisted branches, looked like inverted roots; here and there green leaves still resisted but entangled with parasitic herbs; and the barren bush spread, covering the trunks. Tall, tufted grasses released threads in the wind, and the dense turf, denouncing the weary soil's anemia, waved like a latent fire crackling.
>
> (Coelho Neto 1926, cited in Murari 2009, 232)

Forests fell to appease a growing hunger for coffee, wood, and fuel, and distance and inaccessibility no longer provided any protection. Underwritten mostly by British capital, railroads began to be constructed in the early 1850s. The long-term environmental impacts were anticipated by intellectuals and writers. In 1856, two years after the inauguration of Brazil's first railway, a short line connecting the northern shores of the Guanabara Bay to the foothills of Serra do Mar, José de Alencar warned that "the wagon of progress smokes and shall rush over this immense web of iron rails that will shortly cut [through] your virgin forests" (Alencar 1865; Pádua 2009, 302).

The impacts of the railroad networks in the entire region were immense. They stemmed not only from onsite clearcutting and wood extraction for construction and fuel, but also, more importantly, from the lowered transportation costs for plantation commodities, especially coffee, with the associated slash-and-burn method (Dean 1995; Cribelli 2016). In Mexico, railway tracks grew from 650 kilometers in the 1860s to almost 25,000 kilometers on the eve of the Mexican Revolution in 1910, due mostly to US investors who were eager to have easier access to the country's minerals and forests. In the process, similar access to communal lands was lost forever for thousands of Mexicans (Cariño Olvera and Boyer 2018). In the greater Caribbean, perishable bananas demanded a fast and safe network to connect the plantations to the ports, and this network was built over marshes and communities (Funes Monzote 2018). In Peru, forests fell with the advance of the railroads and their companions—the telegraph lines—to facilitate the production of copper, zinc, lead, and silver. In Colombia, a more modest but extremely transformative 3,000 kilometers of railroads crossed the savanna of Bogota and the whole of Colombia, from mountains to lowlands (Cuvi 2018). In Brazil, where a paltry 1,000 kilometers of iron rails had been laid by 1872, railroad construction accelerated in the 1870s, with another 8,000 kilometers of railway tracks produced by the end of the imperial era (Viotti da Costa 1986). In response to the planter class' demand for lower transportation costs, most railroads were designed to link ports to inland cultivation areas, with few interregional lines. Not surprisingly, railroad construction was concentrated in the coffee-producing areas of the southeast. In 1882, over 70% of the country's total railway tracks were concentrated in the region encompassing the provinces of São Paulo, Rio de Janeiro, and Minas Gerais (Viotti da Costa 1986).

The focus on the railroads also highlights another powerful landscape transformation brought about the agricultural export boom: urbanization. While the old colonial port-cities remained powerful, their role as the gatekeepers between world economy and the hinterland spurred rapid modernization. As export commodities made their way to the ports on railway lines, urbanites benefitted from cheaper access to myriad resources that the same railroads gave them. Though limited, industrialization was unfolding in these urban centers. The changes were also visible in the interior. Fluvial cities were abandoned, while new villages and towns mushroomed along railroads. Easier transportation reinforced the connections from the smallest villages to larger capitals, creating a robust—although unequal—urban network that circulated goods, people, and power (Sedrez and Duarte 2018). As Cuvi (2018, 73) argues, "the trains strengthened cities including Barranquilla, Guayaquil, Lima, Bogotá, Quito, Medellín, Arica, and La Paz, among others."

Thus, cities grew and so did their appetite for energy and water. Their impact on their surrounding regions increased to meet new demands. With over one million residents by the turn of the century, the Brazilian capital of Rio de Janeiro extended to rivers of two different basins, almost one hundred kilometers beyond the city's limits. Some rivers were brought into canals to carry water to the city, and others were dammed to generate power. Rio de Janeiro's growing demand led one smaller city to be abandoned and inundated. The same dam accelerated the proliferation of mosquitoes around the village of São João Marcos, causing an outbreak of malaria that killed more than 1,000 people between 1908 and 1911 (Sedrez and Capilé 2020). The changes in the hydroscape, therefore, brought unequal gains and costs that were enduring.

Landscapes are as much about people's spatial knowledge and imagination as they are about material environments. Therefore, the "second conquest" must also be analyzed in terms of changes in geographic representations. Modern nation-states are strongly linked to territorial sovereignty, which requires mastering geographies to act strategically over space. In Latin America, government mapping initiatives became ever more common from the mid-19th century, covering both basic (topographic) and thematic mapping. These projects were instrumental to providing synoptic, compositional views of natural features and human settlements. Commissioned in 1886 by the provincial government of São Paulo, the *Comissão Geográfica e Geológica* aimed to draw maps in the following way:

> … on the scale of one centimeter per kilometer, which will be at the same time geographical, topographical, itinerary, geological, and agricultural, and in which exactly all the population centers and the industrial and agricultural establishments of certain importance are represented; […] the railways and highways; waterways; mines, etc.; […] the distribution of geological terrains and agricultural lands, as well as the unproductive ones.
>
> (Oliveira 1966, cited in Figueirôa 2008, 766)

Figure 5.1 "Défrichement d'une forêt" (lithograph by Godefroy Engelmann) in Johann Moritz Rugendas, *Malerische reise in Brasilien* (Stuttgart: Daco Verlag Bläse, 1986 [facsimile of the original edition by Engelmann & Cie, Paris, 1835]).

More specialized enterprises were often directed toward the plant kingdom, the greatest source of revenue for most Latin American countries. Cataloging and mapping native plants to produce "national floras" became vital activities in the late 19th and early 21st centuries. Efforts in biogeographic mapping also included what would be called nowadays "land-use/land-cover" maps such as the "Mapa Ecológico de Venezuela," published in 1920 (McCook 2002).

Mixing technoscience and political-economic power, these maps decisively contributed to the refashioning of Latin American forests, both on the ground and in people's minds. In São Paulo, the topographic and thematic maps put together by the *Comissão Geográfica e Geológica* enabled the laying out of railroads, though in some areas the essential work of geographical surveying was accomplished by the railway engineers themselves. Once completed, these infrastructures facilitated a tenfold increase in coffee output between 1920 and 1935, connecting the west of the state—a region dotted with basaltic soils formed through the weathering of igneous intrusions—with the coast (Mattos 1990; Figueirôa 2008). Also fueled by the massive influx of poor European immigrants, this agricultural boom drove the clearing of around 300,000 hectares of forest annually during that period (Victor et al. 2005).

Such unprecedented rates of deforestation contributed to shaping a crucial period in a much longer process through which Latin American tropical forests were reconceived in public perception from menacing jungles to endangered biodiversity hotspots (Pádua 2015; Leal 2018). Heir to a certain Enlightenment tradition of environmental thinking (Pádua 2002), criticism of relentless deforestation was found in brilliant writers like Euclides da Cunha, who wrote vividly about the novel landscapes that were emerging. Echoing earlier writings against deforestation's desiccative effects on the land, Cunha wrote of Brazilians as "desert-makers." "We have been a nefarious geological agent," he eloquently claimed, "and an element of terribly barbaric antagonism to the very nature that surrounds us" (Cunha 1901).

Referring to the landscape-making then taking place in the agricultural frontiers of São Paulo, Cunha remarked the following:

> What does the one who travels on the roads of western São Paulo observe today? He encounters, from moment to moment, along the railway lines, huge amounts of wood in logs, agglomerated in considerable volumes of hundreds of esters, progressing, at intervals, from Jundiaí to the end of all branches. They are the only fuel for locomotives. We evaded the financial crisis and the high price of stone coal by attacking nature's economy and diluting some hectares of our flora every day in the smoke of the boilers. In this way—recurring in the error—to the proven inconvenience of the ultra-extensive crops and the live cautery of burnings, we add the rapid deforestation carried through large-scale clearing.
>
> (Cunha 1901)

Writings like this helped create public awareness of forest exhaustion and its environmental consequences in the first decades of the 20th century. These interventions in the national mindscape would help foster significant legislative and institutional changes regarding the conservation of forests, closely connected to the rise of an intensely nationalist period. In Chile, conservationists sought to restore temperate forest, while in Mexico many national parks were created in the wake of the 1930s social revolutions (Wakild 2018). In Brazil, a highly centralized government, concerned about national natural patrimony, issued a Forest Code (1934) and created the country's first protected area, the Itatiaia National Park (1937), a large forest remnant located on the mountainous triple border of Rio de Janeiro, São Paulo, and Minas Gerais. Elsewhere in Latin America similar processes took place, and the 1920s and 1930s saw the first wave of national park creations as the "second conquest" waned.

Notes

1 For an overview of the growing body of publications, see the Online Bibliography of Latin American Environmental history, http://boha.historia.ufrj.br, with a growing, though not exhaustive list of references.
2 All translations in the text are ours.

References

Baviera, Teresa da Teresa, Princess. [1888] 2013. *Viagem pelos Trópicos Brasileiros [Meine reise in den brasilianischen tropen], translation and notes by Sara Baldus.* Vitória-ES: Arquivo Público do Estado do Espírito Santo.

Bulmer-Thomas, V. 2003. *The Economic History of Latin America since Independence*, 2nd ed. Cambridge: Cambridge University Press.

Cariño Olvera, Micheline and Chris Boyer. 2018. "Mexico's ecological revolutions." In *A Living Past: Environmental Histories of Modern Latin America*, ed. John Soluri, Claudia Leal, and José Augusto Pádua, 23–44. New York and Oxford: Berghahn.

Castro Herrera, Guillermo. 2002. "História Ambiental (feita) na América Latina." *Varia Historia*, v. 18, n. 26 [Accessed on 5 September 2021]: 33–45, http://www.variahistoria.org/s/02_Herrera-Guilherme-Castro.pdf.

Clarence-Smith, William G., François Ruf. 1996. "Cocoa pioneer fronts: The historical determinants." In *Cocoa Pioneer Fronts since 1800: The Role of Smallholders, Planters and Merchant*, ed. W.G. Clarence-Smith, 1–22. London: Macmillan.

Cribelli, Teresa. 2016. *Industrial Forests and Mechanical Marvels: Modernization in Nineteenth-Century Brazil.* New York: Cambridge University Press.

Cunha, Euclides da. [1901] 2020. "Fazedores de desertos." *EUCLIDESITE. Obras de Euclides da Cunha. Contrastes e confrontos*. São Paulo. Available at: https://euclidesite.com.br/contrastes-e-confrontos/fazedores-de-desertos. Accessed January 12, 2021. Originally published in *O Estado de S. Paulo*, 21 out. 1901.

Cuvi, Nicolás. 2018. "Indigenous imprints and remnants in the tropical andes." In *A Living Past: Environmental Histories of Modern Latin America*, ed. John Soluri, Claudia Leal, and José Augusto Pádua, 67–90. New York and Oxford: Berghahn.

Dantas, Marcelo Eduardo and Ana Luiza Coelho Netto. 1996. "Resultantes geo-hidroecológicas do ciclo cafeeiro (1780–1880) no médio vale do rio Paraíba do Sul: uma análise quali-quantitativa." *Anuário do Instituto de Geociências (UFRJ)* 19: 61–78.

De Carvalho Cabral, Diogo. 2014. *Na Presença da Floresta: Mata Atlântica e História Colonial*. Rio de Janeiro: Garamond/Faperj.

De Carvalho Cabral, Diogo and Ana G. Bustamante. 2016. "Introdução: Mudanças na mata." In *Metamorfoses Florestais: Culturas, Ecologias e as Transformações Históricas da Mata Atlântica*, ed. Diogo de Carvalho Cabral and Ana G. Bustamante, 17–33. Curitiba: Prismas.

Dean, Warren. 1995. *With Broadax and Firebrand: The Destruction of the Brazilian Atlantic Forest*. Berkeley and Los Angeles: The University of California Press.

Denevan, William M. 1992. "The pristine myth: The landscape of the Americas in 1492." *Annals of the Association of American Geographers* 82: 369–385.

Dull, Robert A., Richard J. Nevle, William I. Woods, Dennis K. Bird, Shiri Avnery, and William M. Denevan. 2010. "The Columbian encounter and the Little Ice Age: Abrupt land use change, fire, and greenhouse forcing," *Annals of the Association of American Geographers* 100: 755–771.

Eakin, Marshall C. 2008. *The History of Latin America: Collision of Cultures*. New York, St. Martin's Griffin.

Etter, Andres, Clive McAlpine, and Hugh Possingham. 2008. "Historical patterns and drivers of landscape change in Colombia since 1500: A regionalized spatial approach." *Annals of the Association of American Geographers* 98: 2–23.

Federico, Giovanni and Antonio Tena-Junguito. 2018. American divergence: Lost decades and emancipation collapse in Latin America and the Caribbean 1820–1870. *European Review of Economic History* 22: 185–209.

Fernandez-Prieto, Leida. 2018. "Networks of American experts in the Caribbean: The Harvard Botanic Station in Cuba (1898–1930)." In *Technology and Globalisation: Networks of Experts in World History*, ed. D. Pretel and L. Camprubí, 159–187. London: Palgrave Macmillan.

Figueirôa, Silvia F. M. 2008. 'Batedores da ciência' em território paulista: expedições de exploração e a ocupação do 'sertão' de São Paulo na transição para o século XX. *História, Ciências, Saúde – Manguinhos* 15: 763–777.

Fonseca, Antonio Caetano. 1863. *Manual do Agricultor dos Generos Alimnenticios*. Rio de Janeiro: Eduardo & Henrique Laemmert.

Funes Monzote, Reinaldo. 2018. "The greater caribbean and the transformation of tropicality." In *A Living Past: Environmental Histories of Modern Latin America*, ed. John Soluri, Claudia Leal, and José Augusto Pádua, 45–66. New York and Oxford: Berghahn.

Gade, Daniel W. 1999. *Nature and Culture in the Andes*. Madison: University of Wisconsin Press.

Galloway, J. H. 1996. "Botany in the service of empire: The Barbados cane-breeding program and the revival of the Caribbean sugar industry, 1880s–1930s." *Annals of the Association of American Geographers* 86: 682–706.

Graham, Richard. 1972. *Britain and the Onset of Modernization in Brazil*. Cambridge: Cambridge University Press.

Gubert Filho, Francisco A. 2010. "O desflorestamento do Paraná em um século." In *Reforma Agrária e Meio Ambiente: Teoria e Prática no Estado do Paraná*, ed. Claudia Sonda and Silvia Cristina Trauczynski, 15–25. Curitiba: Instituto de Terras, Cartografia e Geociências.

Houghton, R. A., D. S. Lefkowitz and D. L. Skole. 1991. "Changes in the landscape of Latin America between 1850 and 1985. I. Progressive loss of forests." *Forest Ecology and Management* 38: 145–149.

Leal, Claudia. 2000. "Robert West: um geógrafo de la escuela Berkely." *Prologue to Las tierras bajas del Pacífico colombiano*, by Robert West. Bogotá: Instituto Colombiano de Antropología e Historia.

Leal, Claudia. 2018. "From threatening to threatened jungles." In *A Living Past: Environmental Histories of Modern Latin America*, ed. John Soluri, Claudia Leal, and José Augusto Pádua, 115–137. New York and Oxford, Berghahn.

Lewis, Simon and Mark A. Maslin. 2018. *The Human Planet: How We Created the Anthropocene*. London: Pelican Books.

Longchamps, Walthère de Selys. 1875. *Notes d'un voyage au Brésil, [Extract from the Revue de Belgique]*. Brussels: Librairie C. Muquardt.

Maddison, Angus. 2001. *The World Economy: A Millennial Perspective*. Paris: OECD.

Marquese, Rafael B. and Dale Tomich. 2020. "Slavery in the Paraiba Valley and the formation of the world coffee market in the nineteenth century." In *Atlantic Transformations: Empire, Politics, and Slavery during the Nineteenth Century*, ed. Dale Tomich, 193–224. Albany: State University of New York Press.

Mattos, Odilon N. 1990. *Café e Ferrovias: A Evolução Ferroviária de São Paulo e o Desenvolvimento da Cultura Cafeeira*. Campinas-SP: Pontes.

McCook, Stuart. 2002. "'Giving plants a civil status': scientific representations of nature and nation in Costa Rica and Venezuela, 1885–1935." *The Americas* 58: 513–536.

McCook, Stuart. 2011. "The Neo-Columbian Exchange: The second conquest of the Greater Caribbean, 1720–1930." *Latin American Research Review* 46 (Special Issue): 11–31.

McMichael, Philip. 2013. *Food Regimes and Agrarian Questions*. Halifax and Winnipeg: Fernwood Publishing.

Melville, Elinor Gordon Ker. 1994. *A Plague of Sheep: Environmental Consequences of the Conquest of Mexico*. Cambridge: Cambridge University Press.

Miller, Shawn W. 2007. *An Environmental History of Latin America*. Cambridge: Cambridge University Press.

Mintz, Sidney W. 1986. *Sweetness and Power: The Place of Sugar in Modern History*. New York: Penguin.

Murari, Luciana. 2009. *Natureza e Cultura no Brasil (1870–1922)*. São Paulo: Alameda.

Pádua, José Augusto. 2002. *Um Sopro de Destruição: Pensamento Político e Crítica Ambiental no Brasil no Brasil Escravista*. Rio de Janeiro: Zahar.

Pádua, José Augusto. 2009. "Natureza e sociedade no Brasil monárquico." In *O Brasil Império*, ed. K. Grinberg and R. Salles, vol. III, 313–365. Rio de Janeiro: Civilização Brasileira.

Pádua, José Augusto. 2015. "Tropical forests in Brazilian political culture: From economic hindrance to endangered treasure." In *Endangerment, Biodiversity and Culture*, ed. F. Vidal and N. Dias, 148–171. London, Routledge.

Papavero, Nelson, Dante M. Teixeira and William L. Overal. 2001. "Notas sobre a história da zoologia do Brasil. 2. As viagens de Francisco de Melo Palheta, o introdutor do cafeeiro no Brasil." *Boletim do Museu Paraense Emílio Goeldi (série Zoologia)* 17: 181–207.

Paty do Alferes, Barão do. [1847] 1878. *Memoria sobre a Fundação e Costeio de uma Fazenda na Província do Rio de Janeiro*. Rio de Janeiro: Eduardo & Henrique Laemmert.

Ross, Carey. 2017. *Ecology and Power in the Age of Empire: Europe and the Transformation of the Tropical World*. Oxford: Oxford University Press.

Sedrez, Lise. 2009. "Latin American environmental history: A shifting old/new field." In *The Environment and World History*, ed. Edmund Burke III and Kenneth Pomeranz, 272–290. Berkeley, CA: University of California Press.

Sedrez, Lise and Bruno Capilé. 2020. "Os Modernos Rios Cariocas." In *Os Rios do Rio*, ed. Lorelai Kury, Lise Sedrez, Bruno Capilé and Marcelo Motta, 72–129. Rio de Janeiro: Andrea Jakobsson Studio.

Sedrez, Lise and Regina Horta Duarte. 2018. "The Ivy and the wall: environmental narratives from an urban continent." In *A Living Past: Environmental Histories of Modern Latin America*, ed. John Soluri, Claudia Leal, and José Augusto Pádua, 138–162. New York and Oxford: Berghahn.

Soluri, John, Claudia Leal, and José Augusto Pádua. 2018. "Finding the 'Latin American' in Latin American environmental history." In *A Living Past: Environmental Histories of Modern Latin America*, ed. John Soluri, Claudia Leal, and José Augusto Pádua, 1–23. New York and Oxford: Berghahn.

Stein, Stanley J. 1957. *Vassouras: A Brazilian Coffee County, 1850–1900*. Cambridge-MA: Harvard University Press.

Topik, Steven C. and Allen Wells. 1998. "Introduction: Latin America's response to international markets during the export boom." In *The Second Conquest of Latin America: Coffee, Henequen, and Oil during the Export Boom*, ed. Steven C. Topik and Allen Wells, 1–36. Austin: University of Texas Press.

Tucker, Richard P. 2000. *Insatiable Appetite: The United States and the Ecological Degradation of the Tropical World*. Berkeley, Los Angeles and London: University of California Press.

Victor, Mauro Antônio Moraes, Antônio Carlos Cavalli, João Regis Guillaumon and Renato Serra Filho. 2005. *Cem Anos de Devastação – Revisitada 30 Anos Depois*. Brasília-DF: Ministério do Meio Ambiente.

Viotti da Costa, Emilia. 1986. Brazil: "The age of reform, 1870–1889." In *The Cambridge History of Latin America*, ed. Leslie Bethell, Vol. V, 725–777. Cambridge: Cambridge University Press.

Vitale, Luis. 1983. *Hacia una Historia del Ambiente en América Latina: De las Culturas Aborígenes a la Crisis Ecológica Actual*. Mexico City: Nueva Sociedad / Editorial Nueva Imagen.

Wakild, Emily. 2018. "A panorama of parks: Deep nature, depopulation, and the cadence of conserving nature." In *A Living Past: Environmental Histories of Modern Latin America*, ed. John Soluri, Claudia Leal, and José Augusto Pádua, 246–265. New York and Oxford: Berghahn.

Watts, David. 1987. *The West Indies: Patterns of Development, Culture, and Environmental Change Since 1492*. New York: Cambridge University Press.

Williams, Michael. 2003. *Deforesting the Earth: From Prehistory to Global Crisis*. Chicago: University of Chicago Press.

6 Extractivism

The port-a-cathed veins of Guatemala

Liza Grandia

Introduction

Prior to chemotherapy, oncologists typically recommend surgical placement of a "port-a-cath," a medical device that threads a catheter directly into a superior vena cava above the heart. This cyborgian, self-sealing portal gives nurses dual-direction access to a patient's veins. With a simple stab, nurses can administer a chemical infusion designed to kill the cancer (but that itself may poison the body or cause new mutations) or draw blood to determine whether one's cell counts can withstand further toxicity. The port-a-cath may remain in place for many weeks—even years—or, in case of the dual extraction/poisoning of Guatemala, five centuries.

Central to this metaphor is Eduardo Galeano's (1940–2015) remarkably resilient "open vein" thesis on how Spanish lust for silver, gold, and coffee drained the region's lifeblood for world elites. Although Spain provided the firepower, as Galeano aptly put it, "Latin America was a European business." Dutch and Flemish financial firms captured almost a third of the American plunder, with the French, Italian, and English siphoning off a quarter, a fifth, and a tenth respectively (Galeano 1973, 24). More than seven million pounds of silver entered Seville, tripling European supplies between 1503 and 1660, but Spain was so militarily overextended (Wolf 1982) that the Hapsburgs remained mired in debt. As Seneca historian John Mohawk elaborated (2000, 135), colonial "money is not wealth but rather a claim on death, and unless that claim is exercised in some sustainable way, the wealth can evaporate."

Only a cunning few of the Iberian mercenaries like Pedro de Alvarado transformed their spoils into heritable businesses. The return on Alvarado's personal 20,000 peso "investment" to invade Guatemala left him probably the second-richest man in the Americas next to Hernán Cortés himself. Through strategic marriages and court politics, Alvarado acquired numerous "encomiendas" in Mexico and Guatemala, collecting taxes and labor from some 23,000 Indigenous tributaries. He diversified into many other enterprises like slave trading, mining, transport, fishing operations, expansion of the cattle frontier (bovines, goats, but primarily sheep), and expeditions into South America. His gold and silver enterprises stretched from Chiapas to Honduras, with at least 530 enslaved laborers. Foreshadowing Wall Street speculation, Alvarado was a flamboyant gambler. A 1969 analysis estimated the value of his estate at US$2.5 million, which in today's dollars would be almost US$18 million (Sherman 1969).

The conquistadors' colonial mercantilism was as arguably "transnational" (or trans-jurisdictional) as the corporate-driven extractivist boom underway across the Americas (Wolf 1982). To give but a few anecdotes: the beef tallow from Mexican cattle herds lit the mining caverns of South America; the silver of Bolivia's Potosí became the currency of Carolina slave traders (Weatherford 1988); and the cochineal dye from Mexico colored the uniforms of the British "Redcoats" who marauded Ireland, whose peasantry, in turn, survived their own colonial occupation with the caloric value of potatoes pirated from Peru. Although the "world system"

DOI: 10.4324/9780429344428-8

has fallen from trendy academic grace, remarkably similar kinds of colonial demographic, political, legal, religious, environmental, epidemiological, and financial patterns continue to interconnect contemporary Indigenous and Native struggles for survival and territorial autonomy against comparably brutal plunder of their resources today.

As Galeano emphasized during Cold War times, Guatemala still provides 'lessons painfully learned' (Jonas 2015, 3) for the rest of Latin America. In the present-day pumping of Guatemala's veins, colonial themes of land/labor, debt, dependency, trade relations, infrastructure, corruption, coercion, and elite kinship continue to shape patterns of extraction. As Bastos and de León (2014, 149) emphasize, "these businesses act like sixteenth century invaders and all the invaders that followed them" [translation mine]. Through "recurrent dispossession" (Grandia 2009b), colonial mercantilism metastasized into industrial capitalism which mutated through militarized modernism (financed by US foreign aid) into a repressive system of transnational corporate power (Moreno and Salvadó 2017, 134). Much as the conquistadors concealed their savagery through rhetoric of "civilization," 19th-century liberals through "progress," and 20th-century military goons through "security," 21st-century neoliberal elites today justify their scorched subsistence policies as "economic growth." To paraphrase the late Edward Abbey, however, growth for the sake of growth is the crude ideology of a cancer cell. Now diagnosed with terminal climate change, the planet sputters towards ecological collapse. Although Maya movements resisting this broken extractivist model are gaining momentum, they face an encrusted social structure of corruption, legal chicanery, caudillo politics, and criminalization, as well as toxic trauma through the country's port-a-cathed veins.

Land/labor

In the shifting center and peripheries of colonial times, Guatemala historically was a minor player in the mining economy (Helms 1975). Criollo elites instead built their fortunes in agricultural exports—from cochineal, coffee, rubber, cotton, cattle, bananas, and sugar—and most of these agroindustries remain central to the new extractive model (Moreno and Salvadó 2017). Cattle continues to serve as an idiom of power into the present as a low-cost business that allows elites to reconcentrate land holdings in an inflationary economy until investment for other extractive industries becomes possible (Grandia 2009a). This model of development renders the Indigenous subsistence economy, women's reproductive labor, and the natural world as "unproductive" because they fail to produce a cash flow.

By expelling subsistence farmers into the formal economy, land grabbing is effectively labor control grabbing. I have described this elsewhere as a contemporary process of "enclosure" (Grandia 2012), Alonso-Fradejas (2015) as "agrarian extractivism," Solano (2008) as "reconversion," Hurtado (2008) as "recolonato" (recolonization), and Batz (2017) as the "fourth invasion." In this newest cycle of accumulation and dispossession, palm and sugar pose intensified threats to the subsistence economy because as "fields of gold" (Fairbairn 2020) or "flex crops," they can either be sold as raw agricultural commodities; reduced into alternative energy as ethanol or biodiesel; traded in carbon markets (Borras Jr. et al. 2012); or last, but not least, fronted as businesses for laundering drug profits (Anonymous 2011). Some 40% of Guatemalan's territory and its most arable lands for subsistence maize production are now designated as suitable for African palm. As one oil palm magnate derided, "We bring employment and wealth; how are they going to progress with those little maize plants [maicitos]? Who else would be willing to invest US$ 50 million in this petty valley [vallecito de pipiripau]?" (Alonso-Fradejas 2015, 510).

Plantation jobs come at a cost to community, however. While subsistence maize cropping would generate 145 labor days per *manzana* (a colonial land measurement, roughly an acre), sugar and palm offer, respectively, only 23 and 66 days of work on that same amount of land. Local community leaders dole out scarce labor opportunities with prebendalist logic of corruption and clientelism (Alonso-Fradejas, Alonzo, and Dürr 2008). Despite promises of generous wages, palm companies notoriously underpay or withhold wages (Hurtado and Sánchez 2011). As one palm Q'eqchi' laborer remarked:

> Sometimes I wonder if we have advanced at all. As colonos in the estate the patron forced us to work for almost nothing. Still, we had a patch of land to grow our maize. If anything went wrong with the harvest he would never let us starve. He looked after us because he needed us. Now we are free labor, they say. Free to starve, I say. The rich people do not need us and so do not care about us anymore.
>
> (Alonso-Fradejas 2015, 498)

Little wonder another Q'eqchi' leader should remark, 'Before they killed us with the rifle. Today the rifle that kills us is sugar cane and palm' (Alonso-Fradejas, Alonzo, and Dürr 2008, 110).

Debt

Despite the neo-plantation boom, Guatemala remains mired in debt to transnational entities in a pattern that former "economic hit men" (EHM) have publicly confessed was a Cold War strategy exercised throughout Latin America to create dependency on the U.S. (Hiatt 2007; Perkins 2004). Even with the public exposure of geopolitical conspiracy by EHMs, a new generation of unwitting policy-makers have internalized the growth dogma. From five star hotels, these "lords of poverty" (Hancock 1989) continue to induce state officials to accept loans to build the infrastructure, legal stability, and property registries that foreign business interests demand.

Since 2001, the Inter-American Development Bank (the IADB) has loaned Mexico and Central America multi-millions for a coordinated series of "megaprojects" (highways, pipelines, hydroelectric dams, electrical grids, airport development, port construction) to support extractivist industries (Grandia 2013). In tandem with the IADB's "Mesoamerica Project," the US government aggressively imposed a new regional trade agreement in 2004: the DR-CAFTA, whose fine print protects transnational businesses from any environmental or social legislation by guaranteeing them the right to sue for *future lost profits*. Using Guatemala as a "pilot" project based on market theories by Peruvian economist Hernando de Soto (with an aptly inherited conquistador surname), the World Bank has intervened throughout Latin America to create stable land registries to stimulate investment—even though the Guatemalan program clearly catalyzed wide-scale land grabbing and evictions of small farming families (Grandia 2012). Were that not enough, the World Bank's private lending wing, the International Finance Corporation (the IFC), continues to subsidize controversial corporate ventures—for example, a loan to Glamis Gold for $45 million to jumpstart the Marlin Mine (Urkidi 2011). Today almost all of rural Guatemala is threatened by licenses, soil surveys for plantations, biodiversity inventories, and transportation grids (Bastos and de León 2014; Alonso-Fradejas 2015, 491)— all neatly organized onto World Bank-funded cadastral maps.

Ostensibly all these projects are meant to generate cash flow to pay back prior loans. However, given ridiculously low rates of corporate taxation and corruption throughout the system, the Guatemalan government is unlikely to ever recover its investment. In fact, the country's debt obligations ($8.7 billion as of 2019) are rising geometrically—from 18% of the gross domestic product (GDP) in 2000 to 26% in 2020 and expected to reach 32% by 2030. Per capita, Guatemala's newborn citizens now owe foreign creditors $1,158 upon birth. Some 12% of

Guatemala's GDP evaporates into interest payments on previous debts while spending for health is but a paltry 5.7% and education 2.3% of GDP.

When the coronavirus epidemic almost put Guatemala into loan default, the International Monetary Fund (IMF) prescribed an emergency shot of $594 million into the country's leaky (read: corrupt) IV bag in June 2020 (thereby increasing the accumulated debt to over $9 billion). Two hurricanes and 120,000 COVID-19 cases later, the Guatemalan Congress passed a budget in November 2020 that further cut health and education, whilst raising their own per diem stipends (a retro-colonial perk known as a "viático," whose etymology ironically derived from the Catholic sacrament of communion given to those at death's door). To be sure, the IMF could have written off this debt and the defaulted payments of 72 other countries—simply by selling a mere 7% of its personal stash of 2.8 metric tons of gold bars (some of which is surely derived directly from colonial booty), while still reaping a net $62 million due to rising gold prices (Elliott 2020). Instead of providing needed debt relief to Latin America, the tenfold rise in gold prices since 2000 has instead catalyzed an extractivist assault on Indigenous lands previously considered too remote, costly, or risky for mining.

Strategies of resistance

While Maya movements share a number of political strategies within other Indigenous movements for self-determination within the western hemisphere, the grassroots Guatemalan critique of growth- and extractivist-driven development manifests uniquely Maya values of the sacred, balance, consensus, and dignity. After centuries of "counting for nothing" (Waring 1988), above all, they demand "to be taken into account" ("*tomado en cuenta*" in its colloquial expression). The 1991 International Labor Organization (ILO) Convention 169 upholds this right as "Free Prior Informed Consent" (FPIC). While many Indigenous groups now reference the 2007 United Nations Declaration on the Rights of Indigenous Peoples (UNDRIP), the earlier ILO 169 treaty (ratified by an impressive 70% of Central and South American countries, compared with just 6% worldwide) remains an essential legal tool for Latin America's Indigenous human rights claims.

Following ILO 169-inspired FPIC referendum models throughout South America (Peru, Colombia, Chile, and Argentina), anti-mining movements in Guatemala organized the first "community consult" in Sicacapa in 2005 to evaluate community opinion on the Marlin gold mine, whose operations apparently began under the false pretense of being an orchid export business (Nelson 2013). Buttressed by a new Municipal Code passed in 2002 that allows mayoral administrations to seek input on specific administrative decisions affecting their constituents (Urkidi 2011), grassroots Guatemalan leaders have organized more than 85 such community referenda on extractive projects, involving at least one million citizens. Writ large, the majority (99%) voted against mining (Dougherty 2019, 165).

Even if these referenda are not yet legally binding, they have bolstered the visibility and unity of resistance movements and generated important discussions about Maya visions for self-determined government. While post-peace neoliberalism initially segregated Indigenous rights into permitted "folkloric" domains (Hale 2004), cultural rights cannot exist with control over territorial resources and strong local governance systems historically centered at the municipal level (Wolf 1957). The current environmentalization of territorial struggles (Urkidi 2011) has refocused the Maya movement into demands for broader political and economic reform. Beyond township associations, anti-extractivist movements coordinate through a variety of departmental "*mesas*" (regional negotiating tables), horizontal exchanges, and national coalitions/fronts. Emulating the successful popular Chilean uprising in 2019 that led to a constitutional assembly led by a Mapuche woman, Guatemalan social movements have regularly spilled into the streets in 2020 and 2021 to denounce corruption and call for the formation of a new plurinational state (Reina 2008).

In the same way that Indigenous colonial subjects appealed to the Spanish Crown for protection against their criollo overlords, afflicted communities continue to file strategic multi-scalar litigation. Through safeguard protocols like the World Bank's Operational Directive 4.20, Maya claimants have demanded consultations and investigations to halt intrusive infrastructure projects like dams and roads. The Organization of American States' (OAS) International Appeals Court for Human Rights (IACHR) has intervened with advisory rulings on a growing number of cases. In 2009, community groups secured a country visit by S. James Anaya, then United Nations Special Rapporteur on the Rights of Indigenous Peoples (Bastos and de León 2014) who integrated a number of cases (Hudbay, Marlin, San Marcos hydroelectric dam, and the Sacatepequez cement factory) into his global 2010 report to the General Assembly (Anaya 2010). Another rising force is the Maya Lawyers guild (Nimapuj, Asociación de Abogados y Notarios Mayas de Guatemala) established in 2004 by twenty Indigenous lawyers throughout the country (Walter and Urkidi 2016). Finally, the Indian Law Resource Center's international office has accompanied carefully-selected communities as conceptual test cases for international law.

Building on the "conflict" monitory framework established by the United Nations' monitoring team on implementation of the Peace Accords (MINUGUA) in the 1990s, several international databases continue to inventory these sites of resistance. The Guatemalan government itself maintains a registry of conflicts: with 1,214 cases by Alonso-Fradejas's (2015) count. The more iterative Environmental Justice Atlas created at the Universidad Aútonoma de Barcelona refreshes detailed information from the grassroots about more than two dozen cases (Martinez-Alier et al. 2016). Another vital case source are the meticulously detained special reports and "conjunctural notes" put out by "The Observer" (El Observador) (https://elobservadorgt.org) in which Luis Solano's genealogies of corporate oligarchies remain the proverbial "gold standard" of this genre. Courageous Guatemalan scholars have also assembled a number of other timely anthologies (Bastos and de León 2014; Moreno and Salvadó 2017). Most recently, a network of computer-savvy researchers in a new "Observatory on Extractive Industries" is flooding Twitter with maps and alerts of the audacious extent of extractivist industries, whose licensed concessions look like Spanish encomiendas of yore.

These monitoring efforts are necessary in Latin America (with the exception perhaps of Brazil) because environmental journalism is typically relegated to editorial commentary rather than a rigorous and consistent news beat (Takahashi et al. 2019). Filling this vacuum in Guatemala, however, is a new genre of independent and/or Maya-led journalism: *Prensa Comunitaria, Nómada, Plaza Pública*, and the vibrant social media networks of a dizzying number of grassroots associations. A radio-school FGER, which gave apolitical literacy lessons by air in the 1990s, is now one of the most important national outlets for raising consciousness about extractive industries, food sovereignty, and restorative alternatives. Amplifying these stories internationally are teams like the "Green Blood" project plus a cadre of photojournalists, freelancers, and nonprofit writers (James Rodriguez, Sandra Cuffe, Jeff Abbott, Grahame Russell, Rob Mercatante, to name a few).

Maya women journalists like Andrea Ixchíu and Dr. Irma Alicia Velásquez Nimatuj (both profiled in Pamela Yates' documentary *500 Years*) are at the forefront of many of these struggles. Since 15,000 people marched on the Guatemalan capital in the 2012 "Popular, Peasant, Women and Indigenous March," (Alonso-Fradejas 2015), women's groups have mobilized repeatedly, often with savvy street theatre, to denounce femicide (the disproportionate rape and murder of women) around extractivist industries (Arnaud Brandt 2020). As with Native American opposition to the Dakota Access pipeline or the "Idle No More" First Nations uprising, the fundamental human right to water is a central theme of women-led organizing against the toxicity of extractivism.

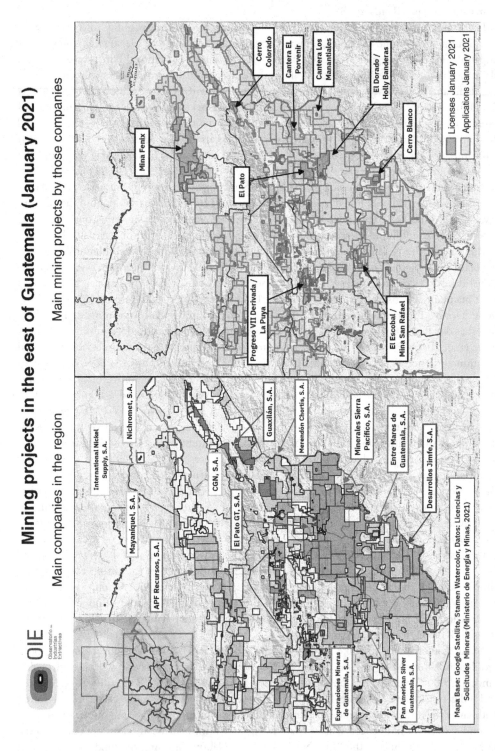

Figure 6.1 Observatory of extractive industries (OIE), 2021.

Port-a-cath veins

Although oncology nurses pump fluids into their patients' port-a-caths during infusions, chemotherapy poisons nevertheless cause profound dehydration, much like the drying and poisoned rivers of Guatemala. Without sufficient fluids to flush chemo drugs from the body, toxicity can linger in the liver, kidneys, and other organ systems for a lifetime of secondary health problems. World-systems analysis and scholarship have tended to focus on the scale and scope of extraction, but less on the toxic residues left behind and natural resources available for community resiliency (Boudia et al. 2018). Denunciations of extractivism have likewise tended to focus more on the financial flows of visible projects and infrastructure (Grandia 2013) and less on the corrosion of social, legal, and ecological systems. Yet extractivism inherently generates contamination (Lopez and Navas 2019) that may last over generations in subtle ways. For example, mothers' observations about rashes on their children's skin mobilized early opposition to the Marlin gold mine; subsequent worries about household access water have sustained the struggle, as the mine consumes more water in an hour than a Mam family might use in twenty years (Haines and Haines 2013).

Although the 1944 revolution ended the transgenerational inheritance of labor debts from the plantation company store, future generations will epigenetically inherit the "ecological debt" (cf. Martinez-Alier 2002) of toxic-bonded labor in these environmental sacrifice zones. Toxic mining sites remain policed and/or militarized long after production has ceased, making it impossible for Indigenous and small farming communities to ever return to the land (Nolin and Stephens 2010). As a community leader succinctly put it, 'In exchange for maintaining their wealth, we have to die' (Fitzpatrick Behrens 2009). Sugarcane workers in Panzós have high rates of fetal mortality (Alonso-Fradejas 2015), and untold numbers of neighboring banana workers have been left sterile from DBCP (Lopez and Navas 2019). With no toxic tracking systems, much less remediation programs, how much it would cost to make these sites safe for human health is anyone's guess. One courageous Q'eqchi' Maya mayor of a frontier town at the crossroads of oil, narcos, palm, and cattle who attempted to fine palm oil companies to 'recover part of the huge expenses we have to pay to restore what they destroy and pollute' (interview with Raxruhá mayor, August 2013) was counter-sued by Guatemala's oligarchic Chamber of Commerce in Constitutional Court (Alonso-Fradejas 2015).

Beyond chemical toxicity associated with extractive industries, neoliberalism inserts a *social* toxicity of corruption into the veins of the political and judicial system. Dishonesty drips down to community relations, pitting neighbors and kin against one another in what were previously egalitarian social structures (Caxaj et al. 2014). How and why Guatemala persists in awarding licenses for a paltry 1% royalty on mining profits begs investigation of who benefits in the shadows at all levels. One of Guatemala's earlier environmental justice advocates, the newspaper columnist Magalí Rey Rosa, summed it up thus: 'The laws are loose, the land cheap, the labor cheap and the politicians cheaper' (Nolin and Stephens 2010, 58). When community leaders, nonprofit people, even priests can be flipped for the right price or with the right threat (Caxaj et al. 2014), fear remains the fuel of neoliberalism (Arnaud Brandt 2020)—making it hard for the stressed social body to heal or respond to other threats. As a grassroots leader recently emailed me, "It's not just the pandemic that is killing us, but the corruption, the lack of medicine in hospital centers, the lack of employment that expels migration to the north, [and] the evictions, assassinations, persecution, jailing of land defense leaders and human rights defenders" (Author fieldnotes 2021).

The intimidation, harassment, and murder of Maya leaders and journalists opposed to extractivist industries perpetuate Cold War patterns of military violence (Solano 2005). To give but one example, a Q'eqchi' Mayan journalist, Bernardo Caal Xol, spent more than four years as a political prisoner in a frame-up for his reporting on hydroelectric threats to the Cahabón River. Like other dissidents, he has been characterized as an "economist terrorist." The criminalization of social movements as "internal enemies" accelerated under the presidential administration of Otto Perez Molina (2012–2015), a former military commander during the civil war genocide. Yet starched-shirt business leaders also endorse violence, as millionaire president Oscar Berger revealed when he remarked to the press, 'We have to protect investors' in 2005 before authorizing an attack on rural communities opposed to the Escobal Mine (Solano 2015; Nolin and Stephens 2010; Grandia 2012). Even were the Guatemalan government willing to protect their own citizens against business bullies, transnational corporations can sue them with trade lawsuits like the DR-CAFTA agreement, which functions as a modern-day Treaty of Tordesillas for US corporations (Solano, Moore, and Moore 2020; Grandia 2014). A Nevada-based mining corporation is using this trade agreement to sue Guatemala for $400 million in future lost profits (Sandoval 2022). The world system lumbers onward.

To counter these complex criollo/corporate collusions, Indigenous environmental defenders and the movements they lead need sustained ethnographic, human rights, journalistic, and hard scientific studies that anticipate and marshal comparative lessons throughout the hemisphere to better understand the toxic legacies of the extractive rush. Unfortunately, whilst Guatemala is besieged by extractive industries, much of the scholarly community continues its own extractivism. As Galeano (2010) himself quipped in an editorial on the "Open Veins of Climate Change,"

> The wordy inflation, which in Latin America is more damaging than monetary inflation, has done us, and keeps inflicting, grave damages. And also, and above all, we are fed up with the hypocrisy of the rich countries, which is leaving us without a planet while it delivers pompous discourses to conceal the hijacking.

Guatemala's northern Petén region where I do research, for example, has hosted hundreds of extractive theses and theoretically-driven projects on what is called green colonialism, environmental colonialism, carbon colonialism (Lopez and Navas 2019, 1) or, more awkwardly, framed as "green settler colonialism" (Ybarra 2017). Theoretically ambitious foreign scholars routinely extract (i.e. steal) data from local NGOs and intellectuals for their own academic advancement and/or move citations from one wordy waste pit to another. Technically-informed studies of environmental racism remain missing from the overwhelmingly forest-focused conservation literature. Among the thousands of studies of flora and fauna sponsored by foreign donors in the Maya Biosphere Reserve and Guatemala's other protected areas, I have yet to find *even one* analysis of the effects of pesticides on wildlife, much less human populations. Although mining contracts, trade agreements, pesticide shipments, and foreign aid bills may not be as alluring to graduate students as Foucauldian discourses of ontology, we need more scholarship of resistance on Guatemala's port-a-cathed veins of commerce like the courageous work cited herein.

References

Alonso-Fradejas, Alberto. 2015. "Anything But a Story Foretold: Multlple Politics of Resistance to the Agrarian Extractivist Proejct in Guatemala." *Journal of Peasant Studies* 42 (3–4):489–515.

Alonso-Fradejas, Alberto, Fernando Alonzo, and Jochen Dürr. 2008. *Caña de Azucar y Palma Africana: Combustibles para un Nuevo Ciclo de Acumulación y Dominio en Guatemala*. Edited by IDEAR. Guatemala City: CONGCOOP y Magna Terra Editores.

Anaya, S. James. 2010. Report by the Special Rapporteur on the Situation of Human Rights and Fundamental Freedoms of Indigenous People, James Anaya addendum. United Nations.

Anonymous. 2011. *Grupos de Poder en Petén: Territorio, Política, y Negocios.* Email Circulation: Unknown publisher.

Arnaud Brandt, Gala 2020. *No TengoTtierra, No Tengo Nada. Violencias Compartidas: Una Etnografía Sobre las Mujeres Q'eqchi'es en Comunidades del Norte de Chisec, Alta Verapaz, Antropología,* Universidad del Valle, Guatemala City.

Bastos, Santiago, and Quimy de León. 2014. *Dinámicas de Despojo y Resistencia en Guatemala. Comunidades, Estado y Empresas.* Guatemala City: Serviprensa.

Batz, Giovanni. 2017. *The Fourth Invasion: Development, Ixil-Maya Resistance, and the Struggle against Megaprojects in Guatemala,* Anthropology: University of Texas Austin.

Borras Jr., Saturnino M., Jennifer C. Franco, Sergio Gómez, Cristóbal Kay, and Max Spoor. 2012. "Land Grabbing in Latin America and the Caribbean." *Journal of Peasant Studies* 39 (3–4): 845–72.

Boudia, Soraya, Angela N. H. Creager, Scott Frickel, Emmanuel Henry, Nathalie Jas, Carsten Reinhardt, and Jody A. Roberts. 2018. "Residues: Rethinking Chemical Environments." *Engaging Science, Technology, and Society* 4:165–78.

Caxaj, C. Susana, Helene Berman, Colleen Varcoe, Susan L. Ray, and Jean-Paul Restoulec. 2014. "Gold Mining on Mayan-Mam Territory: Social Unravelling, Discord and Distress in the Western Highlands of Guatemala." *Social Science & Medicine* 111:50–57.

Dougherty, Michael L. 2019. "How Does Development Mean? Attitudes toward Mining and the Social Meaning of Development in Guatemala." *Latin American Perspectives* 46 (2):161–81.

Elliott, Larry. 2020. Campaigners Urge IMF to Sell Gold to Provide Debt Relief. *The Guardian,* October 11.

Fairbairn, Madelein. 2020. *Fields of Gold: Financing the Global Land Rush.* Ithaca, NY: Cornell University Press.

Fitzpatrick Behrens, Susan. 2009. Nickel for Your Life: Q'eqchi' Communities Take On Mining Companies in Guatemala. *NACLA Report on the Americas.*

Galeano, Eduardo. 1973. *Open Veins of Latin America: Five Centuries of the Pillage of a Continent.* Translated by C. Belfrage. 25th anniversary edition ed. New York, NY: Monthly Review Press.

Galeano, Eduardo. 2010. The Open Veins of Climate Change. *YES! Magazine.*

Grandia, Liza. 2009a. "Raw Hides: Hegemony and Cattle in Guatemala's Northern Lowlands." *Geoforum* 40 (5):720–31.

Grandia, Liza. 2009b. *Tz'aptz'ooqeb': El Despojo Recurrente al Pueblo Q'eqchi'.* Guatemala City: AVANCSO.

Grandia, Liza. 2012. *Enclosed: Conservation, Cattle, and Commerce among the Q'eqchi' Maya Lowlanders.* Seattle: University of Washington Press.

Grandia, Liza. 2013. Road Mapping: "Megaprojects and Land Grabs in the Northern Guatemalan Lowlands." *Development and Change* 44 (2):233–59.

Grandia, Liza. 2014. "Modified Landscapes: Vulnerabilities to Genetically Modified Corn in the Political Economy of Maize Production in Northern Guatemala." *Journal of Peasant Studies* 41 (1):79–105.

Haines, J. T., and Tommy Haines. 2013. Gold Fever. Northland Films, 1 hour and 24 minutes.

Hale, Charles. 2004. "Rethinking Indigenous Politics in the Era of the "Indio Permitido"." *NACLA Report on the Americas* 38 (2):16–21.

Hancock, Graham. 1989. *Lords of Poverty: The Power, Prestige, and Corruption of the International Aid Business.* New York: Atlantic Monthly Press.

Helms, Mary W. 1975. *Middle America: A Culture History of Heartland and Frontiers.* Englewood Cliffs, NJ: Prentice-Hall.

Hiatt, Stephen, ed. 2007. *A Game as Old as Empire: The Secret World of Economic Hit Men and the Web of Global Corruption.* San Francisco: Berrett-Koehler Publishers, Inc.

Hurtado, Laura, and Geisselle Sánchez. 2011. *¿Qué Tipo de Empleo Ofrecen las Empresas Palmeras en el Municipio de Sayaxché, Petén?* Guatemala City: Action Aid.

Hurtado Paz y Paz, Laura. 2008. *Dinámicas Agrarias y Reproducción Campesina en La Globalización: El Caso de Alta Verapaz, 1970–2007, Grupo Pop Noj, Seva Foundation, ProPetén, and Action Aid.* Guatemala City: F&G Editores.

Jonas, Susanne. 2015. Eduardo Galeano, Latin America's Social Justice Laureate. *NACLA Report on the Americas.*

Lopez, Gustavo A., and Grettel Navas. 2019. "Eco-Imperial Relations: The Roots of Dispossesive and Unequal Accumulation." In *Palgrave Encyclopedia of Imperialism and Anti-Imperialism*, edited by I. Ness and Z. Cope. New York: Palgrave Macmillan.

Martinez-Alier, Joan, Leah Tempe, Daniela Del Bene, and Arnim Scheidel. 2016. "Is There a Global Environmental Justice Movement?" *Journal of Peasant Studies* 43 (3):731–55.

Mohawk, John C. 2000. *Utopian Legacies: A History of Conquest and Oppression in the Western World*. Santa Fe, NM: Clear Light Publishers.

Moreno, Sonia Elizabeth, and Camilo Salvadó. 2017. *Industrias y Proyectos Extractivos en Guatemala: Una Mirada Global*. Guatemala City: AVANCSO.

Nelson, Diane M. 2013. "Yes to Life = No to Mining:" Counting as Biotechnology in Life (Ltd) Guatemala. *Scholar and Feminist Online* 11 (3): 1–29. http://sfonline.barnard.edu/life-un-ltd-feminism-bioscience-race/yes-to-life-no-to-mining-counting-as-biotechnology-in-life-ltd-guatemala/0/.

Nolin, Catherine, and Jaqui Stephens. 2010. "'We Have to Protect the Investors:' Development & Canadian Mining Companies in Guatemala." *Journal of Rural and Community Development* 5 (3): 37–70.

Perkins, John. 2004. *Confessions of an Economic Hit Man*. San Francisco: Berrett-Koehler.

Reina, Carmen. 2008. "Retos de la Participación Ciudadana en la Construcción Democrática." *El Observador: Análisis Alternativo Sobre Política y Economía* 3 (14):3–21.

Sandoval, Ana. 2022. "A Mining Lawsuit in Guatemala Shows How Trade Courts Put Big Corporations First." *Inequality.org*, December 5.

Sherman, William L. 1969. "A Conqueror's Wealth: Notes on the Estate of Don Pedro de Alvarado." *The Americas* 26:199–213.

Solano, Luis. 2005. *Guatemala: Petróleo y Minería en las Entrañas del Poder*. Guatemala City: Inforpress Centroamericana.

Solano, Luis. 2008. "Reconversión Productiva y Agrocombustibles: La Nueva Acumulación Capitalista en el Agro Guatemalteco." *El Observador: Análisis Alternativo Sobre Política y Economía* 3 (14):31–61.

Solano, Luis 2015. Guatemala: How a Pseudo-Military Project was Created to Protect the Escobal Mine. *Upside Down World*, April 9.

Solano, Luis, Ellen Moore, and Jen Moore. 2020. *Mining Injustice Thorugh International Arbitration*. Washington, D.C.: Institue for Policy Studies and Earthworks.

Takahashi, Bruno, Juliet Pinto, Manuel Chavez, and Mercedes Vigón. 2019. *News Media Coverage of Environmental Challenges in Latin America and the Caribbean: Mediating Demand, Degradation and Development*. New York: Palgrave Macmillan.

Urkidi, Leire. 2011. "The Defence of Community in the Anti-Mining Movement of Guatemala." *Journal of Agrarian Change* 11 (4):556–80.

Walter, Mariana, and Leire Urkidi. 2016. "Community Consultations: Local Responses to Large-Scale Mining in Latin America." In *Environmental Governance in Latin America*, edited by F. de Castro, B. Hogenboom and M. Baud. New York: Palgrave Macmillan.

Waring, Marilyn. 1988. *If Women Counted: A New Feminist Economics* San Francisco: Harper & Row.

Weatherford, Jack. 1988. *Indian Givers: How Native Americans Transformed the World*. New York: Three Rivers Press.

Wolf, Eric R. 1957. "Closed Corporate Peasant Communities in Mesoamerica and Central Java." *Southwestern Journal of Anthropology* 13 (1):1–18.

Wolf, Eric R. 1982. *Europe and the People without History*. Berkeley: University of California Press.

Ybarra, Megan. 2017. *Green Wars: Conservation and Decolonization in Guatemala's Maya Forest* Berkeley, CA: University of California Press.

7 Environmental colonialism and neocolonialism in Latin America

Chris O'Connell and Rocío Silva-Santisteban

Introduction

From its inception, the European conquest of the Americas was absolute. The impact of colonialism extended beyond territorial control to the fundamental re-shaping of the natural environment and alteration of the lives and livelihoods of its inhabitants beyond recognition. Colonialism swept away centuries of the "ecological and social co-evolution" of the region's indigenous populations (Boyer 2016, 1). The arrival of European settlers, and the plants, animals, and disease they brought with them, devastated the indigenous population of the region and permanently altered its environment. Such was the impact that approximately 90% of the pre-existing population died and an estimated 56 million hectares of land were abandoned, causing the temperature of the entire planet to drop (Koch et al. 2019).

As such, environmental colonialism is woven into the history of the region and its inhabitants, and continues today in both old and new forms. As this chapter outlines, although the forms taken by environmental colonialism and neo-colonialism vary over time, their impact remains consistent and significant. Furthermore, we also demonstrate that environmental colonialism as a mindset has become hegemonic in the region, with the result that many of its human 'victims' have over time become victimizers themselves. In the words of Peruvian thinker José Carlos Mariátegui, contemporary society "suffers from the original sin of the conquest. This is the sin of having been born and having been formed without the Indian and against the Indian" (1925).

In this context, we understand environmental colonialism (and neo-colonialism) as colonial practices that impact the natural environment, as well as indigenous and racialized peoples. On one level, this involves the exploitation and looting of natural assets and the extraction of wealth, along with the impacts these activities produce within colonized territories. The primary beneficiaries of these practices are external actors, but also include national economic and political elites whose "consent and participation" is essential (Atiles-Osoria 2014, 8). At a deeper level, environmental colonialism involves a wider ideological system of control over territories that takes on a range of forms, some more apparent than others.[1] Similarly, the impacts of environmental colonialism do not stop at the exploitation of humans and Nature, but shape the cultural and ontological connections between people and their ecosystems; in other words, their understanding of the world (Alimonda 2011; De Souza Santos 2011).

This chapter outlines the evolution of this phenomenon in a regional context. The first section focuses on the conquests and direct colonial rule, which heralded an ontological break in terms of environmental management in Latin America. The next section examines the post-independence era, in which a decentralized environmental neo-colonialism emerged to expand colonial practices and deepen the logic of coloniality. The third section takes this analysis from the Great Depression to the present day, noting how geopolitical and ideological

DOI: 10.4324/9780429344428-9

changes have altered—but rarely lessened—the extractivist consensus. Finally, the chapter also highlights that resistance by indigenous and other groups has been and remains a constant factor across the periods examined.

An ontological break: Conquest, colonialism and the environment in Latin America

Pre-Columbian societies modified their environments in many ways, but the speed, scale, and sustained nature of the changes under colonial rule was on another level, permanently re-ordering the region's environmental and social systems. Key to these changes was what Europeans brought with them: diseases which devastated native populations; livestock that altered landscapes; and technologies that facilitated the subjugation of the region's ecosystems and societies. But the most significant European import was ideological: the concept of ascribing 'value' to everything (Guardiola-Rivera 2010, 54). This mindset was so at odds with that of indigenous peoples that it led them to fatally underestimate the danger posed by the *Conquistadores*. Guardiola-Rivera notes that in trying to buy his freedom with gold and silver, Inca leader Atahualpa failed to understand that "the thirst and hunger of these men was insatiable" (ibid., 71).

An examination of the dominant economic activities in Latin America during the period of direct rule by the Spanish crown illustrates this point. The most important activities were mining, export-oriented agriculture, and sugar production. Each case demonstrates how colonialism and its systems have left their mark on the environment, as well as on the bodies of the men, women, children and natural beings of the region.

The concept of land as a commodity was alien to the native peoples, whose traditions were based around communal use and occupation. For European settlers, however, Nature was a source of bounty to be managed and manipulated (Gudynas 2003). In particular, land in the 'New World' was viewed as property, which provided both social status and wealth. Justification was found in the view of the region as a "conquerable space", as wild lands in need of taming (O'Gorman 1986). Accordingly, land was rapidly concentrated into large holdings, or *latifundio*.[2] While formal enslavement of the indigenous population was outlawed as early as 1544—due primarily to high mortality rates—other forms, such as debt peonage, were developed to keep them in conditions of "near slavery" (Meade 2016, 79).

Figure 7.1 A mural in the Amazonian town of Nauta in Peru's Loreto province depicts the horrors of the rubber era (O'Connell).

Furthermore, the importation of African chattel slaves meant that most agricultural work was carried out via coerced labor. The *latifundio* system resulted in deforestation, ecological invasion by foreign plants, and the re-shaping of the landscape by European livestock and crops (Boyer 2016, 6–7).

No single crop had a greater impact than "King Sugar" (Galeano 1997, 59). This 'white gold' came to dominate the landscape of north-eastern Brazil, and later much of the Caribbean. Produced as a monocrop, in time sugar devastated soils and water, and left crops vulnerable to disease (Boyer 2016, 7). Sugar production caused the destruction of mangroves in Brazil, and forests in Puerto Rico, Haiti, and Cuba, further degrading soil fertility. Over-reliance on one commodity exposed regions to cycles of boom and bust, as befell Brazil in the 17th century. Furthermore, sugar created an insatiable need for labor, leading to the ramping up of the slave trade. An estimated six million African slaves went to the Caribbean alone, where 70% worked in sugar fields and refineries (Meade 2016, 130). Conditions were brutal, with many enslaved people worked to death within five years (Meade 2016, 124).

Nevertheless, the main social and ecological transformations were linked to mining, the predominant economic activity. While the *Conquistadores* initially sought gold,[3] it was silver that reshaped the contours of Latin America and gave rise to global capitalism. In particular, the mine in Potosí (now Bolivia) provided the basis of a "profoundly unequal ecological exchange" between New and Old Worlds (Moore 2007, 130). During the colonial era, 40,000 tons of silver ore were removed from the mine—around 60–80% of the world's silver (Boyer 2016, 7). According to Guardiola-Rivera, this heralded the invention of globalization through the creation of the first "world money", the silver peso (2010, 55). Yet, like a coin, the story of Potosí had two sides. While Europeans called the mountain that housed the mine '*Cerro Rico*' (Hill of Riches), the native populations knew it by another name: the 'Mountain that Eats Men' (Guardiola-Rivera 2010, 82)—in reference to lives lost in the pursuit of profit. The mine depended largely on the state-organized '*mita*' system of forced labor, which Galeano called "a machine for crushing Indians" (1973, 40). This "awesome operation" of state labor (Williamson 1992, 127) brought upwards of 5,000 '*mitayos*' per year to the mine. Untold numbers died or suffered crippling injuries and disease, with '*mitayos*' given the hardest, most dangerous tasks due to their disposability (Moore 2007, 134). The direct environmental impact of mining was huge: trees were felled for use as fuel and building materials, heavy metal residues and poisonous mercury leached into the soil and water, and animals overgrazed local grasses, leaving "scars" still visible to this day (Boyer 2016, 7).

Silver mining in Potosí also established the model of large-scale industrial mining (Machado-Aráoz 2018, 96). Yet the true impacts on the region's human and natural environment were broader still. The colonial period created three basic conditions that would permit the exploitation of resources, labor, and the environment for centuries to come. The first relates to land occupancy. First, silver mining and the *mita* system led to widespread migration, increased the spread of epidemics, and began a process of urbanization, all contributing to the displacement of indigenous populations and the clearing of land for production (Gil Montero 2011; Moore 2007). These changes ushered in a new period of "ecological imperialism" based on the dominance of European crops (Moore 2007, 136).

Secondly, the colonial period established what Guardiola-Rivera termed the "dual monetary system" to describe the gulf in earnings between Creole elites—mine owners, merchants and landlords—and the rest, many of whom fell into debt peonage (2010, 76). The economic power of elites was mirrored by their social prestige and political clout. It was this power that enabled miners to resist attempts by the clergy to end the *mita* system in the 16th century (Gil Montero 2011, 305–306); and slaveowners to delay the abolition of slavery in Cuba in the

Figure 7.2 The monument to the miners in Oruro, Bolivia (O'Connell).

19th Century (Williamson 1992, 436–437). This "internal colonialism" (Rivera Cusicanqui 2010) by Eurocentric Creoles embedded racialized hierarchies and legitimated the colonial development model (Mignolo 2007, 157–158).

Finally, the subordinate global status of Latin America was clearly established. The unequal exchange with the imperial powers of Spain and Portugal helped to make fortunes in other European countries through trade, finance, and additional conquest (Williamson 1992). Then there was the slave trade. Administered by the Portuguese and Spanish, but benefiting many in Britain, these funds helped catalyze the Industrial Revolution. In these ways, environmental colonialism in Latin America provided Europe with its "comparative advantage" over other regions (Dussel 1998, 5).

Independence and the era of environmental neo-colonialism

The Wars of Independence that liberated the Spanish Americas from direct rule failed to herald a change in social structures. Instead, they led to "colonial nation states", with Creole elites taking power from European monarchies (Mignolo 2007, 157). The "two-tier" colonial system was replaced by a unitary, racialized system of domination by whites over non-whites (Williamson 1992, 115). While slavery was legally abolished in the Spanish Americas, other forms of coerced labor based on racialized categories were established to meet demand (Quijano 2000). Similarly, calls for agrarian reform instead resulted in the consolidation of *latifundios*. The result was perpetual impoverishment for the majority, while large landowners enjoyed "near complete control" over resources (Meade 2016, 182–183).

Indigenous peoples did not meekly accept these changes. A series of rebellions broke out across the region following independence in response to encroachments by the machinery of capitalism onto their territories (Williamson 1992, 246). In turn, the region's fledgling states opted to make war on these groups they branded 'barbarians'.[4] Perhaps the most extreme example occurred in Argentina, where a campaign of "genocidal ethnic cleansing" known as the 'Conquest of the Desert' was waged to clear indigenous Mapuche from the plains (Meade 2016, 288).[5] The land in question was no desert, however, but fertile farmland that was quickly consolidated in the hands of a few loyalists. The campaign was underwritten by British finance, and the ensuing concentration of land ownership "meshed perfectly" with British interests (Meade 2016, 289). Labor needs were met through further colonization by European immigrants, who soon overwhelmed the pre-existing population.

Independence therefore served to deepen the subjugation of land, resources, and people to the colonial matrix of power. This period of environmental neo-colonialism in Latin America had global significance, as it coincided with the shift to carbon-emitting processes by rapidly industrializing countries. Burdened with debt following years of war, these new nations responded by abandoning colonial controls and opening their economies to global commerce (Meade 2016, 225). As a result, the era of the Anthropocene was fed, in large part, by materials from Latin America during the 'primary commodity exporting' phase (Boyer 2016, 8). For some, this period established a "global metabolic rift" whereby the environmental riches of Latin America were appropriated by advanced nations (Clark and Foster 2009). This 'ecological imperialism' created a hierarchical division of nations and asymmetries of wealth, as demonstrated by the history of *guano* (Clark and Foster 2009). Nevertheless, the participation of local elites and the ideological nature of this system of exploitation means it is better categorized as environmental colonialism or coloniality (Atiles-Osoria 2014), as the example of *guano* reveals.

Indigenous farmers in Peru had long known of the benefits of *guano*—sodium-rich droppings of seabirds—for improving soil fertility. But it was not until the 1840s that its global importance became clear, due to the impoverishment of the soil in wheat-growing countries like Britain (Clark and Foster 2009). Peru was indebted to Britain, and *guano* appeared to offer a means to repay its debts. In fact, the forty years of the *guano* boom left Peru even deeper in debt, due to a mix of borrowing and the failure by elites to re-invest their earnings (Meade 2016, 229). The undoubted winners of the boom were the British, who enjoyed a trade monopoly on the product (Clark and Foster 2009, 320). An estimated one million tons of *guano* were exported, leaving the islands bare and the country broke.

Yet the impacts extended beyond a 'boom-and-bust' cycle. Firstly, as Boyer (2016) notes, neo-colonial extractivism resulted in huge damage to the environment. Secondly, the transfer of the riches of the soil not only provided advanced countries with a way out of the "ecological contradiction" of intensive farming, it denied those riches to future generations of Peruvians (Clark and Foster 2009, 319). Thirdly, *guano* came at a huge human cost. Extraction was highly labor-intensive, but slavery had been abolished. A way around the problem was found by importing Chinese 'coolies'. Both Peru and Britain were complicit in this "disguised slavery", with conditions on the guano islands reported as worse than on sugar plantations (Clark and Foster 2009, 322).[6]

From dependent development to the commodities consensus

The patterns of environmental colonialism and neo-colonialism not only continue in contemporary Latin America, but have also deepened and diversified over time. One significant shift came with the end of the 'primary commodity exportation' phase caused by the Great Depression in 1929. This resulted in the "peculiar system" (Quijano 2000, 567) of import-substitution industrialization which saw the state assume a stronger role in economic management. This did not signal the end of environmentally harmful activities in the region, but gave rise to the era of 'dependent development' in which primary products were exported to earn foreign currency to maintain industrialization (Coronil 2000, 99; Cooney, Chap. 16 in this volume). For example, states encouraged the opencast mining of new metals like bauxite by constructing hydro-electric mega-dams to provide them with electricity. This process displaced farming communities and produced toxic "red mud" that contaminated water and soil, however (Dore 2000).

Under military governments in particular, a discourse of resource nationalism was utilized to justify the ongoing deforestation, pollution, and displacement of Indigenous peoples (Boyer 2016, 12–13). Furthermore, transnational corporations—possessed of vital know-how and technology—were subject to lax regulation that resulted in significant environmental and rights issues. Perhaps the clearest example is Ecuador, where Texaco's failure to re-insert toxic waste-water devastated the rainforest and caused a public health disaster among indigenous groups (Acosta 2013). During this period, environmental colonialism also took more nakedly imperialist forms, such as the CIA-sponsored overthrow of Guatemalan President Jacobo Arbenz in 1954, which was done to prevent his planned land reforms from damaging the holdings of the United Fruit Company (Williamson 1992, 353).

The model of dependent development left the region vulnerable to external shocks. This is what occurred during the Debt Crisis of the 1980s, which brought an end to state-led industrialization and began a 'lost decade' of development. The region's vulnerability paved the way for neoliberalization. While neoliberal economics had been instituted by the US-backed Pinochet dictatorship in the 1970s, the crisis opened the possibility of a region-wide ideological shift under conditions of democracy (Silva 2009). In particular, multilateral financial institutions like the IMF and World Bank used 'conditionalities' to pressurize governments to privatize public assets, deregulate industries, and liberalize trade and investment. This period saw global capitalism enter a new era of imperialism that Harvey (2005) termed "accumulation by dispossession". Across Latin America "whole economies were raided" (2005, 66), opening up opportunities for accessing cheap labor and natural resources (Coronil 2000). Under this system, Latin America was reaffirmed as a supplier of primary commodities (Lander 2013, 92).

As nature and humans became 'capital' in themselves, this new colonial era was more socially than geographically defined (Coronil 2000, 101), and resulted in the adoption of neoliberal policies by governments across the ideological spectrum (Silva 2009). The neoliberal consensus produced policies that incentivized extractive activities, with severe environmental consequences. For example, agricultural practices altered drastically during this time, adhering to an industrial model that favored the intensive mono-cultivation of export-oriented crops. Research in Argentina reveals a concentration of land ownership, increased unemployment, and the depression of wages in rural areas, while the heavy use of herbicides polluted soil, water, and air, severely impacting human health (Pengue 2018). Monocropping further damaged the soil, while relying on precarious labor likened to "semi-slavery" (Acosta 2017, 7).

The objective of these reforms was to commoditize all aspects of society and the environment (Silva 2009). However, as the human toll of privatization and deregulation became clear, resistance built gradually. Indigenous and *campesino* movements were at the vanguard of contention, and succeeded in slowing, halting, or rolling back these plans. Water provides the clearest example. As the neoliberal laboratory, Chile was the first country in the region to privatize its water, but others soon followed. Until 2000, that is, when the proposal met with fierce and sustained resistance in the Bolivian city of Cochabamba in what became known as the 'Water War'. There, a coalition of rural and urban social movements successfully opposed the privatization of municipal water supplies by a US-based multinational, creating a "template" for resistance that spread across the region (Silva 2009, 124).

Waves of anti-neoliberal contention obstructed reforms, brought down governments, and elected presidents offering a different path. These were the roots of the 'shift to the left' that included Lula in Brazil, Evo Morales in Bolivia, and Rafael Correa in Ecuador (Silva 2009). The goals of this broadly articulated coalition went beyond mere opposition to questioning "Western patterns of production and knowledge" (Lander 2013, 90). This new thinking was encapsulated by '*sumak kawsay/buen vivir*', a concept drawn from indigenous cosmovision, and which serves as a critique of the development model based on the exploitation of finite resources (Acosta 2017). Versions of '*buen vivir*' were constitutionally enshrined in Ecuador and Bolivia and enabled these governments to challenge the environmental neo-colonialism at the levels of discourse and, to some extent, policy.

In practice, however, no government broke substantively with the extractive model, and in time many came to deepen it (Lander 2013). Gudynas has labelled this "progressive new extractivism" (2009), which combined a stronger state and the redistribution of revenue from extractive activities. Accordingly, extractivism was legitimated as necessary for development, casting indigenous and environmental movements as impediments to progress. Furthermore, the model consolidated dependency, necessitating the ongoing flexibilization of labor and environmental standards (Acosta 2013; Gudynas 2009). In this sense, then, progressive governments differed little from their predecessors (Lander 2013). Svampa termed this the 'commodities consensus': a new "economic and political order" common to both conservative and progressive governments, based on the large-scale exporting of primary products driven by high global demand (2012, 43–44).

New frontiers of environmental colonialism

As existing reserves of minerals and hydrocarbons have depleted, and with world prices receding from their historic highs since 2015, states have gone to ever-greater extremes to exploit Nature's riches. One approach involves new and more extreme forms of extractivism, including open-cast mining, the extraction of heavy crude oils, and hydraulic fracturing ('fracking'). For example, in Argentina governments of both left and right have publicly legitimated fracking as essential to Argentina's economic future, incentivized investment by transnational oil and gas companies, and provided juridical and income security via partial nationalization (Riffo 2017). The industry centers on the *Vaca Muerta* (Neuquén) shale project, dubbed "the capital of unconventional hydrocarbons" (Scandizzo 2017, 45). As in the past, drilling has led to the forced displacement of Mapuche indigenous communities from their lands, with resistance criminalized and repressed, while causing pollution, higher emissions, and tectonic activity (Riffo 2017).

The commodities boom also heralded changes in mining. In Peru, for example, the mega-mining sector is so influential that some see the state as having been 'captured' by corporate interests (Durand 2019). This 'external' colonialism not only weakens state capacity but also helps to foster 'internal' colonialism by conflating mining with 'development'. The rapid growth of alluvial mining in Peru reflects this dynamic. The emblematic case is Madre de Dios, where illegal mining saw over 50,000 hectares deforested by 2016 (de Echave 2016, 136–137), and where mercury was liberally used in a major biodiversity hotspot (Cortés-McPherson 2019, 383). Illegal gold enters global supply chains through the involvement of external actors, including armed groups, criminal gangs, and illicit 'clearinghouses' in countries like Italy, Switzerland, and the US (Verite 2016). Nevertheless, its prevalence in Peru indicates the importance of domestic governance. There, illegal mining 'piggybacks' on the formal sector, benefiting from weakened regulation, state inertia, and the hegemony of the extractive model. In Madre de Dios—where an "illiterate indigenous Quechua-speaker" became the 'Queen of Gold' (Cortés-McPherson 2019, 382)—poor Peruvians not only help to devastate the environment, but also subject others to trafficking and exploitation (Verite 2016).

In the agricultural sector, these environmental neo-colonial traits are evident in the production of genetically-modified (GM) soya. The soya boom has wrought significant and long-lasting changes to environments, societies, economies, and polities, however. First introduced to Argentina by way of a trade window with the European Union (Gudynas 2008), the crop has spread so rapidly across Brazil, Argentina, Paraguay, Uruguay, and Bolivia that agribusiness corporations nicknamed them the 'United Republic of Soya' (Pengue 2018, 26). The advent of soya in Latin America has heralded a 'great transformation' (Gudynas 2008) that amounts to a corporate takeover of agriculture. One feature is the influence of international organizations, including the United Nations, the World Trade Organization, and the World Bank. Another is the active role taken by national governments of all political stripes in promoting the model, with some exceptions such as Ecuador (O'Connell 2021). Thirdly, there is the political influence of agribusiness, both transnational and homegrown (Pengue 2018). A prominent example is the leading role of soya producers in stymying Paraguayan President Fernando Lugo's

Figure 7.3 Indigenous mothers in the "mesa de diálogo," Cuninico, Loreto 2018 (Carrillo).

attempted agrarian reforms, and in his eventual ouster. While large-scale farmers and national conglomerates linked to global supply chains profit handsomely, small farmers and indigenous communities are displaced or are forced to abandon food sovereignty and ancestral knowledge. Meanwhile the environment suffers pollution, soil depletion, and land-use changes that destroy productivity (Gudynas 2008; Pengue 2018).

The pervasive logic of environmental colonialism

The logic of environmental neo-colonialism has been not only deepened but also extended via new forms of territorial control. One example is the unprecedented expansion of infrastructure across the region. Roads, bridges, waterways, and hydroelectric dams have been constructed at record rates, many linked to the 'Initiative for the Integration of the Regional Infrastructure of South America' (IIRSA). According to a 2017 report, a total of almost 600 mega-infrastructure projects had been identified across South America alone, costing an estimated $163 billion (Hildyard 2017). IIRSA was developed by 'neoliberal' governments but enthusiastically embraced by left-leaning administrations (Zibechi 2015).

The impact of these mega-projects in terms of environmental colonialism are manifold. One aspect is the facilitation of extractive activities, with research linking IIRSA to alluvial mining (Cortés-McPherson 2019), soya (Gudynas 2008), fracking (Scandizzo 2017), and human trafficking (Verite 2016). Furthermore, the construction process often causes deforestation, biodiversity loss, and displacement. Linked to this are undemocratic decision-making processes that fail to include affected communities (Svampa 2012). As Zibechi notes, "IIRSA is a neocolonial initiative, a vertical imposition and an aggression" (2015). When communities or organizations question or oppose these projects, they are typically ignored, criminalized, or, if all else fails, violently repressed. Finally, there is the issue that these projects tend to exclusively benefit export-oriented enclaves (Svampa 2012) or other politically influential sectors. Many of these elements were present in the case of the Belo Monte mega-dam in Brazil, which displaced communities, cleared forests, and enabled an alleged 'ethnocide' of indigenous groups (Cardoso and Branford 2019).

Other, less obvious examples of environmental colonialism exist in contemporary Latin America. While tourism (in particular, eco-tourism) is cited as a route to sustainable development, it is also a valuable source of rents for states and profits for the private sector. In this way, some forms of tourism share "analogous features" with extractive industries (Loperena 2017, 620). Some scholars view tourism based on nature and indigenous cultures as a continuation of the colonial dynamic of unregulated commercialization of humans and the environment (Coronil 2000, 98–99). One example is the Tela coast of Honduras, where eco-tourism plans have targeted Afro-descended Garifuna people in a process of "racialized dispossession" and environmental degradation (Loperena 2017, 625). In Colombia's Tayrona National Park, a violent "re-conquest" by military and paramilitary forces rendered the area fit for eco-tourism (Ojeda 2012, 363). This process has led to the privatization of notionally public spaces by a small number of local elites, some of whom trace their titles to royal decrees granted by the Spanish crown (Ojeda 2012, 364).

The use of 'sustainability' as a pretext to dispossess traditional communities connects to the concept of 'conservation colonialism' or 'green colonialism', which describes a system of control within colonized territories that restricts or displaces indigenous populations, disregarding their collective rights. For over a century, this process was violent and openly colonial in nature, involving cultural destruction, forced displacement, and human rights abuses. While large-scale evictions are no longer commonplace, indigenous groups living in protected areas are excluded from decision-making and self-government by states and, increasingly, non-state actors

Figure 7.4 Murals depicting resistance by women and rural communities to the Conga mining project in Peru, taken in the town of Celendin, Cajamarca province (O'Connell).

(Tauli-Corpuz 2016). This often leads to the limiting or outlawing of traditional activities like hunting, fishing, or farming. Failures to award legal titles to indigenous groups or respect good-faith consultation are also common.

Furthermore, conservation schemes have frequently resulted in increased deforestation, where states lack the capacity or political will to protect lands from incursions by extractive activities. Many of these elements have crystallized around global climate governance, and, in particular, the UN's Reducing Emissions from Deforestation and forest Degradation (REDD and REDD+) schemes. REDD and REDD+ are premised on reducing deforestation by offering compensation for conservation through market-based initiatives. In practice, however, the scheme has opened up new possibilities for 'carbon colonialism', which restricts local land rights and permits land grabbing (Lyons and Westoby 2014)—a process referred to as 'green grabbing' (Fairhead et al. 2012). Indigenous and environmental movements denounce what they see as 'false solutions' that allow industrialized nations to use forests as "carbon dumps" for their pollution, while incentivizing land and human rights violations in Latin America (CJA 2019).

Resistance and change

Latin America's history of environmental colonialism is matched at every step by stories of resistance: from Manco Inca leading the Vilcabamba (Cusco) resistance in the 16th century, through the great rebellion of Tupac Amaru II in 1780, to land seizures in Cusco in the 1960s demanding agrarian reform, and the 1977 Mapuche rebellions against the Pinochet dictatorship. More recently, incursions by extractive projects into indigenous territories have provoked

Figure 7.5 The funeral of one of those killed in the Conga Conflict (Chávez).

resistance that draws upon historical and contemporary struggles for inspiration. For example, the 2012 Great March for Water in Peru originated in Cajamarca—the site of the capture of Atahualpa in 1532—and ended in Lima, linking indigenous and peasant groups with urban communities. The march was inspired by Indigenous struggles in Ecuador against oil exploitation in Yasuní National Park (de Souza 2011), as well as by the VIII Indigenous March and the 'Water Wars' in Bolivia (Composto and Navarro 2014).

Nevertheless, many struggles for the vindication of territories, rights, and the environment are treated as political challenges or security threats by governments. Contemporary examples include the resistance of Wayúu women in Colombia to the Cerrejón coalmine (Ulloa 2020) or Bolivian indigenous people defending Isiboro-Sécure Indigenous Territory and National Park (TIPNIS) against a highway championed by Evo Morales, Bolivia's first Indigenous president. In these and many other cases, governments have sought to delegitimize and criminalize protests (O'Connell 2021). In some instances, the treatment of protesters has manifested as violent repression, such as the attack on the TIPNIS marchers or the killing of members of communities resisting the Conga mining project in Peru.

Throughout this history, women have 'woven' multiple decolonial resistances, articulating networks or movements that link the 'right to the body' with the 'right to territories' (Lugones 2008; Mora, Chap. 22 in this volume). Women have questioned both extractivism and epistemic neo-colonialism, criticizing white feminism and organizing new forms of anti-patriarchal struggles in across the region. The case of Berta Cáceres, an indigenous Lenca leader of the resistance to the Agua Zarca hydroelectric project in Honduras and the winner of the Goldman Environmental Prize, is emblematic. Although Berta was murdered by individuals linked to the project in 2016, her legacy lives on. Echoing the final words of Indigenous rebel Tupac Katari ("I will return … and I will be millions"), indigenous and environmental activists around the world now chant: "Berta didn't die, she multiplied."

Conclusion

The colonial and neocolonial control and exploitation of territories and bodies in Latin America is part of a 'continuum of violence' that began with the Conquests and continues to the present as coloniality. As Rivera Cusicanqui (2010) outlines, the struggle for decolonization is dynamic and ongoing. In the contemporary period, the strategies which underpin environmental colonialism have become complex and subtle, as they respond to the different forms of resistance the peoples of Abya-Yala have employed for more than five hundred years. As this overview of environmental colonialism and neocolonialism reveals, the impacts of different development models have been equally harmful to the environment and its Indigenous peoples. Conversely, the Indigenous worldview that sees nature and humanity as inextricably linked continues to inspire rebellion and the search for alternatives. Like the legend of Inkarri, predicted to rise from the earth to re-establish Inca rule, this resistance is therefore rooted in the '*Pachamama*'.

Notes

1 We outline some examples of these forms of control in the chapter, including 'conservation/green colonialism' and colonial forms of tourism. Others not dealt with here due to space restrictions include: energy/fossil colonialism (Smith-Nonini 2020), toxic waste dumping (Thomson and Samuel-Jones 2022), and disaster colonialism (Rivera 2022).
2 Also known as *encomiendas, haciendas, estancias,* or *fazendas*.
3 According to the Florentine Codex the Spaniards were said to "crave gold like hungry swine".
4 Former Argentine President Domingo Faustino Sarmiento's 1845 book "Facundo: Civilization or Barbarism" is an example of how rural and indigenous groups were demonized in this way.
5 A similar campaign took place in Chile between 1861 and 1883, dubbed "the Pacification of the Araucanía".
6 The history of rubber tells a similar story.

References

Acosta, Alberto. "Extractivism and neo-extractivism: Two sides of the same curse." In *Beyond Development: Alternative visions from Latin America*, edited by Miriam Lang and Dunia Mokrani, 61–86. Fundación Rosa Luxemburgo/Transnational Institute, 2013.

Acosta, Alberto. "Post-extractivism: From discourse to practice—reflections for action." *International Development Policy*, no. 9 (2017): 77–101.

Alimonda, Héctor. *La Naturaleza colonizada. Ecología política y minería en América Latina*. CLACSO, 2011.

Atiles-Osoria, José M. "Environmental colonialism, criminalization and resistance: Puerto Rican mobilizations for environmental justice in the 21st century." *RCCS Annual Review. A selection from the Portuguese journal Revista Crítica de Ciências Sociais* 6 (2014): 3–21.

Boyer, Christopher R. "Latin American Environmental History." In *Oxford Research Encyclopedia of Latin American History*. OUP, 2016.

Cardoso Ribeiro, Marilene, and Sue Branford. "The hydroelectric threat to the Amazon basin." In *Voices of Latin America: Social Movements and the New Activism*, edited by Tom Gatehouse, 121–145. LAB, 2019.

Clark, Brett, and John Bellamy Foster. "Ecological imperialism and the global metabolic rift: Unequal exchange and the guano/nitrates trade." *International Journal of Comparative Sociology* 50, no. 3–4 (2009): 311–334.

Climate Justice Alliance (CJA). "Resolution for Climate Justice Alliance on Resistance to Carbon Colonialism in the Amazon." *Indigenous Policy Journal* 30, no. 3 (2019): 293–295. http://www.indigenouspolicy.org/index.php/ipj/article/view/644/626.

Composto, Claudia y Navarro, Mina Lorena. "Introducción". In *Territorios en disputa. Despojo capitalista, luchas en defensa de los bienes comunes naturales y alternativas emancipatorias para América Latina*, 17–32. Bajo Tierra, 2014.

Coronil, Fernando. "Naturaleza del poscolonialismo: del eurocentrismo al globocentrismo." In *La colonialidad del saber: eurocentrismo y ciencias sociales: perspectivas* edited by Edgardo Lander and Santiago Castro-Gómez, 87–111. CLACSO, 2000.

Cortés-McPherson, Dolores. "Expansion of small-scale gold mining in Madre de Dios: 'capital interests' and the emergence of a new elite of entrepreneurs in the Peruvian Amazon." *The Extractive Industries and Society* 6, no. 2 (2019): 382–389.

De Echave, José. "La minería ilegal en Perú: Entre la informalidad y el delito." *Nueva sociedad* 263 (2016): 131–144.

De Souza Santos, Boaventura. "Epistemologías del Sur. Utopía y Praxis Latinoamericana". *Revista Internacional de Filosofía Iberoamericana y Teoría Social* 16, no. 54 (2011): 17–39.

Dore, Elizabeth. "Environment and society: Long-term trends in Latin American mining." *Environment and History* 6, no. 1 (2000): 1–29.

Durand, Francisco. *La captura del Estado en América Latina: Reflexiones teóricas.* Fondo Editorial PUCP, 2019.

Dussel, Enrique. "Beyond eurocentrism: The world-system and the limits of modernity." In *The Cultures of Globalization*, edited by Frederic Jameson and Masao Miyoshi, 3–31. Duke University Press, 1998.

Fairhead, James, Melissa Leach, and Ian Scoones. "Green grabbing: A new appropriation of nature?" *Journal of Peasant Studies* 39, no. 2 (2012): 237–261.

Galeano, Eduardo. *Open veins of Latin America: Five centuries of the pillage of a continent.* Monthly Review Press, 1997.

Gil Montero, Raquel. "Free and unfree labor in the colonial Andes in the sixteenth and seventeenth centuries." *International Review of Social History* 56, no. S19 (2011): 297–318.

Guardiola-Rivera, Oscar. *What if Latin America ruled the world?: How the South will take the North through the 21st century.* Bloomsbury Publishing, 2010.

Gudynas, Eduardo. *Ecología, Economía y Ética del Desarrollo Sostenible.* Abya-Yala, 2003.

Gudynas, Eduardo. "The new bonfire of vanities: Soybean cultivation and globalization in South America." *Development* 51, no. 4 (2008): 512–518.

Gudynas, Eduardo. "Diez tesis urgentes sobre el nuevo extractivismo." In *Extractivismo, política y sociedad,* 187–225. CAAP/CLAES, 2009.

Harvey, David. *The New Imperialism.* OUP, 2005.

Hildyard, Nicholas. "Annihilating Space Through Time." In *Extreme: The New Frontiers of Energy Extractivism in Latin America*, 10–19. Oilwatch, 2017.

Koch, Alexander, Chris Brierley, Mark M. Maslin, and Simon L. Lewis. "Earth system impacts of the European arrival and great dying in the Americas after 1492." *Quaternary Science Reviews* 207 (2019): 13–36.

Lander, Edgardo. "Complementary and conflicting transformation projects in heterogeneous societies." In *Beyond Development: Alternative visions from Latin America*, edited by Miriam Lang and Dunia Mokrani, 87–104. Fundación Rosa Luxemburgo/Transnational Institute, 2013.

Loperena, Christopher A. "Honduras is open for business: Extractivist tourism as sustainable development in the wake of disaster?" *Journal of Sustainable Tourism* 25, no. 5 (2017): 618–633.

Lugones, María. "Colonialidad y género". *Tabula Rasa* 9 (2008): 73–101.

Lyons, Kristen and Peter Westoby. "Carbon colonialism and the new land grab: Plantation forestry in Uganda and its livelihood impacts." *Journal of Rural Studies*, 36, (2014)13–21.

Machado-Aráoz, Horacio *Potosí, el origen: Genealogía de la minería contemporánea.* PDTG, 2018.

Mariátegui, José Carlos. "El hecho económico de la historia peruana". *Revista Mundial*, 1925.

Meade, Teresa A. *History of Modern Latin America: 1800 to the present.* John Wiley & Sons, 2016.

Mignolo, Walter D. "Introduction: Coloniality of power and de-colonial thinking." *Cultural Studies* 21, no. 2–3 (2007): 155–167.

Moore, Jason W. "Silver, ecology, and the origins of the modern world, 1450–1640". In *Rethinking environmental history: World-system history and global environmental change*, edited by Alf Hornborg, J. R. McNeill and Joan Martinez-Alier, 123–142. AltaMira Press, 2007.

O'Connell, Chris. "Civil society and left-wing governments in Latin America: The limits of influence". In *Barriers to Effective Civil Society Organizations*, edited by Ibrahim Natil, Vanessa Malila and Youcef Sai, 111–129. Routledge, 2021.

O'Gorman, Edmundo. *La invención de América*. Fondo de Cultura Económica, 1986.

Ojeda, Diana. "Green pretexts: Ecotourism, neoliberal conservation and land grabbing in Tayrona National Natural Park, Colombia." *Journal of Peasant Studies* 39, no. 2 (2012): 357–375.

Pengue, Walter. "La Republica Unida de la Soja: Concentración y Poder Basado en la Monocultura de Exportación." In *Atlas del Agronegocio*, edited by Dietmar Bartz, 26–27. Fundación Heinrich Böll, 2018.

Quijano, Anibal. "*Coloniality of power, Eurocentrism, and Latin America.*" *Nepantla: Views from South* 1, no. 3 (2000): 533–580.

Riffo, Lorena. "Fracking and resistance in the land of fire: Struggles over fracking in Northern Patagonia, Argentina, highlight the need to decommodify and democratize energy resources and seek alternatives." *NACLA Report on the Americas* 49, no. 4 (2017): 470–475.

Rivera, Danielle Zoe. "Disaster Colonialism: A Commentary on Disasters beyond Singular Events to Structural Violence." *International Journal of Urban and Regional Research* 46, no. 1 (2022): 126–135.

Rivera Cusicanqui, Silvia. *Ch'ixinakax utxiwa. Una reflexión sobre prácticas y discursos descolonizadores.* Retazos, 2010.

Scandizzo, Hernán. "An Iceberg called 'Vaca Muerta'". In *Extreme: The New Frontiers of Energy Extractivism in Latin America*, 44–53. Oilwatch, 2017.

Silva, Eduardo. *Challenging Neoliberalism in Latin America*. Cambridge University Press, 2009.

Smith-Nonini, Sandy. "The debt/energy nexus behind Puerto Rico's Long Blackout: From Fossil Colonialism to new energy poverty." *Latin American Perspectives* 47, no. 3 (2020): 64–86.

Svampa, Maristella. "Resource extractivism and alternatives: Latin American perspectives on development." *Journal fur Entwicklungspolitik* 28 (2012): 43–73.

Tauli-Corpuz, Victoria. "Rights of indigenous peoples". *Report of the Special Rapporteur of the Human Rights Council on the rights of indigenous peoples*. United Nations General Assembly, 2016.

Thomson, Ryan and Tameka Samuels-Jones. "Toxic colonialism in the territorial Isles: A geospatial analysis of environmental crime across US territorial Islands 2013–2017." *International Journal of Offender Therapy and Comparative Criminology* 66, no. 4 (2022): 470–491.

Ulloa, Astrid. "The rights of the Wayúu people and water in the context of mining in La Guajira, Colombia: demands of relational water justice." *Human Geography* 13, no. 1 (2020): 6–15.

Verite. *The Nexus of Illegal Gold Mining and Human Trafficking in Global Supply Chains: Lessons from Latin America*. Verite, 2016.

Williamson, Edwin. *The Penguin History of Latin America*. Penguin, 1992.

Zibechi, Raul. "Interconnection Without Integration in South America: 15 Years of IIRSA", *Upside Down World*, 2015.

8 Water scarcity in Latin America

Maria Fragkou, Natalia Dias Tadeu, Vanessa Empinotti,
Rodrigo Fuster, Maria Teresa Oré, Facundo Rojas,
Anahí Urquiza, and Lucrecia Wagner

Introduction

Latin America has witnessed struggles over its rich biodiversity and natural resources ever since its plundering in colonial times (Ansaldi and Giordano 2012). Attempts at territorial, political, and environmental domination that continue under different forms of capitalist appropriation of nature have transformed the continent into a hotspot for environmental conflicts, registering 937 out of 3,375 environmental conflicts worldwide (Environmental Justice Atlas 2021). This fact is intrinsically linked to the region's position in global production networks, as an indispensable provider of low-cost raw materials, making its economies acutely dependent on international markets and restricted to predetermined development paths. The situation is aggravated by internal socio-political strife related to the coexistence of rural indigenous communities alongside industrial agriculture and large-scale mining; the marked urban/rural dichotomy driven by the expansion of metropolitan areas; weak democratic traditions that permit limit public participation; and the imposition of aggressive neoliberal policies in countries like Chile, Brazil, and Peru. Each of these factors helps to shape a heterogeneous community with varying degrees of understanding and appreciation of nature.

Extractive industries in the region are intensive users of water resources, competing with ecosystems and human consumption (Svampa 2016). This has resulted in the accumulation of water by powerful economic groups and productive sectors and the dispossession of local communities of traditional water sources through contamination and depletion, creating scarcity even in areas of abundant freshwater. In socially-unequal and culturally-diverse lands, the dispossession of natural resources has profound socioeconomic and political causes and implications. As such, water scarcity, especially in Latin America, needs to be addressed as a complex, multidimensional problem, as the result of both physical and social conditions.

In this chapter, we analyze the availability and scarcity of water in Latin America from a critical standpoint, considering the physical processes, economic interests, and political decisions involved. We present an overview of the concept of scarcity in order to demonstrate how our theoretical understanding of the issue has evolved from a hydrometeorological to a political question, focusing in particular on the political ecology literature. We then describe hydroclimatic conditions in Latin America in order to explain the physical distribution of water availability across the region, before contrasting this with an overview of how scarcity has been tackled in Latin American countries. We complement this review by presenting six emblematic case studies of water scarcity production, revealing how this occurs on different scales and through varying relationships between economic and political power. We conclude the chapter with reflections on how water scarcity is generated and manifests in Latin America, as well as its main implications.

DOI: 10.4324/9780429344428-10

Scarcity in theory

Popular and mainstream approaches usually present water scarcity as a solely natural phenomenon, attributed to physical causes of reduced water availability. Quantitative representations of water scarcity based on volumetric accountings of water reserves with the use of physical indicators are simplistic, obscuring the complexity of the issue (Wolfe and Brooks 2003) and its linkage to ecological, socio-political, temporal, and anthropogenic dimensions (Mehta 2003).

The political ecology literature follows this critical stance with regard to the causes of water scarcity, acknowledging its physical dimension before stressing the relevance of anthropogenic factors (Budds 2013). As such, water scarcity is not taken for granted as a phenomenon that is "absolute and nature-given" (Bakker 2000), but is considered to be "socially produced" (Naredo 1997). Scholars in the field have pointed out the discursive, political, and social dimensions of water scarcity (Kaika 2003; Swyngedouw 2009). A major line of research focuses on the role that privatization of water resources and the interests of powerful actors play in the production of water scarcity (Bakker 2000), while a second puts special emphasis on the political nature of water scarcity, analyzing "how social and political institutions, cultural norms, and property rights shape individuals' access to water" (Fragkou and McEvoy 2016; Empinotti et al. 2019). These lines of research concur that scarcity is generated through "socio-political processes, through exclusion, biases, and discrimination" (Mehta 2014, 61) and that these processes have unequal impacts across social groups.

With regard to socially-constructed scarcity, various frameworks have been developed to classify different forms (Fragkou and McEvoy 2016). Classifications include economic or infrastructural scarcity, resulting from conditions of underdevelopment and a lack of financial investment in infrastructure and technology to meet growing water demands (Wolfe and Brooks 2003); institutional or management scarcity, driven by conditioning of water management and, consequently, of user access and control by political, institutional, and administrative factors, including inappropriate land-use practices (Mehta 2007); and quality or perceptual scarcity in contexts where, despite adequate water treatment, bottled water is perceived as safer (Robbins et al. 2014, 269).

This literature offers two important key contributions to the understanding of water scarcity in Latin America. First, it identifies the actors most affected by scarcity, revealing the social inequalities that underlie issues of water access and control. Second, it highlights that the study of possible anthropogenic causes of scarcity can help us to understand the root causes behind the phenomenon and offer realistic solutions under the auspices of social and environmental justice. In the next section, we examine the physical and political dimensions of water scarcity in the Latin American region, exemplified by six case studies that represent the main types of socially-produced scarcity described above.

Physical water availability in the Latin American region: the geographic distribution of climatic and hydrologic conditions

Although water availability in Latin America is considered in general terms to be relatively high, the region presents a complex and diverse distribution of natural water due to its varied orographic and climatic conditions, resulting in heterogeneous hydrologic realities across this vast continent of more than 22,000 square kilometers.

Precipitation is scarce in the north of Mexico, but abundant in the southeast (FAO 2000) and studies on the effects of climate change in the country estimate a temperature increase of around three degrees Celsius, as well as a reduction of around 15% in average precipitation nationally and a decrease in runoff and recharge of aquifers (Martínez-Austria and Patino-Gómez 2012). In Central America, encompassing Costa Rica, Belize, El Salvador, Guatemala, Honduras,

Nicaragua, and Panama, the climate is tropical and influenced by altitude, land relief, and proximity to the sea. Humidity is higher on the Atlantic Coast than on the Pacific and precipitation increases from west to east and from north to south. The greater Antilles region (Cuba, Haiti, and the Dominican Republic) also has a tropical marine climate with defined dry and wet seasons (FAO 2000). The hydroclimatic models for Central America show significant decreases in precipitation during the dry season (Hidalgo et al. 2013).

Moving down to South America, atmospheric circulation provides abundant rainfall in the world's largest rainforest of the Amazon basin. The northern coast of Peru and southern Ecuador are the areas most directly affected by a rise in sea temperature during the El Niño-Southern Oscillation (ENSO) event, which produces intense rainfall in these areas. The same phenomenon generates droughts in other places, such as in the Mexican and South American highlands, as well as in northeastern Brazil (Caviedes 2001). According to Grimm and Sampaio (2020), reductions in springtime rainfall averages are observed in southeastern Brazil during ENSO events, while precipitation tends to increase in Paraguay, northeastern Argentina, central Chile, and parts of Brazil (i.e., southern, central-western, and, in the summer, southeastern regions) (Marengo et al. 2016).

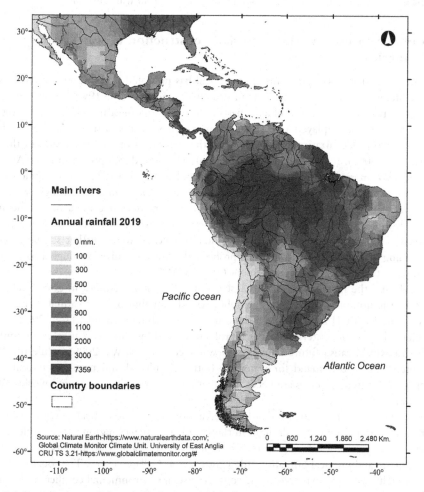

Figure 8.1 Precipitation and rivers in Latin America. Authors, 2020.

In addition to the strong influence of the ENSO in the tropical and subtropical Andes, a persistent decrease in average rainfall has been measured, along with an associated decline in the accumulation of snow and formation of glaciers (Segura et al. 2020). There has also been a progressive decrease in snowfall along the Pacific coast south of 35° latitude, in particular since the 1970s (Quintana and Aceituno 2012). This has resulted in clear glacier recession in Chile and Argentina, as well as significant loss of glacier mass (Bown et al. 2008). Glaciers are important sources of water in the Andean region, regulating river flows, especially in times of drought and reduced snowfall. Several studies have shown increased frequency and intensity of droughts in the central Argentine-Chilean Andes since the mid-20th century (Morales et al. 2020) and the majority of authors attribute these processes to climate change.

The above overview depicts Latin America as a climatically and hydrologically diverse region with variations in water availability across the content. Abundant rivers, glaciers, and great underground water reserves are its main water resources, yet all three are threatened by both climatic and anthropogenic phenomena. Declines in snowfall in the Andes, coupled with higher temperatures, are reducing snow and glacier accumulation, affecting present and future supply to many communities (IPCC 2014).

In what follows, we analyze how the issue of water availability has been tackled by policy and in scientific debate, and its transition from a technical to a political question.

From physical to social: the study and construction of water scarcity in Latin America

The assumption that water problems are technical has prevailed within government agendas, academic literature, and water management principles, and, until the 1990s, water scarcity was defined primarily as a physical problem that should be overcome through technical solutions (Golte 1980). Engineers played a key role, promoting new forms of surface and groundwater irrigation. Such modernizations were linked to the consolidation and expansion of the State apparatus (Oré 2005). It was hoped that they would promote development in Latin American countries. Thus, regional discussions over water problems concerned irrigation practices, tap water connectivity, and efficient water uses, with a special focus on rural settings. Water governance, legislation reforms, and updated management practices became a new trend in the 2000s, as Latin America experienced water reforms and attempts to privatize water services (Rebouças 2003). Such attention to management aspects of water further strengthened the understanding of water scarcity as a natural problem that must be solved through a combination of innovation, technology, and management (Tundisi 2003).

It was during the first decade of the 21st century that South American governments first highlighted the notion of scarcity on the public agenda, attributing it to natural events such as those associated with El Niño (droughts and floods) and climate change. From 2003, scarcity was presented in world water forums as a problem caused by climate change and population growth that would cause future scenarios of water conflict (WWF 2009), while the World Bank stated that "the demand for water for both agricultural and non-agricultural uses is increasing, and water shortages are becoming acute in much of the developing world" (World Bank 2008).

The first Latin American studies that heralded a move to a more critical view of water issues offered an understanding of the drought in the Brazilian semiarid region as the result of social and power structures, introducing an understanding of water and food scarcity that would go beyond climate and physical limitations (de Castro 2004; Callado 1960). The social dimension of water scarcity was then explored in the context of socioenvironmental conflicts and struggles

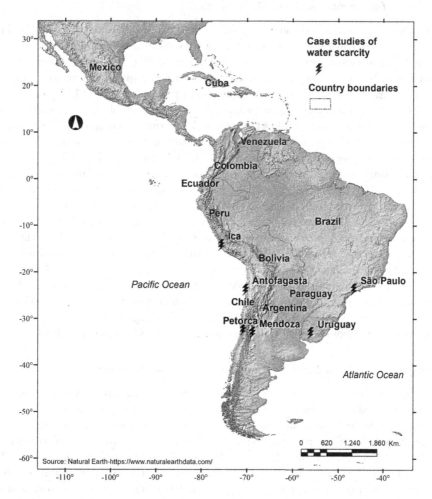

Figure 8.2 Location of the six case studies of socially constructed scarcity in Latin America. Authors, 2020.

over water access resulting from the expansion of agricultural and extractive frontiers, energy production, the resignification of water as a commodity (Martín and Larsimont 2016). Scarcity was also examined in terms of unjust neoliberal regulatory frameworks, such as the infamous Chilean Water Code (Fragkou and Budds 2019).

Recent recognition of the way in which water scarcity has been manufactured has led Latin American governments and academics to reflect upon the increased importance of access to water in the region, as well as to recognition of scarcity not only as a physical and management problem, but as a social problem. Such a critical turn also reflects the consequences of the neoliberal economic model and attempts to commodify water and promote universal access through market tools. In what follows, we analyze six emblematic cases of water scarcity that demonstrate the different ways in which management, regulation, technology, and infrastructure produce water scarcity through accumulation and social exclusion across the wider Latin American region.[1]

Construction of urban water scarcity: São Paulo, Brazil

By 2015, megacities such as São Paulo and Rio de Janeiro in Brazil, Santiago in Chile, and Lima in Peru were experiencing water scarcity, with water reservoirs at their lowest levels historically. Combined with excessive rates of urbanization and inequality, the decrease in precipitation exposed how urban water supply systems in Latin America respond to growing demand for water. In a mega metropolis such as São Paulo, with a population around 12.4 million people in 1,521,110 square kilometers, seven water systems are responsible for supplying domestic (55%) and industrial (6%) uses (CBHAT 2016). The governance system is based on participatory institutions and relies on two legal bodies: the Water Law and the Sanitation Law. These laws are responsible for structuring the management system, policies, and fees for water resources and sanitation services, respectively. The drought that occurred in São Paulo in 2014 and 2015 was denied by the governor of São Paulo state and the CEO of the São Paulo State Water System, even though the reservoirs that supply the city reached 8% capacity, the lowest level ever recorded. The case of water scarcity in São Paulo revealed how centralized decision-making processes prevailed in a participatory institutional process as the result of unequal governance practices that favored the water supply company and the government. Flows of power and water generated inequality in water access, compounded by political denial of the water scarcity issue (Empinotti et al. 2019). At the same time, research has shown how diagnosis of water scarcity as a technical issue generated a less contentious discussion and reinforced the dominance of technical solutions (Quintslr 2018; Lukas and Fragkou 2014), while also discouraging the population from viewing water scarcity as a disaster (Anazawa et al. 2019).

Producing water scarcity in Latin American oases: Mendoza, Argentina

The province of Mendoza is located in central-western Argentina. The power of certain actors over river water and its distribution goes some way toward explaining the geography and social organization of the area (Montaña 2007). Construction of irrigation has created three oases covering around 4% of the province, which are home to more than 95% of its population (around 1,990,338 inhabitants). The General Water Department is the government body responsible for the distribution of surface and subterranean water. The average annual flow rate formed by the five main rivers of Mendoza is around 256.5 cubic meters per second and there are around 20,000 groundwater wells. Decisions as to which actors, territories, and activities are considered subject to water politics is based on the official conception of water scarcity (Grosso 2014). Mendoza provides an example of socially-produced water scarcity and, specifically, of institutional or management scarcity. The semiarid characteristics of the region are combined with state politics that determine actions and omissions. From 2010, Mendoza has suffered the same megadrought as central Chile. Water scarcity has become central to the establishment of consensus and social hegemonies in the province through a process termed "administration of scarcity" (Prieto et al. 2021).

However, water scarcity discourses have been re-signified and adopted by neighborhood organizations, like water assemblies, who have demanded that discussions surrounding extractive activities, mainly mega-mining, consider conditions of water scarcity (Wagner 2014). Concerns about the scarcity and fragility of water resources have been voiced by opponents of mining projects in both Argentina and Chile (Bottaro et al. 2014).

Qualitative water scarcity: Uruguay

Uruguay has a population of about 3.5 million people and relatively high water availability (Achkar et al. 2004). Here, water scarcity commonly can be defined in qualitative rather than quantitative terms. Water pollution is closely tied up with land-use patterns and, particularly, the agricultural sector (Aubriot et al. 2017). Grain and meat production for export has been increasing since 1990, leading to changes in regulatory frameworks. New irrigation legislation has generated significant public debate. Several actors denounced the privatization component of the law, which allows for the construction of dams to promote the expansion of irrigation and intensification of agricultural and livestock activities across the country, thus putting greater pressure on ecosystems and, consequently, on water quality (García-Alonso 2017).

Degradation of water quality can have other knock-on effects, including potabilization problems in the supply of drinking water, as observed in some Departments of Uruguay (Giordano et al. 2020). The conflict illustrates the social production of qualitative water scarcity, primarily in the form of "management scarcity" resulting from modes of appropriation and use of water and territory. Social production of water scarcity associated with water quality degradation and new state policies has also been identified in Brazil (Anazawa et al. 2019) and Mexico (Esparza 2014), respectively. Moreover, public trust in water quality in Uruguay has been damaged and there has been a noticeable increase in consumption of bottled water, mainly in the last decade, indicating a "quality" or "perceptual" scarcity. Similar cases of this scarcity category in Chile are discussed by Fragkou and McEvoy (2016).

Producing scarcity by producing water: desalination in the Antofagasta desert, Chile

The Antofagasta region in northern Chile is the driest place on the planet and the world's largest source of copper (Atienza et al. 2020). Since the 1990s, when a mining sector boom drove rapid economic development and uncontrolled urban expansion in the region, seawater desalination has been a central strategy in both public and private efforts to secure industrial and domestic water supply in this vast desert. Although desalinated water was initially directed primarily toward mining operations, in 2003 Latin America's largest seawater desalination facility for the production of potable water was opened in the regional capital of Antofagasta and currently produces 850 liters per second. The gradual supply of desalinated water to urban hubs allowed the region's water company, *Aguas Antofagasta*, to sell its freshwater surplus to mining companies through private contracts to provide untreated mountain water.

Desalinated water production thus increased overall water availability in this arid area, but it also altered the circulation of water flows by depriving coastal urban hubs of freshwater, and these will soon be supplied with 100% desalinated water. Paradoxically, desalination has also led to the accumulation of freshwater by mining companies, which have taken advantage of legal loopholes (Campero and Harris 2019) and stabilized the contested national water code by masking its shortcomings (Fragkou and Budds 2019). Moreover, the transition to desalinated water provision for urban hubs did not succeed in overcoming pre-existing economic scarcity and negative perceptions of tap water quality among urban dwellers (Fragkou and McEvoy 2016). Similar cases of the use of water production technologies to overcome physical scarcity to the benefit of dominant economic sectors— while availability for general consumption is either maintained or decreases, can also be found in Mexico, where desalination favored the tourism industry (McEvoy 2014) and influenced Mexican–US water relations (Wilder et al. 2016).

Constructing discursive water scarcity and abundance: Ica Valley, Peru

Since Peru created a system of regional governments in 2004, the water governance of the Ica Valley, located on the Peruvian coast, has become more complex. In 1950, an irrigation project had diverted the waters from the Choclococha lagoon, which is located in another basin at 12,000 feet above sea level. Since 2004, this area has been under the control of the Huancavelica regional government. Huancavelica, one of the poorest areas in the country, irrigates the rich Ica valley. In 1992 a neoliberal economic policy began throughout the country. It included the promotion of new crops, such as asparagus, avocado, and grapes, for international markets (Oré 2005). Ica valley became the most successful case of this policy, demanding a large number of workers from many other regions of the country, as well as water resources (Oré et al. 2013). Currently, the valley has a population close to 500,000 people.

In particular, agro-export has used groundwater with an intensity that has put the valley's aquifer at risk. In 2010, the Ica regional governor decreed a "water emergency", and demanded support from the national government to transfer water from various lagoons in Huancavelica. However, the Huancavelica government, jointly with the communities, refused this project, as it would harm local production. The conflict was taken to the Latin American Water Tribunal, which ruled against the project, so it had to be abandoned.

Conclusions

Despite conflicting evidence, states and international organizations continue to define water problems as the result of climate change and population growth, while considering water as an economic and industrial input. Along these lines, Latin American governments have relaxed environmental policies and created incentives for water-intensive economic activities, even in semiarid regions, provoking conflict and inequality. The case of Mendoza in Argentina shows how institutional arrangements are legitimized through the discourse of water scarcity, while social movements that reject mining projects base their opposition on that same scarcity. The Ica Valley in Peru suffers water scarcity as the result of state incentives to promote intensive agricultural production by international corporations. Furthermore, the impact of such decisions on water access is unequal, with low-income urban populations, small farmers, and traditional communities being the most affected.

This neo-Malthusian view reinforces understandings of water scarcity as an apolitical problem in which inequality and power asymmetries are not part of the equation. The case of Antofagasta shows how new institutional practices allow multinational companies to appropriate water through diverse routes, such as water supply providers and mining activities. The case of São Paulo is similar, where governance structures and practices have enabled the state and supply companies to gain control of the participatory decision-making process, thus guaranteeing the supply of water for wealthy sectors. For their part, the cases of the Ica Valley and Uruguay also highlight the ways in which the agricultural sector has been able to secure access to water at the cost of greater scarcity for other areas and actors. Water scarcity is more than simply a physical and technical issue. The six cases presented in this chapter illustrate how flows of power and inequality are factors in the construction of water scarcity and point to the need for our understanding of the issue to incorporate and acknowledge the complexity and multi-dimensionality of hydrosocial relations.

Note

1   To review some cases in Central America see Van Dusen, Richelle, "The Politics of Water in Mexico City" (2016). https://doi.org/10.15760/honors.343. Middeldorp et al., "Social mobilisation and violence at the mining frontier: the case of Honduras". *Extract. Ind. Soc.*, 3 (4) (2016), pp. 930–938.

References

Achkar, Marcel, Ricardo Cayssials, Ana Domínguez and Fernando Pesce. 2004. *Hacia un Uruguay Sustentable Gestión Integrada de Cuencas Hidrográficas*. Montevideo: Programa Uruguay Sustentable, REDES-AT.

Anazawa, Tathiane M., Roberto Luiz do Carmo and Antonio Miguel Vieira Monteiro. 2019. "A escassez hídrica no município de Campinas, Brasil (2013-2015), a partir das percepções da população: Crise hídrica ou desastre socialmente construído?" *Antropologia Americana* 4, no. 8: 61–85.

Ansaldi, Waldo and Verónica Giordano. 2012. *América Latina. La construcción del orden.* Buenos Aires: Ariel.

Atienza, Miguel, Martín Arias-Loyola and Marcelo Lufin. 2020. "Building a case for regional local content policy: The hollowing out of mining regions in Chile." *The extractive industries and society* 7, no. 2: 292–301, https://doi.org/10.1016/j.exis.2019.11.006.

Aubriot, Luis, Lucía Delbene, Signe Haakonsson, Andrea Somma, Federica Hirsch and Sylvia Bonilla. 2017. "Evolución de la eutrofización en el Río Santa Lucía: influencia de la intensificación productiva y perspectivas". *INNOTEC, Revista del Laboratorio Tecnológico del Uruguay*, no. 14: 7–16, https://doi.org/10.26461/14.04.

Bakker, Karen. 2000. "Privatising water, producing scarcity: The Yorkshire drought of 1995." *Economic Geography* 76, no. 1: 4–27, https://doi.org/10.1111/j.1944-8287.2000.tb00131.x.

Bottaro, Lorena, Alex Latta and Marian Sola Alvarez. 2014. "La politización del agua en los conflictos por la megaminería: Discursos y resistencias en Chile y Argentina." *European Review of Latin American and Caribbean Studies*, no. 97: 97–115, http://doi.org/10.18352/erlacs.9798.

Bown, Francisca, Andrés Rivera and César Acuña. 2008. "Recent glacier variations at the Aconcagua Basin, Central Chilean Andes". *Annals of Glaciology* 48: 43–48, https://doi.org/10.3189/172756408784700572.

Budds, Jessica. 2013. "Water, power and the production of neoliberalism in Chile, 1973-2005." *Environment and Planning D: Society and Space* 31, no. 2: 301–318, https://doi.org/10.1068/d9511.

Callado, Antônio. 1960. *Os industriais da seca e os Galileus de Pernambuco: aspectos da luta pela reforma agrária no Brasil.* Rio de Janeiro: Civilização Brasileira.

Campero, Cecilia and Leila M. Harris. 2019. "The legal geographies of water claims: Seawater desalination in mining regions in Chile". *Water* 11, no. 5: 886, https://doi.org/10.3390/w11050886.

Caviedes, César N. 2001. *El Niño in History: Storming Through the Ages.* Gainesville: University Press of Florida.

CBHAT – Comitê de Bacia Hidrográfica do Alto Tietê. 2016. Relatório – I: Plano de Bacia Hidrográfica do Alto Tietê - UGRHI 06 - Ano Base 2016/2035. São Paulo: FABHAT – Fundação Agência Da Bacia Hidrográfica Do Alto Tietê. http://www.sigrh.sp.gov.br/public/uploads/documents/CBH-AT/11958/relatorio-i_plano_final-rev2.pdf.

De Castro, Josué. 2004. *Geografia da Fome. O dilema brasileiro: pão ou aço.* Rio de Janeiro: Civilização Brasileira.

Empinotti, Vanesa, Jessica Budds and Marcelo Aversa. 2019. "Governance and water security: The role of the water institutional framework in the 2013-15 water crisis in São Paulo, Brazil." *Geoforum* 98: 46–54. https://doi.org/10.1016/j.geoforum.2018.09.022.

Environmental Justice Atlas. 2021. "EJAtlas - Global Atlas of Environmental Justice". Accessed March 18, 2021. https://ejatlas.org/.

Esparza, Miguel. 2014. "La sequía y la escasez de agua en México: Situación actual y perspectivas futuras." *Secuencia* 89: 193–219.

FAO. 2000. *"El riego en América Latina y el Caribe en cifras."* Roma: Organización de las Naciones Unidas para la Alimentación y la Agricultura (FAO).

Fragkou, María C. and Jessica Budds. 2019. "Desalination and the disarticulation of water resources: Stabilising the neoliberal model in Chile." *Transactions of the Institute of British Geographers* 45, no. 22: 448–463, https://doi.org/10.1111/tran.12351.

Fragkou, María C. and Jamie McEvoy. 2016. "Trust matters: Why augmenting water supplies via desalination may not overcome perceptual water scarcity." *Desalination* 397: 1–8, https://doi.org/10.1016/j.desal.2016.06.007.

García-Alonso, Javier. 2017. "Las modificaciones a la Ley de Riego y sus impactos ambientales". Centro Universitario Regional del Este (CURE), Universidad de la República. Observatorio del Agua en Uruguay. http://www.observatoriodelaguaenuruguay.com/efectos-de-las-modificaciones-a-la-ley-de-riego/.

Giordano, Gabriel, Natalia Dias Tadeu, and Micaela Trimble. 2020. "Análisis de la gobernanza y aprendizajes de las crisis en las Cuencas de Laguna del Sauce (Maldonado) y Laguna del Cisne (Canelones), Uruguay". Informe Técnico en el marco del Proyecto GovernAgua (SGP-HW 056). Instituto SARAS, Bella Vista-Maldonado, Uruguay. http://saras-institute.org/wp-content/uploads/2020/09/Resumen-Informe-CLC.pdf.

Golte, Jürgen. 1980. Notas sobre la agricultura de riego en la costa peruana. *Allpanchis*, 12 (15): 57–67. https://doi.org/10.36901/allpanchis.v12i15.1151.

Grimm, Alice M. and Gilvan Sampaio. 2020. "Capítulo 2. Observações Ambientais Atmosféricas e de Propriedades da Superfície". In: *Base Científica das Mudanças Climáticas – Painel Brasileiro de Mudanças Climáticas*, edited by Tércio Ambrizzi y Moacyr Araujo, 25–63. Río de Janeiro: COPPE, Universidade Federal do Rio de Janeiro.

Grosso, Virginia. 2014. "La escasez hídrica en tierras secas. Un estudio territorial sobre la apropiación, gestión y uso del agua en la cuenca del río Mendoza, argentina". PhD dissertation, Universidad de Buenos Aires.

Hidalgo, Hugo G., Jorge A. Amador, Eric J. Alfaro and Beatriz Quesada. 2013. "Hydrological climate change projections for Central America". *Journal of Hydrology* 495: 94–112, https://doi.org/10.1016/j.jhydrol.2013.05.004.

IPCC 2014. "Climate Change 2014: Synthesis Report". Contribution of Working Groups I, II and III to the Fifth Assessment Report of the Intergovernmental Panel on Climate Change, edited by Rajendra K. Pachauri and Leo Meyer. Geneva: IPCC. https://www.ipcc.ch/site/assets/uploads/2018/05/SYR_AR5_FINAL_full_wcover.pdf.

Kaika, Maria. 2003. "Constructing scarcity and sensationalising water politics: 170 days that shook athens." *Antipode* 35, no. 5: 919–954. https://doi.org/10.1111/j.1467-8330.2003.00365.x.

Lukas, Michael and María C. Fragkou. 2014. "Conflictividad en construcción: Desarrollo urbano especulativo y gestión de agua en Santiago de Chile." *Revista Ecología Política*, no. 47: 67–71, https://www.ecologiapolitica.info/?p=1636.

Marengo, José A., Ana P. Cunha and Lincoln M. Alves. 2016. "A seca de 2012-15 no semiárido do Nordeste do Brasil no contexto histórico." *Climanálise* 3: 1–6.

Martín, Facundo and Robin Larsimont. 2016. "Agua, poder y desigualdad socioespacial. Un nuevo ciclo hidrosocial en Mendoza, Argentina (1990–2015)". In *Cartografías del conflicto ambiental en Argentina 2*, edited by Gabriela Merlinsky, 31–56. Buenos Aires: Fundación CICCUS.

Martínez-Austria, Polioptro and Carlos Patino-Gómez. 2012. "Efectos del cambio climático en la disponibilidad de agua en México." *Tecnología y Ciencias del Agua* 3, no. 1, 5–20.

McEvoy, Jamie. 2014. "Desalination and water security: The promise and perils of a technological fix to the water crisis in Baja California Sur, Mexico." *Water Alternatives* 7, no. 3: 518–541.

Mehta, Lyla. 2003. "Contexts and constructions of water scarcity." *Economic & Political weekly* 38, no. 48: 5066–5072, https://www.jstor.org/stable/4414344?seq=1.

Mehta, Lyla. 2007. "Whose scarcity? Whose property? The case of water in western India." *Land Use Policy* 24, no. 4: 654–663, https://doi.org/10.1016/j.landusepol.2006.05.009.

Mehta, Lyla. 2014. "Water and human development", *World Development* 59: 59–69, https://doi.org/10.1016/j.worlddev.2013.12.018.

Montaña, Elma. 2007. "Identidad regional y construcción Del territorio En Mendoza (Argentina): Memorias y Olvidos Estratégicos", *Bulletin de l'Institut Français d'études Andines* 36, no. 2: 277–297, https://doi.org/10.4000/bifea.3908.

Morales, Mariano et al. 2020. "Six hundred years of South American tree rings reveal an increase in severe hydroclimatic events since mid-20th century". *Proceedings of the National Academy of Sciences (PNAS)* 117, no. 29: 16816–16823, https://doi.org/10.1073/pnas.2002411117.

Naredo, José Manuel. 1997. *La economía el agua en España*. Madrid: Fundación Argentaria-Visor.

Oré, María Teresa. 2005. *Agua Bien Común y Usos Privados. Estado, riego y conflictos en La Achirana del Inca*. Lima: Fondo Editorial de la Pontificia Universidad Católica del Perú, Universidad de Wageningen y Soluciones Prácticas.

Oré, María Teresa, David Bayer, Javier Chiong, Eric Rendon. 2013. "Water emergency in oasis of the Peruvian coast. The effects of the agro-export boom in the Ica Valley". Colloque "Oasis dans la mondialisation: ruptures et continuités". París, France, 167–176. https://hal.archives-ouvertes.fr/hal-01024460.

Prieto, María del Rosario, Facundo Rojas, Facundo Martín, Diego Araneo, Ricardo Villalba, Juan Rivera and Salvador Gil Guirado. 2021. "Sequías extremas en Mendoza durante el siglo XX y principios del XXI. Administración de la carencia y conflictos socio-políticos". In *Medio Ambiente y transformación rural en la Argentina contemporánea*, edited by Adrián G. Zarrilli and Marta Ruffini. Buenos Aires: Universidad Nacional de Quilmes. In Press.

Quintana, Juan and Patricio Aceituno. 2012. "Changes in the Rainfall Regime along the Extratropical West Coast of South America (Chile): 30-43° S". *Atmosfera* 25, no. 1: 1–22. http://www.scielo.org.mx/pdf/atm/v25n1/v25n1a1.pdf.

Quintslr, Suyá. 2018. "As "duas faces" da crise hídrica: escassez e despolitização do acesso à água na Região Metropolitana do Rio de Janeiro." *Sustentabilidade e Debate* 9, no. 2: 88–101, https://doi.org/10.18472/SustDeb.v9n2.2018.26702.

Rebouças, Aldo da C. 2003. "Água no Brasil: abundância, desperdício e escassez." *Bahia Análise & Dados* 13: 341–345.

Robbins, Paul, John Hintz and Sarah A. Moore. 2014. "Bottled water." In: *Environment and Society: A Critical Introduction*, edited by Paul Robbins, John Hintz and Sarah A. Moore, 259–278. UK: Wiley Blackwell.

Segura, Hans, Jhan Carlo Espinoza, Clementine Junquas, Thierry Lebel, Mathias Vuille and Rene Garreaud. 2020. "Recent changes in the precipitation-driving processes over the southern tropical Andes/western Amazon." *Climate Dynamics* 54: 2613–2631, https://doi.org/10.1007/s00382-020-05132-6.

Svampa, Maristella. 2016. *Debates latinoamericanos. Indianismo, desarrollo, dependencia y populismo*. Buenos Aires: Edhasa.

Swyngedouw, Erik. 2009. "The political economy and political ecology of the hydro-social cycle." *Journal of Contemporary Water Research & Education Issue* 142, no. 1: 56–60, https://doi.org/10.1111/j.1936-704X.2009.00054.x.

Tundisi, Jose Galizia. 2003. *Água no século XXI: enfrentando a escassez*. São Carlos: Rima.

Wagner, Lucrecia. 2014. *Conflictos socioambientales: la megaminería en Mendoza, 1884-2011*. Buenos Aires: Editorial de la Universidad Nacional de Quilmes.

Wilder, Margaret, Ismael Aguilar-Barajas, Nicolás Pineda-Pablos, Robert G. Varady, Sharon B. Megdal, Jamie McEvoy, Robert Merideth, Adriana A. Zúñiga-Terán and Christopher A. Scott. 2016. "Desalination and water security in the US–Mexico border region: Assessing the social, environmental and political impacts." *Water International* 41, no. 5: 756–775, https://doi.org/10.1080/02508060.2016.1166416.

Wolfe, Sarah and David B. Brooks. 2003. "Water scarcity: An alternative view and its implications for policy and capacity building". *Natural Resources Forum* 27, no. 2: 99–107, https://doi.org/10.1111/1477-8947.00045.

World Bank. 2008. "*Informe sobre Desarrollo mundial del 2008*". Bogotá: Banco Mundial, Mayol Ediciones y Mundi-Prensa. http://documents1.worldbank.org/curated/es/747041468315832028/pdf/414550SPANISH0101OFFICIAL0USE0ONLY1.pdf.

WWF - World Water Forum. 2009. Final Report of the 5Th World Water forum. March, Istanbul, Turkey.

Part III

Latin American environmental issues in political–economic context

9 The political economy of the environment in Latin America

Amalia Leguizamón

Introduction

The extraction of natural wealth to sustain political and economic systems has a long history in Latin America: since colonial times Latin American societies have been "nature-exporting" societies (Coronil 1997). Development projects rooted in the extraction of natural resources for exportation have recently sprouted with new vitality, infused by foreign investment and the adoption of new technologies. Since the turn of the 21st century, governments across the political spectrum have promoted extractivism to increase economic growth and social wellbeing, expanding the oil and gas, mining, and agribusiness frontiers across the region. Increased extraction has brought economic growth to many Latin American countries, particularly at times of high commodity prices. But this prosperity has come at great human and ecological cost, as well as growing social conflict, hanging a question mark over the long-term sustainability of extractivism as a development model.

This chapter explores the political economy of the environment in Latin America. First, I summarize main sociological theories on the political economy of the environment as they apply to the study of extractivism in the region. These theoretical approaches explain why developmental pursuits lead to social and ecological disorganization at local and global levels. Then, I give an overview of the history of extractivism in the region, looking into the power dynamics that fuel natural resource extraction for exportation as a development project and its uneven impact on society and the environment. To understand the context of present-day extractivism, I highlight historical patterns and continuities as well as key dilemmas and contradictions of relying on nature-exporting for socio-economic development. I conclude by reflecting on possible environmental futures for the region given the political economic processes and dynamics analyzed.

The crisis of resource extractivism

Sociological perspectives on the political economy of the environment study the power dynamics surrounding the control of material resources that sustain life and the implications of economic development on society and the environment.[1] Broadly, this scholarship focuses on the question of how the normal workings of capitalism transform ecosystems. Critical theories, such as the treadmill of production theory, ecological Marxism, and ecologically unequal exchange, highlight the intrinsic contradiction between capitalist accumulation and environmental protection, arguing that the pressure for sustained economic growth in a finite planet is the source of environmental decline (Bunker 1985; O'Connor 1998; Schnaiberg 1980). Ecological modernization theorists, on the contrary, argue that capitalist growth can be reconciled with environmental protection, chiefly through corporate investment in technological innovation aimed at reducing pollution and resource extraction (Buttel 2000).

DOI: 10.4324/9780429344428-12

Treadmill scholars argue that economic elites (capitalists and investors) pursue profitability by constantly increasing production through investment in new technologies (Gould et al. 2004; Schnaiberg 1980). Economic expansion requires an ever-increasing extraction of natural resources to be used as matter and energy, generating pollution and overall environmental decline. Technological innovation to increase productivity is often more energy- and chemical-intensive, speeds up resource depletion, and returns more toxic elements into the environment. Labor-saving technologies result in fewer and more poorly-paid jobs, while the environments where we live, work, and play become more toxic. The pursuit of economic growth thus leads to decreasing social benefits and increasing ecological disorganization. We can see an example of the treadmill at work with the expansion of the monocultures of genetically modified (GM) soybeans in Argentina, where the adoption of the biotechnology of GM crops has brought economic growth to the country at the expense of endangering the lives and livelihoods of rural communities (Leguizamón 2014).

Political elites (the state) promote the treadmill of economic growth, as tax revenues provide funds for public services (including support for those displaced by labor-saving technologies) and for infrastructure development. The state may also seek to regulate economic activity to protect public health from environmental degradation, in order to get the public support necessary to maintain its authority. Civil society actors (workers, students, Indigenous Peoples) can organize to collectively apply pressure on the state to demand social and environmental protections. Foreign actors, particularly transnational funding organizations such as the World Bank, the US Agency for International Development (USAID) or even Greenpeace, can influence socio-environmental activism in ways that can dramatically alter the state's development trajectory, as seen in the case of Ecuador (Lewis 2016).

The expansion of industrial production in Europe and the United States, as I explain in the following section, has historically depended on the extraction and importation of raw materials from Latin America. The treadmill of political capitalism has always operated on a global scale and has had significant global implications, dynamics exacerbated by the transnationalization of production since the 1980s (Gould et al. 2008). To increase returns on investments, all the while circumventing growing pressure from environmental movements and environmentally-conscious consumers in northern countries, US investors and managers have shifted toxic production abroad—as exemplified by the expansion of *maquiladoras* (assembly plants) in free trade zones (Green 2003). They have also increased capital investment in extractive sectors, expanding the extractive frontier with specialized technologies that increase socio-ecological risk. One such example is the 2012 opening of PEMEX (Petróleos Mexicanos, Mexico's state-owned petroleum company) to private investment to exploit unconventional oil reserves, such as deep water, shale, and tar sands, with hydraulic fracturing (Valdivia and Lyall 2018). Latin America is the second largest oil-producing region in the world while the United States is the world's largest oil consumer and the largest purchaser of Latin American oil. Repeating the pattern of resource dependency I trace in the following section, the US imports crude oil from Latin America and ships back refined fuels.

The theory of ecologically unequal exchange complements the treadmill of production theory by looking directly into the extractive economies of the global South and their historically uneven—and ecologically interdependent—relationship with the global North. The treadmill of production has its theoretical origins in the analysis of industrial development in the United States post-World War II (Gould et al. 2004). The history of Latin America's development is peripheral to this industrial expansion. This is the key structural problem identified by Raúl Prebisch in 1949: as Latin America exports raw materials and imports manufactured goods, over time there is a deterioration in the terms of the trade that leads to the underdevelopment of the commodity-exporting nation and the development of the manufacturing one (Kay 1989). Latin American extractivisms are, as Eduardo Gudynas (2019, 391) argues, in strict terms, neither "productive" nor an "industry." "To consider the extraction and exportation of iron, for example, as 'production' is a crude distortion," he writes, "for nothing is being 'produced': it is being extracted (and, therefore, amounts to a net loss of natural heritage). Neither does it make sense to qualify it as an 'extractive industry,' for no manufacturing process is involved."

The theory of ecologically unequal exchange, originally developed by Steven Bunker (1985), serves well as a complement to analyze how the treadmill operates at its extractive stage, compared to the manufacturing stage of global production. Bunker (1985) argues that economies based on the extraction of value from nature—compared to manufacturing economies that create value from labor, as in the United States and Europe—suffer net environmental losses through the transfer of matter, energy, and wealth. In addition to ecological devastation, extractive economies lead to unequal internal social dynamics. In seeking to "develop" the Amazon, the Brazilian state allied with large-scale capitalist groups to rapidly exploit the region's resources (mining, lumbering, agriculture, and cattle ranching), treating the Amazon as an empty frontier in need of civilization (Bunker 1985). Extractive activities are capital-intensive and have low labor requirements, and thus have little capacity for job creation. In addition, the expansion of extractive frontiers leads to the violent displacement of peasant and Indigenous communities and to land grabbing. Rather than bringing in economic growth and social wellbeing, the result is the underdevelopment of the resource-exporting region, while the resource-consuming nations gain value and their economies accelerate.[2]

In short, focusing a theoretical lens on the political economy of the environment requires questioning the decision-making processes of state and corporate actors over nature and natural resource use. It also calls for an examination of the disparate social and environmental consequences of these decisions, as well as the socio-environmental movements that organize in response to them (see the chapters contained in Part IV, this volume). In the following section, I review the history of extractivism in Latin America following these main theoretical insights. I show how political and economic elites in the region have often embraced a traditional vision of development that quantitatively equates economic growth with social wellbeing, and qualitatively defines development as a linear evolution toward progress, civilization, and Western modernity via constant industrialization and mechanization. This pursuit of economic expansion through the exploitation of nature has had uneven consequences, with powerful actors reaping the benefits of extractivism while poor and Indigenous communities bear the burden of the toxic impact of production and extraction practices.

A brief history of extractivism

Latin America's natural wealth is extraordinary. The wide variety of ecosystems that expand over the Latin American and Caribbean region—tropical and lowland forests, mangroves, coral reefs and wetlands, grasslands and mountain ranges—host over half of the world's biodiversity (UNEP 2016). This lushness has attracted the desire to dominate, control, and exploit nature ever since Columbus landed in the New World. The pillage mentality that characterized the Spanish conquest is embodied in the myth of El Dorado, which the conquistadors thought they had found in Lake Guatavita, in the Colombian Andes—the lake promised to hold, at bottom, the golden treasures of the Muisca. The Spaniards accordingly set about draining the lake to extract the gold, melt it, and ship the ingots away. Across five centuries, the same lust for gold and silver has decimated landscapes and cultures across the region, from Guatavita to Potosí (Galeano 1974).

The Spanish colonies set a pattern of extraction of the region's abundant natural resources for the benefit of foreigners and local elites at the expense of communities and the environment that lasts until today. At the height of the colonial period, in the 18th century, the Spanish—alongside the Portuguese, the French, the Dutch, and the British—had established a system of extraction that delivered Latin American and Caribbean gold, silver, indigo, cotton, sugar, coffee, and cacao to European markets (Restall and Lane 2018). Indigenous peoples were drafted to work in mines and plantations; when bodies were depleted in the Americas, slaves were imported from Africa to replace them. The "triangle of trade" fed workers in British factories while providing an overseas market for British manufactured goods, creating a system of unequal ecological exchange, where the land and labor of the New World subsidized the Industrial revolution, the rise of capitalism, and 19th-century British hegemony (Mintz 1986).

In the 19th century, the commodity trade consolidated, adding a new set of key exports: rubber, nitrates, copper, grains, and beef (Green and Branford 2013). With independence, *criollo* elites—American-born men of Spanish descent—gained political power over the Spanish Crown; but far from changing colonial institutions, they legitimized them. The large estates, *haciendas*, and plantations producing for the export sector, remained in the hands of a few and subordinated to European needs. White wealthy criollos embraced liberal economic theory, calling for economic growth via comparative advantage. Culturally, they espoused the values of European modernity and the Enlightenment. The modern mindset of 19th-century Latin American elites gave birth to a centuries-long quest to bring modernity and progress to the region by taming "savage" landscapes and peoples with science and technology; a civilizing project that continues today with the implementation of genetically modified seeds, hydraulic fracturing, open-pit mining, and other technological inventions for large-scale extractivism (Leguizamón 2020).

The commodity trade that boomed in the late 19th century had, however, busted by the early 20th, following declining demand for Latin American exports in the aftermath of World War I and the Great Depression (Green and Branford 2013). Global crises like these highlight the contradictions and dilemmas of relying on the extraction and exportation of nature for development. Agricultural and mineral commodities are inherently unstable. In addition to sudden drops in demand, economies based on exploiting nature run the risk of resource depletion. Extractive economies are also vulnerable to extreme weather events that can, by force of wind, flood, or drought, wipe out entire harvests. Busts may also result from technological substitution, as exemplified by the demise of the saltpeter industry in Chile and Peru in the mid-19th century. Saltpeter (sodium nitrate) was mined from the Atacama Desert and exported to produce gunpowder and fertilizer. Demand ended when German chemists Fritz Haber and Carl Bosch learned how to synthesize ammonia. The Haber-Bosch process replaced, once and for all, mineral fertilizers with chemicals, leading to the bust of the Chilean and Peruvian economies and to the rise of industrial agriculture (Patel and Moore 2017).

By the mid-20th century, structuralists and dependency theorists began to interpret Latin America's recurrent crises in the light of this history (see the work of Raúl Prebisch, Osvaldo Sunkel, Fernando H. Cardoso, Enzo Faletto, and Andre Gunder Frank, among others). They argued that Latin America's dependent position as an exporter of raw materials and importer of manufactured goods, conditioned the region's "underdevelopment" as well as the economic expansion (or "development") of the global North (Kay 1989). To break free from the colonial legacy of commodity exports, they recommended state-promoted industrialization to replace foreign imports with domestic products. Policies for import-substitution industrialization (ISI) were implemented by populist leaders across Mexico, Argentina, and Brazil (Bértola and Ocampo 2012).

During the ISI period, social conditions improved as a consequence of industrialization policy. The growth in industrial employment changed power relations: as a growing number of unionized workers demanded social security coverage, governments responded by developing welfare states (Huber and Ponce de León 2019). Environmental protection, however, was not a matter of policy concern. On the contrary, the expansion of social benefits—in a pattern repeated by Pink Tide governments in the early 21st century—was funded by resource nationalism (Ellner 2021). During ISI, governments nationalized key resources, like oil, gas, and minerals, and exploited them to develop domestic industries and to promote economic growth. Urban populations benefited at the expense of the rural sector. Natural environments, smallholder peasants, and Indigenous Peoples continued to be perceived as unproductive and in need of modernization. In turn, countries like Mexico and Argentina, also promoted the industrialization of their agricultural sectors through the adoption of the technologies of the Green Revolution, such as hybrid seeds, agricultural machinery, and chemical pesticides (McMichael 2016).

Central American and the Caribbean countries, without the political and economic power necessary to promote domestic industrialization, remained under the grip of the commodity trade, transnational companies, and American hegemony. In Honduras, for example, three large fruit companies (Cuyamel, United, and Standard) monopolized land and transportation routes as well as the banana export farms. As Soluri (2005) argues, the expansion of banana monocultures in Honduras brought extensive social and environmental transformations. In the late 19th century and across the 20th, fruit companies' engineers drained swamplands and cut down forests to make room for banana farms. Monocultures of a single banana variety, the Gros Michel, spread Panama disease across banana plantations. To control plant epidemics, fruit companies funded expensive breeding programs to develop pest-resistant varieties. When those efforts failed, scientists resorted to chemical treatments and, as a last resort, to abandoning the infested farms and expanding the agrarian frontier. Workers suffered from exposure to fumigations and grueling labor conditions; when they dared to strike to contest labor conditions, protests were crushed with violence. This pattern of socio-environmental change has repeated with the expansion of export crop monocultures across the region. Today, Cavendish bananas have mostly replaced the Gros Michel variety, but they are themselves threatened by Black Sigatoka and crown rot disease (Coleman 2020). Producers continue to grow bananas in large plantations and to use chemical cocktails while moving into new territories to escape disease. The banana market is still monopolized by three transnational corporations: Chiquita, Dole, and Del Monte. Latin America produces 80% of the bananas traded globally, bananas which American and European consumers have come to rely on at less than a dollar a pound—bananas made cheap thanks to pesticide-intensive monocrop plantations that continue to harm environments and workers' health. Cheap bananas, in this sense, are not really low-cost, but rather the consequence of a long history of "cheapening," a central strategy for capital accumulation (Patel and Moore 2017).

The making of a global commodity requires significant changes in transportation. In turn, expanding global transport networks have drastic planetary impacts. Fossil fuels are extracted and burned to power ships, trucks, and planes in order to transport commodities across the world, contributing to carbon emissions that lead to climate change. Almost 3% of total global emissions come from shipping alone (Olmer et al. 2017). The capitalist drive to carry larger volumes at a faster speed has also led to the massive redrawing of natural landscapes and the displacement of local communities, as exemplified by the creation of the Panama Canal in 1914 (Carse 2014). Today, probably the most ambitious transportation infrastructure plan for Latin America is the Initiative for the Integration of the Regional Infrastructure of South America, developed by the South American Council of Infrastructure and Planning, or COSIPLAN.

COSIPLAN is set to build dams, waterways, highways, and ports across the Amazon river basin to bring agricultural and mineral commodities of the Andes-Amazon region to oceanic ports. The cost for local ecologies and Indigenous communities of this mega-development project has only worsened with the rollback of social and environmental protections under the Bolsonaro administration in Brazil (Bebbington et al. 2018).

Extractivism intensified across the Latin American region from the 1990s. Burdened by debt and under pressure from international financial institutions like the World Bank and the International Monetary Fund, Latin American governments implemented neoliberal structural adjustment programs focused on exporting commodities for economic growth. Neoliberalism—a political-economic ideology adopted as a narrow set of economic policies including privatization, fiscal and monetary reform, and the deregulation of labor markets and environmental protections—took hold across the region from the 1990s through the early 2000s (Harvey 2005). The neoliberal export drive in Latin America is oriented toward the establishment of maquiladoras in free trade zones (the result of US businesses outsourcing jobs to lower wages and cut costs) and the specialization in nontraditional commodities. Chile, for example, sought to reduce its dependency on traditional copper exports by moving into exporting fresh fruit, timber, and salmon to the US market. By 2000, the country managed to reduce its dependency on copper exports to 40%, down from 79% in 1970 (Green 2003, 120). The impact of Chile's neoliberal agrarian transformation has been uneven (Kay 2002). The fruit export boom concentrated in the Central Region, and its benefits in the hands of large capitalist farmers and agro-industries. Peasant farmers without the capital to shift production had to sell their lands, leading to land concentration and increased dependency on wage labor. The agro-export boom took its heaviest toll on rural workers, and on women in particular. *Temporeras* (female seasonal workers) account for about half of the total temporary agricultural labor force in Chile (Bee 2000, 256). Their health has been severely compromised due to sustained exposure to chemical pesticides, both in the fields and in the packing plants. In addition to the social and ecological injustices created by the agricultural export boom, Chile has not yet shed its dependency on copper: copper exports still accounted for over half of total Chilean exports in 2018 (OEC 2020).

Neoliberalism failed to achieve its promise of socio-economic development. Through the 1990s, poverty and inequality rose across the region, as did social discontent. Social movement organizations—Indigenous and peasant movements in Ecuador, Bolivia, and Venezuela, the unemployed in Argentina—led protests, marches, and riots against neoliberalism (Silva 2009). In Bolivia, struggles against the privatization of water and gas mobilized hundreds of thousands of people in the early 2000s. Protestors demanded the state reclaim public control over natural resources (Spronk and Webber 2007). The region-wide challenge to neoliberalism led to the "pink tide," an electoral wave of leftist leaders, including Hugo Chavez in Venezuela (1999–2013), Luiz Lula da Silva in Brazil (2003–2011), Néstor and Cristina Kirchner in Argentina (2003–2015), Rafael Correa in Ecuador (2007–2017), and Evo Morales in Bolivia (2006–2019).[3]

Pink Tide leaders were elected on social justice platforms, promising social inclusion and economic redistribution to address poverty and inequality. They proposed—and delivered—strong state intervention and increased social spending to address the needs of the most marginalized (Grugel and Riggirozzi 2012). The turn of the 21st century thus brought important political changes to the region, yet there were also important continuities. Most significantly, progressive governments continued to rely on the extraction and exportation of nature for development; and, with that reliance, the negative socio-environmental impacts of extractivism, also continued. Scholars refer to this latest phase of Latin American extractivism as "neo-extractivism," to highlight the significance of state control over natural resources and the social policy component for a more inclusive type of development (Gudynas 2009; Svampa 2019).

Under neo-extractivism, progressive governments sought to capture a larger share of the profits generated by the extractive sectors, through nationalization or export taxation, to finance social spending (Farthing and Fabricant 2018). The commodity boom of the 2010s was key to their success. From 2004 to 2012, global commodity prices rose steadily, as did Chinese investment and demand for Latin American resources (Arsel et al. 2016). In this favorable context, extractivism ramped up. By 2012, oil exports constituted 50% of Ecuador's export earnings and a staggering 97% of Venezuela's. That same year, close to 60% of Peruvian and Chilean exports were minerals and precious metals. Over 50% of Argentina's foreign income was due to agro-food exports, and 23% from soybeans and soybean-derived products alone (OEC 2020).

The commodity boom gave Pink Tide governments the capacity to expand social welfare programs to address the needs of the poor, with remarkable results. From the late 1990s to the late 2000s, income inequality in Latin America declined steadily and significantly; an impressive fact considering inequality rose globally during this time period. Poverty rates also dropped, from 41.5% of the total population down to 29.6%, as government transfers lifted roughly 49 million Latin Americans out of poverty (Lustig et al. 2016). Reductions in poverty and inequality rates took place across the region, as both progressive and conservative governments spent some of the financial windfall of increasing extractivism on social welfare. Data shows that, compared to conservative governments, countries under the Pink Tide had a steeper reduction in both poverty and inequality, reflecting a greater emphasis on redistribution (Feierherd et al. 2020).

By financing social programs and infrastructure development through the export of natural resources, Pink Tide governments gained substantial support and legitimacy for extractivism as a development model (Gudynas 2009; Leguizamón 2020). Addressing social needs with ecological devastation has its limits, however. The expansion of large-scale development projects accelerates the treadmill of production, causing resource depletion and pollution, social inequality, and violence. Thus, in the long term, profits fueled by environmental damage limit the sustainability of neo-extractivism as a development model. In the short term, the fall in commodity prices put an end to welfare politics and the Pink Tide. After 2015, in a context of economic weakness and rising social polarization, center-right coalitions gained power across the region, beginning with Argentina, Chile, and Brazil (Anria and Vommaro 2020). This might not mark "the end of the left turn" for Latin American politics, as countries like Mexico and Argentina have already elected leftist coalitions (Anria and Roberts 2019). Yet despite the turns of the political tide, the patterns of resource dependency and resource depletion are likely to continue.

Environmental futures

Developmental pursuits in Latin America have relied on the appropriation of nature, resulting in social inequality and environmental decline. Learning from the historical, political, and economic dynamics analyzed in this chapter, we can foresee three potential paths to Latin America's environmental future: a continuation of the status quo, a market-based technological transition, or a post-extractivist alternative.

In the short and medium term, a most likely environmental future is path-dependent, thus continuing to rely on extractivism. For as long as external debt repayments continue to consume national budgets and the global North does not step up their international commitment to protect natural ecosystems, national governments will continue to feel the pressure to extract, justifying extractivism as necessary to meet the needs of the poor with the natural wealth of the nation. Ecuador is such an example. High foreign debt (as low as 20% of its GDP in 2012 and as high as 68% in 1992) and the failure of the international community to commit to the Yasuní-ITT initiative (a project to "keep oil in the soil" in return for international payments) led then-president Rafael Correa to famously say "we can't be beggars sitting on a pot of gold," and set to digging Amazonian oil in the name of the development (Lewis 2016).

In the medium and long term, as the negative impacts of climate change intensify, extractivism will hit its ecological limits. Extreme weather events, loss of biodiversity, reduction of available water for energy, agriculture, and drinking, severe soil degradation and desertification, rising sea levels, and other extreme events associated with climate change will force a reckoning with capitalist, extractivist practices (Rojas Hernández 2016). To avert this crisis, it is imperative to reconcile society–nature dynamics. A likely environmental future is reformist: a technological transition toward efficiency improvements. The ecological modernization path relies on corporate investment in green technologies that allow the pursuit of economic growth but with minimal ecological impact (Buttel 2000). This is the discourse behind the expansion of genetically modified soybeans in Argentina: the promise that the adoption of cutting-edge technologies such as genetically-engineered seeds will lead to sustainable development (Leguizamón 2020). This technological optimism is also represented in the promotion of lithium mining in Chile, Bolivia, and Argentina, as corporations like Tesla invest in the production of lithium-ion batteries, which promise to make electric cars and renewable energy both feasible and affordable (Barandiarán 2019).

This sustainable development path proposed by ecological modernization theorists barely displaces—temporally and spatially—the worst impacts of an accelerated treadmill. Market-based solutions do not address the core issue of power imbalances that determine the current nature of society–environment relations. It is thus paramount to pursue an environmental future that transcends extractivism. Post-development transitions in Latin America include *Buen vivir*, the Rights of Nature, and other "alternatives to development" (Escobar 2015). These alternatives require a rethinking of the Eurocentric development project, prioritizing environmental protection and social justice over profitability and growth. *Buen vivir* (living well), for example, sets the focus on the quality of life while recognizing the intrinsic rights of nature, and proposes society–ecosystem relations based on care and harmony instead of domination over people and nature (Gudynas 2011). The post-extractivist path is certainly the most difficult as it meets resistance from established political and economic elites, yet it is necessary for a just, livable future.

Notes

1 For a review of sociological scholarship on the political economy of the environment, see Rudel et al. 2011. As these authors argue, there are no clear boundaries to this field, and at least three other disciplines (political science, geography, and economics, in addition to sociology) claim it as a substantial area of disciplinary study. My approach is sociological and draws from theoretical perspectives on environmental sociology and development sociology (see also Givens et al. 2016).

2 Recent work on ecologically unequal exchange theory further explores how global political-economic factors structure the globally uneven distribution of environmental harm and human well-being. This research documents how the global South bears a disproportionate burden of environmental degradation due to the withdrawal of natural resources (the vertical flow of matter and energy) and the pollution resulting from extractive activities and waste imports from the global North (Givens et al. 2019; Jorgenson and Clark 2009). By displacing the harmful consequences of economic growth—both spatially and temporally—nations of the global North have been able to "green" their environments while maintaining high consumption levels.

3 The Latin America Left in the 21st century is, though far from uniform, united by a common objective to reduce social and economic inequalities—and no longer inspired by a Marxist ideology for public ownership of means of production and centralized planning (Roberts and Levitsky 2011).

References

Anria, Santiago, and Kenneth M. Roberts. 2019. "The Latin American Left isn't Dead Yet." *The Conversation*. October 9: 2019. http://theconversation.com/the-latin-american-left-isnt-dead-yet-124385.

Anria, Santiago, and Gabriel Vommaro. 2020. "En Argentina, un 'giro a la derecha' que no fue y el improbable regreso del peronismo de centro-izquierda." *Más poder local*, 40: 6–10.

Arsel, Murat, Barbara Hogenboom, and Lorenzo Pellegrini. 2016. "The Extractive Imperative in Latin America." *The Extractive Industries and Society* 3 (4): 880–87. https://doi.org/10.1016/j.exis.2016.10.014.

Barandiarán, Javiera. 2019. "Lithium and Development Imaginaries in Chile, Argentina and Bolivia." *World Development* 113 (January): 381–91. https://doi.org/10.1016/j.worlddev.2018.09.019.

Bebbington, Denise Humphreys, Ricardo Verdum, Cesar Gamboa, and Anthony J. Bebbington. 2018. "The Infrastructure-Extractives-Resource Governance Complex in the Pan-Amazon: Roll Backs and Contestations." *European Review of Latin American and Caribbean Studies*. 106: 183–208.

Bee, Anna. 2000. "Globalization, Grapes and Gender: Women's Work in Traditional and Agro-Export Production in Northern Chile." *The Geographical Journal* 166 (3): 255–65. https://doi.org/10.1111/j.1475-4959.2000.tb00024.x.

Bértola, Luis, and José Antonio Ocampo. 2012. *The Economic Development of Latin America since Independence. Initiative for Policy Dialogue*. Oxford, New York: Oxford University Press.

Bunker, Stephen G. 1985. *Underdeveloping the Amazon: Extraction, Unequal Exchange, and the Failure of the Modern State*. Urbana and Chicago: University of Illinois Press.

Buttel, Frederick H. 2000. "Ecological Modernization as Social Theory." *Geoforum* 31 (1): 57–65.

Carse, Ashley. 2014. *Beyond the Big Ditch: Politics, Ecology, and Infrastructure at the Panama Canal*. Cambridge, MA: MIT Press.

Coleman, Kevin. 2020. "Banana Industry in Central America." In *Oxford Research Encyclopedia of Latin American History*. https://doi.org/10.1093/acrefore/9780199366439.013.605.

Coronil, Fernando. 1997. *The Magical State: Nature, Money, and Modernity in Venezuela*. Chicago and London: University of Chicago Press.

Ellner, Steve. 2021. "Introduction: Rethinking Latin American Extractivism." In *Latin American Extractivism: Dependency, Resource Nationalism, and Resistance in Broad Perspective*, edited by Steve Ellner, 1–27. Lanham, Maryland: Rowman & Littlefield.

Escobar, Arturo. 2015. "Degrowth, Postdevelopment, and Transitions: A Preliminary Conversation." *Sustainability Science* 10 (3): 451–62.

Farthing, Linda, and Nicole Fabricant. 2018. "Open Veins Revisited: Charting the Social, Economic, and Political Contours of the New Extractivism in Latin America." *Latin American Perspectives* 45 (5): 4–17.

Feierherd, G., Nora Lustig, W. Long, and S. Quan. 2020. "The Pink Tide and Inequality in Latin America (Working Paper)." Presented at the Seminar of Department of Economics, Tulane University.

Galeano, Eduardo. 1974. *Open Veins of Latin America: Five Centuries of the Pillage of a Continent*. New York: Monthly Review Press.

Givens, Jennifer E., Brett Clark, and Andrew Jorgenson. 2016. "Strengthening the Ties between Environmental Sociology and Sociology of Development." In *The Sociology of Development Handbook*, edited by Gregory Hooks, 69–94. Oakland, California: University of California Press.

Givens, Jennifer E., Xiaorui Huang, and Andrew K. Jorgenson. 2019. "Ecologically Unequal Exchange: A Theory of Global Environmental Injustice." *Sociology Compass* 13 (5). https://doi.org/10.1111/soc4.12693.

Gould, Kenneth A., David N. Pellow, and Allan Schnaiberg. 2004. "Interrogating the Treadmill of Production: Everything You Wanted to Know about the Treadmill but Were Afraid to Ask." *Organization & Environment* 17 (3): 296–316.

———. 2008. *Treadmill of Production: Injustice and Unsustainability in the Global Economy*. Boulder, CO: Paradigm.

Green, Duncan. 2003. *Silent Revolution: The Rise and Crisis of Market Economics in Latin America*. 2nd edition. New York: Monthly Review Press.

Green, Duncan, and Sue Branford. 2013. *Faces of Latin America.* New York: Monthly Review Press.

Grugel, Jean, and Pía Riggirozzi. 2012. "Post-Neoliberalism in Latin America: Rebuilding and Reclaiming the State after Crisis." *Development and Change* 43 (1): 1–21. https://doi.org/10.1111/j.1467-7660.2011.01746.x.

Gudynas, Eduardo. 2009. "Diez Tesis Urgentes Sobre el Nuevo Extractivismo: Contextos y Demandas bajo el Progresismo Sudamericano Actual." In *Extractivismo, Política y Sociedad*, 187–225. Quito: Centro Andino de Acción Popular.

———. 2011. "Buen Vivir: Today's Tomorrow." *Development* 54 (4): 441–47. https://doi.org/10.1057/dev.2011.86.

———. 2019. "Development and Nature: Modes of Appropriation and Latin American Extractivisms." In *The Routledge Handbook of Latin American Development*, edited by Julie Cupples, Marcela Palomino-Schalscha, and Manuel Prieto, 389–99. Routledge International Handbooks. London; New York: Routledge.

Harvey, David. 2005. *A Brief History of Neoliberalism.* Oxford: Oxford University Press.

Huber, Evelyne, and Zoila Ponce de León. 2019. "The Changing Shapes of Latin American Welfare States." *Oxford Research Encyclopedia of Politics*, June. https://doi.org/10.1093/acrefore/9780190228637.013.1656.

Jorgenson, Andrew K., and Brett Clark. 2009. "Ecologically Unequal Exchange in Comparative Perspective: A Brief Introduction." *International Journal of Comparative Sociology* 50 (3–4): 211–14.

Kay, Cristóbal. 1989. *Latin American Theories of Development and Underdevelopment.* New York: Routledge.

———. 2002. "Chile's Neoliberal Agrarian Transformation and the Peasantry." *Journal of Agrarian Change* 2 (4): 464–501. https://doi.org/10.1111/1471-0366.00043.

Leguizamón, Amalia. 2014. "Modifying Argentina: GM Soy and Socio-Environmental Change." *Geoforum* 53: 149–60.

———. 2020. *Seeds of Power: Environmental Injustice and Genetically Modified Soybeans in Argentina.* Durham, NC: Duke University Press.

Lewis, Tammy L. 2016. *Ecuador's Environmental Revolutions: Ecoimperialists, Ecodependents, and Ecoresisters.* Cambridge, MA: MIT Press.

Lustig, Nora, Luis F. Lopez-Calva, and Eduardo Ortiz-Juarez. 2016. "Deconstructing the Decline in Inequality in Latin America." In *Inequality and Growth: Patterns and Policy*, 212–47. London: Palgrave Macmillan.

McMichael, Philip. 2016. *Development and Social Change: A Global Perspective.* 6th edition. Thousand Oaks: Sage Publications.

Mintz, Sidney Wilfred. 1986. *Sweetness and Power: The Place of Sugar in Modern History.* New York: Penguin.

O'Connor, James R. 1998. *Natural Causes: Essays in Ecological Marxism.* New York and London: Guilford Press.

OEC. *The Observatory of Economic Complexity.* 2020. https://oec.world/. Accessed September 25, 2020.

Olmer, Naya, Bryan Comer, Biswajoy Roy, Xiaoli Mao, and Dan Rutherford. 2017. *Greenhouse Gas Emissions from Global Shipping, 2013–2015.* Report for the International Council on Green Transportation. https://theicct.org/publications/GHG-emissions-global-shipping-2013-2015.

Patel, Raj, and Jason W. Moore. 2017. *A History of the World in Seven Cheap Things: A Guide to Capitalism, Nature, and the Future of the Planet.* Oakland, California: Univ of California Press.

Restall, Matthew, and Kris Lane. 2018. *Latin America in Colonial Times.* 2nd edition. Cambridge: Cambridge University Press.

Roberts, Kenneth M., and Steven Levitsky. 2011. "Introduction: Latin America's 'Left Turn': A Framework for Analysis." In *The Resurgence of the Latin American Left*, edited by Kenneth M. Roberts and Steven Levitsky, 1–28. Baltimore: Johns Hopkins University Press.

Rojas Hernández, Jorge. 2016. "Society, Environment, Vulnerability, and Climate Change in Latin America: Challenges of the Twenty-First Century." *Latin American Perspectives* 43 (4): 29–42.

Rudel, Thomas K., J. Timmons Roberts, and JoAnn Carmin. 2011. "Political Economy of the Environment." *Annual Review of Sociology* 37: 221–38.

Schnaiberg, Allan. 1980. *The Environment, from Surplus to Scarcity.* New York: Oxford University Press.

Silva, Eduardo. 2009. *Challenging Neoliberalism in Latin America*. New York: Cambridge University Press.

Soluri, John. 2005. *Banana Cultures: Agriculture, Consumption, and Environmental Change in Honduras and the United States*. Austin, Texas: University of Texas Press.

Spronk, Susan, and Jeffery R. Webber. 2007. "Struggles against Accumulation by Dispossession in Bolivia The Political Economy of Natural Resource Contention." *Latin American Perspectives* 34 (2): 31–47. https://doi.org/10.1177/0094582X06298748.

Svampa, Maristella. 2019. *Neo-Extractivism in Latin America: Socio-Environmental Conflicts, the Territorial Turn, and New Political Narratives*. Cambridge, UK: Cambridge University Press.

UNEP. United Nations Environment Programme. 2016. "Biodiversity in Latin America and the Caribbean." http://www.pnuma.org/forodeministros/20-colombia/documentos/Background_Biodiversity_Document_26_02_16.pdf.

Valdivia, Gabriela, and Angus Lyall. 2018. "The Oil Complex in Latin America: Politics, Frontiers, and Habits of Oil Rule." In *The Routledge Handbook of Latin American Development*, edited by Julie Cupples, Marcela Palomino-Schalscha, and Manuel Prieto, 458–68. Routledge.

10 Ecological debt and extractivism

Tatiana Roa Avendaño

Introduction

A Spanish publication asked about the ecological debt: "Who owes whom?" To which the Catalan professor and ecologist Joan Martínez Alier replied:

> The Ecological Debt of the North to the South is much larger than the financial External Debt of the South to the North. This reality is, however, difficult to quantify, since in many of its aspects this Ecological Debt, added to the historical debts after centuries of colonialism and exploitation, cannot be valued in money.
>
> (Martínez-Alier 2003, 7)

While much has been said about the large external monetary debt owed by Latin American countries to international banking institutions such as the World Bank and the International Monetary Fund, if we count the colonial and ecological debt—despite the challenge of quantifying it—the reality seems to be the other way around: it is the North that owes the South.

In his book *Potosí, el origen. Genealogía de la minería contemporánea*, Horacio Machado (2018) concludes that the 15th-century invasion of America by Europeans and colonial mining extraction were the midwives of capitalism as an economic, political, social, and cultural system that still exploits human beings and nature. With the insertion of the region into the global system in the 19th century, the extraction of its natural resources was crucial to shaping its social, political, and economic development (De Castro, Hogenboom, and Baud 2015, 14). In other words, the ecological debt has its origins in colonial plunder—extraction of minerals, pillaging and destruction of its forests, exploitation of animals—and the usufruct of its natural resources. The debt is projected both in ecologically-unequal exchange and in the impoverishment of Latin America's peoples, as well as in the occupation of environmental space as a result of the industrialized countries' lifestyles, with their associated forms of production and consumption.

Extractive cycles have been recurrent. They may change the natural goods, the places of extraction, or the techniques, but they always have one purpose: to feed the growing appetite of the so-called "powers." And each time, this extraction is accompanied by a dose of plunder, dispossession, violence, and destruction. The ecological debt reflects structures of domination, social inequalities and inequities, and environmental degradation that, as already mentioned, began to take shape several centuries ago.

Between 2004 and 2014, Latin America experienced the so-called "commodity boom." It was caused by high commodity prices, generated by the growing demand from emerging economies, especially China. In these years, the region facilitated the opening up to foreign capital and significantly increased Foreign Direct Investment (FDI) in extractive projects, mainly mining, hydrocarbon, and agro-industrial projects, as well as infrastructure, mostly in support of extractive

DOI: 10.4324/9780429344428-13

developments. But extractivism has not only directed financial flows between the North and the South: it has also generated flows of materials and energy, and left an asymmetrical distribution of ecological costs. This means that Latin America has continued to subsidize European, North American, and Asian countries energetically and materially. In this way, the industrialized countries have increased their ecological debt to the region, expressed in socioenvironmental liabilities, ecologically unequal trade, and the destruction of territories and ways of life of local communities.

The concept of ecological debt is thus a proposal from the struggles and reflections of the Latin American environmental movement that counteracts the concept of external debt, which is typical of the countries that consider themselves to be "creditors"—the countries of the global North. Using the same language, the creditors of the ecological debt are the countries of the South. In recent decades, the ecological debt claim has become increasingly relevant as the extractive model has deepened. This chapter analyses the relationship between ecological debt, foreign debt, and extractivism, and shows the ways in which social and environmental organizations have used the concept as a political tool in their struggle for environmental justice.

Origin, definition, and application of the concept

Aurora Donoso, an Ecuadorian activist, defines the ecological debt as the debt

> ... the North, especially the most industrialised countries, has been accumulating in favour of the peripheral nations through the plundering of natural [goods], through their undervalued sale, environmental pollution, the free use of their genetic resources or the free occupation of their environmental space for the deposit of greenhouse gases or other waste accumulated and disposed of by the industrialised countries.
>
> (Donoso 2000)

This novel concept emerged in the early 1990s, as the Latin American environmental movement was preparing for the 1992 *UN Conference on Sustainable Development* in Rio de Janeiro (better known as the *Rio Summit*), and at a time when the foreign debt crisis was affecting various countries of the South. In 1990, the Instituto de Ecología Política, a Chilean organization, produced a report on the production of chlorofluorocarbon gases (CFCs) used in household appliances, aerosols, and also in various industrial processes in the industrialized countries of the North. In their report, the Chileans warned about the effect of these gases on the ozone layer, and argued that this production and consumption relationship had given rise to an ecological debt owed by the countries of the North to the countries of the South, given that the consequences and damage would affect the entire human population as well as animals. Two years later, a document by environmental organizations, resulting from the alternative Rio Summit, recognized that there is an ecological debt due to the systematic plundering of nature and the practice of ecologically-unequal trade. The Latin American environmentalists' document concluded that the ecological debt is the result of the historical flow of materials and energy, the asymmetric form of ecological costs, and the disproportionate use of environmental space to deposit greenhouse gases (Martínez-Alier 2003). The ecological debt thus contrasts with the North's demand on the South for payment of the external debt.

In the 1990s, the debate and theoretical and political reflections on ecological debt continued to gain momentum. In 1994, the Colombian organization, Fundación para la Investigación para la Protección del Medio Ambiente, FIPMA, published the book *Deuda Ecológica: Testimonio de una reflexión*, an important effort by its director, José María Borrero, to further deepen the concept. In 1997, the Ecuadorian environmental organization Acción Ecológica promoted an international forum on the subject in Ecuador. From that moment on, the ecological debt became one of the most important banners of struggle for the renowned Latin American environmental organization. From 1999 onwards, Acción Ecológica, together with others, promoted the Campaign for the Recognition and Reclamation of the Ecological Debt at the international level, whose aims included "establishing the responsibility and obligation of the industrialised countries of the North to repair and stop the damage caused to the biosphere and the countries of the Third World by the Ecological Debt" (Acción Ecológica 1999, 132) and highlighting the illegitimacy of the foreign debt, as well as halting the flow of material, energy, food, and finance from Latin America outwards. At the turn of the century, other Latin American organizations (e.g. Censat Agua Viva (Friends of the Earth Colombia)), African organizations (e.g. Environmental Rights Action (Friends of the Earth Nigeria)), Asian organizations (Wahli (Friends of the Earth Indonesia) and Jubilee South Asia-Pacific), and European organizations (Observatorio de la Deuda en la Globalización [ODG] of Spain and VODO, the Flemish Platform on Sustainable Development, Belgium) took up the issue as a political tool and started to build alliances, articulations, and political actions around the ecological debt claim.

Undoubtedly, ecological debt has been a political tool of Latin American environmentalism. One of its debates has been around how to quantify the debt, without "reducing the social metabolic analysis to a monocriterial indicator such as money" (Zuberman 2019, 91). Ecological debt, as Joan Martínez Alier rightly says, can be quantifiable not only monetarily but also in "non-monetary units—for example, so many cancer patients, so many hectares deforested, so many tons of materials exported and so much biodiversity lost to pay the external debt" (Martínez-Alier 2007, 24). Ecological economics offers various "indicators of social metabolism that allow us to account for the degree of appropriation that societies make of their natural resources" (Zuberman 2019, 83), for example, the virtual soil or water, or the ecological footprint. However, in some extractive projects, the debt has been quantified in money (Roa-Avendaño & Navas, 2001), for example, to force a company to compensate for damage it has caused, as in the cases of Texaco's oil exploitation in the Ecuadorian Amazon, or Shell's extractive activities in the Nigerian Delta.

In other cases, quantification has also served to highlight the ecologically- and economically-unequal exchange between countries or regions. One example is the work of Mario Pérez-Rincón (2007) on the relationship between international trade and the environment in Colombia, in which the Colombian researcher concludes that the terms of trade of international trade, together with the unequal power relations between North and South, mean that more natural goods need to be exported and exploited to obtain the same amount of imported goods. In Argentina, Federico Zuberman analysed the export of virtual land linked to soybean cultivation over a period from 1970 to 2015. Rather than focusing on an economic estimate, his work is carried out in terms of nutrients exported in chemical fertilizer equivalents. He concludes that "the export of nutrients is not only a debt to future generations, but is part of the ecological debt that, at the global level, the countries of the North continue to incur at the cost of undervalued imports of natural resources from the countries of the South" (Zuberman 2019, 91). Maria Cristina Vallejo has done similar work on Ecuador's banana exports (Martínez-Alier 2007, 24).

Ecological debt, foreign debt and extractivism

In Latin America, the extractive model has been a determinant of its entire history of dependence, exploitation, colonialism, exclusion, racism, and marginalization (Pérez-Rincón 2016b, 3). These extractive processes have generated an ecological debt. However, since the 1990s, extractive processes have accelerated, as large-scale projects using new technologies, more capital, and less employment began to be installed across the continent (Alimonda 2011;

Bebbington and Bury 2013; Hogenboom 2015; Toro-Peréz et al. 2012). This situation also translated into a change in the economic matrix of these countries.

In the last three decades, Latin America has undergone a strong process of primarization, leaving behind the import-substitution industrialization process that prevailed in the post-war period until the late 1970s and early 1980s. The loss of importance of manufacturing industry in gross domestic product (GDP) is a good indicator that confirms the region's reprimarization. Between 1975 and 2006, the GDP of the manufacturing sector fell from 12.7 to 6.4%. In the same period, Argentina's manufacturing GDP falls from 43.5 to 27%, Ecuador's from 19 to 10% of GDP, and Brazil's from 28 to 24% of GDP (Muñiz 2018). In Colombia, "there is a structural change in the primary sector: agricultural exports fall and mining-energy exports increase ostensibly. The former went from 60% to 4% between 1975 and 2014 and the latter increased in the same period from 9% to 68%" (Pérez-Rincón 2016b, 114–115). The trajectory of the leap in the extractive model cannot be understood without taking into account the region's market liberalization policies initiated in the 1990s, which emerged as a result of the so-called Washington Consensus (De Castro, Hogenboom, and Baud 2015; Hogenboom 2015; Toro-Pérez et al. 2012). This liberalization led to accelerated reprimarization and the concentration and foreignization of the economies of these countries (Machado Araoz and Rossi 2017; Pardo Becerra 2018; Pérez-Rincón 2016b; Roa-Avendaño and Navas 2014; Toro-Pérez et al. 2012), as well as the negative growth of industry, manufacturing, and traditional agriculture (Pardo Becerra 2018, 3).

The intensive and extensive exploitation of natural goods for export in this century is related to the increase in social metabolism and the high prices of energy and raw materials. All this leads to a new international economic and political order that Maristella Svampa (2013, 35) calls the *commodity consensus*, determined by the high demand for consumer goods by central and emerging economies. Research reinforces Svampa's argument, making it clear that this new order is characterized by basing the growth of Latin American economies on the extraction of primary goods on a large scale, and their export without greater added-value to the economic powers, taking advantage of the increase in mineral prices at the beginning of the century (Pardo Becerra 2014, 2018; Pérez-Rincón 2016b). From the perspective of Arsel, Hogenboom, and Pellegrini (2016), an *extractive imperative has been* established that has created its own dependencies and legitimization mechanisms.

Mario Pérez-Rincón's (2016b, 16) biophysical analysis of the Colombian economy—which he also carries out for the Andean and Central American countries—provides a better understanding of the model's environmental pressures. Between 1970 and 2013, Colombia multiplied its domestic material extraction (DE) by 350%, from 130 million metric tons (MTM) in the first year to 459 MTM in 2013.[1] Even more telling is the behavior of Colombia's biophysical exports. Between 1990 and 2014, primary sector exports, including natural resource-based manufactures, accounted for 82% of the country's total exports. As evidenced by various studies (Göbel and Ulloa 2014; Pérez-Rincón 2016a and 2016b; Roa-Avendaño and Navas 2014; Toro-Pérez et al. 2012), the intensification of the model has caused profound territorial transformations, serious damage to important ecosystems, and impacts on people's health and lives, increasing the ecological debt that industrialized countries owe to Latin American countries.

But while the extraction of natural resources—most of which are exported to international markets—has grown, the foreign debt has also increased. In less than a decade, this debt has almost doubled from US$1.06 trillion in 2010 to US$1.87 trillion (World Bank 2020), a situation denounced by social and activist organizations and research groups. In 2008, Latin America's public debt represented 40% of the region's Gross Domestic Product (GDP), and currently, according to estimates by the Inter-American Development Bank (IDB), it represents 62% of GDP (BBC 2020), generating serious economic and social implications for the region.

The growth of external debt in Latin America in recent years has been associated with the increase in Foreign Direct Investment (FDI) destined for extractive projects in the region. This began during the aforementioned "commodity boom" (2004–2014), during which extractive activities expanded in most of the continent. Indebtedness not only eroded the public finances of Latin American countries, but has also led to the imposition of structural adjustment policies, strengthening the domination of international financial institutions (IFIs) and transnational corporations. Foreign investment and financial indebtedness are the tools for exercising their control. With the upper hand, the IFIs imposed conditions on the countries of the South to make labor and environmental processes more flexible, and to grant greater incentives and tax exemptions in favour of foreign investment.

In order to meet the commitments imposed by the financial debts, the countries of the region had to deepen and broaden the extractivist model (mining, forestry, oil, fishing, agro-industry). By doing so, they were able to increase exports and, with these resources, service the debt. An emblematic case is that of Ecuador, where "oil has been at the center of external indebtedness, in fact, indebtedness was born from the 'oil boom' in 1972" (Donoso 2017, 51). Between 1971 and the end of 1981, the foreign debt grew 22 times, from US$260.8 million to US$5,869.8 million. The loans were made to build the oil infrastructure that Texaco required to get its oil to international markets. And oil ended up being "the guarantee of indebtedness, since it is used to pay the foreign debt" (Donoso 2017, 51). It is a vicious cycle: the more financial debt, the more extractivism. And the more extractivism, the greater the loss and plundering of the continent's natural heritage, deepening the ecological debt of which our countries are the creditors. In other words, the prioritization of extractive activities has deepened Latin America's structural dependence on industrialized countries and has led to a process of economic indebtedness to these countries. Foreign debt and ecological debt are two sides of the same coin.

On the other hand, it is necessary to emphasize that, although the development of extractivist projects has meant GDP growth in some countries in the region, debt continues to grow. And the region is considered to be the most socially unequal in the world and the extreme poverty of a portion of the Latin American population is significant. A recent report by the Economic Commission for Latin America and the Caribbean (ECLAC) indicates that in 2019, 30.8% of Latin America's population was below the poverty line and 10.7% lived in extreme poverty, rates that increased with respect to 2018 (ECLAC 2019). In other words, decades of extractivism have not managed to reverse the enormous poverty in the region: rather, they have deepened it and are evidence of a vicious cycle: more extractivism, more foreign debt, more ecological debt, and more poverty. Although in the second decade of this century, the debate on ecological debt diminished, the boom in extractivism in recent years, the reprimarization of the economy, and the increase in external debt has revived the discussion on ecological debt and the urgency of quantifying the debt as well as specific environmental liabilities. This relationship between foreign debt, extractivism, and ecological debt makes ecologically-unequal trade more evident,

> where the economies of the periphery are forced to export ever greater quantities of their natural resources in order to gain access to goods supplied almost exclusively by the central countries. This is, in short, an ecological correlate of 'Prebisch's (1979; 1986) theory of deteriorating terms of trade.
>
> (Zuberman 2019, 90–91)

The "cursed gifts" of extractivism

Undoubtedly, one of the areas in which the ecological debt owed by the countries of the North to the South is manifested is the environmental deterioration caused by extractive projects. Three examples illustrate this situation very well.

On Friday, January 25, 2019, at 12 noon local time, three waste containment dams belonging to the Vale S.A. mining company collapsed in Brumandinho in the State of Minas Gerais, Brazil. The tragedy left 270 people dead, another 11 missing, and incalculable environmental, social, cultural, and economic losses (Mendoça 2020). The collapse of the dams caused 13 million cubic meters of sludge to spill in a few seconds (France 24 2020), immediately burying the facilities of the company Vale—one of the main producers and exporters of iron ore in Brazil—as well as several houses in the surrounding area.[2] The contaminated sludge destroyed infrastructure, covered crops, razed trees, and displaced hundreds of families living in the Paraopeba river basin. The waste also contaminated more than 300 kilometers of the river. Water consumption was banned for an indeterminate period of time because of the contamination of the river, and thousands of families in the basin who depended on the river were left without access to water. In addition, hundreds of hectares of land were contaminated, much of it gallery forests and agricultural production areas.

The Brumandinho disaster is not the only one to have occurred in Brazil. Three years earlier, on November 5, 2015, in the same state of Minas de Gerais, other tailings dams from the iron ore mining operations of Samarco Mineração S.A.—also owned by Vale S.A. and BHP Billiton—collapsed. The broken dams released almost 40 million cubic meters of highly polluting sludge that wiped the towns of Bento Rodrigues and Paracatu de Baixo off the map. The tragedy left 19 people dead and contaminated the waters of the Doce river, which supplied water to around 230 municipalities in the states of Minas Gerais and Espírito Santo. Other towns in the Gualaxo river valley were affected to a lesser extent, and contaminated water and sludge even reached the sea.

Decades of extractivism in Latin America have left hundreds of mining waste dams, many of which could collapse at any moment and cause a new disaster. This dammed sludge is an expression of the ecological debt created by mega-mining in Latin America, mostly developed by North American, South African, or Chinese companies that have recently set up operations on the continent.

In Colombia, on March 2, 2018, in La Fortuna, in the municipality of Barrancabermeja, the Lizama 158 well of the company Ecopetrol exploded. Thousands of gallons of crude oil spilled into the Lisama stream, contaminating its waters and the surrounding area. Hundreds of families were temporarily displaced, 25 kilometres of oil flowed before the waters of the stream reached the Sogamoso river. The damage is incalculable and may be irreparable. Months before the accident, the Comptroller of the Republic had warned Ecopetrol about the irregularities of abandoned or closed oil wells, which could lead to situations such as those that occurred with the Lizama well. In a document addressed to the Council of State, the Fracking-Free Colombia Alliance and some congressmen denounced the frequent contingencies related to oil spills and leaks associated with inadequately abandoned wells. This is just one of the environmental liabilities suffered by oil-producing regions, which see their territories devastated by oil developments. The book *Como el agua y el aceite: Conflictos Socio-ambientales de la extracción petrolera*, written by a group of Colombian women social leaders, activists, and academics, and published by Censat Agua Viva, narrates in the form of testimonies, stories, and reflections, the processes of transformation suffered by the territories where hydrocarbons are extracted, including the adverse impacts and forms environmental damage, as well as the conflicts, that this activity generates.

To the extent that extractivism has become a pattern of development in the region, the ecological debt has grown. Because extractivism not only demands the natural goods that are extracted: it also demands land, water, and human labor, causes irreversible impacts and damage, and brings enormous social costs for communities in the present and in the future. The economic income from these commodities does not compensate for the damage and impacts caused by them. On the contrary, the traces of extractivism are increasingly evident: villages razed and buried in contaminated sludge from the mines, abandoned houses, rivers damaged with toxic substances, forests destroyed and transformed into soya plantations, swamps turned into oil sludge, animals and ecosystems contaminated, displaced populations. The Argentinean artist Dana Prieto calls these extractive projects *"regalos malditos"* (damned gifts) because they promise progress, structural improvements, environmental and social benefits, but turn out to be more of a curse.[3]

The Southern Peoples' Ecological Debt Creditors Alliance (Apsade)

As mentioned above, ecological debt is a political tool that has contributed to building agendas, debates, reflections, actions, and articulations. It can be said that an important milestone in the international movement to reclaim the ecological debt was the creation of the Alliance of the Peoples of the South Creditors of the Ecological Debt (hereafter referred to as the Alliance or Apsade). This articulation was created by a group of organizations, networks, and social movements from the South, among others: the International Network Jubilee South/Americas, the Latin American Network against Dams (Redlar), the Action Network on Pesticides and their Alternatives for Latin America (Rapal), Oilwatch, the World Rainforest Movement, the Latin American Observatory of Mining Conflicts in Latin America (Ocmal), the Bolivian Forum on Environment and Development (Fobomade), and Acción Ecológica (Apsade et al. 2004). In September 2000, these organizations met in Prague in the framework of the alternate activities of the annual meetings of the International Monetary Fund (IMF) and the World Bank (WB). This became a space of articulation to unify and strengthen the demands and positions of social and environmental organizations in the face of debates such as foreign debt and ecological debt. The Jubilee South Americas International Network is one of the main promoters of this initiative.

In 2002, the Earth Summit on Sustainable Development was convened by the United Nations in Johannesburg, South Africa. Ten years after the Rio de Janeiro Summit, economic globalization was advancing at a very fast pace, triggering conflicts in a large part of the world. The political debate around the ecological debt gained momentum, and environmental organizations considered it fair to demand "the repayment of the ecological debt, not only in economic terms, but through the full restoration of ecosystems so that communities can recover their sustainable ways of life" (FoEI 2002). In September 2005, at the II South–South Summit of the International Network Jubilee South/Americas, in Havana, the Alliance, together with Jubilee South, the Cuban Chapter of the Hemispheric Social Alliance, and other networks demanded "the cancellation of the illegitimate, inhuman and immoral foreign debts, to stop the looting of natural resources, the destruction and pollution of nature, the violation of the rights of peoples, to prevent the increase of foreign, social-ecological debts and to demand environmental restoration and compensation for the victims" (Apsade 2005). The Havana declaration denounced the way in which external debt has become an instrument of control that favors the deepening of the neoliberal model and destroys life on the planet, while the climate crisis rages against the poorest and most vulnerable populations.

In 2007, in the midst of the great expectations generated by the social mobilization around the debates of the National Constituent Assembly of Ecuador, the Latin American and

Caribbean Meeting of the Alliance was held in Quito (Vittor 2008). There was a great expectation that the proposals worked on during this nascent century could materialize in the social or the governmental sphere.[4] For years, Latin American organizations and social movements had demanded official audits of the foreign debt. In 2007, the government of Rafael Correa decided to promote such an audit applied to a period of 30 years, from 1976 to 2006, and by executive decree created the Commission for the Integral Audit of Public Credit (CAIC). This commission was made up of "representatives and professionals from national and international organisations, social movements and academics, experts on the issue of foreign debt" (Donoso 2009, 8). The final report was delivered at the end of 2008. Acción Ecológica and the Alliance analysed the central aspects that link the foreign debt with the ecological debt, finding a close link between the payment of the foreign debt and the exploitation of natural resources. They also highlighted the conditions imposed by multilateral banks on Ecuador to access new loans and "which paved the way for the control and exploitation of natural resources by transnational companies and, at the same time, guaranteed the payment of the foreign debt" (Donoso 2009, 7).

At the turn of the century, mobilization around the ever-more evident climate crisis became increasingly important, and the call for climate debt or carbon debt—another expression of ecological debt—emerged in force. At the Conference of the Parties to the United Nations Framework Convention on Climate Change (Cop-14), held in Copenhagen, environmentalists and social organizations arrived with high expectations of making progress in the negotiations. The governments of

> Bolivia, Ecuador, Paraguay, Venezuela, Honduras, Costa Rica, El Salvador, Nicaragua, Dominican Republic, Panama, Guatemala, Cuba, Belize, Dominica, St. Vincent and the Grenadines, Antigua and Barbuda, Sri Lanka and Malaysia, [proposed that] the Copenhagen Accords include the recognition and full reparation of the ecological debt owed by the countries of the North to the countries of the South due to climate change.
>
> (Apsade 2009)

The Climate Summit, however, ended without the expected results and from that year onwards, the Alliance lost momentum. Despite this, the concept remained relevant to the demands of environmental justice movements. In this sense, climate debt, a derivation of the concept, has been used by organizations and platforms fighting for climate justice.[5] Their claim demands that industrialized countries, the main contributors to climate change, assume their historical and current responsibility for the unjust and unequal appropriation of the environmental space of developing countries and demand that this must be returned, "decolonized" (Warlenius 2017). The movements consider the recognition of climate debt as a first step in seeking just and effective solutions to the climate crisis (Borrás 2017, 117).

To conclude

With the COVID-19 pandemic, the external debt crisis has become even more evident. The immense economic burden of servicing foreign debt has made it difficult for many governments to provide the concrete social responses that people are demanding to address the health and economic crisis. Today, ecological debt is re-emerging as a banner of Latin American environmentalism. On June 2, 2020, a group of social organizations, activists, and academics from the region launched the Ecosocial Pact of the South, backed by some 570 Latin American organizations and 3,100 individuals. The pact calls for the cancellation of the foreign debt as a step towards "historical reparation for the ecological and social debt contracted by the central countries since colonial times" (Pacto Social Sur 2020). In Argentina, the renowned academic Maristella Svampa and the activist lawyer Enrique Viale propose "remaking the world economic order, which would even promote a debt jubilee" (Svampa and Viale 2020). Ecological debt and external debt are closely related to the extractivist model that has been imposed in the region.

The planet and its territories are crying out for justice in the face of the aggressive extractive processes that destroy ecosystems and unbalance natural cycles. The ecological debt is the debt of an economic model and an unequal global order that has ravaged human and non-human life, but it should be noted that it is the industrialized countries and the richest societies that have caused the most damage. However, it is not a question of calculating the economic value of the ecological debt, but rather of demanding the cancellation of odious foreign debt. It is about transforming ways of life, relationships, and forms of production and consumption—and restoring the possibilities for the human species and other living beings to live in harmony. As activist Ivonne Yanez says in the prologue of the book *No more plunder and destruction: We, the Peoples of the South, are Creditors of the Ecological Debt*, the ecological debt "constitutes a real and concrete solution that is not only economic, but also legal, ethical, and above all political—to environmental impacts, and allows us to envision just societies, with sovereign peoples and nations" (Apsade et al. 2004, 10).

Notes

1 The sectors that grew the most during this period were fossil fuels, which grew 9.6 times, increasing its share of total domestic extraction from 12% to 33%; metallic minerals, which grew 8.9 times, with a share of 7.8%; and construction minerals, which grew 4.3 times, with a share of 25% of total domestic extraction.
2 Following the spill, the Brazilian government "issued a decree to prevent future disasters and ordered the dismantling within three years of mining dams built with the same technique" (Romero 2020).
3 See: https://www.facebook.com/watch/?v=2186690561360500.
4 At the governmental level, the *Caic* was set up to review Ecuador's indebtedness during the period from 1976 to 2006.
5 Climate debt is considered "the debt of emissions and adaptation that industrialized countries owe to poor countries for their excessive emissions—in the past and today—and, on the other hand, for their disproportionate contribution to the effects of climate change" (Borrás 2017, 101).

References

Acción Ecológica. 1999. "No more plunder, they owe us the ecological debt." *Revista Ecología Política*, 18. Barcelona: Editorial Icaria.

Alliance of Southern Peoples Crediting the Ecological Debt, Apsade, Acción Ecológica and Institute oth Third World Ecology Studies. 2004. *No more plunder and destruction. We, the Peoples of the South, are Creditors of the Ecological Debt*. Quito: Ediciones Abya-Yala.

Alliance of Southern Peoples Crediting the Ecological Debt, Apsade. 2005. *Position paper*. Cuba: Havana.

Alimonda, Héctor. 2011. *La Naturaleza colonizada. Ecología política y minería en América Latina*. Buenos Aires: CLACSO.

Arsel, Murat, Barbara Hogenboom, and Lorenzo Pellegrini. 2016. "The extractive imperative in Latin America." *The Extractive Industries and Society* 3: 880–887. Elsevier Ltd.

Bebbington, Anthony and Jeffrey Bury. 2013. "Political Ecologies of the Subsoil." In *Subterranean Struggles: New dynamics of mining, oil, and gas in Latin American*, edited by Anthony Bebbington, and Jeffrey Bury. Texas: University of Texas Press.

BBC. 2020. "Coronavirus: how it will affect Latin America's most indebted countries and where they can get money to fund the fight against the pandemic." *BBC*. https://www.bbc.com/mundo/noticias-52306376.

Borrás, Saturnino. 2017. "Movements for global climate justice: reframing the international climate change scenario." *International Relations*, 33. Grupo de Estudios de Relaciones Internacionales (GERI) - UAM. https://repositorio.uam.es/bitstream/handle/10486/676959/RI_33_6.pdf?sequence=1.

Committee for the cancellation of Third World Debt, Cadtma. 2005. *External Debt, who owes whom.* https://www.cadtm.org/Deuda-Externa-Quien-debe-a-quien.

Economic Commission for Latin American and the Caribean (ECLAC). 2019. *Social Panorama of Latin America.* Santiago. https://repositorio.cepal.org/bitstream/handle/11362/44969/5/S1901133_es.pdf.

De Castro, Fabio, Barbara Hogenboom, Michiel Baud. 2015. *Gobernanza ambiental en América Latina.* Clacso: Buenos Aires: Clacso.

Donoso, Aurora. 2000. *Deuda externa, mecanismo de dominación y saqueo.* Quito: Acción Ecológica.

Donoso, Aurora. 2009. *Ecological Debt. Impacts of foreign debt on communities and nature. Resultados de la auditoría integral ecuatoriana al crédito público.* Quito: Acción Ecológica and Alianza de los Pueblos del Sur Acreedores de Deuda Ecológica.

Foei. 2002. *Clashes with giant companies. 22 campaigns for biodiversity and community.* https://www.foei.org/wp-content/uploads/2015/05/clashes_corporate_giantsesp.pdf.

Göbel, Barbara, and Astrid Ulloa. 2014. *Extractivismo minero en Colombia y América Latina.* Universidad Nacional de Colombia, Bogotá: Universidad Nacional de Colombia (Sede Bogotá). Faculty of Human Sciences. Grupo Cultura y Ambiente / Berlin: Ibero-Amerikanisches Institut.

Hogenboom, Barbara. 2015. "Latin America's Transformative New Extraction and Local Conflicts." *European Review of Latin American and Caribbean Studies*, 99: 143–151.

Machado Araoz, Horacio and Leonardo Javier Rossi. 2017. "Extractivismo minero y fractura metabólica. El caso de Minera Alumbrera Ltd., a veinte años de explotación." *Revista IISE*, 10: 273–286.

Machado Araoz, Horacio 2018. *Potosí, el origen. Genealogía de la minería contemporánea.* Quito: Editorial Abya Yala.

Martínez-Alier, Joan. 2003. "Prefacio." In *Deuda ecológica: ¿quién debe a quién?* edited by Daniela Russi, Ignasi Puig Ventosa, Jesús Ramos Martín, Miquel Ortega Cerdán, and Paula Ungar, 7–9. Barcelona: Icaría Editorial.

Martínez-Alier, Joan. 2007. "Cuantificación de la deuda ecológica." *Revista Reflexión*, Volume 10, No. 3.

Mendoça, Heloísa. 2020. "Mourning and economic euphoria coexist in Brumadinho." *El País*, January 25, 2020. https://elpais.com/internacional/2020/01/25/america/1579983322_525610.html.

Muñiz, Antonio. 2018. "Reprimarización de la economía y dependencia: un modelo perverso." *Revista Zoom*, December 18, 2018. https://revistazoom.com.ar/reprimarizacion-de-la-economia-y-dependencia-un-modelo-perverso/.

Pacto Ecosocial del Sur. 2020. https://pactoecosocialdelsur.com/.

Pardo Becerra, Luis Álvaro. 2014. "Proposals to recover the governance of the Colombian mining sector." In *La Minería en Colombia*, edited by Luis Jorge Garay Salamanca. Bogotá: Contraloría General de la República.

Pardo Becerra, Luis Álvaro. 2018. *Los quince mitos de la gran minería en Colombia.* Bogotá: Heinrich Boll Foundation – Bogotá Office.

Pérez-Rincón, Mario Alejandro. 2007. *Comercio Internacional y Medio Ambiente en Colombia. Mirada desde la Economía ecológica.* Cali: Universidad del Valle.

Pérez-Rincón, Mario Alejandro. 2016a. "Metabolismo social y conflictos ambientales en países andinos y centroamericanos." Presentation at the Latin American Congress on Environmental Conflicts. San José, Costa Rica.

Pérez-Rincón, Mario Alejandro. 2016b. "Characterizing environmental injustices in Colombia: A study of 115 cases of socio-environmental conflicts." Working paper, Universidad del Valle, CINARA Institute, Cali, Colombia.

Roa-Avendaño, Tatiana, and Luisa M. Navas. 2001. *A demand from the South: recognising the ecological debt.* Bogotá: Editorial Bochica/Censat Agua Viva.

Roa-Avendaño, Tatiana, and Danielo Urrea. 2015. "La cuestión ambiental: asunto clave en el proceso de paz." In *Negociaciones gobierno-ELN y sin embargo, se mueve*, edited by Victor De Currea Lugo. Bogotá: Editorial Antropos.

Roa-Avendaño, Tatiana, and Luisa Navas. 2014. *Extractivismo, conflictos y resistencias*. Bogotá: Difundir Ltda.

Romero, Mar. 2020. "Brumadinho under the mud: one year after the tragedy in Brazil." *France 24*, January 25, 2020. https://www.france24.com/es/20200125-brumadinho-bajo-el-lodo-se-cumple-una%C3%B1o-de-la-tragedia-en-brasil.

Svampa, Maristella. 2013. "*Commodity* Consensus and languages of valuation in Latin America," *Nueva Sociedad*, 244: 30–46.

Svampa, Maristella and Enrique Viale. 2020. "Our Green New Deal." *Revista Anfibia*, April 29, 2020. http://revistaanfibia.com/ensayo/green-new-deal/.

Toro-Peréz, Catalina, Julio Fierro-Morales, Sergio Coronado, and Tatiana Roa-Avendaño. 2012. *Mining, territory and conflict in Colombia*. Bogotá: Universidad Nacional de Colombia, Censat Agua Viva.

Vittor, Luis. 2008. "Encuentro latinoamericano y caribeño de la alianza de pueblos acreedores de deudas históricas, sociales-ecológicas." *Ecología Política*, January 1, 2008. https://www.ecologiapolitica.info/?p=5681.

Wartlenius, Rikard. 2018. "Decolonizing the Atmosphere: The Climate Justice Movement on Climate Debt." *Sage Journal*, Vol 27, Issue 2. https://doi.org/10.1177/1070496517744593.

World Bank. 2020. *International Debt Statistics*. http://datatopics.worldbank.org/debt/ids/regionanalytical/LAC.

Zuberman, Federico. 2019. "Suelo Virtual y deuda ecológica. Un cálculo para la expansión de la soja en Argentina." *SaberEs*, 11, No. 1: 81–95.

11 Trajectories of adaptation to climate change in Latin American cities

Climate justice blind spots

María Gabriela Merlinsky and Melina Ayelén Tobías

Introduction

Much has been discussed in recent decades about climate change. The relevance of the issue has permeated political debates at the international and local levels, giving rise to the formation of a global climate change agenda governed by the Kyoto Protocol and the Paris Agreement,[1] both drawn up within the framework of the United Nations Framework Convention on Climate Change (UNFCCC). Central to this debate is the notion of the Anthropocene, the period whose beginning is placed toward the end of the 18th century with the development of the Industrial Revolution (Crutzen 2002).

Underlying the idea of the Anthropocene is an imaginary of human nature as destructive and conquering and of technology as the only possible way out. However, this is only one of the forms that conflicting ideas about nature and what is natural has taken throughout history. From a critical perspective, the main cause of climate change is not the human species as a whole, but the capitalist production model that has prevailed in much of the world in recent centuries. For that reason, the Capitalocene as a label more clearly identifies the great challenge of our time (Malm 2016; Moore 2016).

The effects of climate change are not uniform across the planet, nor are the capacities of different countries and populations to cope with it. In this sense, least developed countries and most vulnerable social groups are most affected by the local effects of climate change. Therefore, we need to build political imaginaries from a climate justice perspective, something that implies a new alliance of peoples and species to revitalize a planet devastated by patriarchal capitalism (Briones, Lanata, and Monjeau 2019; Ulloa 2017).

This is a major challenge for Latin American countries, which must prepare themselves to reduce fossil fuel consumption, minimize damage to their infrastructure, promote gradual adaptation to climate change and avoid irreversible losses of ecosystems and species, while facing serious balance of payments crises and economic dependence rooted in old patterns of colonial domination. Thus, they have few resources to align their policies with a global climate architecture. If we consider that the countries of the region have failed to implement lasting redistributive policies aimed at reducing historical inequalities between regions, social classes and genders, this represents a major dilemma in terms of global climate justice.

Adaptation to climate change in cities is the great challenge of the global South (Dodman et al. 2019) and Latin America does not escape this challenge, as 79.5% of the population lives in urban centers. This is a major crossroads for public policies due to the high density of population and infrastructure, such as high paving and building, which results in increased flooding and heat islands, and the dependence on rural areas for food supply, which makes them vulnerable to possible climate impacts in distant regions (Schaller et al. 2016).

DOI: 10.4324/9780429344428-14

Latin America is a region in which the inhabitants living in areas of high marginalization represent almost a third of the population (27% on average), although this is variable as it is officially estimated at 19.6% for the case of Mexico; 36.6% for Brazil; 33.1% for Argentina; and 68% for Peru, to give some examples (Davis 2006, cited in Delgado Ramos et al. 2015, 37). Inequalities are expressed territorially in terms of both income and access to services. This implies a type of segregation associated with urban informality, where it is concentrated the population most vulnerable to the effects of climate change (Gonzalez de la Rocha 2007; Satterthwaite et al. 2018).

In this chapter we develop a critical analysis of climate justice challenges in Latin American cities. Our central argument is that the dominant global discourse of climate change adaptation invisibilizes socio-environmental inequalities at regional and local scales. Therefore, it is necessary to pay attention to forms of popular urbanization that condemn the poorest and the social classes without resources to live in environmentally disadvantaged sites, which imply higher levels of exposure to climate change. To bring this issue to the center of the political and academic debate, it is necessary to position ourselves in a climate justice approach which serves as an articulating framework between activist and academic knowledge in a perspective that integrates both the biophysical causes and the social and political aspects of vulnerability.

From this perspective we analyze the adaptation trajectories to climate change in Latin American cities, considering three main aspects: a) the sociopolitical component of adaptation practices in terms of the translation of measures in time and space; b) the consideration of socio-spatial inequality as an invisible issue in climate change adaptation policies; and c) the role played by institutional inertia in promoting and sustaining certain urban development traces, given their strong influence on current and future adaptation options. Finally, we will return to an example related to adaptation measures implemented in the Buenos Aires metropolitan region.

Capitalist urbanization, socio-environmental inequality and climate justice

In the last century, the processes of capitalist urbanization have led to a profound reconfiguration of the urban question. The integration of Latin American metropolises into global economic regions is a markedly unequal process leveraged by a flexible urbanism, whose main objective is to territorialize large-scale investments in the built environment, something that plays the role of channeling capital flows within the transnational space. This has very important consequences in terms of land use and in the generation and distribution of rents arising from urbanization processes (Brenner 2013; De Mattos 2017).

In Latin American cities, the geographies of urbanization have created new urban densities, intermediate cities, and a profound reconfiguration of natural spaces. A peculiarity of this process is the growth of the informal city, with settlements that concentrate high rates of poverty and population growth, flood risks, degraded soils, an intensely polluted environment, and infrastructure deficits (Merlinsky 2013; Quimbayo Ruiz 2020).

Along with these transformations, important mobilizations have emerged that are based on the right to the city, in terms of its consideration as a "space of the commons" and as a territorial basis for collective action (Calderón Cockburn 2019; Pradilla Cobos 2016). If the social production of habitat is a vector of urban struggle, climate change adaptation strategies are embedded in human conflicts. Proof of this is the importance that different notions of environmental justice "from below" have acquired in the last two decades, a concept that denounces a situation of injustice evidenced in the unequal geographical distribution of urban income and environmental hazards (Acselrad et al. 2009; Merlinsky 2017; Porto-Gonçalves 2011). What is characteristic of these manifestations is that the notion of environmental justice cuts across the defense of living conditions and health to incorporate the discussion on access to different resources (water, land, infrastructure, among others) in the framework of economic disputes, but also in terms of cultural and identity values (Merlinsky 2021). It has been the circulation of ideas, alliances, and environmental justice movements in the global South that have allowed the re-scaling of this notion in the key of climate justice (Martinez-Alier et al. 2016).[2]

The notion of climate justice re-politicizes the debate in two fundamental ways. On the one hand, by pointing out that those most affected by the effects of climate change are the most vulnerable and the least responsible for greenhouse gas emissions, and at the same time those who have the least influence on decision-making (Buendía and Ortega 2018). On the other, by highlighting the role played by local governments in the reproduction of vulnerabilities in contexts of urban informality. In this sense, the analysis of the literature on cities in the global South shows results ranging from convenient absence to unwitting failure (Anguelovski et al. 2016). Convenient absence refers to political-economic trajectories in which local governments are unable to address adaptation measures in a context of growing socio-economic and spatial inequalities, amid exogenous pressures (Gore 2015; Satterthwaite et al. 2018), lack of support at the national level, and global market pressure (Warner and Kuzdas 2017). Unintended failure draws attention to those adaptation programs that fail to materialize or where fragmented and inadequate local government interventions may create or exacerbate vulnerabilities (Paiva Henrique and Tschakert 2020, 10).

If at the global scale climate change has been presented as a "neutral and apolitical universal imaginary projected by climate sciences and disconnected from local responses to climate" (Jasanoff 2010, 235), it is important to territorialize this discussion and consider different scales in a framing that takes into account the "emerging geographies" of climate justice (Fisher 2012). As O'Brien and Leichenko (2000) summarize, the impacts of climate change—especially in the global South—are intertwined with the effects of economic globalization, accounting for a process of "double exposure" (to climate and neoliberal capitalism) to which certain sectors, regions, ecosystems, and social groups are mostly exposed, thus configuring a new scenario of winners and losers.

Towards a denaturalization of approaches to climate change adaptation

From a decolonial perspective, scholars have criticized the dominant vision of adaptation, that is imposed from the top down, associated with an end state to be achieved, and that ignores the organizational processes of communities. These works question an immanent definition of climate change, as a key element of a neocolonial global order that installs it as a single, future-oriented problem and where "the deaths from disasters of third world inhabitants are more acceptable, more justifiable, than the potential future deaths of people in the first world who have not yet been born". Thus, "climate change is entangled with the contemporary dynamics of both capitalism and development and cannot be considered in isolation from these dynamics" (Cupples 2012, 13).

In their work on disasters and climate change in the Caribbean, Gahman and Thongs (2020) have shown that the imprint of colonial underdevelopment, racial capitalism, and neoliberal extraction has eroded disaster response practices throughout the region. This is due to a structural pattern: adaptation to climate change must contend with austerity measures and/or economic dependence on private capital that offers credit and financial assistance conditional and contingent on prospective profits. This leads to an important observation regarding the relationship between development justice and climate justice.

In this chapter, we consider that adaptation must simultaneously address the root causes of vulnerability in its relation to capitalist urbanization processes within a framework of social transformation guided by a climate justice perspective (O'Brien 2017; Pelling 2010; Ziervogel et al. 2016). To achieve this goal, it is necessary to analyze the urbanization trajectories in which climate change adaptation policies are inserted and to pay attention to the power relations around the processes of unequal appropriation of nature. It involves a dispute with a teleological vision that considers adaptation as an inevitable process, depoliticized and exempt from social conflict.

In this sense, adaptation measures, in addition to offering tools to cope with, anticipate, and recover from impacts, should promote transformative changes in lifestyles, combat poverty, strengthen collective actions, as well as protect people's resources (Forsyth 2010). Adaptation strategies should simultaneously help people to achieve their own goals rather than trying to impose objectives that are not part of their culture or worldview (Ensor 2009). As indigenous climate studies have recently shown, the knowledge carried by communities encompasses information about the nature of ecological change and does so from stories, lessons, and narratives that show a high level of organization to adapt to environmental change, especially in periods prior to conquest and colonization (Chief et al. 2014; Pulido 2018; Whyte 2017).

Toward a re-politicization of the concept: the analysis of adaptation trajectories

Recently, several authors have developed an analysis model that allows evaluating public policy interventions in terms of adaptation trajectories. It is a decision-making tool to identify, sequence, and implement adaptive actions in response to dynamic climatic and other conditions (Lawrence and Haasnoot 2017). It starts from a prospective analysis to identify implementation pathways taking into account ongoing trends, turning points, and points of no return (Haasnoot et al. 2013; Reeder and Ranger 2011). A dynamic approach is thus sought to collectively visualize deliberate adaptation options and re-evaluate them as new knowledge becomes available. The study of these trajectories from a critical perspective has shown that the tool's potential to politically guide stakeholder agreements remains a challenge.

First, by recognizing the sociopolitical component of adaptation practices, the interplay between power, knowledge, and agency are decisive in terms of the translation of measures in time and space. This leads to certain adaptation measures being prioritized over others without an open process of deliberation (Fazey et al. 2018; Grandin et al. 2018; Pelling et al. 2016). Thus, the framings of climate change adaptation are dominated by an expert approach coming from transnational and local NGOs, networks of scientists, international cooperation programs, and state agencies. For example, it has been shown that climate action plans at the municipal scale in Mexico have been elaborated by private consultants or by municipal government personnel trained under the ICLEI-British Embassy program (Delgado Ramos et al. 2015, 117). In these epistemic communities, proposals for solutions oriented to the middle and upper classes of urban centers (clean energy, insulating constructions, sustainable mobility) prevail, to the detriment of addressing the problems of access to land and housing in informal settlements. Thus, there are framing problems that imply the invisibility of those aspects that mark the huge gap in infrastructure coverage (such as water and sanitation) at the metropolitan scale.

Second, not all adaptation options are available to all individuals, communities, or interest groups even if theoretically there are many possible alternatives. Dominant trajectories are often delineated according to the adaptive capacity and opportunity of privileged groups and entities –that occupy positions of authority or have the capacity to influence (Ajibade 2017; Archer and Dodman 2015; Pelling 2010; Swyngedouw 2013). It is important to highlight that women, indigenous groups, inhabitants of informal settlements, and, in general, younger

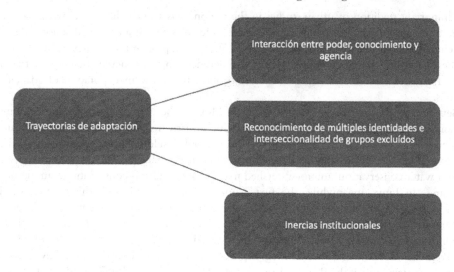

Figure 11.1 Analysis model of adaptation trajectories.
Source: Own elaboration.

generations tend to be excluded from these decision-making processes. This leads to highlight, once again, the importance of considering an approach that takes into account climate justice not only in distributive aspects but also in terms of recognition and procedural ones. Gender-sensitive analyses have shown that women are and will be more affected by climate risks either because they have greater problems of access to land, suffer discrimination within households and communities, or because they are not considered in early warnings (Alston and Whittenbury 2013; Sultana 2013). For that very reason, adaptation plans should not erase or silence their agency in responding to climate change (Arora-Jonsson 2011; Tschakert and Machado 2012; Whyte 2017). In this sense, it is important to recognize the multiple identities and intersectionality in groups that, although disadvantaged in economic terms, exercise leadership positions, work in networks, and are key actors in the production of knowledge (Nightingale 2011).

Third, attention needs to be paid to the institutional inertias that drive and sustain particular urban development trajectories, as these strongly impact current and future adaptation options. Many past decisions can facilitate or inhibit pathways and thus block particular adaptation strategies, something that leads to forms of implementation that sustain the status quo (Fazey et al. 2016; O'Brien 2016; Wise et al. 2014). For example, strategies linked to flood risk management in cities in Argentina tend to prioritize physical works such as dams, coastal defenses, or canalization works at the expense of the environmental management of the territory.

Climate change adaptation programs in Latin American cities

Climate change has traditionally been framed like a national rather than a local problem. Some of this began to change in the global cities of the South with the turn of the millennium, when a transnational agenda emerged, driven by local governments. This happened due to many factors, but international relations undoubtedly played a central role. For example, the emergence of a climate agenda in cities such as Sao Paulo or Mexico was driven by international organizations such as ICLEI—Local Governments for Sustainability (Mauad and Betsill 2019, 5).

The literature coincides in pointing out that, in Latin American cities, climate change adaptation programs have a low level of institutionalization. Sanchez Rodriguez notes that, despite the important tradition of urban studies in the region, it is striking how little work has been done to design interdisciplinary perspectives useful for planning strategies and actions to combat climate change (Sánchez Rodriguez 2013, 15). At best, programs to address climate change in these large cities have been initiated only two decades ago, in an incremental process that has gone from the exclusive consideration of mitigation to a slow incorporation of adaptation actions.

Several cases illustrate these characteristics. In Mexico, the year 2000 marked the beginning of some local responses to climate change with initiatives mainly focused on mitigation. Subsequently, and within the framework of the general development program of the Federal District (2007–2012), the Mexico City Climate Action Program (2008) was promoted, focusing on water conservation, micro-watershed management and soil conservation, among other measures. In Lima, meanwhile, the first actions began in 2011 with the creation of the Metropolitan Environmental Commission, which brings together representatives of regional and municipal governments, as well as members of the private sector, academics, and NGOs. In Santiago de Chile, the adaptation plan was presented in 2012. The main actors involved were the Regional Government and the Regional Secretariat of the Ministry of the Environment, with the participation of the Regional Secretariats of the Ministry of Public Works, the Ministry of Housing and Urban Development, and the Ministry of Health, as well as the municipalities of the Metropolitan Region of Santiago.

The works that have analyzed these programs show that at the regional/local level there is scarce availability of scientific information on climate change, or alternatively that it is characterized by a high level of uncertainty. This situation is due to a general lack of knowledge about climate simulation models and downscaling of global models (Schaller et al. 2016, 267). On the other hand, there are important interjurisdictional and sectoral articulation constraints (Back 2016; Di Giulio et al. 2017), this is due to the fact that adaptation planning is considered primarily as an environmental issue and,—given the low hierarchical status of this portfolio in LA—consequently, receives little political support (Barbi 2015; Schaller et al. 2016).

It is noted that institutional continuity is low due to the replacement of public administration staff during changes of government. Under such circumstances, it becomes difficult to ensure long-term planning, as staff must be constantly trained on the job. In this context, it is external projects and/or NGOs that often provide the information that should normally come from public institutions (Quiroz Benitez 2013; Schaller et al. 2016). All the analyzed works show the existence of financial problems to guarantee the continuity of the actions; this is mainly due to a macroeconomic orientation to combat the fiscal deficit or prioritize social policy actions and the low incidence of these items in the national and local budget (Betsill 2001). In the case of Mexico, Delgado Ramos et al. (2015, 86) have pointed out that, although it is true that mechanisms for financing at the urban scale are beginning to grow, a notable aspect is that the bulk is still channeled to mitigation inasmuch as these types of measures usually allow generating greater profits in the short term through the purchase of technology or services for the adoption of low-carbon infrastructure.

In Brazil, the propositional framework of the National Plan for Adaptation to Climate Change includes practically no reference to climate justice, ignoring the unequal effects of the measures among social groups, classes, and territories (Torres et al. 2020, 4). In this sense, a discursive and superficial consideration of the problem prevails over a disaggregated analysis in terms of socio-spatial inequities.

These considerations can be extended to the analysis of the reports presented by different multilateral organizations, in which the recommendations tangentially touch on the problem of inequality because they only focus on reducing vulnerability and—at best—on sustaining social safety nets (Development Bank of Latin America CAF 2014; Margulis 2016; World Bank

2011). In a recent report for the Economic Commission for Latin America (ECLAC), in the framework of the activities of the EUROCLIMA program—financed by the European Union, Margulis points out that "… any actions to reduce the vulnerability of poor people also represent development actions whether in terms of physical, social or economic capital creation. Adaptation strategies are strictly development strategies, and vice versa" (Margulis 2016, 56). It highlights a vision that tends to naturalize development, assimilating it to progress or overcoming poverty, when in fact in Latin America there is a broad discussion on "alternatives to development," which questions the export orientation of economies, the problem of external and ecological debt, and other worldviews linked to "good living or living well". These documents also lack references to climate justice and procedural aspects that would allow the voice of those most affected to be taken into account in the implementation of adaptation plans. As long as these plans do not take into account the existing initiatives and knowledge in different spheres of civil society, grassroots organizations, and small municipalities, it will not be possible to ensure an approach that takes into account climate justice.

The challenges of adaptation policies in the Metropolitan Region of Buenos Aires (Argentina)

The Regional of Metropolitan Buenos Aires (RMBA) is a territory with a population of 15 million people, composed by the Autonomous City of Buenos Aires and more than 40 municipalities or districts of the Province of Buenos Aires. In this region, the effects of climate change have been expressed since the 1960s and 1970s in an increase in average rainfall and a trend toward an increase in extreme precipitation events (Rusticucci 2007; SAyDS 2015). These trends in climate have generated visible consequences such as floods, heatwaves, changes in heights considered safe for construction, coastal erosion, and forest and grassland fires—that allow us to say that climate dynamics have become one more forcing factor in the production of risk scenarios (Barros and Camillioni 2016). However, these climate risk scenarios are mounted on territorial, institutional, and social scenarios that deepen inequality.

In Buenos Aires, the first government initiatives in relation to climate change date back to 2008 when the first steps were taken to carry out greenhouse gas inventories. This resulted in a climate change law passed in 2011 (Law of Adaptation and Mitigation to Climate Change of the City of Buenos Aires 3871/11) and the development of a climate change action plan 2020, promoted within the framework of the C40 network (an agreement of large cities around the world aimed at reducing greenhouse gas emissions). It is important to note that in the case of Buenos Aires the measures do not have a metropolitan scope. The main reason is due to the historical political tensions between the jurisdictions that make up the metropolitan region of Buenos Aires (National, Provincial and Autonomous City governments), something that prevents the development and implementation of a regional plan for adaptation to climate change that understands the complexity and interrelationship of the metropolitan territory (Merlinsky and Tobías 2016).

Since Argentina is a federal country, the national government is limited in its ability to provide for actions related to climate change (or any other topic related to the environment) to be implemented by local governments. This situation ends up causing the latter to be the ones to promote (or not) measures to address climate change, and this in turn depends on the economic and technical resources available to each jurisdiction. While the Autonomous City of Buenos Aires has a law sanctioned by the legislature, a Climate Change Action Plan, the Province of Buenos Aires does not have specific provincial regulations on the subject, despite having a provincial Environmental Agency for Sustainable Development that has lines and programs associated with climate change.

The urbanization pattern that characterizes the metropolitan region poses great challenges for the implementation of effective adaptation measures in the region. This is because the RMBA has growth in size over time in a heterogeneous manner, leading to an increase in socio-territorial and environmental inequalities. While the northern axis consolidated as the most favored region in terms of economic resources, services and urban infrastructure, the southern zone—historically associated with the industrial development of the country—shows high levels of poverty, lack of public services, and environmental degradation.

The lack of urban land access policies in the region has led to the settlement of popular sectors in peripheral areas, lacking services and environmentally degraded with high sanitary risks. These areas coincide, to a large extent, with the location of the main metropolitan basins that cross the region: the Matanza Riachuelo basin, the Reconquista River basin, and the Luján basin, the first being the most densely populated with nearly eight million inhabitants. In these territories, frequent flooding implies pollution, as the waters of the metropolitan rivers show high levels of environmental degradation, with the presence of heavy metals and untreated organic effluents.

Recently, in the metropolitan region of Buenos Aires an ambitious program aimed at the environmental recomposition of the Matanza-Riachuelo basin established priorities to rehabilitate the riverside landscape and this implied the displacement of the low-income population living on the riverbanks, something that shows a (historical) way of dealing with the urban social issue in terms of the segregation of popular housing spaces toward peripheral territories of the metropolis (Merlinsky 2018; Merlinsky, Scharager and Tobías 2017).

Floods affect the vulnerable population to a greater extent and the responses have not followed a policy linked to risk management on a metropolitan scale. While floods in the Autonomous City of Buenos Aires receive attention and there are even subsidy programs for those affected, in the metropolitan municipalities the response actions are of a welfare nature and once the disaster has occurred.

Figure 11.2 Floods in the southern area of the RMBA (Matanza Riachuelo Basin).

Source: Hugo Partucci (UBANEX Project 2014).

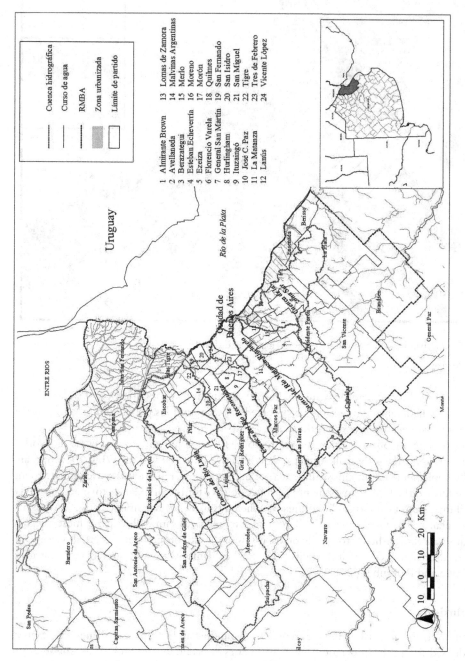

Figure 11.3 Rivers in the Buenos Aires metropolitan watersheds.

Source: Fernández and Garay (2013).

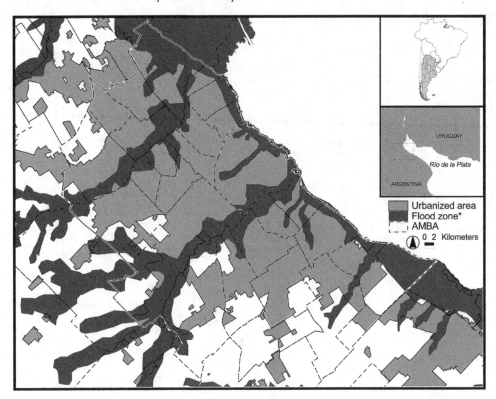

Figure 11.4 Urban growth and flood-prone territories of the AMBA.

Source: Merlinsky and Tobías (2021).

In terms of risk flood management, a culture of canalization and rectification of water courses sponsored by hydraulic engineering has prevailed. The continuous population growth and the absence of maintenance works on the canals led to the obstruction and deterioration of the system (Viand and González 2012) and the response is the projection of mega hydraulic works, whose costs and large scale make difficult their effective implementation. While the works are partially implemented in the most affluent areas of the city, this represents a cycle of convenient absence (due to the lack of investment in habitat policies for the popular sectors) and involuntary failure (given the impossibility of meeting the costs of large-scale infrastructure works on a metropolitan scale). Given the lack of resources to recover from a flood, the most affected popular sectors are pushed into a cycle of dispossession (loss of property and assets, displacement to more disadvantaged sites in the city) that further damages their future adaptive capacity (Hallegatte and Rozenberg 2017).

Conclusions

Adaptation to climate change in cities represents a multiple and complex challenge that cannot be addressed from a technocratic perspective. Cities concentrate the bulk of greenhouse gas emissions, have high rates of energy and material consumption, and are the place of residence of the world's population most vulnerable to the impacts of climate change. In this chapter we wanted to contribute to a discussion from the perspective of climate justice, something that implies a double dilemma. On the one hand, the people and regions most vulnerable to the impacts of climate change are not the main generators of carbon emissions; on the other hand, adaptation policies depend heavily on the income level of countries and the resources available to local governments.

As we have shown, the existence of urban adaptation policies is a necessary but not sufficient condition to face these dilemmas and it is necessary to ask: adaptation for whom, decided by whom, and at what cost? The answer to these questions requires identifying "blind spots" linked to power asymmetries in the management, appropriation and distribution of natural resources, as well as in terms of the reproduction of persistent urban inequalities.

The analysis of adaptation trajectories in Latin American cities shows a lack of prioritization of climate change policies in the urban policy agenda. This has repercussions on the fragmentation of measures, due to funding limitations, institutional discontinuity, and interjurisdictional coordination problems. In terms of a research agenda, this shows that it is essential to incorporate the temporal and spatial dimension in the analysis. In the same direction, a key point in this article has been to show that, in Latin American cities, adaptation programs fail to overcome implementation barriers in contexts of high urban informality.

In reference to the case of Buenos Aires, we have seen how the trajectory of flood risk management policies oscillates between convenient omission and involuntary failure. Risk prevention infrastructures based on a hydraulic approach only reach the population with greater resources, which is concentrated in central areas, and relegate the most vulnerable population to the impacts of a recurrent cycle of disasters. All this implies a greater dispossession of the popular sectors, who see their response capacity weakened, thus reproducing intergenerational inequalities.

Disputes over access to urban land represent one of the most important aspects of the political struggle for the right to the city in Latin American cities. For this reason, adaptation plans should not erase or silence the agency capacity of social groups living in informal areas. It is not enough to incorporate a climate justice approach to adaptation; it will also be necessary to alter the balance of power and consider aspects of distributive justice, the dilemmas of participation that are problems of recognition and political representation.

Notes

1 The Kyoto Protocol was developed under the United Nations Framework Convention on Climate Change in 1997 to address the causes and problems associated with climate change. The signatory countries undertook to reduce the level of greenhouse gas emissions into the atmosphere. It came into force in 2005 and was extended until 2020, although throughout the period countries such as the United States and Australia (both with high greenhouse gas emission rates) refused to participate, which strongly weakened its effectiveness. For its part, the Paris Agreement was negotiated at the XXI Climate Change Conference (COP21) in 2015 to enter into force at the time the Kyoto Protocol ends. Its objective is to "strengthen the global response to the threat of climate change, in the context of sustainable development and efforts to eradicate poverty," although its ability to achieve the goal was limited by the withdrawal of the United States in 2017.

2 The term began to gain prominence in 2000, at the first Climate Justice Summit held in The Hague, in parallel to the Sixth Conference of the Parties (COP 6). Two years later, in 2002, the Second Earth Summit (also called Rio + 10) was held in Johannesburg, where international environmental groups established the Bali Climate Justice Principles.

References

Ajibade, Idowu. 2017. "Can a future city enhance urban resilience and sustainability? A political ecology analysis of Eko Atlantic city, Nigeria." *International Journal of Disaster Risk Reduction* 26: 85–92.

Alston, Margaret, and Kerri Whittenbury. 2013. *Research, Action and Policy: Addressing the Gendered Impacts of Climate Change*. Dordrecht, Netherlands: Springer.

Anguelovski, Isabelle, Linda Shi, Eric Chu, et al. 2016. "Equity impacts of urban land use planning for climate adaptation: critical perspectives from the global North and South." *Journal of Planning Education and Research* 36(3): 333–348.

Archer, Diane, and David Dodman. 2015. "Making capacity building critical: power and justice in building urban climate resilience in Indonesia and Thailand." *Urban Climate* 14: 68–78.

Arora-Jonsson, Seema. 2011. "Virtue and vulnerability: discourses on women, gender and climate change." *Global Environmental Change* 212: 744–751.

Ascelrad, Henri, Cecilia Mello, and Gustavo Bezerra. 2009. *O que é Justiça Ambiental. Rio de Janeiro.* Brazil: Garamond.

Back, Adalberto Gregório. 2016. "Urbanização, planejamento e mudanças climáticas: desafios da capital paulista e da região metropolitana de São Paulo." PhD. thesis, Universidade Federal de São Carlos.

Barbi, Fabiana. 2015. *Mudanças climáticas e respostas políticas nas cidades: os riscos na Baixada Santista.* SP: Editora da Unicamp, Campinas.

Barros, Vicente, and Inés Camillioni. 2016. *Argentina and climate change. From physics to politics.* Buenos Aires: Eudeba.

Betsill, Michele M. 2001. "Mitigating climate change in US cities: opportunities and obstacles." *Local Environment* 6(4): 393–406.

Brenner, Neil. 2013. "Theses on urbanization." *Public Culture* 25(1): 85–114.

Briones, Claudia, José Luis Lanata, and Adrián Monjeau. 2019. "The future of the Anthropocene." *Utopía y Praxis Latinoamericana* 24(84): 19–31.

Buendía, Mercedes P., and Jordi Ortega. 2018. "Environmental justice and climate justice: the slow road, but with no return, to just sustainable development." *Barataria. Revista Castellano-Manchega De Ciencias Sociales* (24): 83–100.

Calderón Cockburn, Julio. 2019. "The right to the city in Latin America and approaches to the informal city." In *Derecho a la ciudad: una evocación de las transformaciones urbana en América Latina*, edited by Fernando Carrión Mena, and Manual Dammert-Guardia. Lima: CLACSO, Flacso – Ecuador, IFEA.

Chief, Karletta, John J. Daigle, Kathy Lynn, and Kyle Powis Whyte. 2014. "Indigenous experiences in the U.S. with climate change and environmental stewardship in the Anthropocene." In *Forest conservation and management in the Anthropocene: Conference proceedings*, edited by Alaric V. Sample, and Patrick R. Bixler. USDA, Forest Service, Rocky Mountain Research Station, RMRS- P-71: 161–176. Rocky Mountain Research Station Publishing services. Washington, DC.

Crutzen, Paul J. 2002. "Geology of mankind." *Nature* 415: 23–23.

Cupples, Julie. 2012. "Wild Globalization: The Biopolitics of Climate Change and Global Capitalism on Nicaragua's Mosquito Coast." *Antipode* 44: 10–30.

Delgado Ramos, Gian Carlo, Ana De Luca Zuria, and Verónica Vázquez Zentella. 2015. *Adaptación y mitigación urbana del cambio climático en México.* México City: CLACSO-UNAM.

Development Bank of Latin America, CAF. 2014. *Climate change adaptation program lima*, Lima Peru.

De Mattos, Carlos Antonio. 2017 "Financiarización, valorización inmobiliaria del capital y mercantilización de la metamorfosis urbana." *Revista Sociologías* 42: 24–52.

Di Giulio, Gabriela et al. 2017. "Relatório Técnico-Científico Workshop Mudanças climáticas e o processo decisório na megacidade de São Paulo: análise das discussões promovidas." *Sustentabilidade em Debate* 8(2): 75–87.

Dodman, David, Diane Archer, and David Satterthwaite. 2019. "Editorial: Responding to climate change in contexts of urban poverty and informality." *Environment and Urbanization* 31(1): 3–12.

Ensor, Jonathan. 2009. "Biodiverse Agriculture for a Changing Climate", *Practical Action*. The Schumacher Centre for Technology and Development, Bourton on Dunsmore, UK. 1–27.

Fazey, Ioan, Peter Moug, Simon Allen, Kate Beckmann, David Blackwood, Mike Bonaventura, Kathryn Burnett, Mike Danson, Ruth Falconer, Alexandre S. Gagnon, et al. 2018. "Transformation in a changing climate: a research agenda." *Climate and Development* 10(3): 197–217.

Fazey, Ioan, Russell M. Wise, Christopher Lyon, Claudia Câmpeanu, Peter Moug, and Tammy E. Davies. 2016. "Past and future adaptation pathways." *Climate and Development* 8(1): 26–44.

Fernández, Leonardo, and Diego Garay. 2013. *Biodiversidad Urbana: apuntes para un sistema de áreas verdes en la región metropolitana de Buenos Aires.* Buenos Aires: Prometeo - UNGS.

Fisher, Susannah. 2012. "The emerging geographies of climate justice." *Centre for Climate Change Economis and Policy*, Working Paper No. 94.

Forsyth, Tim. 2010. "Climate change: is Southeast Asia up to the challenge? Forest and climate change policy: what are the costs of inaction?" IDEAS reports - special reports, Kitchen, Nicholas (ed.) (SR004). LSE IDEAS, London School of Economics and Political Science, London, UK.

Gahman, Levi, and Gabrielle Thongs. 2020. "Development justice, a proposal: reckoning with disaster, catastrophe, and climate change in the caribbean." *Transactions of the Institute of British Geographers* 45(4): 763–778.

Gonzalez de la Rocha, Mercedes. 2007. "The construction of the myth of survival." *Development and Change* 38(1): 45–66.

Gore, Christopher. 2015. "Climate Change Adaptation and African Cities. Understanding the Impact of Government and Governance on Future Action." In *The Urban Climate Challenge: Rethinking the Role of Cities in the Global Climate Regime*, edited by Craig Johnson, Noah Toly, and Heike Schroeder, 205–226. Abingdon: Routledge.

Grandin, Jakob, Håvard Haarstad, Kristin Kjærås, and Stefan Bouzarovski. 2018. "The politics of rapid urban transformation." *Current Opinion in Environmental Sustainability* 31: 16–22.

Haasnoot, Marjolijn, Jan H. Kwakkel, Warren E. Walker, and Judith ter Maat. 2013. "Dynamic adaptive policy pathways: A method for crafting robust decisions for a deeply uncertain world." *Global Environmental Change* 23(2): 485–498.

Hallegatte, Stephane, and Julie Rozenberg. 2017. "Climate change through a poverty lens." *Nature Climate Change* 7(4): 250–256.

Jasanoff, Sheila. 2010. "A new climate for society." *Theory Culture & Society* 27(2–3): 233–253.

Lawrence, Judy, and Marjolijn Haasnoot. 2017. "What it took to catalyse uptake of dynamic adaptive pathways planning to address climate change uncertainty." *Environmental Science & Policy* 68: 47–57.

Lin, Ning, and Eric Shullman. 2017. "Dealing with hurricane surge flooding in a changing environment: part I. Risk assessment considering storm climatology change, sea level rise, and coastal development." *Stoch Environ Res Risk Assess* 31: 2379–2400.

Malm, Andreas. 2016. *Fossil capital. The rise of steam power and the roots of global warming*. London: Verso.

Margulis, S. 2016. *Climate change studies in Latin America. Vulnerability and adaptation of Latin American cities to climate change*. Santiago de Chile, ECLAC, EUROCLIMA Project Documents.

Martinez-Alier, Joan, Leah Temper, Daniela Del Bene, and Arnim Scheidel. 2016. "Is there a global environmental justice movement?" *The Journal of Peasant Studies* 43(3): 731–755.

Mauad, Ana Carolina, and Michele Betsill. 2019. "A changing role in global climate governance: São Paulo mixing its climate and international policies." *Brazilian Journal of International Politics* 62(2): 1–17.

Merlinsky, Gabriela. 2013. *Politics and environmental justice in the Buenos Aires metropolis. El conflicto del Riachuelo*. Buenos Aires: Fondo de Cultura Económica.

Merlinsky, Gabriela. 2017. "Environmental justice movements and the defense of the commons in Latin America. Five theses in progress." In *Latin American Political Ecology*, coordinated by Héctor Alimonda, Catalina Toro Pérez, and Facundo Martín. Autonomous City of Buenos Aires, CLACSO; Mexico: Universidad Autónoma Metropolitana; CICCUS.

Merlinsky, Gabriela. 2018. "Environmental justice and recognition policies in Buenos Aires." *Perfiles Latinoamericanos* 26(51): 241–263.

Merlinsky, Gabriela. 2021. *All ecology is politics. Struggles for the right to the environment in search of alternative worlds*. Buenos Aires: Siglo Veintiuno Editores.

Merlinsky, Gabriela, and Melina Tobías. 2016. "Inundaciones y construcción social del riesgo en Buenos Aires: acciones colectivas, controversias y escenarios de futuro." *Cuadernos del CENDES* 91: 45–63.

Merlinsky, Gabriela, and Melina Tobías. 2021. "Water conflicts in the Matanza Riachuelo and Reconquista river basins. Keys to think about water justice at a metropolitan scale." *Punto Sur Magazine*, No. 5 July–December (in press).

Merlinsky, Gabriela, Andrés Scharager, and Melina Tobías. 2017. "Environmental recomposition and population displacement in Buenos Aires. Controversies over the liberation of the towpath in the Matanza Riachuelo watershed." *Cuaderno Urbano* 22(June): 53–72.

Moore, Jason. 2016. *Anthropocene or Capitalocene? Nature, history and the crisis of capitalism*. Oakland: PM Press.

Nightingale, Andrea J. 2011. "Bounding difference: intersectionality and the material production of gender, caste, class and environment in Nepal." *Geoforum* 42(2): 153–162.

O'Brien, Karen. 2016. "Climate change and social transformations: is it time for a quantum leap?" *Wiley Interdisciplinary Reviews: Climate Change* 7(5): 618–626.

O'Brien, Karen. 2017. "Climate change adaptation and social transformation." In *International encyclopedia of geography*, edited by Douglas Richardson, Noel Castree, and Michael F. Goodchild MFm 1–8. Atlanta, GA: American Cancer Society.

O'Brien, Karen, and Robin M. Leichenko. 2000. "Double exposure: assessing the impacts of climate change within the context of economic globalization." *Global Environmental Change* 10(3): 221–232.

Paiva Henrique, Karen, and Petra Tschakert. 2020. "Pathways to urban transformation: from dispossession to climate justice." *Progress in Human Geography* 45(5): 1–23.

Pelling, Mark. 2010. *Adaptation to climate change: From resilience to transformation.* Abingdon: Routledge.

Pelling, Mark, Thomas Abeling, and Matthias Garschagen. 2016. "Emergence and transition in London's climate change adaptation pathways." *Journal of Extreme Events* 3(3): 1–25.

Porto-Gonçalves, Carlos Walter. 2011. *A globalização da natureza e a natureza da globalização.* 2nd edition. Rio de Janeiro: Civilização Brasileira.

Pradilla Cobos, E. (2016). The transformations of conflicts and social movements in Latin American cities. In *El derecho a la ciudad en América Latina: visiones desde la política*, coordinated by Fernando Carrión and Jaime Erazo. Mexico: UNAM, IDRC/CRDI, CLACSO.

Pulido, L. 2018. "Racism and the Anthropocene." In *The remains of the Anthropocene*, edited by Gregg Mitman, Robert S. Emmet, and Marco Armiero, 116–128. Chicago: University of Chicago Press.

Quimbayo Ruiz, Germán. 2020. "Territory, sustainability, and beyond: Latin American urbanization through a political ecology." *Environment and Planning E: Nature and Space* 3(3): 786–809. doi: 10.1177/2514848619887933.

Quiroz Benítez, Diana Esmeralda. 2013. "Cities and climate change: the case of Mexico City's climate policy." *Demographic and Urban Studies* 28(2): 343–382.

Reeder, Tim, and Nicola Ranger. 2011. "How do you adapt in an uncertain world? Lessons from the Thames Estuary 2100 project." *World Resources Report*, Washington, DC, USA.

Rusticucci, Matilde. 2007. "IPCC. State of the Situation." *Revista Encrucijadas* n° 41. Magazine of the University of Buenos Aires. July.

Sánchez Rodriguez, Roberto. 2013. Urban responses to climate change in Latin America. Inter-American Institute for Global Change and the Economic Commission for Latin America and the Caribbean United Nations, Santiago de Chile.

Satterthwaite, David, Diane Archer, Sarah Colenbrander, David Dodman, Jorgelina Hardoy, and Sheela Patel. 2018. "Responding to climate change in cities and in their informal settlements and economies." Paper prepared for the *International Scientific Conference on Cities and Climate Change*, Edmonton, March.

Schaller, Nathalie, Alison Kay, Rob Lamb, and Neil Massey. 2016. "Human influence on climate in the 2014 southern England winter floods and their impacts." *Nature Climate Change* 6(6): 627–634.

Secretaría de Ambiente y Desarrollo Sustentable - SAyDS. 2015. *Tercera Comunicación Nacional de la República Argentina a la Convención Marco de las Naciones Unidas sobre el Cambio Climatico.* Office of the Chief of Cabinet of Ministers and SAyDS.

Sultana, Farhana. 2013. "Gendering climate change: geographical insights." *The Professional Geographer* 66(3): 372–381.

Swyngedouw, Erik. 2013. "The non-political politics of climate change." *ACME: An International Journal for Critical Geographies* 12(1): 1–8.

Torres, Pedro H., Ana Lia Leonel, Gabriel Pires de Araujo, and Pedro Jacobi. 2020. "Is the Brazilian National Climate change adaptation plan addressing inequality? Climate and environmental justice in a Global South perspective." *Environmental Justice* 13(2). doi: 10.1089/env.2019.0043.

Tschakert, Petra, and Mario Machado. 2012. "Gender justice and rights in climate change adaptation: opportunities and pitfalls." *Ethics and Social Welfare* 6(3): 275–289.

Ulloa, Astrid. 2017. "Environmental and extractive dynamics in the 21st century: is it the age of the anthropocene or the capitalocene in Latin America." *Desacatos* 54: 58–73.

Viand, Jessica and Silvia González. 2012. "Crear riesgo, ocultar riesgo: gestión de inundaciones y política urbana en dos ciudades argentinas." *Primer Encuentro de Investigadores en Formación de Recursos Hídricos.* Buenos Aires: Instituto Nacional del Agua.

Warner, Benjamin, and Christopher Kuzdas. 2017. "The role of political economy in framing and producing transformative adaptation." *Current Opinion in Environmental Sustainability* 29: 69–74.

Whyte, Kyle. 2017. "Indigenous Climate Change Studies: indigenizing futures, decolonizing the Anthropocene." *English Language Notes* 551–2: 152–162.

Wise, R., I. Fazey, M. Stafford Smith, S. E. Park, H. C. Eakin, E.R.M. Archer Van Garderen, and B. Campbell. 2014. "Reconceptualising adaptation to climate change as part of pathways of change and response." *Global Environmental Change* 28: 325–336.

World Bank. 2011. *Guide to climate adaptation in cities.* Washington, DC.

Ziervogel, Gina, Anna Cowen, and John Ziniades. 2016. "Moving from adaptive to transformative capacity: building foundations for inclusive, thriving and regenerative urban settlements." *Sustainability* 8: 1–26.

12 Environmental disasters and critical politics

Alejandro Camargo and Juan Antonio Cardoso

Introduction

In a time of climate crisis, Latin America and the Caribbean has stood out as the "second most disaster-prone region in the world" (OCHA 2020, 2). From hurricanes and storms, to earthquakes and floods, environmental disasters have disturbed environmental, social, economic, and political orders in the region. Between 2000 and 2019, for instance, around 152 million people were affected by nearly 1,205 disastrous events (OCHA 2020, 2). These catastrophes have a material basis in the geographical and biophysical features of the subcontinent. Three of these features are particularly significant for this chapter. First, the Pacific seaboards of México, Central América, and South América are the areas most prone to earthquakes, because they are located in the "ring of fire," a path running along the Pacific Ocean; as "one of the most geologically active areas on Earth," this "ring of fire" is characterized by active volcanoes and recurrent earthquakes (Masum and Akbar 2019, 1). Secondly, floods are the most common disaster in Latin America and the Caribbean (OCHA 2020, 2) and can be a consequence of a variety of factors, such as urbanization and land-use changes (Fernández Illescas and Buss 2016). The El Niño Southern Oscillation (ENSO) cold phase, known as La Niña, has significantly contributed to flood-induced disasters in the region, as this phenomenon has abruptly altered precipitation patterns (Fernández Illescas and Buss 2016). Thirdly, Mexico, Central America, and the Caribbean are located in a hurricane area, which translates into significant tropical storm activity each year. Global warming is expected to increase precipitation extremes and, consequently, the severity of floods and Atlantic hurricanes (O'Gorman 2012; Spencer and Strobl 2020).

This physical reality often intersects with different forms of inequality and power imbalances. In fact, the magnitude and impact of a disaster is commonly assessed through issues of vulnerability and risk, which in turn consider variables such as dependency, institutional and infrastructural development, access to health care, and food security. From this perspective, Guatemala, Haiti, and Honduras experience the highest levels of disaster risk and vulnerability in the region, given their disadvantaged and marginal position regarding the aforementioned variables (Unión Europea et al. 2018). Unequal access to land is particularly illustrative of this phenomenon. It is estimated that by 2030, 84% of the people in Latin America and the Caribbean will be living in cities. Limited land for urban growth, uneven development, and inequality force vulnerable families to settle in at-risk areas such as flood-prone zones and on slopes where they face the threat of landslides. Furthermore, land insecurity in rural areas hinders post-disaster recovery (IFRC 2011, 11). These intersections of inequality, power imbalances, and disasters account for the politics of disaster, and mean that disturbing nature's spaces also exposes the tensions and contradictions of society.

DOI: 10.4324/9780429344428-15

In this chapter we present a reflection on the politics of environmental disasters in Latin America and the Caribbean. A crucial antecedent for this reflection is the critique of the naturalness of the so-called "natural" disasters. At least since the 1970s, scholars have questioned the idea that environmental disasters are a manifestation of biophysical processes. For instance, in their classic article, "Taking the Naturalness out of Natural Disasters," Phil O'Keefe, Ken Westgate and Ben Wisner (1976) observed that "underdeveloped" countries around the world had experienced more recurrent disasters. However, no significant geological or climatological changes had occurred. The authors suggested that the increasing number of disasters at the time should therefore be explained from the perspective of people's growing vulnerability, marginalization, and the unequal dynamics of development. But while this assertion was novel among academics at the time, it was not necessarily the case for the survivors of disasters: The authors remind us that Guatemalan peasants used to refer to earthquakes as "classquakes," thereby undermining their naturalness. Put differently, while scholars usually take credit for denaturalizing disasters, communities affected by these calamities have long ago come to this conclusion through their own experience.

The politicization of disasters in Latin America and the Caribbean has inspired a large body of literature. Topics addressed in this literature include how disasters intertwine with the violation of human rights during states of emergency (Marchezini 2014), the violence inherited in the formation of Latin American states (Fonseca 2020), the production of collective memory (Ullberg 2013), and the relationships between affect and neoliberalism (Barrios 2017).[1] In line with the broader literature on disaster capitalism, scholars have also studied how disasters become opportunities for the expansion of capitalist relations (Stonich 2008). Despite the variety of topics and approaches to politics, however, the study of how disasters become scenarios for social activism and unrest has not been sufficiently developed. Scholars have pointed out the importance of community action, social networks, resilience, and solidarity in the context of disasters (e.g., Larenas et al. 2015; Chandes and Paché 2009; see also Gustavo García-López et al., Chap. 38 in this volume). Numerous cases, however, have also shown that disasters are moments of mobilization against broader and more structural conditions of inequality and power imbalances. In this chapter we focus on this dimension of disaster politics. We borrow Tania Li's (2019) concept of "critical politics" to refer to those modalities of collective action that confront prevailing power relations and hierarchies. This concept allows us to differentiate our approach from other ideas of politics in disaster studies, which refer to the international politics of aid, development, and humanitarianism (Hannigan 2012), or broader frameworks considering changing political regimes, histories, causes, consequences, and impacts (Pelling and Dill 2010). Furthermore, approaching disasters from the critical politics of activism and grassroots mobilization allows us to decenter the denaturalization of disasters from academia.

Therefore, we show that the modalities of social mobilization and unrest that emerge in the context of environmental disasters transcend the event *per se* to make claims regarding the inequalities produced in capitalism and colonialism. We concentrate on the ways in which actors and communities produce critical politics in the aftermath of disaster. This form of politics is far from continuous, homogeneous, or unidimensional, however. Although some forms of mobilization may emerge from the disaster itself, others were born before, in different contexts and for other reasons, in which case the disaster constitutes an arena to broaden and resignify those claims. An analysis of different cases, therefore, reveals how mobilization unfolds around a variety of specific issues, and different claims intersect in moments of disaster and risk. In particular, we present here distinct instances in which disasters intertwine with mobilizations against urban inequalities, the state, colonialism, and environmental injustice. We dedicate one section to each of these three phenomena, followed by some concluding remarks. We use the term environmental disaster to avoid the depoliticized idea of "natural" disasters, which is still used in policy arenas, despite long-standing criticism; in contrast to "natural," the adjective "environmental" encompasses the multiple and profound interactions between biophysical and social forces.

The right to remain

On the morning of September 18, 1985, an 8.1-magnitude earthquake hit central and western Mexico. Although the epicenter of the earthquake was on the Pacific coast of the state of Michoacán, the devastating effects of this phenomena were experienced with greater intensity four hundred kilometers away, in the heart of Mexico City. The construction of this urban space on lakes and its high population density—especially in the center of the city—were crucial factors that contributed to the dramatic human and material impacts of the earthquake. Yet the economic crisis which began in 1982, and the expansion of neoliberal policies during the administration of President Miguel de la Madrid, also played a crucial role (Anderson 2011, 148), as this crisis had manifested in growing unemployment, lower wages, and limited access to housing, thereby increasing vulnerability. In this context, the neighborhoods in the center of Mexico City became the setting for the emergence of various popular movements. Even though these movements initially focused on recovery in the aftermath of the disaster, they became an important platform for mobilization around issues of housing during the rest of the decade (Haber 2006, 177).

While it took days for the government to make it to the poorest areas affected by the disaster, tourists and businesspeople lodged in hotels were prioritized and received immediate attention. Nevertheless, this unequal response in the management of the disaster was not an obstacle to community action; volunteers, victims, and their relatives organized rescue activities, traffic control, neighborhood surveillance, created unofficial shelters, and collected food and water for their neighbors. Most of this aid and resources had been allocated through grassroots neighborhood groups, some of which were created prior to the disaster, while others came into being in the aftermath.[2] On some occasions, the claims of the former—around the excessive increase in rents and the lack of access to housing and public services—coincided with the claims of the latter, who opposed eviction, denounced the gentrification of the historic center, and demanded housing reconstruction by the state. The Coordinadora Nacional del Movimiento Urbano (CONAMUP) brought together many of these movements, not only with the purpose of coordinating and promoting rescue and care activities for the victims, but also with the goal of promoting democratic participation in the reconstruction of the city (Moctezuma 1986, 36). As Ramírez Saiz (1986) observed, social mobilization in the aftermath of the earthquake was part of broader urban struggles in Mexico City. Despite the constant announcements by the government of plans to make collapsed buildings into green spaces, as well as its failure to grant collective titles as proposed by neighborhood organizations, between 1986 and 1987 404,000 homes were built, of which many tenants became owners (Haber 2006, 183).

In southern Chile, communities also mobilized around social and infrastructural reconstruction in the aftermath of a disaster. During the first months of 2008, earthquakes in the Los Lagos region reactivated the eruptive activity of the Chaitén volcano, thereby forcing people to evacuate the entire city of Chaitén. For Mandujano, Rodríguez, Reyes and Medina (2015), this case involved a double catastrophe: First, the eruption of a volcano on May 2 produced a large ash cloud— more than 25 kilometers high—and caused the overflow of the Río Blanco, which buried the southern part of the town of Chaitén on May 12. Second, the political and technical decisions to mitigate the environmental phenomenon led to the evacuation of the entire population, the disintegration of the community, and the drastic demographic decline of the town.

A few weeks after the eruption, some people created "Hijos y Amigos de Chaitén" (Sons/ Daughters and Friends of Chaitén), a community organization to support reinhabiting the town they had been forced to leave. This "illegal" occupation not only produced violent responses on the part of the police, but also reinforced, among the Chaiteninos, the common goal of recovering their territory (Berezin 2015). Through take-overs and *funas*,[3] the "rebels"— as the press and government authorities called them—persisted for four years in their struggle to be part of the reconstruction of the town, to live there peaceably, and to demand the restoration of access to electricity, water, health care, and education. It was not until 2011 when President Sebastián Piñera ordered the reconstruction of the northern part of Chaitén in its original location, by which time many families had given up living in the town: only 10% of the initial population returned.[4]

Colonialism and the state in question

On September 20, 2017, Hurricane Maria hit Puerto Rico, triggering one of the most devastating disasters in the recent history of the island and of the Caribbean as a whole. Sustained winds of more than 64 mph (103 kmph) and torrential rains that caused floods up to a depth of 15 feet (4.6 m) destroyed thousands of homes, and severely damaged infrastructure, including the entire electricity and telephone network of the island. Two weeks after the storm, President Donald Trump argued that the hurricane had not been a "real catastrophe," since only sixteen deaths (out of the 2,975 that would be reported later) had been acknowledged at that time. Most of the deaths, however, were not due to strong winds of floods, but to structural failures after the storm:

> uncleared roads that did not allow ambulances to arrive, lack of water distribution that led residents to contaminated water sources, lack of generators in hospitals, and more than half a year without electricity to power medical equipment, refrigerate lifesaving medications such as insulin and provide public lighting and traffic lights to prevent deadly accidents.
>
> (Bonilla and LeBrón 2019, 3)

Bureaucratic and institutional negligence ended the lives of thousands of people in a context of economic recession, rising debt, deep cuts in public budgets, and, in general, Puerto Rico's condition as a colony of the United States.

Puerto Ricans decided to take the recovery of the island into their own hands, given the poor response on the part of the US government, as well as the multiple problems concerning the allocation of aid from the national and the local state. Centros de Apoyo Mutuo (Mutual Support Centers, or CAMs), a grassroots development initiative, emerged out of pre-existing networks of activism and solidarity, and became pivotal in the assistance and support to communities (Vélez-Vélez and Villarrubia-Mendoza 2018). Others, especially women, organized themselves into "brigades" to clean roads, distribute water and food, provide health care for the wounded, and build shelters and houses. They also promoted or created groups to help the victims and the community in specific areas. For example, Colectiva Feminista en Construcción (Feminist Collective in Construction)—which existed well before the disaster—connected pregnant women with medical and legal services during the emergency, while promoting spaces for discussion about different forms of oppression on the island (Bustos 2020, 92). Ayuda Legal Puerto Rico (Legal Aid Puerto Rico) is a group of lawyers who have offered free legal services and advice to the poorest inhabitants regarding access to justice and the defense of fundamental rights for a just recovery—the organization is in fact the first legal assistance

Figure 12.1 "The disaster is the colony". This was a slogan coined by activists to highlight the colonial roots of the hurricane Maria disaster. Art by Francesco "Cesco" Lovascio di Santis (2018) as part of "The Portrait-Story Project", in collaboration with the Mutual Aid Project of Mariana, Humacao, Puerto Rico.

initiative explicitly focused on cases of disasters (Bustos 2020, 95). El Departamento de la Comida (The Department of Food) built upon its prior experience to promote food sovereignty among Puerto Ricans in the aftermath of the disaster. They teach about and build gardens for self-determination and sustainability through food, which became crucial ideas since most crops were destroyed in the wake of Hurricane María (Bustos 2020, 99). In the context of the Covid-19 pandemic, these forms of solidarity and activism have again been crucial for the allocation of soap, thermometers, and face masks, as well as the implementation of prevention measures (Rodríguez-Soto 2020, 307).

The community organizing and activism that emerged in reaction to Hurricane María in Puerto Rico was not limited to the immediate response to the emergency. Their mutual aid efforts and actions have also been mobilized as survival mechanisms in the face of a broader and long-standing fight against centuries of colonialism and its deleterious effects on economic and legal autonomy, environmental sustainability, and justice (Atiles-Osoria 2013). "Tired of watching people die after Hurricane María, Puerto Ricans needed to do two things: fill the immediate gaps in order to keep people alive and, at the same time, continue to demand accountability and action from governments" (Rodríguez-Soto 2020, 302). The idea that "solo el pueblo salva el pueblo" (only the people save the people), with which many activists explained their mobilization, reflects the failure of the state to significantly improve the lives of its citizens, as well as the need for decolonization as a pivotal step towards reparation and justice. These claims resulted in eleven days of protests in 2019, which triggered the resignation of Governor Ricardo Rosello. Rosello was accused of homophobic and misogynistic attacks and of being the head of a corrupt and inefficient government in the context of the hurricane disaster. These struggles for self-determination are manifested beyond the legal and political spheres, as they permeate everyday practices and discourses of reparation. For Bustos (2020, 101), these struggles constitute a third space of sovereignty that allows Puerto Ricans to be critical and imagine their lives outside the domination and sovereignty of the USA.

Disastrous infrastructure and environmental justice

Disasters often expose the failure of infrastructure and its role in the production of environmental injustice. In 2010, catastrophic floods ravaged Colombia and caused serious infrastructural, social, environmental, and economic damage. In this context, Uprimmy (2010) pointed out that the poorest and most marginalized people are the ones who suffer the most from the effects of disasters. For him, it is no coincidence that governments have even promoted the settlement of poor people in at-risk areas, as was the case for the Southern Atlantico in Northern Colombia, where the heavy rains of 2010 caused the collapse of a canal and the flooding of rural villages and farmland. Even though this flooding seemed atypical and unforeseen, the Southern Atlántico was in fact a floodplain until the 1950s. At that time, the national government built a dike to progressively drain the area and convert it into a large agricultural field. With the financial and technical support of the World Bank, the Colombian government undertook an agricultural modernization and agrarian reform project in this area during the 1960s and 1970s. Population density in the area increased significantly, as rural workers migrated from neighboring areas to participate in the growing agricultural project. Nevertheless, despite the dike, flooding was never fully controlled; on various occasions, floodwaters from the Magdalena River devastated crops and hindered agricultural development (Camargo 2020). The 2010 floods occurred in what used to be a floodplain, but instead of marshes the floodwaters met poor peasants. In the aftermath of the disaster, peasants created the Comité Central del sur del Departamento del Atlántico (Central Committee of the Atlantic Department), whose objectives went well beyond the phase of emergency management. Although the Colombian government blamed this disaster on global climate change, peasants knew it was a consequence of infrastructural abandonment and political negligence. Therefore, for the Comité the solution was to recover and create new agricultural infrastructures for large-scale crop production. They mobilized to participate in the economic transformation of the area and proposed the implementation of a commercial agriculture scheme to participate in the global food economy. Through this project, they expected to overcome poverty, marginalization, and vulnerability.

Dams have increasingly become disastrous infrastructures where social mobilization intensifies. Since the 1980s, the Movimento dos Atingidos por Barragens (Movement of the Affected by Dams) in Brazil, for instance, has advocated for the human rights of those affected by the expansion and collapse of dams and other mega-projects. Colombia has also taken part in this growing problem: Hidroituango is the largest dam in the country, and has been presented as a landmark infrastructure to mitigate climate change and promote sustainable development through clean energy generation. Yet in 2018, blockages in the dam's tunnels caused flooding upstream and downstream in the Cauca River, leaving approximately 600 people homeless and hundreds evacuated due to flood risk (Gill 2018). This disaster added to a broader conflict between the dam and the surrounding communities. Ríos Vivos (Living Rivers), a network of grassroots organizations, has denounced how communities along the Cauca River have been the victims not only of the dam-related disaster, but also of human and environmental rights violations, dispossession, and displacement. The Asociación de Mujeres Defensoras del Agua y de la Vida (Association of Women Defenders of Water and Life), one of the network members, has exposed the differential effects and vulnerabilities caused by the dam. The Cauca River has been a zone of conflict in which illegal armed actors have threatened fishers, farmers, women, and other vulnerable subjects. In addition, gold mining in this area is constantly threatening people and the environment through the mismanagement of mercury and other pollutants. By way of strikes, forums, and other forms of mobilization, Rios Vivos has framed the 2018 disaster within a broader set of devastating consequences of hydropower and mining development.[5]

In times of human-induced climate change, it is likely that extreme weather events will exacerbate (Herring et al. 2020). This in turn creates new scenarios for social mobilization and reconfigures previous demands by linking them with the challenges and inequalities of climate change. As Svampa (2020) observes, today climate justice constitutes a critical space for action which has been revitalized by a more radical youth, the "heat" of denialism, and ecological disasters. That space includes, according to Svampa, strikes and demonstrations led by young people, as well as grassroots movements, transnational networks, and social movements that confront political and economic elites and institutions. In vulnerable areas such as the island states of the Caribbean, these spaces are increasingly growing as climate change, the green economy, and extractivism intersect (see Sheller 2020). In Jamaica and the Dominican Republic, for instance, communities have already mobilized against the expansion of extractivism and its ecological and social consequences (Sotolongo Gutiérrez 2020). As climate change intensifies, critical politics can be expected to evolve as well.

Conclusions

While disaster scholars have long called attention to the unnaturalness of so-called natural disasters, but communities mobilizing in disastrous contexts have made that idea an everyday principle, upon which they have built their claims, interpretations, and agendas. In this process, new claims intertwine with long-standing struggles, and old movements diversify their goals. Their practices of care and aid contributed to the recovery of those affected by calamity and have also created spaces for political interpellation. Critical politics, therefore, constitutes both spaces of hope and solidarity, as well as platforms to politicize and destabilize the naturalness of disasters.

The lens of critical politics, however, must also be attentive to the shortcomings and tensions within and among forms of activism. Although grassroots mobilization emerges in multiple disasters, their continuity and success are not always guaranteed. Some movements are stronger and more visible than others, while others may remain unnoticed by academia and policy-makers. The politics of disasters are therefore not confined to the actions and responses of governments or policy-makers. On some occasions, these actors resist recognizing such events as disasters, or frame them as inevitable tragedies for which no specific agents can be held responsible (Taddei 2019). The politics of disasters is found as well in the multiple initiatives and actions led by communities and individuals, who seize these opportunities to revitalize their claims and imagine new possibilities of life. A careful analysis of the political in the context of disasters must then go beyond the specific event to consider the ruptures left behind by the situation, the political spaces for struggle, and the demands that congregate different actors around broader inequalities, power imbalances, and hierarchies.

Understanding environmental disasters through critical politics is not only a way to contribute to the expansion of the study of the politics of the environment in Latin America and the Caribbean. It is also an opportunity to fathom the heterogeneity among the so-called "victims" of disasters, including their expectations and own conceptualizations of how their lives have been affected. This endeavor necessarily demands an intersectional approach. In the mobilization against the Hidroituango dam, for instance, rural and urban women have played a pivotal role. In fact, the work of Asociación de Mujeres Defensoras del Agua y de la Vida has been illustrative of this intersectionality. Furthermore, "solo el pueblo salva el pueblo" also refers to a *pueblo* (people) which is mostly saved by women, who are also situated along axes of class, race, and colonialism. Recognizing the heterogeneity within critical politics is a necessary step toward a more inclusive response to disaster through collaborative action and adaptation to climate change.

Notes

1 The review literature on disasters in Latin America and the Caribbean is beyond the scope of this this chapter. For a recent review with a special focus on anthropological approaches, see the introduction of the volume, "The Anthropology of Disasters in Latin America: State of the Art" by Virginia García-Acosta (2019).

2 Grassroots organizations such as Unión Popular de Inquilinos de la Colonia de Morelos, Unión de Vecinos de la Colonia de Guerrero and Coordinación de los Cuartos de Azotea de Tlatelolco were established in the early 1980s to denounce excessive rent costs in a context of high inflation rates. While Unión de Vecinos y Damnificados del Centro, Unión de Vecinos y Damnificados 19 de Septiembre, Sindicato de Costureras 19 de septiembre, were some of the grassroots organizations which emerged in the aftermath of the earthquake, during the recovery period.

3 Funa is a Chilean term referring to marches and demonstrations intended to publicly denounce, repudiate, or expose a person, institution, or group whose acts are considered unjust. In the case of Chaitén, for four years the "rebels" held several funas in the towns to which they were relocated, public roads, and state institutions.

4 See Letelier and Irazábal 2018 for a similar instance in Talca, Central Chile.

5 Visit https://riosvivoscolombia.org/en/ for more information.

References

Anderson, Mark. 2011. *Disaster writing: The cultural politics of catastrophe in Latin America*, 145–189. Charlottesville, VA: University of Virginia Press.

Atiles-Osoria, José M. 2013. "Colonialismo ambiental, criminalización y resistencias: Las movilizaciones puertorriqueñas por la justicia ambiental en el siglo XXI." *Revista crítica de ciências sociais* (100): 131–152.

Barrios, Roberto E. 2017. *Governing affect: Neoliberalism and disaster reconstruction.* Lincoln, NE: University of Nebraska Press.

Berezin, Alan D. 2015. "Chaitén: una historia en el lugar." *Magallania (Punta Arenas)* 43 (3): 91–106.

Bonilla, Yarimar, and Marisol LeBrón, eds. 2019. *Aftershocks of disaster: Puerto Rico before and after the storm.* Chicago, IL: Haymarket Books.

Bustos, Camila. 2020. "The third space of Puerto Rican sovereignty: Reimagining self-determination beyond state sovereignty." *Yale JL & Feminism* 32: 73.

Camargo, Alejandro. 2020. "Aguas indomables: vulnerabilidad y transformaciones hidrosociales en el sur del departamento del Atlántico." In *Fragmentos de historia ambiental colombiana*, edited by Claudia Leal, 145–168. Bogotá: Ediciones Uniandes.

Chandes, Jérôme, and Gilles Paché. 2009. "Pensar la acción colectiva en el contexto de la logística humanitaria: Las lecciones del sismo de Pisco." *Journal of Economics, Finance and Administrative Science* 14 (27): 47–61.

Fernández Illescas, Coral, and Stefan Buss. 2016. *Ocurrencia y gestión de inundaciones en América Latina y el Caribe: Factores claves y experiencia adquirida.* Washington, DC: Banco Interamericano de Desarrollo.

Fonseca, Carlos. 2020. *The literature of catastrophe: Nature, disaster and revolution in Latin America.* London: Bloomsbury Academic.

García-Acosta, Virginia. 2019. *The anthropology of disasters in Latin America: State of the art.* London: Routledge.

Gill, Stephen. 2018. "Colombia's Hidroituango dam floods: hundreds left homeless." *Colombia Reports.* Available at https://colombiareports.com/colombias-hidroituango-dam-floods-hundreds-left-homeless/.

Haber, Paul Lawrence. 2006. *Power from experience: Urban popular movements in late twentieth-century Mexico.* University Park, PA: Pennsylvania State University Press.

Hannigan, John. 2012. *Disasters without borders: The international politics of natural disasters.* Cambridge, UK: Polity.

Herring, Stephanie, Nikolaos Christidis, Andrew Hoell, Martin Hoerling, and Peter Stott, eds. 2020. "Explaining extreme events of 2018 from a climate perspective." *Bulletin of the American Meteorological Society* 101 (1): S1–S128.

IFRC. 2011. *Desastres en América: argumentos para la preparación jurídica.* Ginebra: Federación Internacional de Sociedades de la Cruz Roja y de la Media Luna Roja.

Larenas, Jorge, Marcela Salgado, and Xenia Fuster. 2015. "Enfrentar los desastres socionaturales desde los capitales y recursos comunitarios: el caso de la erupción volcánica de Chaitén, Chile." *Magallania (Punta Arenas)* 43 (3): 125–139.

Letelier, Francisco, and Clara Irazábal. 2018. "Contesting TINA: Community planning alternatives for disaster reconstruction in Chile." *Journal of Planning Education and Research* 38 (1): 67–85.

Li, Tania Murray. 2019. "Politics, interrupted." *Anthropological Theory* 19 (1): 29–53.

Mandujano, Fernando, Juan Carlos Rodríguez, Sonia Reyes, and Patricio Medina. 2015. "La erupción del volcán Chaitén: voyerismo, desconfianza, academia y Estado. Consecuencias urbanas y sociales en la comunidad." *Universum (Talca)* 30 (2): 153–177.

Marchezini, Victor. 2014. "La producción silenciada de los "desastres naturales" en catástrofes sociales." *Revista mexicana de sociología* 76 (2): 253–285.

Masum, Mohammed, and Md. Ali Akbar. 2019. "The Pacific ring of fire is working as a home country of geothermal resources in the world." In *IOP Conference Series: Earth and Environmental Science* 249, 012020.

Moctezuma, Pedro. 1986. "La CONAMUP." *Estudios Políticos* 4-5 (4–1): 30–37.

OCHA. 2020. *Desastres Naturales en América Latina y el Caribe, 2000–2019.* Available at https://reliefweb. int/sites/reliefweb.int/files/resources/OCHA-DESASTRES_NATURALES_ESP%20%281%29.pdf.

O'Gorman, Paul A. 2012. "Sensitivity of tropical precipitation extremes to climate change." *Nature Geoscience* 5 (10): 697–700.

O'Keefe, Phil, Ken Westgate, and Ben Wisner. 1976. "Taking the "naturalness" out of "natural disaster"." *Nature* 260: 566–567.

Pelling, Mark, and Kathleen Dill. 2010. "Disaster politics: tipping points for change in the adaptation of sociopolitical regimes." *Progress in Human Geography* 34 (1): 21–37.

Ramírez Saiz, Juan Manuel. 1986. "Organizaciones populares y lucha política." *Cuadernos Políticos* 45: 38–55.

Rodríguez-Soto, Isa. 2020. "Mutual aid and survival as resistance in Puerto Rico." *NACLA Report on the Americas* 52 (3): 303–308.

Sheller, Mimi. 2020. *Island futures: Caribbean survival in the Anthropocene.* Durham, NC: Duke University Press.

Sotolongo Gutiérrez, Rosabel. 2020. "El cambio climático, los impactos en el Caribe y las luchas socio–ambientales." In *Cambio climático y sus impactos en el gran Caribe*, edited by Jaqueline Laguardia Martínez, 167–185. Buenos Aires: CLACSO.

Spencer, Nekeisha, and Eric Strobl. (2020). "Hurricanes, climate change, and social welfare: Evidence from the Caribbean." *Climatic Change* 163 (1): 337–357.

Stonich, Susan. 2008. "International tourism and disaster capitalism." In *Capitalizing on catastrophe: Neoliberal strategies in disaster reconstruction*, edited by Nandini Gunewardena and Mark Schuller, 47–68. Lanham, MD: AltaMira Press.

Svampa, Maristella. 2020. "¿Hacia dónde van los movimientos por la justicia climática?" *Nueva Sociedad* 286: 107–121.

Taddei, Renzo. 2019. "The field of anthropology of disasters in Brazil: Challenges and perspectives." In *The Anthropology of disasters in Latin America*, edited by Virginia García-Acosta, 45–62. London: Routledge.

Ullberg, Susann. 2013. *Watermarks: urban flooding and memoryscape in Argentina.* Stockholm: Acta Universitatis Stockholmiensis.

Union Europea, INFORM, CEPREDENAC, and UkAid. 2018. *Índice de gestión de riesgos para América Latina y el Caribe.* Available at https://www.unicef.org/lac/media/1601/file.

Uprimmy, Rodrigo. 2010. *Justicia ambiental e inundaciones. Dejusticia.* Available at https://www.dejusticia. org/justicia-ambiental-e-inundaciones/.

Vélez-Vélez, Roberto, and Jacqueline Villarrubia-Mendoza. 2018. "Cambio desde abajo y desde adentro: Notes on Centros de Apoyo Mutuo in post-María Puerto Rico." *Latino Studies* 16 (4): 542–547.

13 Latin America in the chemical vortex of agrarian capitalism[*][†]

María Soledad Castro-Vargas and Finn Mempel

Introduction

In this chapter, we explore how the conceptualization and use of chemical substances are related to the development of agrarian capitalism in Latin America. Our analysis incorporates a chemical geographies lens, which examines chemistry and chemical substances to rethink the human relationships with the living and more-than-living world. These insights seek to uncover how chemicals are embedded in the co-production of socio-natures (Romero et al. 2017).

Patterns of capitalist accumulation have determined particular configurations of the global food system, including their relationship to off-farm inputs, especially chemicals and fertilizers. Friedmann and McMichael (1989) conceptualize these temporarily stable configurations as distinct food regimes, a concept we deploy with respect to Latin America. During the first food regime (1870–1930s), Latin America constituted a source of cheap food and raw materials that supported the industrialization of Europe. The second food regime (1950s–1970s), characterized by United States (US) hegemony, involved the mechanization and "chemicalization" of agriculture production through the imposition of industrial agriculture, and saw a movement of 'surplus' food from the US to Latin America (Bernstein 2016). The contemporary food regime has been determined by the mechanisms of agricultural financialization that emerged in the 1970s, together with the structural adjustment programs of the 1980s that dismantled the state-led institutional arrangements built during the former regime. The role of the private sector became central, placing the state on a secondary plane supporting 'market rules' (Patel 2013; Werner 2021). Governments reduced social services and agrarian subsidies while they promoted non-traditional export agriculture. In contrast to the previous regime dominated by the US, the neoliberal policies of the World Trade Organization have yielded a multipolar political arrangement dominated by a small number of multinational corporations (McMichael 2020).

One can think chemically about the evolution of Latin America's role in global food regimes from various angles and relate to different aspects of agri-food value chains (see Figure 13.1). Our analysis explores how different types of industrial off-farm capital have appropriated agricultural production through chemicalization. Firstly, we examine how patterns of boom and bust relate to agricultural items' chemical properties, defining their appropriation in industrial circuits of accumulation through processing and final applications. We then analyze the role that pesticides have played in the development of industrialized agriculture in Latin America. It will become apparent, through our analysis, that the function of pesticides or the processing and final use of commodities transcend that of a mere material input or output. Rather, each of the

[*] Both authors contributed equally to this chapter.
[†] Instituto de Ciencia y Tecnología Ambiental. Universitat Autònoma de Barcelona.

DOI: 10.4324/9780429344428-16

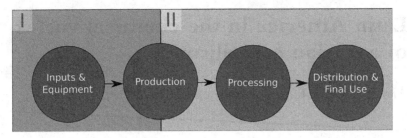

Figure 13.1 Industrial appropriation by off-farm capital. I. Inputs. II. Processing and final use.

nodes presented in Figure 13.1 constitutes a coupling between nested value chains in a complex network, through which industrial capital operates by transforming each node with possible effects on all connected parts of the network.

Commodity booms and the dynamics of sociometabolic circuits

Modern agriculture and the chemical gaze

Links between agriculture, food, and industrial chemistry have become ubiquitous since the early 20th century, accompanied by an emerging set of perceptual patterns, which Landecker has dubbed the "chemical gaze" (Landecker 2019). This gaze is a quintessential modernist approach to nature that dissects agricultural products and processing wastes into their biochemical components and creatively remakes them within industrial metabolic circuits. It thereby recasts agricultural products as substitutable carriers of matter.

A given configuration or alteration of these metabolic pathways is intrinsically linked to a mode of production and social relations. This becomes obvious when examining the Industrial Revolution and its dependence on colonized territories, slave labor, and rural dispossession enabling flows of cotton wool feeding the textile industry and cotton seed processed into vegetable oil and animal feed. The history of food and agricultural systems can then be read through the way in which these sociometabolic pathways have been initiated, altered, or rearranged over time and how these processes relate to broader social structures.

Industrialization, protectionism, and substitution: rubber

An early example of the tight coupling between the chemical industry and agriculture was the rubber booms and busts in the Amazon region during the first food regime. Natural rubber became an increasingly important raw material through the scientific discoveries of various rubber solvents and, most importantly, the discovery of sulfur vulcanization in the 1830s (Guise-Richardson 2010). These advances in the chemical industry allowed for a diverse range of new applications for rubber and led to its ubiquity in the industrial world.

Brazil was the dominant source for wild rubber from the native *Hevea brasiliensis*, which in turn evolved into a mainstay of the country's economy, surpassed only by coffee in terms of export revenue (Dean 1987). This first rubber boom (1879–1912) was characterized by enormous wealth accumulation in the hands of rubber barons, the exploitative subcontracting of *seringueiros*, and the enslavement and brutal extermination of Indigenous communities. It would shape social relations in the regions for subsequent commodity booms (de la Rosa 2004). Plantation-style rubber cultivation in East Asia brought the first rubber boom to an end as new supplies and production methods depressed prices (Garfield 2011, 71–72).

The 1920s also saw the emergence of the Chemurgy movement in the United States. The movement's goals were threefold: to find new processing pathways for agricultural commodities in industrial products, to revive the domestic agricultural economy and to decrease dependence on imported items (Finlay 2004). Given the dependence on rubber imports from Brazil and East Asia for US industry and the military, the movement pushed for a biomass-based rubber substitute. Hope was placed on the rubber extraction from the guayule plant, native to deserts in the US West, and on the development of synthetic rubber from biobased alcohol.

When the US was cut off from the East Asian rubber supply during World War II, the Rubber Agreement between the Brazilian Vargas government and the United States led to a second, short-term rubber boom (1942–1945) in the Amazon region (Garfield 2011, 226–27). The Chemurgy movement was ultimately unsuccessful in its advocacy for a biomass-based rubber substitute and, in the long run, petroleum by-products, such as styrene and butadiene, became the dominant feedstock for new synthetic rubbers. These began replacing natural rubber for many applications in the postwar period. More generally, the petrochemical boom, with its center in the US Gulf Coast, became the basis of modern organic chemistry while Chemurgy's dream of biomass-based feedstocks was reduced to micro-level applications in the postwar economy and only to be revived decades later (Finlay 2004).

Evolving metabolic pathways in the post-war era: soybeans

The demise of the first rubber boom coincided with the general collapse of the 19th-century wave of globalization, after World War I. Across Latin America, widespread adoption of import-substitution industrialization (ISI) policies followed, which sought to end dependence on the exportation of primary commodities and led to a net decrease in the region's share of global agricultural trade (Baraibar Norberg 2020, 79).

In the search for new markets for the large surplus of US agricultural produce following World War II, the US government and private companies sought new industrial applications for individual biochemical components of various agricultural commodities or their processing wastes. Some of these applications would become the basis of commodity booms in Latin America during the contemporary food regime. Soybeans, sugarcane, maize, or oil palms are grown not merely as food items, but as raw materials for a complex network of metabolic pathways, which expanded rapidly after the postwar era (Borras et al. 2016).

A good example of this development is the soybean boom across Brazil and much of the Southern Cone countries. Like rubber, soybeans first made their way into global industrial metabolic circuits in the first wave of globalization before World War I. Starting in the 1890s, Japan used its imperial enterprise in northeastern China to import soybean meal to serve as nitrogen fertilizer in its bid to rapidly intensify agriculture (Hiraga and Hisano 2017). Soybean oil, an excess product of soybean meal processing, would soon be shipped to Europe and North America, where it was embraced as an ingredient for margarine, shortening, soap and paints at a time of great demand for vegetable oils for industrial applications (Prodöhl 2013).

This separation into oil and protein fractions has remained the basis for the successful commercialization of soybean products and has generated other by-products, which have found their own applications (e.g., soy lecithin). After World War II, when the United States had become the largest producer of soybeans, soymeal became the excess product of oil extraction, since it had lost its prior role as a meat substitute in civilian diets during World War II (Prodöhl 2013; Du Bois 2018). After a series of breakthroughs in making soybean cake more suitable for animal diets by removing anti-nutritional factors (Dourado et al. 2011), soybean protein rapidly evolved into a cornerstone of surging mass poultry and swine production systems in the US and Europe.

The rapid expansion of large-scale export-oriented soybean production across Brazil, Argentina, Paraguay, and Bolivia has occurred in the contemporary food regime in a context of deregulation, and the general embrace of biotechnology and large agribusiness. However, as shown above, the soybean boom is a product not only of neoliberal policies, but also of the restructuring of industrial metabolic pathways since the postwar era, which has resulted in a dominant "industrial grain-oilseed-livestock complex" (Weis 2013) at the global scale. In the case of China, imported soybeans are a cornerstone of the recent surge in concentrated animal feeding operations. The radical transformation from net exporter to largest importer of soybeans was made possible by state-led agroindustrialization and an adjustment of China's policies relating to self-sufficiency in staple crops. These have redefined soybean meal as an industrial rather than agricultural product (Schneider 2017). The large biomass flows from Latin America to China mirror the reorientation of the former's trade relations toward the Pacific for its other raw materials, particularly minerals.

The age of chemical versatility: flex crops

The focus of capital accumulation in the "primary" sector of agriculture and mining, called *reprimarización* in Spanish, has become dominant since the beginning of the neoliberal wave. This shift inherited many metabolic patterns and industrial applications for cash crops from the postwar era. The new commodity boom began to feed into global industrial circuits, just as the US agricultural surplus had been doing since World War II. However, the contemporary food regime also came with its own restructuring of these metabolic patterns. Since the late 1990s, the *bioeconomy* as a policy agenda has revived some of Chemurgy's earlier ideas to substitute fossil resources with renewable biomass sources. The use of renewable feedstocks has also become one of the principles of *green chemistry*, which aims to reduce the environmental impacts of the chemical industry (Anastas and Warner 1998).

While the Chemurgy movement had advocated for plant-based industrial applications mainly to reduce dependence on imports, the logic behind this new push for biomass-based organic chemistry has been quite the opposite. The chemical industry's hunger for plant-based feedstocks in the neoliberal era has resulted in a new coupling between the sourcing of agricultural items, financialization, and free trade ideology. The basic ingredients for metabolic pathways are increasingly sourced according to the comparative advantage of the production of certain crops, the quality of their biochemical components, and their respective maximum yields per hectare (e.g., corn for starch, soybeans for protein, oil palms for oil). The ever-increasing versatility of these *flex crops* in terms of their possible end-uses, mediated through the chemical industry, has also rendered them attractive objects of speculation through financial derivative markets and land deals (Oliveira and Schneider 2016).

The dynamic reconfigurations in metabolic patterns in the last decades have once again remade the boundaries between processing products as either waste or commodities. In the case of soybeans, the protein meal, which was mainly the excess product of oil extraction in the mid-1940s, became the economically most relevant component due to the grow in concentrated animal feeding operations, most recently in East Asia. While the oil component has long been important in industrial applications and as a cooking oil, its recent application for biodiesel fuel production through the chemical process of transesterification has further diversified its end-uses and greatly increased its value.

With the expansion of oil palm plantations, the increasing availability of a cheap and easily fractionated vegetable oil has positioned palm and palm kernel oil as extremely versatile raw materials in a myriad of metabolic pathways with different end-uses since the 1980s. In the case of personal care items, a shift in consumer preference towards plant-based ingredients led to a widespread application of palm oil to replace animal tallow as feedstock for synthetic surfactants. Tallow had often been locally available as processing waste from the meat industry, while oil palm plantations are concentrated in a limited number of tropical regions, including parts of Latin America.

Nutritionism and superfoods

Thinking chemically opens up additional perspectives on the evolution of nutritional science and dietary advice and their links with food marketing. Following Gyorgy Scrinis in his characterization of the ideology of "nutritionism" (Scrinis 2008), we can observe how a reductionist understanding of food to the biochemical or nutrient level has become dominant. Nutritionism has evolved from a mere focus on the quantification and sufficient intake of core nutrients (e.g. calories and protein) to include concerns over excess consumption and finally into a new form of functional nutritionism. This latest stage seeks to optimize health and bodily performance by emphasizing beneficial properties of ever more subcategories of biochemical components (e.g. omega-3 fats) for specific needs or conditions.

Through this focus, nutrients become isolated from traditional food items in the form of synthesized supplements or additives, but certain niche food items also become fetishized due to their biochemical composition in line with a belief in associated health benefits. This is particularly visible in the discourse on *superfoods*, which regularly uses specific health claims along with a focus on non-Western, 'authentic,' traditional, or Indigenous food items to open new niche markets (Loyer 2016). Latin American traditional and Indigenous food items, such as açai berries, chia seeds, maca, or quinoa have been at the forefront of this trend, which sits uncomfortably between ethical, environmentally conscious, fair-trade consumption culture on the one hand, and racist cultural essentialism and commodification on the other hand (Loyer and Knight 2018).

Latin America in the global pesticide complex

A brief history of synthetic pesticides

When analyzing food regimes from a chemical perspective, pesticides are a crucial element of this history, since they have facilitated the transformation of agrarian landscapes to the rhythm of capital accumulation patterns. The history of pesticides can be analyzed through three main phases: (1) Use of non-synthetic pesticides (prior to the first food regime); (2) the production of inorganic synthetic pesticides (during the first food regime); and (3) the production of organic synthetic pesticides (starting with the transition to the second food regime) (Zhang, Jiang, and Ou 2011). This last phase began with the development of dichlorodiphenyltrinchloroethane (DDT) during the World War II, which became available for agricultural purposes in 1945. It inaugurated an era of synthetic organic pesticides (Özkara, Akyıl, and Konuk 2016).

Four key aspects stand out in the history of pesticides. First, since the beginning of industrialized agriculture, pest control has been inextricable from, and has co-evolved with, chemical warfare ideologically, techno-scientifically, and organizationally (Russell 2001; Romero 2016). Second, science has played a crucial and ambiguous role, not only collaborating in the production and dissemination of these substances, but also by contributing to parameters and accepted ideas about their regulation (Boudia and Jas 2014). Third, the use of certain substances has been a cyclical process between their creation, the rise of concern about their effects and resistance-related issues, and their consequent regulation or prohibition. This has occurred in the case of insecticides with the shift from organochlorines to organophosphates and then to pyrethroids and to neonicotinoids (Davis 2017). Fourth, the intensive use of pesticides, their regulatory policies and socio-ecological effects have been historically and geographically shaped unevenly across the globe (Bertomeu-Sánchez 2019; Shattuck 2021).

From the circle of poison to the global pesticides complex

The institutional and technological groundwork for the second food regime lies in the Green Revolution, inaugurated in the 1940s when the Rockefeller Foundation launched a process of transformation of agriculture, seeking to develop varieties of high-yielding grains (Galt 2014). The concept of the "Green Revolution," first coined by William Gaud, described the results that the US had achieved by funding developing countries for crop production, combining the use of fertilizers and agrochemicals with hybrid seeds, irrigation, and soil management techniques (Picado 2012). The US government, with the collaboration of recipient governments and the World Bank, performed this large-scale philanthropic intervention, aimed to fight Communism (Patel 2013). The project was consolidated in the 1960s, when the agro-industrial production system was exported to the Third World by private foundations and international development agencies (Clapp 2012).

The relationship between global North and global South in terms of production, commercialization, and socio-environmental exposure to pesticides has been the subject of much debate. The *circles of poison* concept, introduced by Weir and Schapiro in 1981, described uneven relationships in pesticide geopolitics. It captured how particular pesticides are banned in the global North, but still produced for export to the global South. Due to the promotion of export-led agriculture in the global South, pesticides would then return to their manufacturing place as pesticide residues in fruits and vegetables (Weir and Schapiro 1981). Although 'circles of poison' constitutes an umbrella term relevant to a body of scholarship in political ecology and environmental justice, it is not in itself sufficient to explain the intricacy of pesticide networks. The concept of the 'pesticide complex' proposed by Galt (2008) emerged from these discussions, highlighting the need for new empirical data and global information about pesticide use, trade, regulation, exposure, and effects. This framework refers to "all aspects of pesticides' life-cycles from conception to environmental fate and arises from overlapping spatial patterns of pesticides in the world" (Galt 2008, 786).

The geometry of the pesticide complex has been rearranged due to three major shifts in the agrochemical industry in the first two decades of this century: First, a decrease in the number of new patents along with the expiration of old patents; second, the consolidation of the generics industry; and, third, the positioning of China as a global leader in the pesticide industry (Shattuck 2021; Werner, Berndt, and Mansfield 2022). Despite the increasing use of pesticides at a global level, the number of new patents for active ingredients (AIs) has steadily declined from 13 new patents of AIs per year during the 1990s to 7 per year in the 2010s (Haggblade et al. 2017). Additionally, the cost of research and the registration process for new AIs in pesticides is estimated to be more than $286 million, taking an average of 11 years until they are released to the market (McDougall 2016). Consequently, the pesticide industry has focused on the development of new pesticide formulations that mix off-patent AIs with co-formulants to respond to pest resistance. The boom in off-patent herbicide formulations and the fact that they are the most widely used type of pesticide across the globe, especially in the Global South, is referred to as the herbicide revolution (Haggblade et al. 2017).

While the global pesticide complex has penetrated into deeper layers of human and non-human lives, its causes and consequences are more indecipherable than ever. According to Shattuck (2021), three aspects stand out as a result of this continuously transforming process. First, pesticide use and dependence has increased, particularly in the global South. Second, its production core has shifted from the global North to the global South, with the predominance of China and India in the market. Third, although the current configuration of the pesticide complex touches more people, it is even more difficult to study due to lack of access to information, particularly for those most affected. China's rising centrality to the global food system, with its particular state-led participation in global markets, deserves attention (McMichael 2020). As Werner, Berndt, and Mansfield (2022, 14) point out, "in the new map of chemical ubiquity, middle-income countries are also principal producers, exporters, and end markets in a geography characterized by new south–south dynamics.". These large-scale market dynamics have transformed relationships between chemicals and socio-natures, into new spatial configurations and power relationships that require a hard look. Interactions between the state and agrarian capital and within each of these spheres deserve special attention to comprehend changes (e.g., regulatory tension in Costa Rica) (Jansen 2017; Castro-Vargas and Werner 2022).

Tracking the pesticide treadmill

Latin America and Asia have been the regions of the world where pesticide use has grown the most during the past decades. From the 2000s to 2015, the Latin American pesticide market grew from a value of $4 billion to $12 billion (Shattuck 2021). This dramatic increase has led to several problems, such as the pesticide exposure of workers and rural communities; the presence of pesticide residues in food, soil, and groundwater; and environmental degradation (Bertomeu-Sánchez 2019). Unfortunately, the issue remains understudied, resulting in a lack of knowledge about the use and effects of these substances (United Nations 2017).

The Green Revolution and its implications remain salient to our understanding of the contemporary pesticide complex for several reasons. First, the Green Revolution laid the foundations for the 'pesticide treadmill,' an ecological feedback loop that involves the disruption of agro-ecosystems with pesticide use, and the subsequent development of biocide resistance and pest resurgence, which results in a constantly increasing dependence to pesticides (Murray 1994; Nicholls and Altieri 1997). Second, the Green Revolution constituted an *agri-biopolitics*, understood as "a political technique that made certain populations of humans thrive alongside companion crops" (Hetherington 2020, 1). Third, from its early inception, the Green Revolution has been a violent process. Pre-existing colonial structures were re-organized and reformulated through development policies, providing the conditions for the expansion of the agricultural frontier, and displacing Indigenous peoples and peasants. This reinforced the racist, enslaving and settler colonial logic of plantations (Murray 1994; Williams 2018; Hetherington 2020). In short, the Green Revolution is best understood as a long, enduring process, with multiplate configurations, that continues to shape the present, as proposed by Patel when he asks when, exactly, was the Green Revolution (Patel 2013).

During the second food regime, Central American banana and cotton plantations were characterized by a very high use of pesticides that affected the socio-environmental health of rural communities and workers. Cotton cultivation during the twentieth century, promoted by US development agencies using state subsidies, contributed to position Latin America as the Third World's leader in intensive pesticide use. Banana plantations abruptly transformed landscapes and the interaction among people, plants, and pathogens, developing a tropical-based production system, fully based on chemical inputs (Soluri 2005). From the 1940s to the 1960s, more than 12,000 workers each year applied intensively copper sulphate to combat Sigatoka disease. The fight against this fungus was arbitrarily dictated by techno-scientific specialists, at the expense of the health of workers or *periqueros*, so-called because of the blue tone of their skin from chemical applications (Marquart 2003). Later, in the 1960s and 1970s, the toxic nematicide DBCP produced by Dow and Shell, was broadly applied in banana plantations. Its widespread use led to severe health consequences, from permanent infertility to psychological trauma, for more than 30,000 workers. As Mora Solano (2014) points out, the environmental suffering of workers affected by DBCP is a direct consequence of production dynamics and it exemplifies how environment and bodies are inseparable. In both cases, the *agri-biopolitics* of the Green Revolution established a phytosanitary protection regime safeguarding certain plants for the abstract wellbeing of selected people, at the expense of the lives of others (Hetherington 2020).

In the contemporary food regime pesticide use intensified, due to the growth of global food market and the availability of fresh products all year round. Moreover, the number of pesticide users increased, since small-scale farmers not only began to participate in the export crop market, but also to incorporate chemical inputs in their own farming (Murray 1994; Nicholls and Altieri 1997; Galt 2008). Pesticides have become pervasive in agricultural landscapes, carrying with them logics of violence while also interacting with colonial and racist legacies. In Colombia, the use of the herbicide glyphosate for coca eradication campaigns demonstrates the persistent links between pesticides and chemical warfare, constituting an ecocide through instantaneous, structural, and slow violence (Meszaros Martin 2018). Furthermore, the regulatory process of glyphosate shows the agri-biopolitical nuances around its discussion, which have managed to separate health concerns of individual bodies from those of social and political bodies (Silva 2017).

Pesticide pollution has also constituted a mechanism of dispossession. For Camacho (2017), Martínez Sánchez (2019) and Hurtado and Vélez-Torres (2020), pollution represents a subtle and violent process that undermines social production and reproduction of daily life and, at the same, time re-signifies it. As part of this re-signification, communitarian feminism has consolidated the concept of the body/territory as a unity, which understands health and healing as a shared path. According to Cabnal (2018, 103), "defending the territory/body entails to assume the body as a historical territory in dispute with ancestral and colonial patriarchal power, but we also conceive it as a vital space for the recovery of life." Even though the global South holds multiple *sacrifice zones*, conceived as spaces of environmental suffering which are intrinsic to the logic of uneven development, these have also represented places to weave resistances into the administration of death (Olmedo and Ceberio De León 2021). Pesticide pollution both exacerbates other oppressions and injustices and extends and amplifies fields of struggle that have been established in the defense of life (Martínez Sánchez 2019).

Conclusions

In this contribution we have analyzed how industrial capital has increasingly appropriated agricultural production in Latin America through chemicalization. This process recasts plants, seeds, trees, and their products as carriers of chemical components, initiating new sociometabolic pathways and thereby new circuits of capital accumulation. Further, the incursion of industrial capital into pest management has increased the use and dependence on synthetic pesticides and propagated a model of input-intensive agriculture.

We have shown how this chemicalization has left long-term legacies, such as the industrial grain–oilseed–livestock complex and the pervasiveness of synthetic pesticides in Latin American agri-food systems. However, we have also pointed out highly dynamic processes, such as the shifting geographies of pesticide production and formulation as well as the ever-changing roles of chemical components derived from agricultural items and their fluid existence between waste, by-product, and commodity. These legacies and dynamics have shaped agricultural landscapes throughout Latin America by initiating commodity booms, favoring particular cash crops, and affecting social relations as well as socio-environmental health in rural communities.

References

Anastas, Paul, and John Warner. 1998. *Green Chemistry: Theory and Practice.* Oxford: Oxford University Press.

Baraibar Norberg, M. 2020. *The Political Economy of Agrarian Change in Latin America. Argentina, Paraguay and Uruguay.* Cham: Palgrave Macmillan.

Bernstein, Henry. 2016. "Agrarian Political Economy and Modern World Capitalism: The Contributions of Food Regime Analysis." *The Journal of Peasant Studies* 43 (3): 611–47.

Bertomeu-Sánchez, José Ramón. 2019. "Introduction. Pesticides: Past and Present." *HoST-Journal of History of Science and Technology* 13 (1): 1–27.

Bois, Christine M Du. 2018. *The Story of Soy.* London: Reaktion Books.

Borras, Saturnino M, Jennifer C Franco, S Ryan Isakson, Les Levidow, and Pietje Vervest. 2016. "The Rise of Flex Crops and Commodities: Implications for Research." *Journal of Peasant Studies* 43 (1): 93–115. https://doi.org/10.1080/03066150.2015.1036417.

Boudia, Soraya, and Nathalie Jas. 2014. "Introduction." In *Powerless Science?: Science and Politics in a Toxic World*, edited by Soraya Boudia and Nathalie Jas, 1–26. Oxford: Berghahn Books.

Cabnal, Lorena. 2018. "Tzk'at, Red de Sanadoras Ancestrales Del Feminismo Comunitario Desde Iximulew- Guatemala." *Ecología Política* 54: 100–104.

Camacho, Juana. 2017. "Acumulación Tóxica y Despojo Agroalimentario En La Mojana, Caribe Colombiano." *Revista Colombiana de Antropología* 53 (1): 123–50.

Castro-Vargas, María Soledad, and Marion Werner. 2022. "Pesticide Regulation and the State in Costa Rica." Manuscript Submitted for Publication.

Clapp, Jennifer. 2012. *Food.* Cambridge: Polity.

Davis, Frederick Rowe. 2017. "Insecticides, Agriculture, and the Anthropocene." *Global Environment* 10 (1): 114–36.

Dean, Warren. 1987. *Brazil and the Struggle for Rubber: A Study in Environmental History.* Cambridge: Cambridge University Press.

Dourado, Leilane Rocha Barros, Leonardo Augusto Fonseca Pascoal, Nilva Kazue Sakomura, Fernando Guilherme Perazzo Costa, and Daniel Biagiotti. 2011. "Soybeans (Glycine Max) and Soybean Products in Poultry and Swine Nutrition." In *Recent Trends for Enhancing the Diversity and Quality of Soybean Products*, edited by Dora Krezhova. InTechOpem. https://doi.org/10.5772/18071.

Finlay, Mark R. 2004. "Old Efforts at New Uses: A Brief History of Chemurgy and the American Search for Biobased Materials." *Journal of Industrial Ecology* 7 (3–4): 33–46. https://doi.org/10.1162/108819803323059389.

Friedmann, Harriet, and Philip McMichael. 1989. "Agriculture and the State System: The Rise and Decline of National Agricultures, 1870 to the Present." *Sociologia Ruralis* 29 (2): 93–117. https://doi.org/10.1111/j.1467-9523.1989.tb00360.x.

Galt, Ryan E. 2008. "Beyond the Circle of Poison: Significant Shifts in the Global Pesticide Complex, 1976–2008." *Global Environmental Change* 18 (4): 786–99. https://doi.org/10.1016/j.gloenvcha.2008.07.003.

———. 2014. *Food Systems in an Unequal World: Pesticides, Vegetables, and Agrarian Capitalism in Costa Rica.* Tucson, AZ: University of Arizona Press.

Garfield, S. 2011. *The Devil's Milk: A Social History of Rubber.* New York: Monthly Review Press.

Guise-Richardson, Cai. 2010. "Redefining Vulcanization: Charles Goodyear, Patents, and Industrial Control, 1834–1865." *Technology and Culture* 51 (2): 357–87.

Haggblade, Steven, Bart Minten, Carl Pray, Thomas Reardon, and David Zilberman. 2017. "The Herbicide Revolution in Developing Countries: Patterns, Causes, and Implications." *The European Journal of Development Research* 29 (3): 533–59.

Hetherington, Kregg. 2020. "Agribiopolitics: The Health of Plants and Humans in the Age of Monocrops." *Environment and Planning D: Society and Space* 38 (4): 682–98.

Hiraga, Midori, and Shuji Hisano. 2017. "The First Food Regime in Asian Context? Japan's Capitalist Development and the Making of Soybean as a Global Commodity in the 1890s–1930s." Kyoto.

Hurtado, Diana, and Irene Vélez-Torres. 2020. "Toxic Dispossession: On the Social Impacts of the Aerial Use of Glyphosate by the Sugarcane Agroindustry in Colombia." *Critical Criminology* 28 (4): 557–76.

Jansen, Kees. 2017. "Business Conflict and Pesticide Risk Regulation in Costa Rica: Supporting Data on Laws and Instructive Events, 1998–2014."

la Rosa, Francisco Javier Ullán de. 2004. "La Era Del Caucho En El Amazonas (1870–1920): Modelos de Explotación y Relaciones Sociales de Producción." In *Anales Del Museo de América*, 183–204.

Landecker, Hannah. 2019. "A Metabolic History of Manufacturing Waste: Food Commodities and Their Outsides." *Food, Culture and Society* 22 (5): 530–47. https://doi.org/10.1080/15528014.2019.1638110.

Loyer, Jessica. 2016. "Superfoods." In *Encyclopedia of Food and Agricultural Ethics*, 1–7. Dordrecht: Springer Netherlands. https://doi.org/10.1007/978-94-007-6167-4_574-1.

Loyer, Jessica, and Christine Knight. 2018. "Selling the 'Inca Superfood': Nutritional Primitivism in Superfoods Books and Maca Marketing." *Food, Culture and Society* 21 (4): 449–67. https://doi.org/10.1080/15528014.2018.1480645.

Marquart, Steve. 2003. "Pesticidas, Pericos y Sindicatos En La Industria Bananera Costarricense, 1938–1962." *Revista de Historia* 47: 43–95.

Martínez Sánchez, Gloriana. 2019. "La Piñera Nos Contaminó El Agua: Mujer, Trabajo y Vida Cotidiana En Comunidades Afectadas Por La Expansión Piñera En Costa Rica." *Revista Latinoamericana de Geografía* 10 (2): 3–23.

McDougall, Phillips. 2016. "The Cost of New Agrochemical Product Discovery, Development and Registration in 1995, 2000, 2005–8 and 2010–2014. R&D Expenditure in 2014 and Expectations for 2019. A Consultancy Study for CropLife International, CropLife America and the European Crop Pro."

McMichael, Philip. 2020. "Does China's 'Going out' Strategy Prefigure a New Food Regime?" *Journal of Peasant Studies* 47 (1): 116–54. https://doi.org/10.1080/03066150.2019.1693368.

Meszaros Martin, Hannah. 2018. "Defoliating the World' Ecocide, Visual Evidence and 'Earthly Memory.'" *Third Text* 32 (2–3): 230–53.

Mora Solano, Sindy. 2014. "Agroquímicos y Sufrimiento Ambiental: Reflexiones Desde Las Ciencias Sociales." *Revista Reflexiones* 93 (1): 199–206.

Murray, Douglas L. 1994. *Cultivating Crisis: The Human Cost of Pesticides in Latin America*. Austin, TX: University of Texas Press.

Nicholls, Clara Ines, and Miguel A Altieri. 1997. "Conventional Agricultural Development Models and the Persistence of the Pesticide Treadmill in Latin America." *International Journal of Sustainable Development and World Ecology* 4 (2): 93–111. https://doi.org/10.1080/13504509709469946.

Oliveira, Gustavo de L.T., and Mindi Schneider. 2016. "The Politics of Flexing Soybeans: China, Brazil and Global Agroindustrial Restructuring." *Journal of Peasant Studies* 43 (1): 167–94. https://doi.org/10.1080/03066150.2014.993625.

Olmedo, Clara, and Iñaki Ceberio De León. 2021. "Zonas de Sacrificio y Sufrimientos Invisibles. El Caso de Nonogasta, Provincia La Rioja, Argentina." *Revista Austral de Ciencias Sociales* 2021 (40): 161–78. https://doi.org/10.4206/rev.austral.cienc.soc.2021.n40-09.

Özkara, Arzu, Dilek Akyıl, and Muhsin Konuk. 2016. "Pesticides, Environmental Pollution, and Health." In *Environmental Health Risk-Hazardous Factors to Living Species*, edited by M. Larramendy and S. Soloneski, 1–27. London: IntechOpen.

Patel, Raj. 2013. "The Long Green Revolution." *Journal of Peasant Studies* 40 (1): 1–63. https://doi.org/10.1080/03066150.2012.719224.

Picado, Wilson. 2012. "En Busca de La Genética Guerrera. Segunda Guerra Mundial, Cooperación Agrícola y Revolución Verde En La Agricultura de Costa Rica." *Historia Agraria* 56: 107–34.

Prodöhl, Ines. 2013. "Versatile and Cheap: A Global History of Soy in the First Half of the Twentieth Century." *Journal of Global History* 8 (3): 461–82. https://doi.org/10.1017/S1740022813000375.

Romero, Adam M. 2016. "Commercializing Chemical Warfare: Citrus, Cyanide, and an Endless War." *Agriculture and Human Values* 33 (1): 3–26.

Romero, Adam M., Julie Guthman, Ryan E. Galt, Matt Huber, Becky Mansfield, and Suzana Sawyer. 2017. "Chemical Geographies." *GeoHumanities* 3 (1): 158–77. https://doi.org/10.1080/2373566X.2017.1298972.

Russell, Edmund. 2001. *War and Nature: Fighting Humans and Insects with Chemicals from World War I to Silent Spring*. Cambridge: Cambridge University Press.

Schneider, Mindi. 2017. "Wasting the Rural: Meat, Manure, and the Politics of Agro-Industrialization in Contemporary China." *Geoforum* 78: 89–97. https://doi.org/10.1016/j.geoforum.2015.12.001.

Scrinis, Gyorgy. 2008. "On the Ideology of Nutritionism." *Gastronomica* 8 (1): 39–48. https://doi.org/10.1525/gfc.2008.8.1.39.

Shattuck, Annie. 2021. "Generic, Growing, Green?: The Changing Political Economy of the Global Pesticide Complex." *The Journal of Peasant Studies* 48 (2): 231–53. https://doi.org/10.1080/03066150.2020.1839053.

Silva, Diego. 2017. "Security and Safety in the Glyphosate Debate: A Chemical Cocktail for Discussion2." *Alternautas* 4 (2): 46.

Soluri, John. 2005. *Banana Cultures: Agriculture, Consumption, and Environmental Change in Honduras and the United States.* Austin: University of Texas Press.

United Nations. 2017. "*Informe de La Relatora Especial Sobre El Derecho a La Alimentación.*" Ginebra: Naciones Unidas.

Weir, David, and Mark Schapiro. 1981. *Circle of Poison: Pesticides and People in a Hungry World.* San Francisco, CA: Institute for Food and Development Policy.

Weis, Tony. 2013. *The Ecological Hoofprint – The Global Burden of Industrial Livestock.* London: Zed Books.

Werner, Marion. 2021. "Placing the State in the Contemporary Food Regime: Uneven Regulatory Development in the Dominican Republic." *The Journal of Peasant Studies* 48 (1): 137–58. https://doi.org/10.1080/03066150.2019.1638367.

Werner, Marion, Christian Berndt, and Becky Mansfield. 2022. "The Glyphosate Assemblage: Herbicides, Uneven Development, and Chemical Geographies of Ubiquity." *Annals of the American Association of Geographers* 112 (1): 19–35. https://doi.org/10.1080/24694452.2021.1898322.

Williams, Brian. 2018. "'That We May Live': Pesticides, Plantations, and Environmental Racism in the United States South." *Environment and Planning E: Nature and Space* 1 (1–2): 243–67.

Zhang, WenJun, FuBin Jiang, and JianFeng Ou. 2011. "Global Pesticide Consumption and Pollution: With China as a Focus." *Proceedings of the International Academy of Ecology and Environmental Sciences* 1 (2): 125.

14 Resource radicalisms

Thea Riofrancos

> It is madness to say no to natural resources, which is what part of the left is proposing—no to
> oil, no to mining, no to gas, no to hydroelectric power, no to roads. This is an absurd novelty,
> but it's as if it has become a fundamental part of left discourse. It is all the more dangerous com-
> ing from people who supposedly speak the same language. With so many restrictions, the left
> will not be able to offer any viable political projects… We cannot lose sight of the fact that the
> main objective of a country such as Ecuador is to eliminate poverty. And for that we need our
> natural resources.
>
> Rafael Correa, "Ecuador's Path"

Introduction

In 2011, the fourth year of the administration of Ecuadorian leftist president Rafael Correa,
more than a hundred social movement organizations and leftist political parties gathered for the
"Meeting of Social Movements for Democracy and Life." According to the manifesto written
at this meeting, these organizations and parties were rooted in diverse experiences of social
mobilization, including anti-mining, environmentalist, public transit worker, feminist, and gender diversity struggles, and "the indigenous and peasant uprising for water and land"[1] (manifesto 2011). They condemned Correa's government for "represent[ing] an authoritarian and
corrupt model of capitalist modernization."

Popular movements had rebuked prior governments for being antidemocratic and neoliberal.
But this document also deployed a new critical category: "the extractivist model," defined as a
political-economic order based on the intensive extraction and export of natural resources
(Gudynas 2009; Svampa 2015). The manifesto stated that this model, with its blatant disregard
for nature and indigenous communities, all the more pernicious for being shrouded in a "sup-
posed image of the left" and "a double discourse," must be as militantly resisted as neoliberalism
had been in the recent past.

A year later, in an interview in the Chilean leftist magazine *Punto Final*, and during protracted
political conflict with many of these same social movements, President Correa charged that rejecting the extractive model was a "colossal error" that was particularly "lethal because it utilizes our
same language, proposes the same objectives and even invokes our same principles" (Cabieses
Martinez 2012). Correa grounded his arguments in appeals to the leftist canon, asking, "Where in
The Communist Manifesto does it say no to mining?" and "What socialist theory says no to mining?"
A few months later, in an interview in *New Left Review*, he expressed exasperation with what he
saw as activists' "absurd" and "dangerous" opposition to resource extraction (Correa 2012).

While Correa and the organizations that signed the manifesto vehemently disagreed over the
model of development, they did agree on one thing: to each, the other represented a perversion

DOI: 10.4324/9780429344428-17

of leftism, a perversion particularly insidious for being cloaked in the language of radical transformation. Each side accused the other of betraying the principles of socioeconomic equality, popular empowerment, and anti-imperialism that have defined the Latin American Left for over a century. Correa identified himself with a regional movement of "socialism for the twenty-first century," named neoliberalism as the cause of myriad social, economic, and political ills, rejected US hegemony, and presided over a state that had dramatically increased social spending and that enjoyed widespread political support among the poor. His discourse resonated with a long history of popular calls for the expropriation and nationalization of natural resources. The anti-extractive social movements that opposed him traced their organizational lineage to worker, peasant, and indigenous struggles, and their critique of the extractive model was indebted to the systematic analysis of imperialism and dependency that characterizes Latin American critical thought. But they voiced a more recent radical demand: an end to the extractive model of development.

Why did activists who had for decades resisted neoliberalism now protest against a leftist government? More generally, what accounts for the emergence of radical anti-extractive movements? And how might they reshape resource politics across the globe?

Dominant approaches to the study of oil or mineral-dependent states focus on how resource dependency shapes regime type or economic development (Karl 1997; Weyland 2009). They conclude that such states tend to be authoritarian and corrupt, and rule over societies that are alternately portrayed as politically quiescent or prone to violent resource-related conflicts. Completing this picture of pathology is economic underdevelopment: "boom-and-bust" price cycles, profligate state spending, and a pervasive "rentier mentality" (Beblawi 1987) are seen to divert investment away from productive sectors and reproduce resource dependency.

My approach rejects such pessimistic determinism and expands the study of resource politics well beyond the halls of the petro-state.[2] In Ecuador, grassroots activists were key protagonists in the contentious politics of oil and mining. In dynamic conflict with state and corporate elites, popular mobilization shaped the political and economic consequences of resource extraction. The stakes of these conflicts were high. Constitutional authority, democratic sovereignty, and the possibility of a "post-neoliberal" state hung in the balance.

In the heat of political struggle, activists craft critiques of extraction and enact processes of resistance. I call these *resource radicalisms*, and I will show how they shape the strategies, identities, and interests of state and movement actors alike. The concept of resource radicalism brings into relief how intellectual production is intertwined with political mobilization. From rallying cries to animated debates to everyday reflection, activists analyze the prevailing order and articulate visions of a world otherwise.

Around the globe, conflict in relation to extraction, energy, and infrastructure has escalated—and it will only continue to do so in a rapidly warming and politically unstable world (Parenti 2011; Welzer 2015). Situated at the frontiers of capitalism's relentless expansion, mining and oil projects are sites of dispossession and contamination. They are structured by local, national, and global scales of political economy and ecology (Bebbington et al. 2017; Bridge and Le Billon 2017; Perreault 2017). As a result, they afford multiple venues of conflict. Due to their uneven geographic distribution, and thus uneven environmental and social impacts, natural resources are "intensely local" (Bebbington et al. 2017). At the same time, they are commodities in international supply chains shaped by the investment decisions of multinational firms and volatile global prices. Dangerous labor conditions and relative worker autonomy have historically made sites of extraction focal sites of class conflict. And these local conflicts also have national significance: governments in the global South have taken an acute interest in regulating oil and mineral sectors since the early twentieth century, including via direct ownership of extractive firms. As a key source of fiscal revenue, these sectors are considered "strategic"—a status justifying the deployment of physical force to protect extractive projects from protest or other disruptions. More fundamentally, in such national contexts, the processes of extraction and state formation have reinforced each other (Karl 1997; Smith 2007). Meanwhile, potent resource imaginaries, developed by movements and institutions, have shaped their political consequences (Bebbington et al. 2017; Kohl and Farthing 2012; Perreault and Valdivia 2010).

In Latin America, the politics of resource extraction are particularly charged. Across the region's diverse histories, resource extraction traces a long arc: colonial plunder, independence-era "enclave economies," mid-century nationalist projects of oil-fueled modernization, subsequent privatization and deregulation of hydrocarbon and mineral sectors, and, most recently, attempts at oil-funded equitable development. Over the course of four centuries, the extraction (or harvesting) and export of primary commodities has relegated the region to "peripheral" status in the global division of labor (Svampa 2015). This status, rooted in colonial domination, places it on the losing end of an unequal exchange of raw goods for refined or manufactured imports. Dependency only intensified after independence, with the proliferation of mines and plantations that functioned as economic enclaves, often foreign-owned and with weak linkages to the rest of the national economy. Although the history of extraction is a history of underdevelopment, natural resource sectors have long inspired developmental ambitions on the part of state officials—and hopes of radical sovereignty on the part of popular movements. Inspired by such visions, in the mid-20th century, several resource-dependent Latin American countries underwent forms of "endogenous development," investing rents in industrial sectors. Their goal was to ultimately diversify economies and export revenues. But ensuing neoliberal reforms of deregulation and market integration reinforced the reliance on primary sectors—a trend only exacerbated by the commodity boom of 2000 to 2014, and trade and financial dependency on China.

Recent leftist administrations in Latin America are ideal sites to explore resource conflict because of this history, and because both policy-makers and social movements have explicitly politicized—and radicalized—the relationship between development and extraction. In the process, they have raised deeper questions about the state, democracy, and the ecological foundations of global capitalism. Ecuador is an especially revealing window into regional, and global, resource radicalisms. It is among the most commodity-dependent economies on the continent, and has seen intense conflict between a leftist government committed to an extraction-fueled, broad-based development model and an array of movements militantly opposed to resource extraction in all forms.

The Ecuadorian dispute over resource extraction between a self-described socialist leader and the social movement activists who helped bring him to power testifies to a unique historical moment. In Latin America, the turn of the millennium was marked by the proliferation of "counter-hegemonic processes" in the halls of state power and in the streets (Escobar 2010). At the height of the Pink Tide in 2009, leftist administrations governed almost two-thirds of the region's population (Levitsky and Roberts 2011). But this moment was also marked by the intensification of an export-oriented, resource-intensive model of accumulation, highly dependent on not only North American and European but also Chinese capital. In Ecuador, activists who had protested decades of neoliberal policies in tandem with the region's leftist, critical, and decolonial intellectuals now resisted a leftist government and what they called "the extractive model" of development.[3]

The region is now home to a variety of resource radicalisms. Depending on the context, activists' grievances and demands center on indigenous rights, environmental contamination, labor exploitation, foreign ownership, territorial autonomy, and local self-determination; or, often, some combination thereof. In some cases, disputes over extraction pit leftists with histories of common political struggle against one another. Leftist governments in Bolivia, Ecuador, and Venezuela espoused a state-centric resource nationalism, while indigenous and popular environmental movements (*ecologismo popular*) struggling against the expanding extractive frontier envisioned a post-extractive future.[4] These movements articulated a novel critical discourse centered on the concept of extractivism that called into question the unity of state, nation, territory, and resources. Although this discourse has circulated transnationally in both activist (Klein 2014) and academic circles (Svampa 2015; Veltmeyer and Petras 2014), in Ecuador the radicalization of resource politics was both particularly acute and historically dynamic (Escobar 2010; Svampa 2015). It was acute because, during the presidential administration of Rafael Correa, the dispute over extraction became *the* primary source of discord between state actors and social movements—and among bureaucrats themselves. And it was historically dynamic because in the space of less than a decade, many popular sector organizations dramatically changed their position on resource extraction. In response to the social and environmental impacts of extractive projects, they abandoned their historic calls for expropriation, nationalization, and the collective ownership of the means and products of extraction—what I call *radical resource nationalism*—and embraced *anti-extractivism*: the militant opposition to all forms of resource extraction. In the streets and in the courts, in popular assemblies in affected communities and on nature walks to the sites of planned extraction, they identified and resisted the disparate nodes of extractivism. From their perspective, each of these nodes reproduced the extractive model—and furnished an opportunity to disrupt its ubiquitous development.

Resource radicalisms

While the ascendency of new leftist governments may have unevenly transformed resource *policy*, it has fundamentally transformed the *politics* of extractive economies (Hogenboom 2012, 151–2). Indigenous, peasant, environmental, and labor movements, among others that had protested against neoliberalism, paved the way for the electoral success of leftist parties. In the wake of electoral victories, these movements demanded a range of deeper initiatives to reorganize the relationship between state, society, economy, and nature—from wholesale nationalization to the construction of a "post-extractive" economy—that leftist governments have not implemented. From the perspective of these movements' activists, such reorganizations are vital to the project of decolonizing a continent in which the history of resource extraction is intimately tied to that of conquest and subjugation. In response to such demands, leftist national governments in countries such as Argentina, Bolivia, Brazil, Ecuador, and Venezuela have often reprimanded indigenous and environmental groups, framing them as obstacles to the national good of resource-funded development. Meanwhile, as the Ecuadorian case reveals, these groups have struggled to organize an anti-extractive mass movement with the size and capacity of the earlier anti-neoliberal popular bloc, a point to which I return in the conclusion.

In Ecuador, both neoliberal and nationalist policies have been unevenly implemented. But as ideologically inflected policy paradigms, they oriented state and corporate actors vis-à-vis resource sectors. They formed part of the political terrain that structured (and was structured by) the interactions between state actors and social movements. And these governance models were imbued with social meaning via the emic categories through which they were apprehended and analyzed, including those articulated by social movements.

Resource radicalisms are articulated by popular organizations and social movements, whether oil and mine workers' unions, urban neighborhood associations, environmental groups, or indigenous federations. Their members, militants, and activists are the architects of these radical critiques of prevailing models of extraction, critiques which not only guide social movement strategy—and, in moments of confrontation, elicit repressive responses from the state—but shape the terms and stakes of political conflict. State actors responded to new critiques of resource extraction by redeploying the terms of critique as justifications for extraction.

In Ecuador, popular movements have articulated two resource radicalisms—radical resource nationalism and anti-extractivism—in the course of struggles over economic development, resource extraction, territorial rights, and democratic sovereignty. These radicalisms map onto two different political periods (1990 to 2006, and 2007 to 2017, respectively), but not neatly or discretely: prior to their bifurcation as two distinct discourses, a nascent rejection of oil-led development coexisted alongside calls to nationalize oil resources. Popular movements consolidated and deployed these resource radicalisms in opposition to the prevailing paradigm of resource governance (that is, neoliberalism and post-neoliberal resource nationalism). And in each period, activists' critiques and processes of resistance also shaped state practices. They forced state actors to adopt new ideological justifications for their promotion of extraction, incited ideological disputes among bureaucrats, and slowed down the development of large-scale mining projects and oil exploration.[5]

During what the social movement manifesto refers to as the "yesterday" of neoliberalism, the same organizations that now fought against extractivism had instead demanded the nationalization of resource extraction. They saw the nationalization of ownership as vital to the recuperation of national sovereignty and the redistribution of national wealth. This was a regional pattern in: Argentina, Bolivia, Brazil, Chile, Venezuela, and elsewhere, Indigenous, peasant, trade union, and environmental organizations resisted the deregulation and privatization of resources such as oil, minerals, water, and natural gas (Kohl and Farthing 2012; Perreault and Valdivia 2010). These groups demanded various forms of popular control over resource extraction, ranging from nationalization to worker control to local management by the Indigenous peoples whose territory overlapped with hydrocarbon reserves. The hegemony of neoliberal policies allowed for this provisional alignment of social movement organizations with such distinct political trajectories and positions on extraction. I call this formation *radical resource nationalism*. As Benjamin Kohl and Linda Farthing (2012) discuss with regard to the case of Bolivia, this popular resource imaginary is firmly "anti-imperialist and proto-nationalist." It is also an emotionally charged appeal that is often "formed around grievances rather than potentialities and focus[ed] on demands to recoup what has been lost and continues to be lost through foreign-controlled extraction" (Kohl and Farthing 2012, 229).

In Ecuador, during that same period and alongside the crystallization of radical resource nationalism, another radical position on extraction was beginning to emerge. In the course of conflictual and sometimes violent encounters between oil companies and Indigenous peoples of the Amazon, the latter articulated a militant defense of territory against oil exploration. The demands voiced by Sarayaku, Achuar, and Shuar leaders provided the discourses and shaped the political strategies that would be subsequently unified under the banner of anti-extractivism.

These intertwined critiques of extraction coexisted until the new political conjuncture of the late 2000s converted them into mutually opposed positions. In this new context, marked by Correa's inauguration (in 2007), a Constituent Assembly (2007–2008) that rewrote the Constitution, and the Correa government's avid promotion of large-scale mining (2009–2017); the first position, radical resource nationalism, became an ideological resource for an administration seeking to take political and economic advantage of soaring global demand for primary commodities. But state actors reinterpreted nationalism as the *redistribution* of resource rents, rather than *expropriation* and national ownership. This was a nationalism amenable to courting foreign capital and deepening global market integration. In response, social movement activists and critical intellectuals abandoned their previous demands for nationalization, and reoriented their resistance to target what they now called the extractive model, amplifying the history of localized opposition to oil extraction in the Amazon into a wholesale anti-extractivism. This model, they argued, pollutes the environment, violates collective rights, reinforces dependency on foreign capital, and undermines democracy. The gravity of the extractive model's political, economic, and environmental consequences is matched by the *longue durée* timescale of its domination: for anti-extractive activists, extractivism originated with European conquest and was only reproduced by the recent turn to post-neoliberal resource nationalism.

Although its elements had existed in an inchoate form prior to Correa's rise to power, the reign of an avowedly post-neoliberal administration was the key historical condition for a mode of critique and resistance that zeroed in on resource extraction itself. Correa spoke of the nation, sovereignty, democracy, a "solidary" economy, equality, citizenship, participation, and, most importantly and poetically, of an end to the "long night of neoliberalism." He emphasized paying off the "social debt" accumulated under decades of austerity and economic crisis. Drawing on a long-established discursive repertoire of social resistance, he identified a cast of political and economic enemies: the international financial system, foreign corporations, domestic oligarchs, and corrupt political parties. In direct response to resounding popular demands, he called for a constituent assembly to refound the state. However, in part because of these clear ideological signals, Correa found himself in heated political conflict with indigenous, campesino, environmental, labor, and feminist social movements. If even a self-identified leftist government could reproduce or, worse, intensify the rapacious exploitation of nature and the subordination of Indigenous communities to a homogeneously defined nation, in the process violating collective rights and centralizing power, then, social movement activists concluded, the root of the problem was not the ideological stripe of elected officials but the "civilizational" model that encompassed socialism and capitalism alike. The crystallization of this discourse in turn fomented a dispute among the Left over whether emancipation lies in an alternative form *of* economic development, or in alternatives *to* the very concept of development, seen as historically rooted in relations of coloniality. In Escobar's (2010) terms, Ecuador exemplified the conflict between "neo-developmentalism and post-development" (see also Svampa 2015).

Most of the extant scholarship employs extractivism as a descriptive or analytical term to refer to extractive activities, the policies and ideologies that promote them, their socio-environmental effects, and the forms of resistance that they provoke (e.g. Acosta 2009; Veltmeyer and Petras 2014). In contrast, my analytic perspective historicizes this critical discourse, and regards social movement activists and intellectuals as protagonists in crafting its conceptual architecture. This mode of analysis does not regard discourse as ontologically distinct from or epiphenomenal of "reality," but rather takes discourse to be the linguistic mediation of social relations and the concrete medium through which we reflect upon, make, and remake our social worlds. Anti-extractive activists aspire to a post-extractive future characterized by a harmonious relationship between humans and nature. In envisioning this future, they drew on the grievances and demands of southeastern Amazonian indigenous communities, which formed the basis for a wholesale rejection of extraction in all forms. Under the rubric of anti-extractivism, a multi-scalar alliance of indigenous and environmental movements enacted new forms of democratic participation, organized outings to the territories slated for extraction, produced their own knowledge regarding socio-environmental impacts, brought cases to the Constitutional Court, and physically blockaded attempts to develop mining or oil projects. The systemic object of their critique was immanent in the spatial contours of their resistance. Traversing mountains, wetlands, and rainforest, they mobilized a network of directly affected communities along the frontiers of extraction, confronting the extractive model at the roots of what they saw as its expansionary imperative.

Conclusion

At this juncture, these novel democratic practices face new threats and challenges. On the one hand, the Latin American Left suffered a series of defeats. In 2013, Hugo Chávez, the first president elected in what would later be named the Pink Tide, died of cancer. The next year, the commodity boom came to a decisive end with a precipitous drop in oil prices, and recessions followed. In quick succession, these dramatic events were followed by the election of the conservative president Mauricio Macri in Argentina, the parliamentary coup that removed Dilma Rousseff from power in Brazil, Bolivian voters' rejection in a popular referendum of Evo Morales's attempt to run for a fourth term (a rejection subsequently overruled by the country's Constitutional Court), Venezuela's descent into seemingly intractable political-economic crisis, and the 2018 defeat of the Brazilian Workers' Party presidential candidate Fernando Haddad by Jair Bolsonaro, an open admirer of the military dictatorship that ruled that country from 1964 to 1985. More recently, Andrés Arauz, former minister under Correa, lost the 2021 presidential elections to right-wing banker and millionaire Guillermo Lasso. Although many factors determined this outcome, among them was the deep split in the Ecuadorian Left, with Arauz failing to win the endorsement of the CONAIE and many environmental and Indigenous activists instead rallying around the candidacy of anti-mining leader Yaku Pérez.

On the other hand, at the time of writing, the Right has not achieved regional political hegemony: in Bolivia, Argentina, Peru, and Chile, Leftists defeated right-wing candidates and now hold national office. In Chile, a massive social uprising that began in late 2019 resulted in a major victory: the country is now undergoing a Constitutional Convention, to rewrite the dictatorship-era constitution. Progressive forces swept the delegate elections, many with ties to feminist, student, environmental, and Indigenous movements.

These political fluctuations are unfolding against a backdrop of mounting and multifaceted crises. At the time of writing, Covid continues to ravage the region and Latin America had the most severe pandemic-related economic fallout of anywhere in the world (Martin 2021). The region is also deeply mired in unsustainable and illegitimate debt, and extremely vulnerable to climate change. Meanwhile, the extractive model of development continues to destroy ecosystems and threaten Indigenous rights.

Amid such turmoil, what is the possibility of Latin American leftists reconstructing a viable political project that can weave together egalitarian and ecological demands? The future is, more than ever, uncertain and unpredictable. But if the past three decades of contentious politics in the region offer any indication, resource extraction will remain a historically specific horizon and a fertile ground for radical politics. In this tumultuous context, we can expect militant activists to refashion their critiques, revise their strategies, and assemble new resource radicalisms.

Notes

1 "Manifiesto del encuentro de movimientos sociales del Ecuador por la democracia y la vida," August 9, 2011 (http://www.inesc.org.br/noticias-es/2011/agosto/manifiesto-del-encuentro-de-movimientos-sociales-del-ecuador-por-la-democracia-y-la-vida).
2 As Michael Watts (2005) argues, the notion that primary commodities, abstracted from social power relations, is a form of "commodity determinism". See also Huber (2013).
3 For the term "decolonial," see Mignolo (2011).
4 The term *ecologismo popular* was developed by Joan Martínez-Alier, and refers to territorialized conflicts that arise in response to the detrimental socio-environmental effects of economic growth, which threaten local means of subsistence. The actors that mobilize in such conflicts may or may not explicitly invoke environmental discourse. See Martínez Alier (2007), Latorre (2009).

5 For example, the Correa administration embraced a variant of resource nationalism that was devoid of much of its radical content (for example, no nationalizations or expropriations) and ideologically repurposed it to delegitimize anti-extractive resistance and promote extraction at all cost.

References

Acosta, Alberto. 2009. *La Maldición de La Abundancia*. Quito, Ecuador: Ediciones Abya-Yala.

Arditi, Benjamin. 2008. "Arguments about the Left Turns in Latin America: A Post-Liberal Politics?" *Latin American Research Review* 43, no. 3: 59–81.

Bebbington, Anthony, Abdul-Gafaru Abdulai, Marja Hinfelaar, Denise Humphreys Bebbington, and Cynthia Sanborn. 2017. "Political Settlements and the Governance of Extractive Industry: A Comparative Analysis of the Longue Durée in Africa and Latin America." *ESID Working Paper* No. 81. Manchester: Global Development Institute, University of Manchester.

Bebbington, Denise Humphreys. 2012. "Consultation, Compensation and Conflict: Natural Gas Extraction in Weekhayek Territory, Bolivia." *Journal of Latin American Geography* 11, no. 2: 49–71.

Beblawi, Hazem. 1987. "The Rentier State in the Arab World." In *The Rentier State*, edited by Hazem Beblawi and Giamcomo Luciani, 49–52. New York: Routledge.

Bowen, James D. 2011. "Multicultural Market Democracy: Elites and Indigenous Movements in Contemporary Ecuador." *Journal of Latin American Studies* 43, no. 3 (August 2011): 451–83.

Bridge, G., and P. Le Billon 2017. *Oil*. Hoboken, NJ: John Wiley & Sons.

Cabieses Martinez, Francisca. 2012. "Revolución ciudadana, el camino del Ecuador," Punto Final, May 25, 2012 (http://www.puntofinal.cl/758/rafael758.php).

Cameron, Maxwell A., and Kenneth E. Sharpe. 2010. "Andean Left Turns: Constituent Power and Constitution-Making." In *Latin America's Left Turns: Politics, Policies, and Trajectories of Change*, edited by Maxwell A. Cameron and Eric A. Hirschberg, 61–80. Boulder, CO: Lynne Rienner Publishers.

Ciplet, David, and J. Timmons Roberts. 2019. "Splintering South: Ecologically Unequal Exchange Theory in a Fragmented Global Climate." In *Ecologically Unequal Exchange*, 273–305. Cham, Switzerland: Palgrave-Macmillan.

Correa, Rafael. 2012. "Ecuador's Path: Interview." *New Left Review* 77: 88–112.

Escobar, Arturo. 2010. "Latin America at a Crossroads: Alternative Modernizations, Post-Liberalism, or Post-Development?" *Cultural Studies* 24, no. 1: 1–65.

Fligstein, Neil, and Doug McAdam. 2011. "Toward a General Theory of Strategic Action Fields." *Sociological Theory* 29, no. 1: 1–26.

Foucault, Michel. 1968. "Politics and the Study of Discourse." In *The Foucault Effect: Studies in Governmentality*, edited by Graham Burchell, Colin Gordon, and Peter Miller, 53–72. Chicago: University of Chicago Press.

———. 1981. "The Order of Discourse." In *Untying the Text: A Post-Structuralist Reader*, edited by Robert Young. New York: Routledge.

———. 2005. *Security Territory Population: Lectures at the College de France, 1977–78*. Translated by Burchell, Graham. New York: Palgrave Macmillan.

Frank, Andre Gunder. 1972. *Lumpen-Bourgeoisie, Lumpen-Development: Dependency, Class, and Politics in Latin America*. New York: Monthly Review Press.

Ghandi, Abbas, and C.-Y. Cynthia Lin. 2014. "Oil and Gas Service Contracts around the World: A Review." *Energy Strategy Reviews* 3: 63–71.

Glaeser, Andreas. 2010. *Political Epistemics: The Secret Police, the Opposition, and the End of East German Socialism*. Chicago: University of Chicago Press.

Gledhill, John. 2011. "The Persistent Imaginary of "The People's Oil": Nationalism, Globalisation and the Possibility of Another Country in Brazil, Mexico and Venezuela." In *Crude Domination: An Anthropology of Oil*, edited by Andrea Behrends, Stephen P. Reyna, and Günther Schlee, 165–89. New York: Berghahn Books Ltd.

Gudynas, Eduardo. 2009. "Diez Tesis Urgentes Sobre El Nuevo Extractivismo." In *Extractivismo, Política y Sociedad*, edited by CAAP, 187–225. Quito: Centro Andino de Acción Popular (CAAP).

Haslam, Paul Alexander, and Nasser Ary Tanimoune. 2016. "The Determinants of Social Conflict in the Latin American Mining Sector: New Evidence with Quantitative Data." *World Development* 78: 401–19.

Haslam, Paul, and Pablo Heidrich. 2016. "From Neoliberalism to Resource Nationalism: States, Firms and Development." In *The Political Economy of Natural Resources and Development: From Neoliberalism to Resource Nationalism*, edited By Paul A. Haslam and Pablo Heidrich, 1–32. New York: Routledge.

Hogenboom, Barbara. 2012. "Depoliticized and Repoliticized Minerals in Latin America." *Journal of Developing Societies* 28, no. 2: 133–58.

Huber, Matthew T. 2013. *Lifeblood: Oil, Freedom, and the Forces of Capital*. Minneapolis: University of Minnesota Press.

Karl, Terry Lynn. 1997. *The Paradox of Plenty: Oil Booms and Petro-States*. Berkeley: University of California Press.

Klein, Naomi. 2014. *This Changes Everything: Capitalism vs. the Climate*. New York: Simon & Schuster.

Kohl, Benjamin, and Linda Farthing. 2012. "Material Constraints to Popular Imaginaries: The Extractive Economy and Resource Nationalism in Bolivia." *Political Geography* 31: 225–35.

Latorre, Sara. 2009. "El ecologismo popular en el Ecuador: pasado y presente." Instituto de Estudios Ecuatorianos. https://www.iee.org.ec/ejes/sociedad-alternativa-2/el-ecologismo-popular-en-el-ecuador-pasado-y-presente.html.

Levitsky, Steven, and Kenneth M. Roberts 2011. "Introduction: Latin America's 'Left Turn': A Framework for Analysis." In *The Resurgence of the Latin American Left*, edited by Steven Levitsky and Kenneth M. Roberts, 1–30. Baltimore: Johns Hopkins University Press.

Luong, Pauline Jones, and Erika Weinthal. 2010. *Oil Is Not a Curse: Ownership Structure and Institutions in Soviet Successor States*. Cambridge: Cambridge University Press.

Mahdavy, Hussein. 1970. "The Patterns and Problems of Economic Development in Rentier States: The Case of Iran." In *Studies in Economic History of the Middle East*, edited by M.A. Cook London: Oxford University Press.

Mahler, Garland. 2018. "What/Where is the Global South?" https://globalsouthstudies.as.virginia.edu/what-is-global-south, accessed December 8, 2018.

Martin, Eric. 2021. "Latin America to Rebound From Worst Recession in Two Centuries Bloomberg." *Bloomberg*, March 20, 2021. Available at https://www.bloomberg.com/news/articles/2021-03-20/latin-america-to-rebound-from-worst-recession-in-two-centuries?sref=apOkUyd1.

Martínez-Alier, Joan. 2007. "El ecologismo popular." *Ecosistemas* 16, no. 3: 148–151.

Mignolo, Walter. 2011. *The Darker Side of Western Modernity: Global Futures, Decolonial Options*. Durham: Duke University Press.

Mouffe, Chantal. 2000. *The Democratic Paradox*. New York: Verso.

Nem Singh, Jewellord T. 2012. "Who Owns the Minerals? Repoliticizing Neoliberal Governance in Brazil and Chile." *Journal of Developing Societies* 28, no. 2: 229–56.

Ordóñez, Andrea, Emma Samman, Chiara Mariotti, and Iván Marcelo Borja Borja. 2015. *Sharing the Fruits of Progress: Poverty Reduction in Ecuador*. London: Overseas Development Institute.

Parenti, Christian. 2011. *Tropic of Chaos: Climate Change and the New Geography of Violence*. New York: Nation Books.

Perreault, Tom. 2017. "Tendencies in Tension: Resource Governance and Social Contradictions in Contemporary Bolivia." In *Governance in the Extractive Industries*, edited by Lori Leonard and Siba N. Grovogui, 17–38. London: Routledge.

Perreault, Tom, and Gabriela Valdivia. 2010. "Hydrocarbons, Popular Protest and National Imaginaries: Ecuador and Bolivia in Comparative Context." *Geoforum* 41, no. 5: 689–99.

Prebisch, Raúl. 1950. "Crecimiento, Desequilibrio y Disparidades: Interpretación Del Proceso de Desarrollo Económico." *Estudio Económico de América Latina y El Caribe* 164, no. 1: 3–89.

Riofrancos, T. 2020. *Resource Radicals: From Petro-Nationalism to Post-Extractivism in Ecuador*. Durham, NC: Duke University Press.

Rosales, Antulio. 2013. "Going Underground: The Political Economy of the 'Left Turn' in South America." *Third World Quarterly* 34, no. 8: 1443–57.

Ross, Michael. 2001. "Does Oil Hinder Democracy?" *World Politics* 53: 325–61.

Ruiz Acosta, Miguel, and Pablo Iturralde. 2013. *La Alquimía de La Riqueza: Estado, Petroleo y Patrón de Accumulación En Ecuador.* Quito: Centro de Derechos Económicos y Sociales.

Shever, Elana. 2012. *Resources for Reform: Oil and Neoliberalism in Argentina.* Santiago: Stanford University Press.

Smith, Benjamin. 2007. *Hard Times in the Lands of Plenty: Oil Politics in Iran and Indonesia.* Ithaca: Cornell University Press.

Svampa, Maristella. 2015. "Commodities Consensus: Neoextractivism and Enclosure of the Commons in Latin America." *The South Atlantic Quarterly* 114, no. 1: 65–82.

Tockman, Jason, and John Cameron. 2014. "Indigenous Autonomy and the Contradictions of Plurinationalism in Bolivia." *Latin American Politics and Society* 56, no. 3: 46–69.

Veltmeyer, Henry, and James Petras, eds. 2014. *The New Extractivism: A Post-Neoliberal Development Model of Imperialism of the Twenty-First Century?* London: Zed Books.

Watts, Michael. 2005. "Righteous Oil? Human Rights, the Oil Complex, and Corporate Social Responsibility." *Annual Review of Environment and Resources* 30: 373–407.

Webber, Jeffery R. 2014. "Revolution against 'Progress': Neo-Extractivism, the Compensatory State, and the TIPNIS Conflict in Bolivia." *Crisis and Contradiction: Marxist Perspectives on Latin America in the Global Political Economy*, 302.

Welzer, Harald. 2015. *Climate Wars: What People Will Be Killed for in the Twenty-First Century.* Hoboken, NJ: Wiley.

Weyland, Kurt. 2009. "The Rise of Latin America's Two Lefts: Insights from Rentier State Theory." *Comparative Politics* 41, no. 2: 145–64.

Woo-Cumings, Meredith, ed. 1999. *The Developmental State.* Ithaca: Cornell University Press.

15 The fruits of labor or the fruits of nature?

Toward a political ecology of labor in Central America

Andrés León Araya

Introduction

Latin American Political Ecology (LAPE) is an expanding field of inquiry. The last few decades have seen a sharp increase in both the number of researchers subscribing to it, and the topics being studied. From indigenous ontologies and practices, through conflicts over water and other resources or commons, to mining and monocrops, Latin American scholars are analyzing the interactions between society and nature, from varied and innovative perspectives. LAPE, as with political ecology in general, is far from a homogenous field. What seems to keep this literature together is the importance given to producing situated knowledges that speak from and to Latin American realities and that promotes and strengthens emancipatory political practices (Alimonda 2017). As Enrique Leff (2017), one of the better-known exponents of LAPE, has argued, the greatest difference between the political ecology coming from the global North and the global South is that, while the former focuses in understanding the impacts of the global unequal ecological distribution, the latter is more interested in the production of more equal and sustainable worlds of life ("*mundos de vida*").

I would argue that another common tendency in most LAPE literature is a generalized disregard for workers, and indeed for the working class, as a focus of analysis. It is important to point out that in this chapter I differentiate between, on the one hand, work and workers; and, on the other hand, labor. The first one refers to those people whose life depends on working for a salary. The second one to the human activity of producing landscapes to secure the production/reproduction of social life (see for example, Barca 2019; Gutierréz Aguilar 2015; Kirsch and Mitchell 2004; Navarro 2015).

This disregard emerges from both LAPE's theoretical foundations and methodological tendencies. Theoretically, one of its immediate antecedents and the source of inspiration for many of its main exponents are the debates regarding imperialism and the place of Latin America within the global economy, which took place in the region from the 1960s onward and that are usually brought together under the mantle of "dependency theory." As we will see, one of the central debates within Latin American dependency theory had to do with whether capitalism in the region should be understood in terms of its articulation to the global market, or through the expansion and deepening of the wage labor relation. In general, the first thesis, based on commercial circulation, became more predominant and intertwined with the post-structuralist and decolonial critique of modernity, deeply influencing the ways in which LAPE scholars think about capitalism and, thus, work and workers.

Methodologically, we find at least three tendencies within the literature: i) the prevalence of "communities" and "peoples" (*pueblos*) as the main subjects of analysis and political transformation; ii) the activist-scholar perspective of many of its practitioners; and iii) the creation of an absolute split between the logics of capital, and those of the communities and peoples. Further,

DOI: 10.4324/9780429344428-18

LAPE scholars tend to prefer participatory and dialogical research techniques, influenced by Freire (1996) and Fals-Borda (1984). Since workers tend to spend less time in community spaces, and might have less time available or interest, they do not tend to be as involved as other people in participatory research processes, thus becoming a sort of blind spot for much of the LAPE.

The result is that a clear spatial division is drawn between the "community" and the "plantation/factory/plant/mine," rendering invisible the experience of those people who continuously traverse between both worlds. Further, by confronting communities and peoples with industrial and extractivist activities, workers are often seen as alienated subjects with little political liberation potential. This is a significant and costly oversight. Workers produce the laboring landscapes of the plantations/factories/plants/mines, and it is the exploitation of their labor what allows the accumulation of capital. At the same time, they play an important role in the dynamics of social reproduction of the communities, either through the salaries that they receive, or through other activities that they partake in.

In this chapter, I argue for a political ecology of labor that reconciles the debates regarding the relation between nature and society with the experience and practices of plantation workers. To do so, I will first present a brief explanation of the oversight of labor and workers within LAPE, by presenting its dependency theory roots. Next, I will propose a political ecology approach to labor and workers that allows us to suture the analytical divide between community and plantations, by thinking about the landscapes that are produced by human labor, using as an example the cases of pineapple production in Costa Rica and oil palm production in Honduras.

Labor and dependency theory

In the aftermath of World War II, the development question came to dominate much of the global political landscape. This question was framed in terms of how the so-called "underdeveloped," or the Third World in the Cold War vocabulary, could reach the levels of technological advance and economic welfare of the developed world. According to the dominant modernization theory, presented by authors such as W.W. Rostow (1990), the underdeveloped societies needed to repeat the path taken by the developed ones, by following a set of steps. Very quickly, critical scholars—many from those 'underdeveloped' regions—began to push back against these theories due to their ethnocentric, teleological,[1] and simplistic approach to the question of development and progress.

One of the principal sites of this contention was Latin America, where, from the 1960s onwards, a combination of development-oriented state officials and escalating levels of worker and peasant organization created the conditions for a vibrant debate regarding the place of the region within the global dynamics of capital accumulation and its uneven and combined character (Grosfoguel 1997). Oversimplifying a complex affair, we find two camps: On the one hand, the "circulationists," for whom the origins of capitalism were to be found in the creation of the world market and the conquest of the Americas.[2] On the other hand, for the "mode of production" camp, what defined capitalism was the dominance of wage labor, as opposed to "free" indentured or subsistence labor, and the production and circulation of surplus value.[3] In general terms, the circulationist position came to dominate the (mainstream) dependency (*dependentista*) school of thought, and with it came the emphasis on forms of (uneven) exchange and rent extraction, while leaving aside or downplaying the attention to labor and class formation.[4]

In the 1990s, when dependency theory, as most modernist theories, came under attack by the postmodern critique, it was the modes of production's emphasis on labor and class that was mainly rejected. Under the coloniality debate, the discussions on circulation moved from commodities to ideas and theories, to propose that the particularity of the Latin American identities were the more than 500 years of coloniality which had led us to a process of estrangement with our own roots and position in the world.[5] As such, the former debates regarding economic policy, class struggle and the state turned into discussions regarding culture and identity, particularly the recuperation of non-European lines of thought in the continent, the struggles for autonomy of indigenous, afro-descendant and peasant groups, and the community as the locus of the set of new possible worlds that were to come. This, of course, is not to say that the discussion on the economy and the state disappeared, but rather that the standpoint from which it is seen, shifted.[6] We find a very good example in LAPE.

LAPE attempts to better understand and transform the relations between society and nature in the region, from a perspective that is closer to those groups that they deem to be in the forefront of the dispute against capitalism. This dispute is not only material, but also symbolic, in the sense of the modes of thinking about nature and the relations between humans and non-humans. What lies in the back of this perspective is an indictment of the modernist project as a whole, including the forms of political organization of the "traditional" left, such as labor unions and peasant centrals. While this anti-neoliberal is not exclusive of Latin America nor of the decolonial perspective,[7] we find a good example in some of LAPE's better known exponents. The late Héctor Alimonda (2017, 40) proposed that we think about LAPE as "… an encounter between the Latin American tradition of critical thought and the vast experiences and strategies of resistance of the peoples [*pueblos*] against the pillage and 'ransacking economy' [*economía de rapiña*]." Arturo Escobar (2017), another of the better-known male proponents of LAPE, is far more lapidary when he claims that "those who still insist in the path of development and modernity are either suicidal, or at least ecocidal, and definitely anachronist" (66). For him,

> The theoretical and political contributions for rethinking the region reverberate across the whole continent, in the gatherings of the *pueblos*, in the thinking *mingas* [collective work], in the debates of the movements and collectives, in the assemblies of the communities in resistance, in the youth, women, peasant and environmentalist mobilizations, and definitely also, within those sectors that have been considered traditionally as the places par excellence of critical thought, such as the academia and the arts.
>
> (52–53)

Also, in the last few decades, following the rise of what came to be known as the "pink tide" (left-leaning) governments in South America, the notion of (neo)extractivism was accrued as a critique of the reliance of these governments on the exploitation of natural resources, particularly oil and gas extraction and mining. According to this line of thought, the logic of extractivism is based on the idea that there is a difference within capitalism between the extraction of value from labor (surplus value), as in the case of industrial activities, and the direct extraction of value from nature. Of course, industrial capitalism is based on the transformation of raw materials into commodities. However, while in industry the transformation of raw materials entails "adding" value, in terms of labor, technology, infrastructure, among others; in extractivist activities, rather than addition, what we find is subtraction, through the destruction of the sources of value: labor and nature.

Authors such as Eduardo Gudynas (2009) and Maristella Svampa (Svampa and Viale 2014) have defined extractivism in terms of a mass of natural resources that are extracted and exported with little added value and minimal linkages to the "national economies." Initially used to think about mining, with the boom in South America of monocrops such as soy in the 1990s, the notion was extended to include forms of agribusiness. Agrarian extractivism argues that production of soy, sugarcane, palm oil or bananas do not entail value-added processing, sectoral linkages, or employment generation (McKay 2017). As such, agrarian extractivism resonates strongly with dependency theory, particularly in arguments about center–periphery relations, the uneven terms of global exchange and the rentist patterns of periphery capitalism.[8]

From a Central American perspective, it is worth wondering when exactly did the regional strategy of articulation to the world market *stop* being agrarian and extractivist. While the isthmus's economic structure has changed significantly in the last few decades, one of the constants is the dependence on exports for the accumulation of capital: from bananas, coffee, and sugar to oil palm, tropical fruits (melon, pineapple, citrus, among others), tubers, and flowers. Further, it is also important to point out a significant difference between the location of those who write about LAPE, and those places that do not seem to fit their definitions. Unlike South America, and particularly soy plantations, in Central America, the continued capture of labor is fundamental for the operations of monocrop plantations.[9] Unlike soy, much more dependent on industrial technology and mechanization, coffee, bananas, oil palms, and pineapples are heavily reliant on cheap and available labor pools. And while the living and working conditions of this labor are very much deplorable, it is clear that an approach to the political ecology of these places that mainly ignores not only these conditions, but also the crucial place of workers in the reproduction of the plantation economies, will simply not do.

Toward a political ecology of work

Since at least the 1970s, feminist scholars have pushed us to think not only about the production process of commodities under capitalism, but also to focus on the dynamics of reproduction: the set of nonmarketized, and thus "invisible," relations, which supply the conditions of possibility for capital accumulation (the "visible" economy), and life in general, to take place (Fraser 2017). Since most of this labor of reproduction is done by women, these activities tend to be feminized, rendering invisible in most cases their contribution to the sustaining of life and thus, as without value.

Something similar happens with nature. One of the central tenets of Western modernity, including capitalism, is the separation between society and nature, with human (masculine) labor as a form of translation between both. This separation is not one between equals; rather, it places nature not only below, but at the service of society. As a result, labor, and not nature, is understood as the source of value.[10] Thus, similar to the case of feminized reproduction labor, nature, and its centrality for the reproduction of life, is rendered invisible and understood as without value.

This has led many ecofeminist and feminist political ecologists to approach capital and capitalism, not from the perspective of production as such, but from that of the production and reproduction of life as a whole, and the centrality of both, feminized labor and nature, in this process.[11] In this way, by exploring how the split between, on the one hand, nature and society, and, on the other hand, between production and reproduction iterated and reiterated, we can then shed light to the ways in which inequalities operate under capital, and where "… ecology is therefore the system of relationships between what is used to produce, what is produced, the waste of production, the bodies of those who produce, and the environment in which production, reproduction and waste take place" (Barca 2012, 66–67). However, although this approach allows us to expand what we understand as working class, to include those people who occupy the lower places in the labor hierarchy—the informal sector, the care economy, among others—and who tend to work in the more polluted places, with the most unhealthy conditions, there has been a tendency in most of the literature to think about these subject through other categories, such as "subaltern," "poor," or "communal," and to ignore, or underplay, the experience of those wage workers who work in capitalist operations, such as industrial plants and plantations (Barca 2014).

The result, as I mentioned before, is a spatial and analytical differentiation between the "community" and the "plantation." I believe this is part of a tendency within much of the environmental justice and LAPE literature to think about the operations of capital only in negative terms. That is, only in terms of what is being destroyed and lost—the forest, the river, the peasant farms, etc.—, not what is created in the process: the monoculture fields, the extraction, processing and packaging plants, the roads and ports, among others. The reasons for this bias are evident and understandable. Communities, particularly those amid environmental conflicts, are more accessible than plantations, and people inhabiting these communities tend to be more willing to talk than the employees and managers of the companies. Also, politically and ethically, working with people and organizations that are struggling against capital and have less chances of having their voice heard, is much more appealing, and urgent, than focusing on the ever-powerful transnational firms that stand against them.

Nonetheless, understanding what is being produced by the operations of capital, in terms of both landscape and subjects, is fundamental for their disruption (Barca 2014; Mitchell 2003). In this line, Neil Smith's (2008) ideas on the production of nature become very useful. For him, the separation between society and nature under capitalism is a false one, since what we tend to take for granted as "natural" is very much the result of the capitalist process of uneven development. In their search for profit, capitalist ventures transform and produce a patchy set of differentiated and unequal landscapes that are, of course, also the result of the dynamics of contestation by different groups. As such, peasant communities and *milpas* are as much the product of this uneven development, as the pineapple or oil palm plantations and their workforces. What we need to make sense of are the relations that articulate these differentiated spaces, and here the strategic position of plantation workers becomes crucial. Not only are they the ones who produced—and reproduce—the plantations with their labor, but they also live, depend upon, and contribute to the production and reproduction of the communities.

In general terms, monoculture plantations in Central America cannot exist without the exploitation of the labor of the neighboring communities. How and why this labor is available, is a question that needs to be answered within the historical conditions of each place. However, what most of them have in common is the destruction of previous communal forms of organizing the reproduction of life. These forms were not fully monetized and rested in complex sets of activities that might have included forms of collective labor, keeping house gardens, growing *milpa*, keeping some small farm animals, and seasonal agricultural work. With the attempts of modernizing the region's countryside in the 1960s and 1970s, these worlds were rattled by state attempts to implement agrarian reforms that would "bring" them into the national economy, by having them produce agrarian commodities: either foodstuffs for the domestic market, or exportable crops (Kay 1998). Thus, when the 1990s rolled in with the imposition of the neoliberal project, the logic of monocrops was already installed, and naturalized, in many Central American communities. Let me present two examples.

In the Bajo Aguán region, in the Honduran North Coast, impoverished peasant households from different parts of the country began arriving in the 1970s as part of an agrarian reform colonization project. Through a painful and long process that I explore in some detail elsewhere (León Araya 2017, 2019), these peasant families transformed the region's landscapes from thick bush, scattered peasant and Garifuna (black indigenous) communities, and extensive cattle ranches, into kilometers upon kilometers of geometrically arranged oil palm plantations. At the same time, households which used to live off their own labor and exchanging or selling the surpluses in their communities became members of oil palm-producing cooperatives, dependent now on a monetary income in the plantations to secure their social reproduction. In the process, these households subvert some of our more rooted understandings on peasants, in both academic discussions and everyday conversations, as smallholding units, producing for their procurement first, for the market later, and dependent on unpaid family labor (Shanin 1982; van der Ploeg 2010).

When in the 1990s a new law made the previously unsellable agrarian reform lands occupied by the cooperatives available in the market, seven out of every ten hectares distributed previously were sold or grabbed and accumulated by a limited set of national large landowners and transnational companies. For most of the families involved, this meant losing their access to the means of production in a moment where their subsistence was all but completely dependent on having a monetary income. As a result, most of them were now forced to work as wage laborers in the same plantations that they, or their parents, had created a few decades before. Thus, in the Bajo Aguán of the 2000s, peasant organizations were struggling to recover, or reconstruct, a way of being in the world where the fruits of their labor needed to be monetized for reproduction to take place. In other words, their struggles were not around what was being produced, palm oil was seen by everyone involved as the main game in town, but about who should keep the lion's share of the wealth produce: the companies, for profit an reinvestment, or the peasant cooperatives and communities to expand their wellbeing.[12] Phrased more schematically, upon the oil palm-dominated landscapes produced during the agrarian reform period, the monetization of reproduction transformed the ways in which households and communities, who identify themselves as "peasants," understood their place in the world and the ways in which they could make a good living. As such, while completely enmeshed in a capitalist economy, neither the agrarian reform cooperatives, nor today's peasant organizations, operate as capitalist ventures, but rather, as what the Indian economist Kalyan Sanyal (2013) calls "need economies." Unlike capitalist ventures, the need economies use just enough of this surplus in restarting the cycle, and the rest is spent in consumption for their members. In this sense, if for the capitalist venture accumulation for accumulation's sake is the final goal, for the need economies it is full employment and making enough money as to assure each member a good enough income to survive.

From this perspective, the "working class environmentalism" (Barca 2012) of these peasant organizations complicates some of our traditional understandings of what peasant and environmentalist activism looks like. For example, in 2012, the Unified Peasant Movement of the Aguán (MUCA), the largest and better-known peasant organization in the Aguán, was awarded the Fourth Annual Food Sovereignty Prize by the organization WHY Hunger, for their struggles to recover the land that they deemed had been taken away illegally in the mid-1990s. In very few words, the food sovereignty movement is a broad global coalition of farmers, growers, consumers, and activists that proposes that communities should have control over the ways in which food is produced, traded, and consumed, and strive for a food system designed to generate healthy communities and protect the environment, rather than generate profit, as is the case of the corporate agribusiness model that is dominant nowadays (Altieri and Nicholls 2016; Edelman 2016; Rosset and Martínez 2014). While this general idea of food sovereignty is applied and understood in different ways by different actors, one thing most of them have in common is a critique of monocultures, understood as an extractivist activity, and the promotion of agroecology and other forms of sustainable agriculture. However, the main economic activity of MUCA is the production of palm oil and most of the estates that they have targeted for "recuperation"—"invasion" from the perspective of the landowners and the state—are mature oil palm plantations already in full production. How do we make sense of this contradiction? How can we have food sovereignty among oil palms? From the perspective of organizations such as MUCA the solution is to continue the production of oil palms but combine it with other activities such as growing staple crops, and raising chickens, as well as using less agro-chemicals in general. This clearly entails a different conversation regarding either what is understood as food sovereignty, or who is said to be part of the movement. In any case, it speaks to the ways in which working-class organizations understand their relations to nature, and food, in landscapes produced by their dead labor and in which reproduction is mainly done through the market.

We find a different situation in the Costa Rican North Region, close to the border with Nicaragua. In the last few years Costa Rica has become the largest exporter of fresh pineapples in the world. This was the result of a spectacular expansion of the crop in the last twenty years. Nowadays, of the more than 65,000 hectares dedicated to the crop, over 60% are located in the North Region (Arguedas González et al. 2020). Initially colonized by poor peasant families and would-be cattlemen from the country's central region, as well as by Nicaraguan rubber tappers, in the 1970s and 1980s migration toward the North Region intensified, as impoverished and landless peasants from other parts of the country began to populate the region with the support of the government. As part of the United States' strategy of containment of the Sandinista regime in Nicaragua, the government received and poured millions of dollars into the region to create a "living frontier": that is, a border region settled by Costa Rican nationals that could work both as a barrier against the Sandinistas, and a support network for the anti-Sandinista irregular forces (Girot 1989; Granados and Quesada 1986).

In the 1990s, after the electoral defeat of the Sandinistas in Nicaragua, and the imposition of the neoliberal project in the region, the Costa Rican North Region became an important site of expansion of agroexports: starting with oranges and tubers and later pineapples. This was the result of what Alberto Cortés-Ramos (2003) calls a "structural coincidence" between the political economy of both countries. In the case of Nicaragua, the aftermath of the Sandinista period came with a set of policies, and lack of them, that deeply impoverished the country's rural areas and unleashed a massive wave of emigration mainly toward Costa Rica, of poor peasant families in search of work. In the case of Costa Rica, the imposition of the neoliberal project took above all the shape of a process of commercial liberalization and the promotion of new exportable crops, such as pineapple. As a result, state support for staple crops diminished significantly, while subsidies and tax breaks were given to exporters (Edelman 1999). These changes hit particularly hard places like the North Region, dominated by peasant communities and middle and large cattle ranchers. As a result, cheap land became available, while inflows of migrants provided a cheap source of labor, creating a truly transnational and transborder economy. Close to the border between both countries, the tensions created by this dynamic are very evident. Every day, Nicaraguan workers cross the border through different "blind sites" to work in the Costa Rican pineapple plantations. This, of course, cheapens the costs of production for the fruit companies, as they are able to pay these migrant workers a salary that would be insufficient to sustain a household in Costa Rica, but enough to do so in Nicaragua with its lower living costs. However, this is not the only advantage of this system of "overexploitation of labor" for the companies. Migrant workers tend to enter the country irregularly, without migratory papers, and, thus, are more vulnerable in legal terms, and less likely to protest their working conditions or take part in the frequent, albeit mostly unsuccessful, attempts to organize labor unions.

Also, from the perspective of the peasant communities that must coexist with the pineapple plantations, these immigrant workers are seen as a problem. Not only their work in the plantations translate into the contamination of their water sources and the poisoning of their animals and plants due to pesticide drift (Carazo et al. 2016; Diepens et al. 2014), they are also seen as foreigners that come into the country to steal their jobs and lower the wage rates. As a result, the shared grievances that they might have against the pineapple plantation economy as such—environmental degradation and hideous working conditions—, are portrayed rather as a conflict between Costa Ricans and Nicaraguans. It becomes evident then, that, understanding the environmental conflicts in Costa Rica due to the expansion of the pineapple plantation economy, must include a discussion regarding how this migrant labor force was produced, and how the fruit of their labor, in the shape of the plantations, is enmeshed into the everyday life of the Costa Rican peasant communities under siege. The work of Nicaraguan migrant laborers is the condition of possibility for both the pineapple plantations in the Costa Rican side, and their hometowns back in Nicaragua. At the crux lies a set of agroexport landscapes, which have been naturalized as the main source of development, and that hide a complex web of more than human relationships that extend both spatially and historically, beyond a particular time and place.

Dead labor and the question of value

One way of thinking about these cases is through the Marxian differentiation within capitalism between dead and living labor. While the latter refers to the process of transforming means of production into commodities, dead labor is past labor ossified and materialized into definite things, including infrastructure and landscapes. Workers—living labor—have an estranged relationship with the world that results from this dead labor. It does not appear to them as the result of the relationship between living laborers, past and present, but rather, as "an immense accumulation of commodities" (Marx 1992, 125). In other words, men and women confront a world that is of their own making, through their interaction with non-humans, but that appears upon their eyes and other senses, as a collection of objects, mostly owned by others (the plantation, the packaging plant, the factory, among others), and ruled by a set of institutions that present themselves more like immutable natural laws, than the social relationships that they are.

What this means for our case in point is that men and women, and the scholars that study them, have an easier time recognizing themselves, their dreams and their life-sustaining activities, in the communities, *milpas*, and the forests and rivers, than in the plantations and packaging plants where they toil for the enrichment of others and the deterioration of themselves, both in terms of their bodies and those same forests, *milpas*, and rivers. At the crux lays the question of value: "… the name for how productive, social activities get divided up within societies, activities—labor, in the very broadest sense—that yield the assemblages of humans and nonhumans that are necessary to sustain life, as well as spark new life" (Henderson 2013, xii). While under capital, this organization of labor, in the very broadest sense, gravitates around monetary profit, separating it from nature in the form of dead labor (commodities), and thus presenting it as the only source of value; for the people that confront its operations it is exactly the opposite: Dead labor is the anathema of value, the material manifestation of their separation from their means of existence.

From this perspective, we can think about the struggles over value, of how we think society should be organized and the place that humans' and non-humans' activities should have in it, as a way of analyzing the complicated relations between nature and society. As such, a political ecology of labor should explore the ways in which nature is being produced in particular settings, paying special attention to the ways in which not only the line between society and nature is drawn, but also between the human and non-human, production and reproduction, men and women, and so on and so forth. Further, it should look at the ways in which both the communities and the plantations are being produced, by whom, and how they relate to each other within larger socioecological contexts, and what types of worlds are produced, and destroyed, in the process.

Conclusions

In this chapter, I have argued for a political ecology of labor. One that looks at both the complicated relations between nature and society, and the role played by labor—understood in the broadest sense—in the creation of the worlds that humans, and non-humans, live in. I begin by highlighting the anti-workers bias found in most LAPE literature and linked it to both its methodological preferences (participatory and popular education methods), and theoretical roots (dependency theory and Latin American decolonial critique).

Then, following the insights of Latin American ecofeminist and feminist political ecologists, I propose that we think about the political ecology of capitalism as the constant process of separating humans from their means of existence. This process is dynamic and ever-changing, with the frontiers between the "visible" and valued (production and society), and the "invisible" and devalued (social reproduction and nature), constantly being contested and rewritten. At the crux we find labor: the world-making, and life-sustaining, activity that mediates between the human and the non-human, producing both nature and society in the process. Using two examples from plantation economies in Central America—oil palm in Honduras, and pineapple in Costa Rica—I showed the role that previous labor has in defining the contexts in which current environmental conflicts take place, and how crucial it is to include workers to make sense of these situations.

Finally, I introduced the notions of the production of nature, dead labor and value, to move away from the sharp differentiations that tend to be drawn between the plantation and the community in the study of environmental conflicts. If we think of both nature and society as produced, with labor playing an important role in the creation of the resulting landscapes, it becomes clear that this differentiation is not a useful one. Phrased differently, if both the communities and the plantations are produced, the result of dead labor, then the question should shift towards the process by which both spaces are produced, and the relationships that link them. Here is where the question of value becomes important.

Broadly, value is the way in which the world-making and life-producing activities of a society are organized. Thus, exploring the conflicts over how it should be defined in particular contexts becomes a powerful window into the ways in which the relationships between nature and society are understood and produced. It is also a way of proposing that we think about humans as part of a larger differentiated whole, where the frontiers between society and nature are all about the power relations that push to define them in certain ways, placing certain actors, human and non-human, in particular positions and patterns of inequality.

Notes

1 Understanding time and history as a single ascending line in which the past is radically different from the present, and where the future is achievable, but at the same time always postponed (Trouillot 2003).

2 Probably the better-known representative and proponent of this thesis was Gunder-Frank (1966, 1967), who had a great influence in Wallerstein (1984). For a review of his controversial legacy in Latin America, see Kay (2011). In Africa, a proponent of this thesis was Rodney (1972).

3 As such, while most of Latin America was clearly linked to a world market since the 1500s, the wage labor relation was not yet dominant and, thus, it could not be said that these were fully capitalist. One of the better-known proponents of this line of thought was Laclau (1971).

4 There were exceptions to the norm. For example, the Brazilian Marini (2008) proposed that the greater accumulation of capital in the central (developed) countries was the direct result of the over-exploitation of labor in the (underdeveloped) peripheries, that is, the extension of the working day, combined with wages below the cost of reproduction of the labor force.

5 Lander's (1993) influential edited volume in the mid-1990s inaugurated a boom of the postmodern decolonial debates in the region. This was criticized by other scholars who pointed out the lack of women, Black, and indigenous perspectives, and representatives, within their discussions (for example, Cumes 2012; Curiel 2007; Rivera Cusicanqui 2010).

6 Also, as authors such as Victoriano Serrano (2010) have argued, to understand what happened to the social sciences in Latin America in the 1980s and 1990s, we have to take into account the terror unleashed against the left by the military regimes during the previous decades. With many thinkers writing from exile, or murdered, and with the generalized persecution of any political thought that seemed communist, class as a concept of analysis was also set aside. When democratization began in the 1980s, in the context of "neoliberal multiculturalism," the intellectual cues to pick up on a class-oriented analysis were not there.

7 Particularly important are the works of Löwy (1996) from Brazil and of Gutierréz Aguilar (2008) and Holloway (2010), working from Mexico.

8 E.g., Silva-Santisteban (2017) talks about "dependent patriarchy" (*patriarcado dependiente*), and Costantino (2013) proposes a "political ecology of dependency."

9 An extensive literature exists on the exploitation of labor in cattle ranching, sugarcane and coffee plantations, and the continuity of unfree forms of labor in these activities (for example, Chonchol 1994; Florescano 1975; Gordillo 2004; Rogers 2010; Schwartz 1978). However, it rarely dialogues with the literature on agrarian extractivism.

10 As Jason Moore (2015) mentions, the idea that labor, not land, was the source of value, was a novel idea within economic thought at the time of the rise of capitalism.

11 For example, Navarro (2019), following De Angelis (2004) and Moore (2015), proposes that we think about capitalism as a form of ecology predicated upon the separation of humans, not only from the means of production, but also, from their "means of existence": all the material and symbolic elements that guarantee subsistence and make possible the reproduction of life. For her, separation entails three different dynamics: 1) guaranteeing the intervention and flow of capital through the web of life; 2) hiding and deforming the interconnections and relations between all the different forms of life that inhabit the planet; and 3) radically transforming the human and non-human metabolisms to satisfy the needs of capital accumulation.

12 This is not to say that there are no conflicts in Central America, and Honduras, regarding the production of oil palms, and the encroachment of the plantation on communal lands (Alonso-Fradejas 2012, 2015; Brondo 2013; Mingorría 2016).

References

Alimonda, Héctor. 2017. "En clave de sur: La ecología política latinoamericana y el pensamiento crítico." In *Ecología Política Latinoamericana. Pensamiento Crítico, Diferencia Latinoamericana y Rearticulación Epistémica*, Vol. I, edited by C. Hector Alimonda, Toro Pérez, and F. Martín, 33–50. Buenos Aires: CLACSO.

Alonso-Fradejas, Alberto. 2012. "Land Control-Grabbing in Guatemala: The Political Economy of Contemporary Agrarian Change." *Canadian Journal of Development Studies/Revue Canadienne d'études Du Développement* 33 (4): 509–528.

Alonso-Fradejas, Alberto. 2015. "Anything but a Story Fortold: Multiple Politics of Resistance to the Agrarian Extractivism Project in Guatemala." *Journal of Peasant Studies* 42 (3–4): 489–515.

Altieri, Miguel Ángel, and Clara Inés Nicholls. 2016. "Agroecología y soberanía alimentaria en América Latina." In *Soberania Alimentar (SOBAL) e Segurança Alimentar e Nutricional (SAN) na América Latina e Caribe*, edited by Islandia Bezerra and Julian Perez-Cassarino, 96–118. Curitiba: Editora UFPR.

Arguedas González, Catalina, Cornelia Miller Granados, and Christian Vargas Bolaños. 2020. *Monitoreo del estado de la piña en Costa Rica para el año 2018*. San José, Costa Rica: Laboratorio Prias, Centro Nacional de Alta Tecnología (Cenat).

Barca, Stefania. 2012. "On Working-class Environmentalism: A Historical and Transnational Overview." *Interface: A Journal for and about Social Movements* 4 (2): 61–80.

Barca, Stefania. 2014. "Laboring the Earth: Transnational Reflections on the Environmental History of Work." *Environmental History* 19 (1): 3–27.

Barca, Stefania. 2019. "Labour and the Ecological Crisis: The Eco-Modernist Dilemma in Western Marxism (s) (1970s-2000s)." *Geoforum* 98: 226–235.

Brondo, Keri Vacanti. 2013. *Land Grab: Green Neoliberalism, Gender, and Garifuna Resistance in Honduras*. Tucson: University of Arizona Press.

Carazo, Eva, Javiera Aravena, Vanessa Dubois, Jorge Mora, Francisco Parrado, and Andrés Mora. 2016. *Condiciones de Producción, Impactos Humanos y Ambientales en el Sector piña en Costa Rica*. San José, Costa Rica: Oxfam Germany.

Chonchol, Jacques. 1994. *Sistemas Agrarios en América latina*. Santiago: Fondo de Cultura económica.

Cortés-Ramos, Alberto. 2003. "Apuntes Sobre las Tendencias Migratorias en América Central en la Segunda Mitad del Siglo XX." *Reflexiones* 82 (2): 8.

Costantino, Agostina. 2013. "Apuntes Para Una Ecología Política de la Dependencia. El Caso del Acaparamiento de Tierras." *Revista Sociedad y Economía* 25: 39–54.

Cumes, Aura Estela. 2012. "Mujeres Indígenas Patriarcado y Colonialismo: Un Desafío a la Segregación Comprensiva de las Formas de Dominio." *Anuario de Hojas de WARMI* 17: 1–16.

Curiel, Ochy. 2007. "Crítica Poscolonial Desde las Prácticas Políticas del Feminismo Antirracista." *Nómadas* 26: 92–101.

De Angelis, Massimo. 2004. "Separating the Doing and the Deed: Capital and the Continuous Character of Enclosures." *Historical Materialism* 12 (2): 57–87.

Diepens, Noël J., Sascha Pfennig, Paul J. Van den Brink, Jonas S. Gunnarsson, Clemens Ruepert, and Luisa E. Castillo. 2014. "Effect of Pesticides Used in Banana and Pineapple Plantations on Aquatic Ecosystems in Costa Rica." *Journal of Environmental Biology* 35 (1): 73.

Edelman, Marc. 1999. *Peasants Against Globalization: Rural Social Movements in Costa Rica*. Redwood City, CA: Stanford University Press.

Edelman, Marc. 2016. *Estudios Agrarios Críticos: Tierras, Semillas, Soberanía Alimentaria y los Derechos de las y los Campesinos*. Quito: Editorial IAEN.

Escobar, Arturo. 2017. "Desde abajo, por la izquierda, y con la tierra: La diferencia de Abya Yala/ Afro/ Latino/ América." In *Ecología Política Latinoamericana. Pensamiento Crítico, Diferencia Latinoamericana y Rearticulación Epistémica: Vol. I*, edited by Hector Alimonda, C. Toro Pérez, and F. Martín, 51. Buenos Aires: CLACSO.

Fals-Borda, Orlando. 1984. "Participatory Action Research." *Development: Seeds of Change* 2: 18–20.

Florescano, Enrique. 1975. *Haciendas, Latifundios y Plantaciones en América Latina*. México: CLACSO, Siglo XXI.

Gunder-Frank, Andre. 1966. *The Development of Underdevelopment*. Boston: New England Free Press.

Gunder-Frank, Andre. 1967. "Sociology of Development and Underdevelopment of Sociology." *Cahiers Internationaux de Sociologie* 42 (JAN-J): 103–131.

Fraser, Nancy. 2017. "Behind Marx's Hidden Abode: For an Expanded Conception of Capitalism." In *Critical Theory in Critical Times: Transforming the Global Political and Economic Order*, edited by Penelope Deutscher, and Cristina Lafont, 141–159. New York: Columbia University Press.

Freire, Paulo. 1996. *Pedagogy of the Oppressed*. New York: Continuum.

Girot, Pascal. 1989. "Formación y Estructuración de una Frontera Viva: El Caso de la Región Norte de Costa Rica." *Geoistmo* 3 (2): 17–42.

Gordillo, Gastón. 2004. *Landscapes of Devils: Tensions of Place and Memory in the Argentinean Chaco*. Durham: Duke University Press.

Granados, Carlos, and Liliana Quesada. 1986. "Los Intereses Geopolíticos y el Desarrollo de la Zona nor-atlántica Costarricense." *Estudios Sociales Centroamericanos* 40: 47–65.

Grosfoguel, Ramón. 1997. "A TimeSpace Perspective on Development: Recasting Latin American Debates." *Review (Fernand Braudel Center)* 20 (3/4): 465–540.

Gudynas, Eduardo. 2009. "Diez Tesis Urgentes Sobre el Nuevo Extractivismo." *Extractivismo, Política y Sociedad* 187: 187–225.

Gutierréz Aguilar, Raquel. 2008. *Los ritmos del Pachakuti: Movilización y Levantamiento Indígena-popular en Bolivia* (Vol. 6). Buenos Aires: Tinta Limón.

Gutierréz Aguilar, Raquel. 2015. "Políticas en Femenino: Transformaciones y Subversiones no Centradas en el Estado." *Movimientos Sociales. Nuevos Escenarios, Viejos Dilemas*, 123.

Henderson, George. 2013. *Value in Marx: He Persistence of Value in a More-Than-Capitalist World*. Minneapolis: University of Minnesota Press.

Holloway, John. 2010. *Crack Capitalism*. London: Pluto Press.

Kay, Cristobal. 1998. "Latin America's Agrarian Reform: Lights and Shadows." *Land Reform, Land Settlement and Cooperatives* 2: 9–31.

Kay, Cristobal. 2011. "Andre Gunder Frank:'Unity in Diversity'from the Development of Underdevelopment to the World system." *New Political Economy* 16 (4): 523–538.

Kirsch, Scott, and Don Mitchell. 2004. "The Nature of Things: Dead labor, Nonhuman Actors, and the Persistence of Marxism." *Antipode* 36 (4): 687–705.

Laclau, Ernesto. 1971. "Feudalism and Capitalism in Latin America." *New Left Review* 67: 19–38.

Lander, Ernesto. 1993. *La Colonialidad del Saber: Eurocentrismo y Ciencias Sociales*. Buenos Aires: CLACSO.

Leff, Enrique. 2017. "Las Relaciones de Poder del Conocimiento en el Campo de la Ecología Política: Una Mirada desde el sur." In *Ecología Política Latinoamericana. Vol. I*, edited by Hector Alimonda, C. Toro Pérez, and M. Facundo, 129–166. Buenos Aires: CLACSO.

León Araya, Andrés. 2017. "Domesticando el Despojo: Palma Africana, Acaparamiento de Tierras y género en el Bajo Aguán, Honduras." *Revista Colombiana de Antropología* 53(1): 151–185.

León Araya, Andrés. 2019. "The Politics of Dispossession in the Honduran Palm Oil Industry: A Case Study of the Bajo Aguán." *Journal of Rural Studies* 71: 134–143.

Löwy, Michael. 1996. De Karl Marx a Emiliano Zapata: La dialéctica Marxiana del Progreso y la Apuesta Actual de los Movimientos Eco-sociales. *Ecología Política* 10: 97–105.

Marini, Ruy Mauro. 2008. *América Latina, Dependencia y Globalización*, edited by C. E. Martins. Bogotá: Siglo del Hombre Editores.

Marx, Karl. 1992. *Capital: Volume 1: A Critique of Political Economy* (B. Fowkes, Trans.; Reprint edition). London; New York: Penguin Classics.

McKay, Ben M. 2017. "Agrarian Extractivism in Bolivia." *World Development* 97: 199–211.

Mingorría, Sara. 2016. "Violencia, Silencio, Miedo: El desvelo del Conflicto de Palma Aceitera y Caña de Azúcar en el Valle del Polochic, Guatemala." *Ecología Política* 51: 73–78.

Mitchell, Don. 2003. "Cultural Landscapes: Just Landscapes or Landscapes of Justice?" *Progress in Human Geography* 27 (6): 787–796.

Moore, Jason W. 2015. *Capitalism in the Web of Life: Ecology and the Accumulation of Capital*. London: Verso Books.

Navarro, Mina Lorena. 2015. *Luchas por lo Común: Antagonismo Social Contra el Despojo Capitalista de los Bienes Naturales en México*. Ciudad de México: Bajo Tierra Ediciones.

Navarro, Mina Lorena. 2019. "Despojo múltiple y separación del capital sobre el tejido de la vida." In *Teoría del valor, Comunicación y Territorio*, edited by Francisco Sierra Caballero, 277–296. Madrid: Siglo XXI de España Editores.

Rivera Cusicanqui, Silvia. 2010. *Ch'ixinakax Utxiwa. Una Reflexion Sobre Prácticas y Discursos Descolonizadores*. Buenos Aires: Tinta limon.

Rodney, Walter. 1972. *How Europe Underdeveloped Africa London*. London: Bogle, L'Ouverture Publications.

Rogers, Thomas D. 2010. *The Deepest Wounds: A Labor and Environmental History of Sugar in Northeast Brazil*. Chapel Hill, NC: University of North Carolina Press.

Rosset, Peter, and María Elena Martínez. 2014. "Soberanía Alimentaria: Reclamo Mundial del Movimiento Campesino." *Ecofronteras* 18: 8–11.

Rostow, Walt Whitman. 1990. *The stages of Economic Growth: A Non-communist Manifesto (Third)*. Cambridge: Cambridge University Press.

Sanyal, Kalyan. 2013. *Rethinking Capitalist Development: Primitive Accumulation, Governmentality and Post-Colonial Capitalism*. New Delhi: Routledge India.

Schwartz, Stuart B. 1978. "Indian Labor and New World Plantations: European Demands and Indian Responses in Northeastern Brazil." *The American Historical Review* 83 (1): 43–79.

Shanin, Teodor. 1982. "Defining Peasants: Conceptualisations and De-Conceptualisations: Old and New in a Marxist Debate." *The Sociological Review* 30 (3): 407–432.

Silva-Santisteban, Rocío. 2017. *Mujeres y Conflictos Ecoterritoriales. Impactos, Estrategias, Resistencias*. Lima: Mega Trazo.

Smith, Neil. 2008. *Uneven Development: Nature, Capital, and the Production of Space*. Athens: University of Georgia Press.

Svampa, Maristella, and Enrique Viale. 2014. *Maldesarrollo: La Argentina del Extractivismo y el despojo* (Vol. 3088). Buenos Aires: Katz Editores.

Tronti, Mario. 2019. *Workers and Capital*. London: Verso.

Trouillot, Michel-Rolph. 2003. *Global Transformations: Anthropology and the Modern World*. London: Palgrave Macmillan.

van der Ploeg, Jan D. 2010. *The Peasantries of the Twenty-First Century: Struggles for Autonomy and Sustainability in an Era of Empire and Globalization*. London: Earthscan.

Victoriano Serrano, Felipe. 2010. "Estado, Golpes de Estado y Militarización en América Latina: Una Reflexión Histórico Política." *Nueva Epoca* 64: 19.

Wallerstein, Immanuel. 1984. *The Politics of the World-Economy: The States, the Movements and the Civilizations*. Cambridge: Cambridge University Press.

16 Transnationals, dependent development and the environment in Latin America in the 21st century

Paul Cooney

Introduction

Since the 1980s, the world economy has entered a phase of neoliberal globalization, reflected in the dominance of free-market policies imposed by international institutions, predominantly the IMF, the World Bank, and the World Trade Organization (WTO), and by the ever-growing hegemony of transnational corporations (TNCs). Given the problems of the neoliberal paradigm a range of critiques from the left have appeared in recent decades. Consider the examples of David Harvey's *The New Imperialism* (2004), Hardt and Negri's *Empire* (2000), Robinson (2004) with his work on the role of transnationals, and also dependency theorists, such as Dos Santos (2000). As critiques against the neoliberal model spread and progressive governments gained traction across Latin America after the turn of the century, there was a resurgence of dependency theory and debates regarding what constitutes development in Latin America, emphasizing the controversial and dialectical relationship between the center and the periphery.

Moreover, as a result of neoliberal globalization in recent decades, most countries of Latin America have experienced some combination of deindustrialization, reprimarization and neo-extractivism.[1] This is evident in the increased extraction of natural resources, be it minerals, fossil fuels, or lumber, and also in the expansion of agroindustry dominated by TNCs: cattle, soy, bio-fuels, palm oil, flowers, and a range of other non-traditional agricultural and primary exports (see Robinson 2008, Ch. 2).

This chapter examines the nature and causes of these shifts in Latin America and presents an evaluation of the environmental and social impacts produced by these new 'development' trajectories. Given the dynamic expansion of the sectors associated with neoextractivism, there has been an increasing number of environmental problems experienced in Latin America. In recent decades, most notable examples are deforestation, especially the Amazon rainforest; air, water, and soil pollution, reduction in biodiversity, soil erosion, desertification, and a range of related health problems, from lung and skin diseases, anencephaly, to cancer, among others. Another socio-environmental issue related to neoextractivism is the construction of dams, often linked to electricity-intensive mining projects, where transnational finance has played a notable role.

In order to understand the current phase of capitalism, namely TNC-dominated neoliberal globalization, the next section discusses recent theoretical debates and challenges. The two main subsections examine the dependency approach and ecologically unequal exchange. The third section presents a more detailed analysis of the trajectories of reprimarization, combined with an assessment of the range of associated environmental problems. Although this chapter presents a range of issues for different countries and parts of Latin America, there will be

DOI: 10.4324/9780429344428-19

particular emphasis on the Amazon rainforest—popularly known as "the lungs of the planet"—since this is not just a problem affecting several countries of the region, but plays a key role with regards to the global climate crisis.

Lastly, in the final section, an attempt is made to synthesize the discussion of recent theoretical advances with the shifting trajectories and associated environmental consequences. It is hoped that a fuller and more critical understanding of the socio-environmental problems faced by the populations of Latin America during the first decades of the 21st century can benefit from recent developments in dependency theory and ecologically unequal exchange, among others. Thus, there is the challenge of combining theory and praxis so as to contribute toward the social and environmental struggles being waged by workers, peasants, feminists, environmentalists, and other social activists.

Theoretical debates and challenges

This section considers the dependency school and also the concept of ecologically unequal exchange as well as the roles of TNCs, the WTO, and ground rent in the current context. Dependency can be seen as the flipside of imperialism, but instead of looking at imperialist policies of the center, it is the view from the dependent countries, namely, an analysis from the viewpoint of the countries in the periphery, which as a result of imperialist policies, are suffering from dependent capitalist 'development'.

Dependency theory[2]

First and foremost, there is no general agreed-upon theory of dependency and any attempt to treat dependency as such is mistaken. The more appropriate term is the dependency school, suggesting an umbrella which encompasses a range of different concepts, orientations and political differences, which all aim to explain the dependent relationship of the periphery in relation to the center. Unfortunately, much of the debate about or within the dependency school, has led to significant confusion, and intense debates and polemics, though more in the past than the present.

The dependency school developed in the 1960s and 1970s and was a critique of both the modernization school and the Economic Commission for Latin America (ECLA) perspective associated with Keynesian development policies in the periphery or the Third World, in particular advocating import-substitution industrialization (ISI) and the development of the internal market. Although authors of ECLA, such as Prebisch, referred to the center and periphery, they did not have a radical critique of the structural relationships associated with imperialism and lacked class analysis. The critiques of ECLA and exposure of its limitations was part of what defined dependency theory: the main argument being that the countries of the periphery were dependent upon the needs of accumulation of the countries of the center, thus lacking autonomy and the possibility of genuine development.

The main contributors of the school in its early days were Theotônio Dos Santos (1978), Ruy Mauro Marini (1973), Fernando Henrique Cardoso (1980), and Andre Gunder Frank (1967) among those working on Latin America, while authors such as Samir Amin (1976), Walter Rodney (1972), and Clive Thomas (1974) used dependency concepts in their work on other regions. Some would argue that Wallerstein should be included, but I would argue his analysis of world-systems theory was developed in parallel. In the case of Frank, he actually crossed paths with Marini and Dos Santos in the late 1960s at the Universidade de Brasília; thus, his influence was clear in the early years of the Brazilian dependency school, though in the following years he came to work more closely with Wallerstein and world-systems theory.

Brazil was clearly the main cradle of the development of the dependency school, with many key authors, however, due to space limitations only two of the main founders, will be considered, Theotônio Dos Santos, and Ruy Mauro Marini; followed by a presentation of more recent contributions by a new generation of scholars, mostly from Brazil.

Dos Santos was one of the founders of dependency, and continued working with this approach until he recently passed away in 2018 and was a mentor of several of the new generation of *dependentistas* in Brazil. Dos Santos's definition of dependency is "a situation in which a certain group of countries have their economies conditioned by the development and expansion of another economy, to which their own is subjected" and that "dependency conditions a certain internal structure which redefines it as a function of the structural possibilities of the distinct national economies" (Dos Santos 1970). Chilcote (1981), in identifying four different formulations of dependency,[3] argued that Dos Santos had a theory of the new dependency, reflecting the technological and industrial dominance established by multinational corporations after WWII. Therefore, according to Dos Santos, the main issue is the interplay between the internal Latin American structures and international structures and that this is the critical starting point for an understanding of the process of development in Latin America (1970). Now let us turn to Marini.

Ruy Mauro Marini was one of the founders of dependency in Brazil, and from an early stage argued that the countries of the periphery constituted another type of capitalist mode of production, distinct from the center. Although over time, his work and his followers have tended to emphasize the issue of super-exploitation and the importance of marxist categories and class. Below is the general view of dependency as presented by Marini.

> … the Latin American countries are connected to the capitalist center countries by means of a structure defined and established and based upon an international division of labor in which the production relations of the periphery are transformed to guarantee the reproduction of dependency and imperialism.
>
> Marini (1973, 109–113)

One major point stressed by Marini is the problem of "unequal exchange" and, as argued by most dependency theorists, Latin American countries predominantly produce primary products or raw materials, with minimal value-added, for export. Marini argues that in the periphery, capitalists do not have access to the most advanced technology and thus have lower levels of productivity as compared to the center. In order to stay competitive, they seek to increase the rate of exploitation (rate of surplus value) as a means of compensation. Therefore, as a survival mechanism, the capitalists of Brazil and elsewhere in the periphery pay wages below the value of labor power and this constitutes super-exploitation.[4]

In the last couple of decades, there has been a resurgence of discussions around dependency theory, especially in Brazil, including some of the original theorists, such as Dos Santos (2000) and Bambirra (1978), but also a new wave of younger scholars, such as Carlos Eduardo Martins (2011), Marcelo Carcanholo (2017) and Correa Prado and Castelo (2013). The new generation considers themselves to be working with a marxist theory of dependency, strongly based on the work of Marini and Dos Santos, with particular emphasis on super-exploitation, but also addressing imperialism and sub-imperialism, a case such as Brazil, subject to the imperialism of the center; yet imposing their economic and political dominance on other weaker countries of the periphery in their region, such as Uruguay, Paraguay or Bolivia.

The main argument continues to be the structural problem of dependency and the necessity of super-exploitation, in a context dominated by TNCs and with even greater inequality than in the past. A recent contribution by Mariano Féliz (2019), is an attempt to use the framework developed by Marini, in addressing super-exploitation, and problems of development in the current period addressing the attempt of new developmentalism (*neodesarrollismo*) in the context of Argentina during the Kirchner governments. Although nature is included in his discussion of super-exploitation, he does not develop this very far.

In addition to Féliz, another Argentine, Claudio Katz (2018), has worked with the dependency approach. Katz is a well-known Argentine Marxist economist, and many consider him to be a dependency theorist, though he may be more of a fellow traveler. In his book evaluating 50 years of dependency, Katz (2018) stresses the importance of this school, which over five decades is one of the few sources of scholars and activists which continue to emphasize the problems of imperialism and underdevelopment and how this has worsened with the period of neoliberal globalization.

In addition to working on issues of imperialism and dependency, Katz has also incorporated other issues, such as the environment, gender, and social movements. In this regard there is a clear confluence of dependency theory, the crises of capitalism, and critical environmental analysis in the work of Katz and thus his research is an example of those working on the environment and engaging critically with the dependency approach (Katz 2018).

There have been a range of criticisms of "dependency theory" given the many different theorists and arguments. Some were clearly legitimate but others reflected dogmatic or rigid views within marxist and leftist intellectual circles. One example is the term 'development of underdevelopment' from Frank which came to be strongly criticized, especially after the success of the Asian Tigers.[5] Nevertheless, in recent discussions in Latin America around dependency this term is rarely mentioned. Marini's argument that dependent countries have a separate mode of production was also strongly criticized, though some still employ it, but do not tend to stress it. Lastly there were critiques, and still are, of the extensive use of the term super-exploitation, especially given minimal attempts to empirically test super-exploitation.

Unfortunately, much of the critical left and Marxist analyses of the past in Latin America, be it theories about imperialism, dependency, or unequal exchange, paid minimal attention to environmental issues. At present, a new generation in Latin America has sought to broaden the discussion beyond classes and nation-states, incorporating critical discussions around the environment, gender, ethnicity, and race, largely as a result of many fervent social movements in the region. Nevertheless, the radical thinkers of dependency are often not at the forefront of many of these debates, though many are active in social movements. Fortunately, there are some working from a dependency approach that have sought to broaden the analysis to include the issue of the environment, such as Féliz and Katz mentioned above. Another example is a Brazilian author, Wendell Ficher Teixeira Assis, who seeks to combine the dependency approach with an analysis of coloniality to address issues of the environment and race.

The author describes environmental problems associated with the process of reprimarization in Brazil, referring to environmental degradation and pollution and refers to "ecological imperialism" given the role of TNCs. Though not common within dependency research, perhaps work by authors such as Teixeira Assis, Féliz, and Katz can encourage others using the dependency approach to incorporate more discussion of the environment and other social issues in the near future.

The dependency school continues with a critical analysis emphasizing the extraction of resources and wealth from the region as the main basis for underdevelopment in Latin America. This approach correctly argues that the accumulation processes in the center is the priority for the global economy, where the 'periphery' is seen as secondary (whether workers or capitalists). The agroexport economy that dominated Latin America at the end of the 19th century was the initial dominant stage of dependency. In the case of countries such as Argentina, Brazil and Mexico, from the 1930s through the 1970s, a serious industrialization via ISI, implied a second stage of dependency,[6] given the shift in production, flows of surplus value and profits, and the operations of multinational corporations within the periphery. The dependency approach is one of the few which still points to imperialism as a major part of the problem, be it debt, or dependence on foreign capital and as discussed below, prevented from breaking away from underdevelopment, which too many academics in the global North continue to be unwilling to recognize.

The present period of neoliberal globalization constitutes a new and third phase of dependent development. In order to comply with WTO rules, countries which experienced industrialization, such as Argentina and Brazil, have been forced to eliminate tariffs and subsidies. As a result, they can no longer compete in manufacturing, and must resort to where they have a natural advantage and ground rent, which is in the primary sector, limiting their options to neoextractivism or reprimarization.

In conclusion, the present period implied a major shift with respect to the possibilities of development for the majority of the Latin American region, such that the only possibility of being competitive is via sectors where natural fertility or productivity is an advantage, namely minerals, petroleum, and agriculture. These conditions of ground rent based on differential fertility in land and differential productivity, in the case of mines or oil at present, is practically the only basis for countries of the periphery to compete.[7] As argued by Robinson (2004), the hegemony of the TNCs in the neoliberal globalization phase is concomitant with the emergence of a transnational capitalist class (TCC) and the decline in the autonomy of the nation-state, and how TNCs come to play a greater role in the formation of national policies, especially in the periphery. Discussion of a new transnational class alliance and analyzing the socio-environmental impacts of several key dynamic sectors is presented below. Though first, the contributions made in the field of ecological economics[8] and the specific theoretical debates around ecologically unequal exchange, will be considered in the next section.

Ecologically unequal exchange

In order to understand the concept of Ecologically Unequal Exchange, it is clearly appropriate to first review the history of the concept of Unequal Exchange; this former concept being strictly economic and in general limited to monetary exchanges of value. In debates on unequal exchange, there has been a lack of consensus, and so far, there seems to be even less consensus with respect to the concept of Ecologically Unequal Exchange. The original concept of unequal exchange was first presented by Arghiri Emmanuel (1972) and further developed by Amin (1976), Mandel (1975), and others in the 1960s and 1970s. The development of this concept can be traced back to the debates from the 1950s associated with ECLA and especially Prebisch, expounding upon declining terms of trade between the center and the periphery and issues of development and industrialization. This came to be strongly associated with the advances of ISI in Latin America from the 1930s through the 1970s.

The concept of unequal exchange is usually defined in terms of the unequal transfers of quantities of labor time embodied in the commodities that were traded between countries. Emmanuel and others sought to demonstrate that the problem of unequal exchange derived from wage disparities between the center and periphery, however there was much debate around this point; As Shaikh (1980) and others correctly argued, underdevelopment was the cause not the consequence of these wage disparities.

In the past two decades, we have observed the extension of the discussion of unequal exchange to the realms of ecology, geography, and ecological sociology, which sought to go beyond just analyzing monetary stocks and flows by incorporating an analysis of material and energy stocks and flows. Therefore, research on ecologically unequal exchange has entered a broader realm than just radical political economy and moved into a realm of new questions and debates. These include proposals of different theories of value, demands to address the problem of unequal flows of embodied land, energy, water, etc., rather than just labor; the emphasis is on the consumption and deterioration of a range of natural resources, required for the production of particular commodities, especially for the First World.

There are a range of concepts and several approaches that have achieved important advances in conceiving of these asymmetric flows without compensatory remuneration, but rather with very serious and notable ecological consequences. Authors that have made major contributions in this debate include: Bunker (1985); Hornborg (2011); Rice (2007); Foster and Holleman (2014); Jorgenson and Clark (2009) and Odum (1996), among many others. An important recent contribution in this debate is that of Foster and Holleman, who argue:

> Unequal economic exchange is mainly concerned with a quantitative value problem related to exchange-value relations, while unequal ecological exchange is chiefly concerned with use-value relations and real wealth (including the contradictions between use value and exchange value).
>
> (Foster and Holleman 2014, 207)

Beyond the importance of Marx's key distinction between use value and exchange, emphasized by Foster and Holleman, other advances conceptually in discussions around ecologically unequal exchange elaborate on the biophysical dimension of such exchanges, drawing on concepts of biocapacity and ecological footprints. The latter is defined as the sum of natural resources, water, clean air, soil, and energy, which a country uses to produce the resources consumed and to absorb wastes. These all need to be taken into account and they do not necessarily enter directly as commodities into the production process, though they may be necessary use values, part of the reproduction behind the production process (Global Footprint Network 2009).

Foster and Holleman make reference to the path-breaking work by Bunker (1985) referring to the issue of unequal ecological exchange as far back as 1985 and arguing that the unequal exchange of energy and materials occurred to the detriment of extractive economies or "extreme peripheries". Bunker, similar to Emmanuel, refers to differential wages leading to unequal exchange between the periphery and center but then includes "the transfer of natural value in terms of raw resources". He argues:

> The outward flows of energy and the absence of consumption-production linkages combine with the instability of external demand and with the depletion of site-specific natural resources to prevent the storage of energy in useful physical and social forms in the periphery, and leave it increasingly vulnerable to domination by energy-intensifying social formations at the core.
>
> (Bunker 1985, 45)

In his attempt to present interdisciplinary analysis, much less common at the time, Bunker made a significant contribution to the field, forcing political economists, among others, to think about environmental and ecological issues.

Another important area of empirical research has been attempts to estimate ecological footprints, or carrying capacity, including several researchers working with the framework of 'ecologically unequal exchange'. Several Latin American authors have made advances in these areas, including Vallejo (2009) and Peinado (2019). These authors and others have produced empirical estimates of ecologically unequal exchange using both monetary and physical categories, and present clear tendencies of imbalances born by the periphery.[9]

As is often the case in economics, physical ratios, such as input–output coefficients, are derived from monetary transaction data, as the explicit details of physical quantities are often unknown or impossible to tease out of the available data. Foster and Holleman (2014) argue that although there remain a number of methodological problems, the main conclusion is still the expected result of the first world countries consuming more of the natural resources and not paying for the environmental consequences, and instead it is the global South employing its natural resources, be it energy, water, soil, and clean air, etc. to accommodate the excessive consumption in the North.

Advances by several authors, including the work associated with Foster and Holleman, Jorgenson, Odum, and Hornborg, are clearly crucial and important. In spite of encouraging advances in recent years, a number of serious problems remain. Therefore, the ecologically unequal exchange framework needs to continue its efforts to move forward, both theoretically and empirically, in the coming future, in order to improve the analysis of value, matter, and energy flows, and estimating international transfers, as such analysis is crucial in addressing a range of problems of wealth and environmental imbalances in the current global economy.

Reprimarization and the environment in Latin America

During the first decade of the 21st century, the *commodities boom* was produced as a result of the surge in demand for primary exports, namely agricultural products, minerals, petroleum, and other raw materials. The dynamics of growth for both the agroindustrial and other primary goods sectors, is often at the expense of manufacturing, constituting (re)primarization in the region. This tendency was a product of the phenomenal growth in China[10] and its demands of a steady flow of primary inputs and food, as well as the further consolidation of neoliberal globalization dominated by TNCs and reinforced via WTO rules. Moreover, China is providing loans and expertise for infrastructure projects, including hydroelectric plants, in addition to purchasing lands and brokering development projects, reflected in deals and treaties in numerous countries of Latin America (e.g., Argentina, Brazil, Ecuador).

There are a range of key environmental impacts from the reprimarization tendencies that deserve examination, the most notable being: (1) deforestation of the Amazon rainforest linked to the expansion of lumber, cattle, soy, mining, and petroleum; (2) air, water, and soil pollution as a result of mining and the use of pesticides in agriculture, especially for soy, (3) soil erosion as a result of mining and petroleum exploration, soy, and other agroindustries. There are also major socio-environmental concerns related to dams, in Brazil, Peru, Ecuador, and elsewhere, and these are often tied to mining. Additional environmental concerns are loss of biodiversity, risks of desertification, landslides, and earthquakes, such as due to fracking, and the concern of climate change in general. Moreover, there are numerous and growing social conflicts, involving human rights abuses.[11]

Table 16.1 below presents a fairly extensive, though not exhaustive, list of the sectors in Latin American countries, reflecting primary sectors, often the Non-Traditional Agricultural Exports (NTAE) category mentioned above, which entail major environmental problems.

Deforestation of the Amazon rainforest

Currently, the Amazon rainforest constitutes a *frontier for global capital accumulation*, with an ever-growing presence of TNCs, between cattle and JBS, Vale and mining, soybeans and Amaggi.[12] In the Brazilian Amazon, a common sequence is the removal of trees for lumber, followed by cattle, and then the planting of an intermediate crop to re-establish the nutrients in the soil lost during livestock production. The final step is the planting of soy, often heralded as

Table 16.1 Primary Sectors in Latin America

Sector	Products	Countries
Agroindustry	GM Soy	Brazil, Argentina, Uruguay and Paraguay
	GM wheat, cotton, corn	Argentina and Brazil
	Cattle/livestock	Argentina and Brazil
	Flowers	Colombia and Ecuador
	Fruits and Vegetables	Latin America (all countries)
	Palm Oil	Brazil, Colombia, Ecuador, Peru and Venezuela
Mining	Lithium (for batteries)	Chile, Bolivia, and Argentina
	Gold	Argentina, Brazil, Ecuador, Mexico
	Iron ore	Brazilian Amazon
	Silver	Argentina, Brazil, Bolivia, Colombia, Ecuador, Chile and Venezuela
	Copper	Chile, Brazil, Peru, Uruguay, Ecuador, Colombia
	Bauxite/Aluminum	Brazil
Fossil Fuels	Petroleum/Natural Gas	Argentina, Brazil, Colombia, Bolivia, Ecuador, Mexico, Venezuela
Biofuels	Wood, corn, sugar cane, vegetable oils	Argentina, Brazil, Chile, Colombia, Ecuador, Peru;

an environmentally-friendly alternative to cattle ranching; yet soy is actually one of the main motors behind the process of deforestation in the Amazon, given its higher profitability. The latter scenario is most common in the Brazilian state of Mato Grosso, but the drive to deforest is connected with clearing the trees for mineral and oil extraction, agriculture, etc. in many parts of Latin America. These cases are by no means limited to Brazil but include the Ecuadoran, Colombian, Peruvian, Bolivian, and Venezuelan Amazon.

Evidently the deforestation of the Amazon is not limited to Brazil, and moreover, it is relevant for the issue of climate change for the planet.[13] During 2019, the issue of forest fires, in particular those promoted by Bolsonaro, was a major environmental and human rights concern (Lappé 2019). For the first 26 days of August 2019, forest fires in the Amazon increased by almost 30%, resulting mainly from the Brazilian President's "call for fires" reflecting his criminal interest to burn down indigenous villages, namely genocide, and demonstrating a total lack of concern regarding the local and global environmental impacts of the fires. Given the major TNCs expansion in the Amazon, there are also many other environmental disasters happening or waiting to happen.

Agroindustry, soy, pesticides, and GMOs

The recent shift toward a more intensive use of biotechnology, far beyond the Green Revolution of the 1970s, reached a more mature stage in the 1990s. This was with the introduction of a technological package including genetically modified zero-tillage seeds (GMOs), and greater usage of agrochemicals. At present, both Brazil and Argentina are among the main producers and exporters of soy with transgenic soy constituting almost all Argentine, and Brazilian production. "Roundup Ready" is a variant of a soy seed produced by Monsanto which, by design, is particularly resistant to glyphosate.[14] As a result, the intensive use of "zero tillage" combined with Roundup Ready increased productivity, however, the intensive use of agrochemicals in the production of soy in Argentina, over time, has led to several problems. In order to maintain productivity, higher dosages, 3 to 5 times more Roundup, have been used over recent decades, due to herbicide-resistant weeds. This has increased the cost of imports for farmers as well as

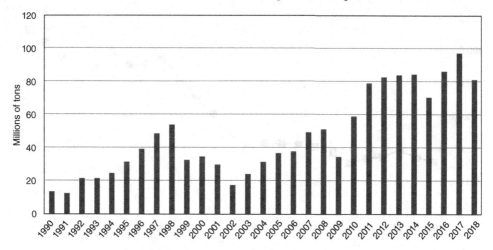

Figure 16.1 Pesticides Imported by Argentina: 1990–2018 (millions of tons).

Source: FAO (2020).

exacerbated health issues (Pengue 2005). The use of pesticides not only affects agricultural workers in the field, but since they are transported through the air, they impact adjacent communities as well, causing general health problems and more serious illnesses, such as cancer or anencephaly, and has led to a water crisis in the province of Entre Rios in Argentina. The growing use of glyphosate due to the expansion of soy and other crops and the need to use ever larger doses for increasingly resistant weeds is reflected in Figure 16.1 for Argentina.

The notable growth of harvested soy in the two most industrialized countries of South America in recent decades, namely, Argentina and Brazil can be observed in Figure 16.2 and presents a clear example of reprimarization. The area harvested for soy in Brazil grew an impressive 284% for between 1991 and 2019, while for Argentina the growth was 231% between 1991 and 2017, with maximums of 29 million hectares in a year for Argentina and almost 37 million hectares in 2019 for Brazil.

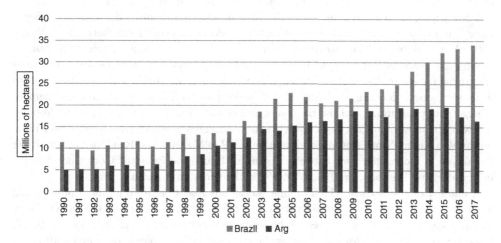

Figure 16.2 Area Harvested for Soy in Argentina and Brazil: 1990–2017 (millions ha).

Source: IBGE/PAM, 2020; INDEC/INTA, 2020.

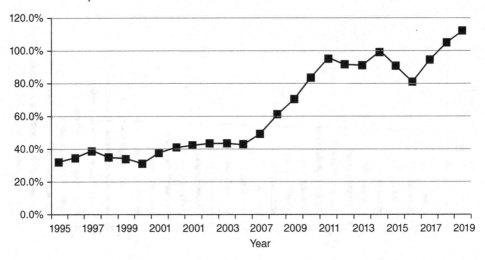

Figure 16.3 Primary to Manufacturing Exports (%) (Brazil: 1995–2019).
Source: MDIC, 2020.

Another example of the reprimarization tendency in Brazil is provided in Figure 16.3, which presents the ratio of primary exports in relation to manufacturing exports. Since 1995, this ratio grew from roughly 33% to approximately 110% in 2019, as primary exports have now surpassed manufacturing exports. Moreover, during the period of ten years (1999–2009), exports overall grew roughly 318%, while exports of primary goods grew at 525%. Beyond Brazil and Argentina, there is a general tendency toward reprimarization in Latin America, especially in the last couple decades, primarily due to the growth of demand from China.

Environmental impacts of mining

In Latin America, at least 14 countries made legal changes to accommodate TNC investment in mining in recent decades. Between 1991 and 1999, Latin America went from being the fourth to the first most common destination for investment in mineral exploration, as a result of a 500% increase in investment. Given the amazing growth of China's demands for minerals there was a surge in the prices of metals at the world level, quadrupling between 2002 and 2007. As a result, many Latin American countries experienced an increased role of the State in mining and associated infrastructure projects, be it for highways, waterways, or dams. The latter are crucial for electricity-intensive mining. The main socio-environmental impacts related to dams are: erosion, flooding, landslides, loss of biodiversity, land dispossession and displacement of significant populations, and also the loss of quality water sources and river-based livelihoods. A few of the most significant cases involving social struggles in recent years were: Belo Monte on the Rio Xingu, Santo Antonio and Jirau on the Rio Madeira, and Rio Doce in Brazil; the Itaipú dam in the frontier zone between Argentina, Brazil, and Paraguay; Coca Codo Sinclair in Ecuador; and Inambari, Ene, and Marañón in the Peruvian Amazon.

A number of countries in which mining was all but absent began to promote projects and granted concession to TNCs; consider the cases of Argentina, Ecuador, and Colombia. There was also significant expansion in established mining countries of the region, namely Brazil, Peru, Bolivia, Chile, and Mexico. Besides the environmental problems associated with dams, mining involves the generation of toxic waste, the processing and chemical treatment of thousands of tons of rock, which lead to substantial quantities of both solid and liquid waste.

There are diverse forms of chronic contamination and accidents involving cyanide, mercury etc., which impact the quality of air, soil, and water, not to mention deforestation, and irreversible changes to the water table, etc., which are all issues generating negative effects for public health (Sacher Freslon and Cooney 2019).

Final considerations

This chapter sought to present the major theoretical contributions related to the dependency school and ecologically unequal exchange, emphasizing Latin American research and authors. The shifts brought about through TNC-dominated neoliberal globalization were also examined. A key result was the decline in autonomy for countries of the periphery, especially due to the disproportionate role of TNCs in national economies, and the role played by the WTO. These 'external' factors, combined with internal shifts in class alliances, led to deindustrialization, reprimarization, and the increasing domination of the primary sector for much of Latin America and a range of social and environmental problems. It was hoped that the wave of 'progressive' governments in Latin America in the first decades of the 21st century would pursue a new trajectory and make a break from the domination by the 1st World and TNCs, however, the end result was that neoextractivism was actually strengthened.

Therefore, Latin America continues to suffer from the major problems of unequal exchange of wealth (in its monetary form) and a worsening ecologically unequal exchange of natural resources and environmental degradation, and given the current conjuncture of the pandemic, the short-term future looks rather bleak. In evaluating the reality of Latin America, and given the range of social and environmental problems, there is a need to consider alternatives. The contributions of dependency and ecologically unequal exchange, in recognizing the unjust economic structure in which the world operates and in which transfers of wealth and resources tend to have a clear flow from the periphery to the center, should be seen as critical tools, among several, in the construction of these alternatives.

Notes

1 See Gudynas (2015) and Bolados in this volume, for the definition and discussion of neoextractivism, which goes beyond just mining and fossil fuels.
2 It is a challenge to present a coherent synopsis of the dependency school, given so many different views and interpretations and switches of perspective among several of the authors. The *Latin American Perspectives* (1981) volume, with an excellent Introduction by Ronald Chilcote, was very useful for this discussion (see LAP (1981), which includes Chilcote (1981)).
3 Chilcote (1981) identified four different formulations of dependency: Frank (1967) development of underdevelopment; the new dependency of Dos Santos (1970); dependent capitalist development by Cardoso (1972); and, lastly, the work of Quijano (1971) dependency as a reformulation of classical theories of imperialism.
4 Super-exploitation is defined by Marx as the price of labor-power (wages) below the value of labor power; a situation which over the long run is not feasible, since wages would often be below subsistence level.
5 The argument of Frank is that the more contact with first world capital the more underdeveloped a country will be, but the examples of Taiwan, South Korea, and the other Asian Tigers arguably provided strong counter examples to Frank's argument.
6 The second stage of dependency is where countries began to produce manufactured goods, though not yet consumer durables, thus no longer needing to import so many basic manufactured goods, but rather needing to import machinery and other capital goods, This was often accomplished through the relocation of first world factories into the peripheral countries, thus avoiding tariffs, but still resulting in problems for balance payments due to trade deficits.
7 For a more developed analysis of the role of ground rent from a marxist perspective and analyzing reprimarization in Latin America, see Cooney (2021).

8   An important collection on political ecology in Latin America has been compiled by Pengue (2017).
9   The research carried out with respect to ecological unequal exchange and ecological footprints are tied to work around the concept of ecological debt (see Roa in this Handbook).
10 Since 2000, China has become the primary or secondary trading partner for the majority of countries in the region (Slipak 2014).
11 See the Global Environmental Justice Atlas (www.ejatlas.org) and also OCMAL (Observatory for Mining Conflicts in Latin America) (www.ocmal.org).
12 JBS is the largest meat product transnational in the world, based in Brazil, very much tied to the expansion of cattle in the Amazon; Vale do Rio Doce (Vale), the second largest mining TNC globally; Amaggi Group, is the largest TNC for soy in Brazil.
13 For more analysis of reprimarization and deforestation of the Amazon, see Rivero and Cooney (2010), or Cooney (2021).
14 Glyphosate is one of the most used herbicides in the world, and is the main active component of their product Roundup. In addition, glyphosate causes cancer among other health problems.

References

Amin, S. 1976. *Unequal Development: An Essay on the Social Formations of Peripheral Capitalism*. New York: Monthly Review Press.

Bambirra, Vania. 1978. *Teoría de la Dependencia: Una Anticrítica*. Mexico City: Ediciones Era.

Bunker, S. J. 1985. *UnderDeveloping the Amazon; Extraction, Unequal Exchange and the Failure of the Modern State*. Chicago: University of Chicago Press.

Carcanholo, Marcelo Dias. 2017. *Dependencia, Super-explotación del Trabajo y Crisis: Una Interpretación Desde Marx*. Madrid: Ediciones Maia.

Cardoso, Fernando Henrique. 1972. "Dependency and Development in Latin America." *New Left Review* 74 (July–August): 83–95.

———. 1980. "As ideias e seu lugar: ensaios sobre as teorias do desenvolvimento." *Cadernos CEBRAP (33)*, Editora Vozes.

Chilcote, Ronald H. 1981. "Issues of Theory in Dependency and Marxism." *Latin American* 8(3–4): 3–16.

Cooney, Paul. 2021. *Paths of Development in the Southern Cone: Deindustrialization and Reprimarization and Their Social and Environmental Consequences*. London: Palgrave.

Correa Prado, Fernando, and Castelo, Rodrigo. 2013. "O Início do Fim? Notas Sobre a Teoria Marxista da Dependência no Brasil Contemporâneo." *Pensata* 3 (1): 10–29.

Dos Santos, Theotônio. 1970. "The Structure of Dependence." *American Economic Review* LX (May): 231–236.

———. 1978. *Imperialismo y Dependencia*. Mexico City: Ediciones Era.

———. 2000. *A Teoria da Dependência: Balanço e Perspectivas*. Rio de Janeiro: Civilização Brasileira.

Emmanuel, A. 1972. *Unequal Exchange*. New York: Monthly Review Press.

Féliz, Mariano. 2019. "Neodevelopmentalism and Dependency in Twenty-first Century Argentina: Insights from the Work of Ruy Mauro Marini." *Latin American Perspectives*, Issue 224, 46 (1): 105–221.

Foster, J. B., and Holleman, H. (2014). "The Theory of Unequal Ecological Exchange: A Marx-Odum Dialectic." *The Journal of Peasant Studies* 41 (2): 199–233.

Frank, André Gunder. 1967. *Capitalism and Underdevelopment in Latin: Historical Studies of Chile and Brazil*. New York: Monthly Review Press.

Global Footprint Network. 2009. "Huella Ecológica y Biocapacidad en la Comunidad Andina." https://www.footprintnetwork.org/content/images/uploads/CAN_Teaser_ES_2009.pdf.

Gudynas, Eduardo. 2015. *Extractivismos. Ecología Economía y Política de un Modo de Entender el Desarrollo y la Naturaleza*. 1st ed. Lima: Cedib y ClAES.

Hardt, M. and Negri, A. 2000. *Empire*. Boston: Harvard University Press.

Harvey, David. 2004. *El nuevo imperialismo*. Madrid: Ediciones Akal.

Hornborg, A. 2011. *Global Ecology and Unequal Exchange*. London: Routledge.

Jorgenson, A.K. and Clark, B. 2009. "The Economy, Military, and Ecologically Unequal Exchange Relationships in Comparative Perspective: A Panel Study of the Ecological Footprints of Nations, 1975–2000." *Social Problems* 56 (4): 621–646.

Katz, Claudio. 2018. *La Teoría de la Dependencia, 50 años después*. Buenos Aires: Batalla de Ideas.

Latin American Perspectives. 1981. "Dependency and Marxism." Issues 30 and 31, Volume VIII, Numbers 3 and 4, Summer and Fall, 1981.

Lappé, A. 2019. "Follow the Money to the Amazon. Who Is Profiting from the Development That Led to These Fires?" *The Atlantic*, September 4.

Mandel, E. 1975. *Late Capitalism*. London: Verso.

Marini, Ruy Mauro. 1973. *Dialéctica de la Dependencia*. México D.F.: Ediciones Era.

Martins, C. E. 2011. *Globalização, Dependência e Neoliberalismo na América Latina*. São Paulo: Boitempo.

Odum, H. T. 1996. *Environmental Accounting: Emergy and Environmental Decision-Making*. New York: John Wiley and Sons.

Peinado, G. 2019. "Economía Ecológica y Comercio Internacional: el Intercambio Ecológicamente Desigual como Visibilizador de los Flujos ocultos del Comercio Internacional." *Revista Economía* 71 (112): 53–69.

Pengue, Walter. 2005. "Transgenic Crops in Argentina: The Ecological and Social Debt." *Bulletin of Science, Technology & Society* 25 (4): 314–322.

Pengue, Walter (editor). 2017. *El Pensamiento Ambiental del Sur*. Buenos Aires: Ediciones UNGS.

Quijano, Aníbal J. 1971. *Nationalism and Colonialism in Peru: A Study in Neo-Imperialism*. New York: Monthly Review Press.

Rice, J. 2007. "Ecological Unequal Exchange." *Social Forces* 85: 1369–1392.

Rivero, S. and Cooney, Paul. 2010. "The Amazon as a Frontier of Capital Accumulation: Looking Beyond the Trees." *Capitalism, Nature, Socialism: A Journal of Socialist Ecology* 21: 30–56.

Robinson, W. I. 2004. *A Theory of Global Capitalism- Production, Class and State in a Transnational World*. Baltimore: The John Hopkins University Press.

———. 2008. *Latin America and Global Capitalism- A Critical Globalization Perspective*. Baltimore: The John Hopkins University Press.

Rodney, Walter. 1972. *How Europe Underdeveloped Africa*. London and Dar es Salaam: Bogle-L'Ouverture and Tanzania Publishing House.

Sacher Freslon, William and Cooney, Paul. 2019. "Transnational Mining and Accumulation by Dispossession". In: *Environmental Impacts of Transnational Corporations in the Global South*, 11–34. Paul Cooney and William Sacher Freslon (eds.). United Kingdom: Emerald Publishing.

Shaikh, Anwar. 1980. *Foreign Trade and the Law of Value*. Part 2. *Science and Society* 44 (1): 27–57.

Slipak, A. 2014. "Un análisis del ascenso de China y sus vínculos con América latina a la luz de la Teoría de la Dependencia." *Realidad Económica*, 282 February 16 to March 31.

Teixeira Assis, W. F. 2014. "Do Colonialismo à Colonialidade: Expropriação Territorial na Periferia do Capitalismo." *Caderno CRH*, Salvador, 27 (72): 613–627, Set/Dez 2014.

Thomas, Clive Y. 1974. *Dependence and Transformation: The Economics of the Transition to Socialism*. New York: Monthly Review Press.

Vallejo, M. C. 2009. *La estructura Biofísica de la Región Andina y sus Relaciones de Intercambio Ecológicamente Desigual (1970–2005). Un estudio comparativo*. Madrid: Fundación Carolina.

17 Challenging the logic of "the open veins"?

The geography of resource rents distribution in Peru and Bolivia

Felipe Irarrazaval

Introduction

Much research regarding the environment in Latin America begins with referencing *Open veins of Latin America* by Eduardo Galeano (1973) to evidence the long-term history of foreign plundering and the many dramatic events surrounding resource extraction that have shaped Latin American geography. Although this historical-geographical pattern arguably remains unbroken—the natural resources sector is still the leading net exporter, and income derived from it remains a fundamental source of public revenue for national government (ECLAC 2014), changes in the political economy of resource extraction at multiple scales ask for a critical appraisal and revisiting of *Open veins of Latin America* (Bebbington 2009; Bury and Bebbington 2013; Farthing and Fabricant 2018). In particular, many forms of social control over and against resource extraction have emerged in Latin America in the last 20 years.

The last cycle of high global demand of natural resources—roughly between the late 90s and 2015—involved a massive increase in resource extraction in Latin America, particularly minerals like silver (Peru), copper (Chile and Peru), lead (Bolivia), gold (Peru and Bolivia), iron ore (Brazil and Mexico), and hydrocarbons, mainly oil (Venezuela, Mexico, Brazil, Colombia and Ecuador) and natural gas (Brazil, Argentina, Bolivia, Colombia and Venezuela) (see Bebbington and Bury 2014 for an overview). Such an increase in resource extraction was massively contested at local and regional levels (Arsel, Hogenboom, and Pellegrini 2016; Bebbington and Bury 2014). Although the unrest of local communities is not necessarily new, political conditions during this cycle—such as the emergence of non-traditional forms of citizenship (Delamaza, Maillet, and Neira 2017; Rich, Mayka, and Montero 2019), decentralization reforms (Brosio and Jiménez 2012; Suarez-Cao, Batlle, and Wills-Otero 2017) and international agreements (such as the ILO Convention 169[1] or EITI[2])—allowed local communities to contest extraction and re-scale their mobilizations at different levels (Haarstad 2014; Riofrancos 2017). This scenario consequently triggered a variety of multi-scalar political projects that looked to shape resource governance in multiple ways (Arce 2016; Fry and Delgado 2018; Rasmussen and Lund 2018). For example, some local communities reject the materialization of extractive projects (Conde and Le Billon 2017), local groups might challenge the distribution of the benefits related to resource extraction (Hinojosa et al. 2015; Arce 2016; Irarrázaval 2018) or political elites might encourage developmental projects that rely on resource extraction (Gudynas 2012; Svampa 2012).

Whereas the current scenario of social mobilization does not break the historical pattern of dependency and inequality, what might be changing is "the governance of nature and the social control and subsequent use of its subsidy" (Bebbington 2009, 9). For example, relevant transfers of resource rents to local communities, such as direct payments to Indigenous communities (Anthias 2018), development funds (O'Faircheallaigh 2013), or corporate social responsibility programs (Suescun Pozas, Lindsay, and du Monceau 2015), are expressions of the mechanisms

DOI: 10.4324/9780429344428-20

by which extractive industries obtain local licenses to operate. Likewise, the massive increases in resource extraction during the last decades not only financed the social programs of progressive governments—within the "new extractivism" (Gudynas 2010; Svampa 2012)—but also involve a massive transfer of resource rents to the local governments of producing areas in many countries (Mejia and Arellano Yanguas 2014; Viale 2015).[3] Altogether, the different forms of resource rents transfer that underpinned the last commodity boom not only reflect the ongoing way in which the benefits of resource extraction are being distributed among different scales, but also call for a more in-depth appraisal regarding how those new scales are challenging, or not, the logic of "open veins".

This chapter analyzes the way in which natural gas rents in Peru and Bolivia have been mobilized at the sub-national level, and the consequences of such a rearrangement for the long-standing issues of dependency and inequality that characterize the 'open veins' of Latin America. My core argument is that the new scenario of contestation merely changed the distribution of rents, reproducing new forms of uneven development rather than confronting the classic problems of dependency and the inequalities related to resource extraction in the region. The chapter is organized as follows: The next section discusses the changes in the geography of natural gas rents distribution in Peru and Bolivia in the last commodity boom. The following section examines how those changes have impacted economic dependency and territorial inequality. Finally, the chapter concludes that resource rent distribution failed in promoting alternative development paths.

The geography of resource rents distribution in Peru and Bolivia

The core of the social science research agenda on extractive industries during the last commodity supercycle has been to discuss the contradictions between the Latin American states' accumulation strategies, based on the rents derived from extractive industries, and the everyday life and local knowledges of communities affected by resource extraction (Gudynas 2010; Svampa 2012; Arsel, Hogenboom, and Pellegrini 2016; Alimonda, Toro, and Martin 2017). However, such a research agenda has neglected the vibrant spatial diversity of actors that influence resource governance in Latin America, and consequently have modified the way in which the benefits of extractive industries are distributed—as other contributions have shown (Berdegué, Bebbington, and Escobal 2015; Hinojosa et al. 2015; Arce 2016; Irarrázaval 2018). Although the scenario is variegated, this chapter mainly addresses sub-national mobilizations that seek to reconfigure the way in which resource rents are distributed, and consequently the effects of such distribution in terms of dependency and inequality.

Even though local groups might share the wealth from natural resources extraction through different processes—such as direct payments to local communities or development funds, the most institutionalized expression of resource rents distribution is revenue-sharing policies. Those policies distribute rents from extractive industries—collected by states through taxes and royalties—to sub-national levels (Mejia and Arellano Yanguas 2014; Viale 2015). As such, revenue-sharing policies are a channel for distributing the value produced by extractive industries, but are limited exclusively to an institutionalized territoriality—i.e., sub-national governments. There are different distribution formulas which benefit producers' areas—i.e., where natural resources are being extracted, depending on how sub-national elites have historically institutionalized their position at the national level (see Irarrázaval 2020). Depending on the political power of the sub-national groups at the national level, as well as on the relation of those groups with the national level, producers and non-producers in sub-national areas define how they divide the extractive wealth (Mejia and Arellano Yanguas 2014). In this regard, it is crucial to grasp the way in which revenue- sharing policies in Bolivia and Peru were arranged.

During the first hydrocarbon nationalization in Bolivia (1937), it was determined that the producer department (the first sub-national level in Bolivia) must receive 11% of the total revenues collected (Anaya Giorgis 2014). Even though there is no clear explanation for why the government defined a revenue-sharing policy,[4] the policy strengthened the political power of Santa Cruz—the leading producer region—and consequently the main rival of La Paz, the country's capital (Pruden 2012). This is critical for Bolivia's sub-national politics because hydrocarbon reserves are in the lowlands. Bolivia's lowlands—the so-called half-moon because of its cartographic form—has a different ethnic and political composition compared to the highlands, where Indigenous identity is stronger (Perreault and Valdivia 2010). As a result, the political struggles for revenue sharing—mainly led by Santa Cruz—reflect the country's political cleavage (Pruden 2012). While the highlands seek to increase control over revenues for developing social policies, as well as for reducing the economic power of the lowlands, the lowlands aim to conserve the sharing policy to secure the strength of their regional power. The result of this scenario is that producer departments receive 11% of the natural gas extracted from its territory through *regalias*, a form of royalty, plus the 12,5% of the 32% through IDH,[5] a complementary tax. After a severe political juncture in 2008, the lowland sub-national governments gave away a portion of IDH to finance a pension program introduced by Evo Morales's government (Humphreys and Bebbington 2010).

In Peru, the sub-national powers have been historically weak vis-à-vis the capital, Lima (Orihuela and Thorp 2012). However, in 1974 there was a great wave of protest in the department of Loreto, demanding increased autonomy (Gruber and Orihuela 2017). This department is the most isolated area of the country, located in the heartland of the Amazon basin and lacking transport connections by land to the rest of the country. Additionally, the then- promising oil industry was actively drilling within its jurisdiction. As a result, a mechanism of redistribution (*canon*) was created, which states that 50% of the taxation and royalties of any natural resource must remain in the producer department (first sub-national level in Peru) (Viale 2015). Although this policy continues to provide the main principle for revenue sharing in Peru, there has been a critical modification regarding how the 50% must be distributed within the country: during Alberto Fujimori's dictatorship (1990–2000), many opposition leaders got positions at municipalities (the second sub-national level in Peru). Subsequently, during the transition period after the dictatorship (2000–2003), those opposition leaders pushed for an increase in the *canon* transfer to municipalities instead of departments (Arellano-Yanguas 2011). Additionally, mining companies also pushed for this reform, because they thought it would be a good path for securing local licenses (Arellano-Yanguas 2011, 2019). As a result, the municipalities in Peru receive a large proportion of the different types of the *canon*.[6]

Against this background, and along with the increase of hydrocarbon extraction and high oil prices, there was a massive transfer of resource rents to the sub-national governments of producing areas in Peru and Bolivia. The data presented in Figure 17.1 reveal that the amount of hydrocarbon revenues distributed to sub-national levels in Bolivia has exploded since 2009, and substantially increased during the same period in Peru. The following section looks more closely at how those rents were invested, the consequences of those expenditures in terms of the dependency of local governments on resource rents, and how those rents challenged uneven development. The analysis focuses on the key producing area of hydrocarbons in Bolivia—the Tarija department—and Peru—the local districts of La Convención province (Figure 17.2).

Challenging the logic of the open veins?

The emergent body of literature on the "local resource curse" (Arellano-Yanguas 2011; Manzano and Gutiérrez 2019; Orihuela and Echenique 2019) has been keener to analyze the relation between sub-national areas and resource rents than other approaches within human

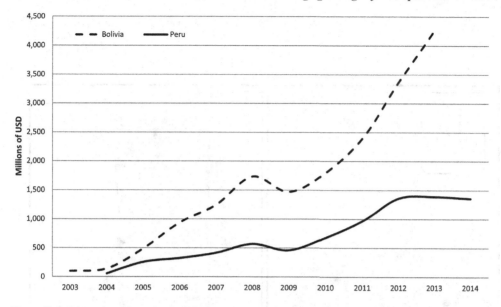

Figure 17.1 Sub-national incomes from hydrocarbon production. Own elaboration based on Viale (2015).

geography, environmental sociology, or political ecology. However, many of these contributions reproduce the "post-political" assumptions of the resource curse literature, which neglects the geopolitical processes through which states produce spaces for extraction (Watts 2004) and the divergent dynamics that shape local development in production networks of resource extraction (Bridge 2008). Whereas the local resource curse literature has mainly analyzed how the distribution of resource rents produce local conflicts (Arellano-Yanguas 2011; Orihuela, Pérez, and Huaroto 2019) and the effects that resource rents have on socio-economic development (Arellano-Yanguas 2019), this section discusses how resource rents have impacted economic dependency and uneven development in Tarija and La Convención.

Concrete, steel and resource dependency

During the most recent commodity supercycle, the local governments of the producing areas increased their budget radically. The population of Tarija department in 2012 was nearly 483,000. In the same year, the hydrocarbon incomes of the department were US$283 million and US$476 million in 2014 (Fundación Jubileo 2017). Even though the revenue-sharing regime in Peru is fragmented among the districts and the department of the producing area, the geographical regime of rent distribution creates significant transfers per inhabitant in the Echarati district. While the population of Echarati was 42,676 in 2007, the income of the municipality, because of the *canon* law, was US$15 million in 2017—nearly a quarter of the municipal income in 2011 (US$65 million).[7]

Considering this scenario, Tarija department and the districts of La Convención—particularly Echarati—started an aggressive investment in specific areas. As Figure 17.3 shows for Echarti, the lead areas of investment—transport, urban development and agriculture—involved an intensive development of construction projects, which were the primary expression of the resource rents in the landscape. As noted by the local population in Echarati, the most noticeable changes were: 'roads and sidewalks, steel and concrete, steel and concrete, that was the main change' (Interviewee, Civil society, Echarati Peru)

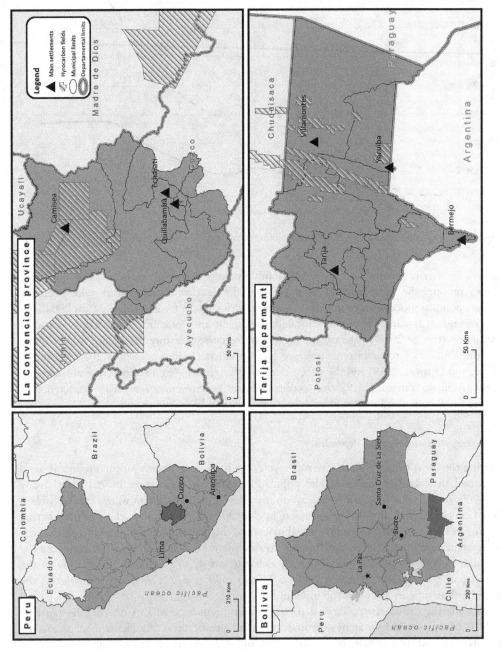

Figure 17.2 Cases studied. Own elaboration.

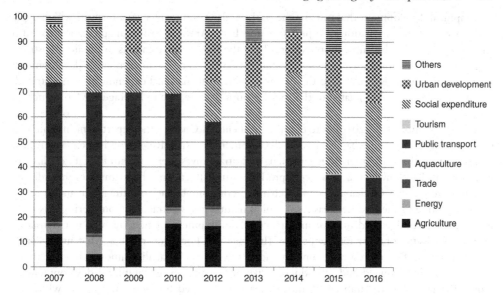

Figure 17.3 Expenditure by area in Echarati (percentage), Peru. Own elaboration based on SIAF.

Resource rents were mainly visible through new buildings—including many white elephants—and road development: 'Road infrastructure growth, the buildings also have grown (…) Finally, we see infrastructure in Tarija; the issue of mobilities was incredible; there is more mobility than people; you do not see bicycles anymore' (Interviewee, Regional assembly, Tarija). The total length of roads in the Tarija department was 5,697km in 2005 but increased to 14,591km by 2014 (INE 2016). Even though this kind of investment in construction looks to cover infrastructure gaps—such as in education, health, or trade infrastructure—there was no clear development project regarding the investment of natural gas rents. As a consequence, local governments failed to boost any other economic sector beyond construction: 'This department has a high dependency on public revenues, and hydrocarbon rents are what mobilize the economy, the infrastructure' (Interviewee, Local government, Tarija Bolivia).

Far from boosting economic diversification, the local private sector has adapted almost wholly to work with the public sector for developing infrastructure. As acknowledged by a member of the local government in Tarija, 'we do not have a self-sustainable model but a paternalistic one that depends on public incomes'. The interviewees acknowledged that the local economy before the commodity boom was small but relatively stable and diversified, with active participation of the agrarian sector. Through the increase in public budgets as a consequence of resource rents, however, 'the only effect are more fiscal incomes, but the economic structure of dependency has been deepened' (Interviewee NGO, Cusco Peru). As such, this scenario in which sub-national actors receive a massive amount of resource rents did not change the pattern of dependency over resource extraction in the short- or middle- term.

Uneven development

The way in which sub-national governments spend resource rents matters not only for analysis of the failure of economic diversification in the context of revenue sharing policies, but also for understanding how those transfers impact on different forms of spatial inequalities. Following Smith (1984), such spatial inequalities are nothing other than a contingent form of uneven geographical development that mirrors the multi-scalar contingent spatial form in which capitalism arranges the circulation of capital (see also Leon Araya 2015 for an application to Latin America). As such, this subsection explores how the massive investments in public infrastructure led by sub-national governments produced two different forms of uneven development: First, a spatial inequality related to the places in which sub-national expenditure concentrated. Second, a class inequality related to the social groups that are able to leverage the 'construction boom'.

In Peru, Muñoz et al. (2016) note that the local elite of Cusco—the department in which La Convencion is located—barely has a development perspective beyond their own city. Consequently, the departmental government prioritizes investments within the city or directly benefiting it. La Convencion province reproduces this pattern, and the main cities that administer resource rents—Quillabamba and Echarati—concentrate the investments in urban areas, barely investing in the remote areas from which natural gas is extracted. Considering this context, the large budget and massive infrastructure development of Echarte 'was not translated into the necessities of those remote people' (Interviewee, Local government, Megantoni Peru).

Whereas transfers of resource rents reduced the spatial inequality among districts in La Convencion, it produced and reproduced a pattern of uneven geographical development in which the towns of Echarati and Quillabamba concentrated the new investments while the remote areas where natural gas is extracted (Bajo Urubamba or Camisea) barely enjoyed the benefits. The former president of Peru—Ollanta Humana—summarized such unevenness by pointing out that, "Echarati is probably the richest district of Peru. They are full of swimming pools and useless stuff. They have forgotten that Bajo Urubamba and the areas near Camisea belongs to Echarati, and those people live in extreme poverty" (El Comercio 2014). As such, the transfer of resource rents barely improved the living conditions near extraction sites (Castro et al. 2014).

Likewise, in Bolivia, the administration of Tarija departments operated within a similar geographical horizon of development and consequently concentrated expenditures in Tarija city. The developmental project of Tarija's elite is grounded on their colonial differentiation from the peasant and Indigenous sectors that live within the department (Vacaflores and Lizárraga 2005; Humphreys and Bebbington 2010), and have historically marginalized the areas in which those people concentrate: "The inequality within the region was incredible, impressive. There was a huge gap, a huge historical debt with the peasants' sector. A historical debt in a territorial sense and also a social one" (Interviewee, former departmental authority, Tarija).

This geographical unevenness was particularly evident between Tarija city, the surrounding area called Cercado, and El Chaco, the area to the east from which resources were extracted. The main settlements of El Chaco (Villamontes, Carapari, and Yacuiba) have stronger functional bonds with Santa Cruz de la Sierra than with Tarija, despite the fact that Tarija is the department's capital. As such: "there were many resources in the department, and those resources came from El Chaco, but El Chaco never improved, and Tarija [the capital] improved" (Interviewee, Local government, Villamontes Bolivia). Despite the extensive road network in the Tarija department, there is no asphalt road between Tarija city and El Chaco. As in Peru, resource rents distribution produced new scales of uneven development within this Bolivian department.

The uneven development mentioned above also has a class expression: not only were investments concentrated at Echarati, Quillabamba, and Tarija cities, but so too were the local elites who leveraged such a scenario. These elites mainly participate in the construction sector, as owners of construction firms or suppliers, because they have contacts with local authorities, capital to invest in construction, and the necessary knowledge to acquire contracts with the state. This new elite is visible in the urban landscape, because "they expend it [the money] very well, they built nice houses, and they bought houses. If you visit Los Parrales or Miraflores, you will find new rich people who take advantage of the moment" (Interviewee, NGO, Tarija Bolivia).

However, this new urban landscape is not only related to the construction firms that take advantage of the moment but also to the local authorities that practised corruption when signing contracts: "Public expenditure is not only a business for firms but also for the politicians (…) then you understand why the interest in offering contracts, the business of the concrete" (Interviewee, NGO, Tarija Bolivia).Three consecutive majors of Echarati have been formally accused of corruption—two are fugitives and one is in jail while two former governors of Tarija face the same situation: one is a fugitive and another under investigation. Thereby, resource rents have not only produced new scales of uneven development within the sub-national areas that administered the revenues, but also across the class structure.

Conclusions

This chapter assesses the way in which the multi-scalar scenario of resource rents distribution might challenge the long-standing patterns of resource dependency and uneven development that characterize the "open veins" *of* Latin America. It discusses how sub-national struggles might modify the geographical distribution of resource rents, and consequently take a larger piece of the pie. Sadly, this chapter shows that the local governments of Tarija and La Convencion did not challenge the historical pattern of economic dependency of resource rents, and instead produced new scales of uneven development. As discussed elsewhere (Arellano-Yanguas 2011; Wilson and Bayón 2018; Irarrázaval 2020), the distribution of resource rents operates as a mechanism for legitimizing resource extraction at the sub-national level rather than as a path to alternative local forms of development. Even though it makes sense that resource rents should be administered closer to the extraction site, the cases analyzed through this chapter show that the pursuit of even a minimum of spatial justice is a much more complex challenge.

Local and isolated communities near where extraction takes place barely participate in the many political debates regarding at which scale resources should be administered. Accordingly, the distribution of resource rents to sub-national or local governments might not be the right path for transferring benefits to local communities because the transfer of rents to sub-national governments does not guarantee spatial justice (see also Watts 2004; Hinojosa et al. 2015; Wilson and Bayón 2018). As noted in the last subsection, resource rents distribution mainly benefits local elites, because they are better prepared to capture the benefits through different mechanisms (local government administration, local construction firms, local suppliers, etc.). These cases do not, therefore, offer hopeful examples that allow us to imagine a post-extractivist future. It is necessary to dig in a different direction.

Notes

1 The International Labour Organisation convention 169, introduced in 1989, is intended to protect Indigenous communities. It has been ratified by 22 countries to date, including Bolivia and Peru.
2 The Extractive Industries Transparency Initiative is a global standard for disclosing information about the contracts signed between firms and states.
3 Depending on the political regime of resource rents distribution—so-called revenue sharing (see Viale 2015; Mejia and Arellano Yanguas 2014). Excluding Chile, every Latin American country distributes a percentage of the royalties captured by the central government to the districts or departments in which the extraction take place (see Viale 2015 for an overview).
4 The most probable explanation is that Bolivia's government followed the Argentine model for creating its hydrocarbon politics. Argentina was a federal country, and its sub-national governments collect and control the revenues from natural resources. In 1937, revenue sharing was not a key issue for the political agenda; thus, Bolivia simply followed the federal guidelines from Argentina.

5 Direct tax to hydrocarbons (IDH according to its Spanish acronym).
6 From the total *canon*, 40% is equally distributed to all the municipalities within the producer department; 25% is equally distributed to all the municipalities within the producer province (an intermediate level); 10% is for the producer municipality; 20% for the producer department; and 5% for the department's public universities.
7 According to Perupetro (Estadística anual de Hidrocarburos).

References

Alimonda, Hector, Catalina Toro, and Facundo Martin, eds. 2017. *Ecologia Politica Latinoamericana. Pensamiento Crítico, Diferencia Latinoamericana y Rearticulacion Epistemica*. Vol. 1. Buenos Aires: CLACSO, Consejo Latinoamericano de Ciencias Sociales.

Anaya Giorgis, Juan José. 2014. 'Estado y Petróleo En Bolivia: 1921–2010'. PhD thesis, Buenos Aires: FLACSO.

Anthias, Penelope. 2018. *Limits to Decolonization: Indigeneity, Territory, and Hydrocarbon Politics in the Bolivian Chaco*. Cornell Series on Land: New Perspectives on Territory, Development, and Environment. Ithaca, NY: Cornell University Press.

Arce, Moises. 2016. 'The Political Consequences of Mobilizations against Resource Extraction'. *Mobilization: An International Quarterly* 21 (4): 469–83. https://doi.org/10.17813/1086-671X-21-4-469.

Arellano-Yanguas, Javier. 2011. 'Aggravating the Resource Curse: Decentralisation, Mining and Conflict in Peru'. *The Journal of Development Studies* 47 (4): 617–38. https://doi.org/10.1080/0022038100370 6478.

———. 2019. 'Extractive Industries and Regional Development: Lessons from Peru on the Limitations of Revenue Devolution to Producing Regions'. *Regional & Federal Studies* 29 (2): 249–73. https://doi.org/10.1080/13597566.2018.1493461.

Arsel, Murat, Barbara Hogenboom, and Lorenzo Pellegrini. 2016. 'The Extractive Imperative in Latin America'. *The Extractive Industries and Society* 3 (4): 880–87. https://doi.org/10.1016/j.exis.2016.10.014.

Bebbington, Anthony. 2009. 'Latin America: Contesting Extraction, Producing Geographies'. *Singapore Journal of Tropical Geography* 30 (1): 7–12. https://doi.org/10.1111/j.1467-9493.2008.00349.x.

Bebbington, Anthony, and Jeffrey Bury, eds. 2014. *Subterranean Struggles: New Dynamics of Mining, Oil, and Gas in Latin America*. Reprint edition. Austin, TX: University of Texas Press.

Berdegué, Julio A., Anthony Bebbington, and Javier Escobal. 2015. 'Conceptualizing Spatial Diversity in Latin American Rural Development: Structures, Institutions, and Coalitions'. *World Development, Growth, Poverty and Inequality in Sub-National Development: Learning from Latin America's Territories*, 73 (September): 1–10. https://doi.org/10.1016/j.worlddev.2014.10.015.

Bridge, Gavin. 2008. 'Global Production Networks and the Extractive Sector: Governing Resource-based Development'. *Journal of Economic Geography* 8 (3): 389–419.

Brosio, Giorgio, and Juan Pablo Jiménez. 2012. *Decentralisation and Reform in Latin America: Improving Intergovernmental Relations*. Northamtpon, MA: Edward Elgar Publishing.

Bury, Jeffrey, and Anthony Bebbington. 2013. 'New Geographies of Extractive Industries in Latin America'. In *Subterranean Struggles. New Dynamics of Mining, Oil and Gas in Latin America*, edited by Anthony Bebbington and Jeffrey Bury, 27–66. Texas: University of Texas Press. https://wordpress.clarku.edu/abebbington/2013/2013/new-geographies-of-extractive-industries-in-latin-america/.

Castro, Gonzalo, Juan Jose Garrido, Richard Korswagen, Patricia Majluf, Miguel Santillana, Glenn Sheppard, and Richard Chase. 2014. *Camisea: Emerging Lessons in Development, First Consolidated Report (2010-2014)*. Lima: South-Central Peru Panel.

Conde, Marta, and Philippe Le Billon. 2017. 'Why Do Some Communities Resist Mining Projects While Others Do Not?'. *The Extractive Industries and Society* 4 (3): 681–97. https://doi.org/10.1016/j.exis.2017.04.009.

Delamaza, Gonzalo, Antoine Maillet, and Christian Martínez Neira. 2017. 'Socio-Territorial Conflicts in Chile: Configuration and Politicization (2005-2014)'. *European Review of Latin American and Caribbean Studies* 104: 23–46. https://doi.org/10.18352/erlacs.10173.

ECLAC. 2014. 'Pactos para la igualdad: Hacia un futuro sostenible'. Text. Trigésimo quinto período de sesiones de la CEPAL. Santiago: CEPAL. http://www.cepal.org/es/publicaciones/36692-pactos-la-igualdad-un-futuro-sostenible.

El Comercio, R. E. C. (2014, septiembre 9). Humala sobre La Convención: "Lo que se están peleando es plata". *El Comercio Perú*. https://elcomercio.pe/peru/ica/humala-convencion-peleando-plata-361590-noticia/.

Farthing, Linda, and Nicole Fabricant. 2018. 'Open Veins Revisited: Charting the Social, Economic, and Political Contours of the New Extractivism in Latin America'. *Latin American Perspectives* 45 (5): 4–17. https://doi.org/10.1177/0094582X18785882.

Fry, Matthew, and Elvin Delgado. 2018. 'Petro-Geographies and Hydrocarbon Realities in Latin America'. *Journal of Latin American Geography* 17 (3): 10–14.

Fundación Jubileo. 2017. 'Un mejor uso y destino de la renta petrolera'. 60. Serie Debate Público. La Paz: Fundacion Jubileo.

Galeano, Eduardo H. 1973. *Las Venas Abiertas de América Latina*. Madrid: Siglo XXI.

Gruber, Stephan, and José Carlos Orihuela. 2017. 'Deeply Rooted Grievance, Varying Meaning: The Institution of the Mining Canon'. In *Resource Booms and Institutional Pathways*, 41–67. Latin American Political Economy. Cham: Palgrave Macmillan. https://doi.org/10.1007/978-3-319-53532-6_2.

Gudynas, Eduardo. 2010. 'Si eres tan progresista ¿por qué destruyes la naturaleza? Neoextractivismo, izquierda y alternativas (Tema central)', April. http://repositorio.flacsoandes.edu.ec/handle/10469/3531.

———. 2012. 'Estado Compensador y Nuevos Extractivismos: Las Ambivalencias del Progresismo Sudamericano'. *Nueva Sociedad* 237: 128–46.

Haarstad, Håvard. 2014. 'Cross-Scalar Dynamics of the Resource Curse: Constraints on Local Participation in the Bolivian Gas Sector'. *The Journal of Development Studies* 50 (7): 977–90. https://doi.org/10.1080/00220388.2014.909026.

Hinojosa, Leonith, Anthony Bebbington, Guido Cortez, Juan Pablo Chumacero, Denise Humphreys, and Karl Hennermann. 2015. 'Gas and Development: Rural Territorial Dynamics in Tarija, Bolivia'. *World Development*, Growth, Poverty and Inequality in Sub-National Development: Learning from Latin America's Territories, 73 (September): 105–17. https://doi.org/10.1016/j.worlddev.2014.12.016.

Humphreys, Denise, and Anthony Bebbington. 2010. 'Extraction, Territory, and Inequalities: Gas in the Bolivian Chaco'. *Canadian Journal of Development Studies / Revue Canadienne d'études Du Développement* 30 (1–2): 259–80. https://doi.org/10.1080/02255189.2010.9669291.

Irarrázaval, Felipe. 2018. 'Metano-territorialidades: La "Era del Gas Natural" En Peru y Bolivia'. *Journal of Latin American Geography* 17 (3): 153–82. https://doi.org/10.1353/lag.2018.0045.

———. 2020. 'Contesting Uneven Development: The Political Geography of Natural Gas Rents in Peru and Bolivia' *Political Geography* 79: 102161.

Manzano, Osmel, and Juan David Gutiérrez. 2019. 'The Subnational Resource Curse: Theory and Evidence'. *The Extractive Industries and Society*, April. https://doi.org/10.1016/j.exis.2019.03.010.

Mejia, Andres, and Javier Arellano Yanguas. 2014. 'Extractive Industries, Revenue Allocation and Local Politics', March. https://kclpure.kcl.ac.uk/portal/en/publications/extractive-industries-revenue-allocation-and-local-politics(62e61387-82e1-4032-a383-425ff199e058).html.

Muñoz, Paula, Martín Monsalve, Yamilé Guibert, César Gudalupe, and Javier Torres. 2016. *Élites Regionales En El Perú En Un Contexto de Boom Fiscal: Arequipa, Cusco, Piura y San Martín (2000–2013)*. Documentos de Investigación 7. Lima: Universidad del Pacífico.

O'Faircheallaigh, Ciaran. 2013. 'Community Development Agreements in the Mining Industry: An Emerging Global Phenomenon'. *Community Development* 44 (2): 222–38. https://doi.org/10.1080/15575330.2012.705872.

Orihuela, José Carlos, and Victor Gamarra Echenique. 2019. 'Volatile and Spatially Varied: The Geographically Differentiated Economic Outcomes of Resource-Based Development in Peru, 2001–2015'. *The Extractive Industries and Society*, July. https://doi.org/10.1016/j.exis.2019.05.019.

Orihuela, José Carlos, Carlos A. Pérez, and César Huaroto. 2019. 'Do Fiscal Windfalls Increase Mining Conflicts? Not Always'. *The Extractive Industries and Society* 6 (2): 313–18. https://doi.org/10.1016/j.exis.2018.07.010.

Orihuela, José Carlos, and Rosemary Thorp. 2012. 'The Political Economy of Managing Extractives in Bolivia, Ecuador and Peru'. *Social Conflict, Economic Development and Extractive Industry: Evidence from South America*, Social conflict, economic development and extractive industry: evidence from South America. London [u.a.]: Routledge, ISBN 978-0-415-62071-0. - 2012, pp. 27–45.

Perreault, Tom, and Gabriela Valdivia. 2010. 'Hydrocarbons, Popular Protest and National Imaginaries: Ecuador and Bolivia in Comparative Context'. *Geoforum* 41 (5): 689–99. https://doi.org/10.1016/j.geoforum.2010.04.004.

Pruden, Hernán. 2012. 'Las Luchas "Cívicas" y Las No Tan Cívicas: Santa Cruz de La Sierra (1957-59)'. *Revista Ciencia y Cultura* 29 (December): 127–60.

Rasmussen, Mattias Borg, and Christian Lund. 2018. 'Reconfiguring Frontier Spaces: The Territorialization of Resource Control'. *World Development* 101 (January): 388–99. https://doi.org/10.1016/j.worlddev.2017.01.018.

Rich, Jessica A. J., Lindsay Mayka, and Alfred P. Montero. 2019. 'Introduction The Politics of Participation in Latin America: New Actors and Institutions'. *Latin American Politics and Society* 61 (2): 1–20. https://doi.org/10.1017/lap.2018.74.

Riofrancos, Thea N. 2017. 'Scaling Democracy: Participation and Resource Extraction in Latin America'. *Perspectives on Politics* 15 (3): 678–96. https://doi.org/10.1017/S1537592717000901.

Smith, N. 1984. *Uneven Development: Nature, Capital and the Production of Space*. London: Blackwell.

Suarez-Cao, Julieta, Margarita Batlle, and Laura Wills-Otero. 2017. 'Presentación: El auge de los Estudios Sobre la Política Subnacional Latinoamericana'. *Colombia Internacional* 90 (April): 15–34. https://doi.org/10.7440/colombiaint90.2017.01.

Suescun Pozas, María del Carmen, Nicole Marie Lindsay, and María Isabel du Monceau. 2015. 'Corporate Social Responsibility and Extractives Industries in Latin America and the Caribbean: Perspectives from the Ground'. *The Extractive Industries and Society* 2 (1): 93–103. https://doi.org/10.1016/j.exis.2014.08.003.

Svampa, Maristella. 2012. 'Extractivismo Neodesarrollista y Movimientos Sociales. ¿Un Giro Ecoterritorial Hacia Nuevas Alternativas?' In *Más Allá Del Desarrollo*, edited by Grupo Permanente de Trabajo sobre Alternativas al Desarrollo, 185–218. Quito: Abya Yala.

Vacaflores, Carlos, and Pilar Lizárraga. 2005. 'La lucha por el Excedente del Gas y la resignificación de las Contradicciones de la Identidad Regional en Bolivia. Proyectos de Dominación y Resistencia en una región Productora de Hidrocarburos'. *OSAL* 17 (May): 21–31.

Viale, Claudia. 2015. 'Distribución de la Renta de las Industrias Extractivas a los Gobiernos Subnacionales en América Latina: Análisis Comparativo y de Tendencias'. 1. Kawsaypacha Digital: Documentos Para el debate Ambiental. Lima: INTE-PUCP. http://repositorio.pucp.edu.pe/index//handle/123456789/52358.

Watts, Michael. 2004. 'Resource Curse? Governmentality, Oil and Power in the Niger Delta, Nigeria'. *Geopolitics* 9 (1): 50–80.

Wilson, Japhy, and Bayón, Manuel. 2018. 'Potemkin Revolution: Utopian Jungle Cities of 21st Century Socialism'. *Antipode* 50 (1): 233–54.

Part IV
Environmental struggles and resistance

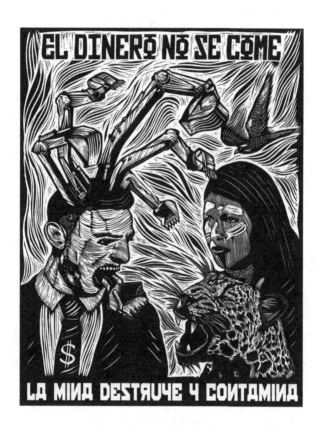

18 Resistance of women from "sacrifice zones" to extractivism in Chile

A framework for rethinking a feminist political ecology

Paola Bolados García

Introduction

In this chapter, we explore the eruption of a territorial political ecology and why that must be read in an ecofeminist key. This upsurge is occurring at the intersection of critical eco-socioenvironmental reflexivity, rooted in the political ecology of extractivisms, and in various experiences situated in the territorial and socio-environmental feminisms that have arisen during recent decades in the Latin American region. The first is understood as an ecoterritorialization of environmental movements and the construction of an alternative expert knowledge by Indigenous and farmer communities and environmental groups, after decades of resistance against the extractivist advance and its multiple disastrous consequences for their lives. The second responds to an ecofeminization of territories, in which women from societies formed under social and sexual inequalities begin a process of denaturalizing the structures of patriarchal domination. We refer to women's organizations in territories affected by various extractivisms—energy/mining, agro-export, forestry—and their articulation in response to the need to renew forms of struggle and defense of their territories, activating collective proposals that question the current forms of politics and knowledge. Without an initial definition of their struggle as feminists, women from these territories initiate responses against the policies of predatory overexploitation of nature, the consequences of which put the lives of their families, localities, and ecosystems at risk.

Although these processes of struggle and resistance have been collected and accompanied through the knowledge production of Latin American political ecology and the extractivism literature in recent decades (Alimonda 2011, 43; Gudynas 2015, 15; Machado 2015, 13), they have not explicitly exposed gender violence as a central part of extractivist violence. In this chapter, I highlight both processes: (1) the emergence of women's movements (feminisms) from a territorial experience without any previous feminist traditions; (2) how the specialized literature in Chile has not paid sufficient attention to the ecofeminization of territories subject to extractivism, in particular the relationship between extractive violence and patriarchal violence.

In Brazil, Argentina, and Chile, feminist perspectives have been slow to converge with socio-environmental ones, unlike in Mexico, Guatemala, Colombia, Bolivia, and Ecuador, where these paths converged at an earlier stage, giving rise to rich collaborative productions (Lugones 2008, 73; Curiel 2009, 8; Ojeda 2012, 65; Espinosa-Miñoso 2014, 7; Cabnal 2017, 100).

Among the perspectives that make explicit and deepen the intersection between gender violence and extractivism in the region are the so-called territorial or socio-environmental feminisms, as well as ecofeminist and feminist political ecology readings and proposals

DOI: 10.4324/9780429344428-22

(Ojeda 2012, 65–70). In Colombia, the work of Astrid Ulloa (2016, 134, 2020, 88), compiles processes of territorialization of feminisms among Indigenous, farmer, and Afro-descendant women's groups in the face of the advance of the mining industry. Further south, an antecedent for the processes of ecofeminization of territory are the contributions of feminist organizations who focus their criticism on the patriarchal and androcentric capitalist model of neoliberalism in the context of the struggles for water in Bolivia (Guzmán 2019, 39–40). These Latin American feminisms or ecofeminisms have increasingly permeated rural areas, pushing for a theoretical and conceptual reworking of Latin American political ecology. The ecoterritorial turn has experienced a sustained feminization and gendering as the production of knowledge from political ecology, southern feminisms, and ecofeminisms experienced a demand from the socio-environmental movements themselves for greater collaboration, interpellation and inter-penetration of each other's contributions.

The articulation between feminisms and environmental movements is more recent in Brazil, Argentina, and Chile, probably due, among other reasons, to the urban and mestizo predominance of their population and/or the racist and whitening policies associated with the emergence of their national states in the 19th century. Some antecedents of the recent visibility of women's protagonism in the defense of territories in the context of environmental conflicts are works such as those of Maristella Svampa (2015, 130) in Argentina, and in Chile, the collaborative production of Bolados and Sánchez (2017, 35–36) and Bolados, Sánchez, Alonso, Orellana, Castillo and Damann (2018a, 86–87). Both evidence an emergent process of reflection and production of feminist knowledge that initiates articulations with political ecology in the context of pressure and protagonism of women in the processes of resistance and production of situated and collaborative knowledge.

Various women's groups in defense of their territories have initiated a process of protagonism and denunciation of the socio-environmental damage and the effects, inequalities and violence that extractivism imposes on women and dissidents. These are increasingly being incorporated in their collective practices of the concept of good living, as the horizon of a new relationship between society and nature. They have also emphasized the rights of nature as a new legal strategy to dismantle the extractivist architecture of current environmental neoliberalism. Rural women have led an epistemic re-elaboration, that denounces environmental violence, deregulation processes, and the impacts of environmental policies in the construction of sacrifice zones.

In Latin America, the struggle for water in Bolivia in 2003, followed by the defense of the Yasuní territory in Ecuador in 2007, were important processes to awaken the hope of a path towards a Latin American political ecology based on the Indigenous practices of good living, and indicate the viability of political projects rooted in a normative regime reflecting the reality of Latin America's of Indigenous-mestizo origins (Acosta and Martínez 2009, 75; De la Cuadra 2015, 7–9). The case of Bolivia marked a path of recognition of the plurinationality, on the one hand, and the recognition of Mother Earth as a subject of rights, on the other. Nevertheless, this process was confronted by extractivist mining policies and other emblematic projects, such as the well-known TIPNIS Project that included the construction of a highway crossing forest-dwelling communities without prior consultation and consent. These communities experienced intense repression that contravened recognized protections embodied in national constitutional legislation and international agreements. In Ecuador during the government of President Correa, after a constituent period that resulted in the first constitution recognizing the Rights of Nature, international organizations promoted the protection of the Yasuní Reserve as a special biodiversity site, protecting it from oil exploitation. In the end, the proposal was not made effective, and the Ecuadorian government was pressured to return to the extractivist path.

In a post-multiculturalist context, Indigenous, farmer, and Afro-descendant communities, have provided an account of the limitations of the multicultural recognition of the 1990s and its perverse capacity to reconcile cultural recognition with the spatialization of capital, particularly in Indigenous territories protected by national and international regulations. The contradictory confluence between democracy and neoliberalism at the end of the 20th century (Dagnino 2004, 98) ran parallel to the problematic confluence between policies of cultural recognition and a deepening of environmental neoliberalism. This paradox implied a geopolitical reorganization of territories, particularly indigenous territories, where a large part of the resources demanded by the capitalist economy reside.

The hidden face of Latin American neo-extractivism: the sacrifice zones

In the period from 2000 to 2010, new export booms were observed in Latin America in sectors such as agro-exports, forestry, fishing, mining, and energy (Bustos, Prieto, and Barton 2015, 15). This export boom expanded not only in neoliberal countries, but also, and even more complexly, under progressive governments, embodying a neo-extractivism of a developmentalist nature.[1] This project marked the Latin American agenda since the beginning of the century and involved actions associated with generating optimal infrastructural conditions for the movement of goods in the region. These included hydroelectric projects, highways and energy corridors, and/or ports to ensure the entry and exit of products and natural resources.[2] These openings reinforced the historical role of raw material suppliers and fueled processes of reprimarization of the region's economies—an economy centered on the intensification of international flows associated with the export of natural resources, particularly due to China's growing demand for these commodities during this period (Svampa and Slipak 2017, 365–366).

The new wave of environmental violence deepened processes of social and environmental degradation in a context of accelerating crisis associated with climate change. Communities throughout the region have been affected by multiple mining projects; real estate extractivism that generated unprecedented loss of forests and changes in land use; overexploitation of water and contamination of soil and water as a result of the expansion of agriculture and forestry; and intensification of energy production based on polluting fossil fuels, particularly coal and petroleum. These activities had as a correlate the proliferation of ecological disasters and the appearance of so-called "sacrifice zones"—territories and communities destined to be sacrificed for national progress and the neoliberal economy—which gave rise to a scenario of explosive growth of socio-environmental conflicts and a process of progressive criminalization of environmental defenders.

The concept of sacrifice zones has gained prominence during the last decade in Chile and other countries in the region, as a result of denunciations by communities and non-governmental organizations of the extreme degradation of conditions in some territories, especially coastal areas (Bolados and Jerez 2019, 149–150). An inheritance from the North American (Lerner 2010, 3) and Brazilian (Acselrad 2004, 12–13) environmental justice movements, the concept refers to places where damage to the ecology and human health are prevalent as a result of the concentration of dangerous, toxic, and polluting activities. These tend to be located in economically poor and culturally Black and Indigenous communities. Whereas the environmental justice movement in the United States emphasizes chemical contamination and the racial dimension, in Latin America the term has been associated more with the economic and political violence of extractivism, as well as the corruption associated with environmental deregulation and the designation of areas for the concentration of investment projects. It has also been used to describe waste dumps or residual waste sites from activities associated with energy—mining, forestry, fishing, etc.; the emergence of the term reveals the paradoxes and contradictions of Latin American neo-extractivism, and, in particular, its exacerbated modality in neoliberal Chile. In Argentina, Brazil, and Chile, the term has been linked to industrial pollution and the environmental liabilities of mining (Svampa and Viale 2014, 84). In particular, they are associated with the construction of energy mining complexes that have been responsible for disasters such as those experienced in 2015 and 2018 in the State of Minas Gerais, Brazil, where the rupture of the Vale tailings dam caused the death of the Rio Dulce and, with it, the death of people, economies and ecosystems (Losekann and Vervloet 2016, 232).

In Chile, the concept of sacrifice zones has been configured from the very beginning as a category that seeks to describe and denounce the consequences of extractive activities associated with mining, energy production, and forestry. These are spaces where the waste from these activities is concentrated, and where disasters are a permanent presence, as a result of deficient environmental legislation. In the context of day-to-day disasters, women articulate collective practices associated to alter the extractivist/patriarchal violence of their territories, materializing in 2015 as the first environmental group of women whose name "Women of Sacrifice Zones in Resistance of Puchuncaví and Quintero" will propose the resistance to assume the victimization of the term.

Eruption of a territorial and feminized political ecology in Chile

Acting from the place assigned to them by energy-mining extractivism—the care of children and the ill—women break into the public dispute to recover their territories and challenge the economic and political decisions that destine them for sacrifice. In the case of Chile, the term has begun to have different meanings in other territories besides the coastal ones. The use of the concept of sacrifice zones to characterize the energy-mining activities at coastal ports, gave rise to the use of the term to describe socio-environmental disasters resulting from agro-export activities as in the case of the Province of Petorca, where communities and groups such as MODATIMA (Movement for the Defense of Water, Land and the Environment) denounce the theft of water—and the violation of the human right to it—as a result of legislation that prioritizes productive consumption, leaving the local communities without access. The Province of Petorca presents one of the extreme examples of the consequences of the privatization and commercialization of water in Chile, and of the monopoly of water in the agricultural and forestry sector assured in the constitution from 1980 as well as in the water code of 1981, where water is established as a public good which serves private property (Bolados, Henríquez, Ceruti y Sánchez 2018b, 169; Mundaca 2014, 43–44). In this context, a women's group, Mujeres MODATIMA emerged, organized not only by women from Petorca but also by women from all over the country. They have articulated as women in the defense and recovery of water as a common good that cannot be privatized, highlighting how the problems of access to water affect women in a more radical way. In this context, they denounce the hydraulic and gender inequalities, associated with care activities and assigned housework, as well as the economic and health effects generated by the lack of water.

Initially, both groups had no clear definition as feminist groups, but since the feminist movement of 2018 their articulation took a more explicit course as criticism of extractivism and the sexual organization of work that it imposes. Faced with the environmental violence they and their families experience in their bodies, they strengthen their links with foundations and other non-governmental organizations, as well as with other groups of women in resistance, such as the NO Alto Maipo women's group, whose work is articulated against the hydroelectric megaproject that changed the course of the Maipo River, the main source of water for the Santiago Metropolitan Region (which concentrates almost 60% of the national population).

From the so-called sacrifice zones, a new understanding of environmental justice and human rights has emerged, based on the care of life and the right to health. In this process, women have acquired relevance and protagonism, denouncing power structures, inequalities, and environmental suffering. A rupture has emerged, and not only with the logic of irrational and destructive accumulation in currently-contaminated territories. Women from the two communes that form part of the Quintero bay-basin that supports energy-mining extractivism, expose gender inequalities and the conditions of double exploitation of women as they are excluded from the labor force of the energy-mining complex, which is mainly made up of men. They must also assume the damage generated by the contamination and the consequences of living in a sacrifice zone through the care of children and the elderly. From the first poisonings of students in 2011 and 2012, the initiative to establish a women's organization was born, which materialized in 2015. In this context, and assuming the denomination of sacrifice zones, they denote their rejection of the victimization that this term involves, calling themselves "Women from Sacrifice Zones in Resistance of Puchuncaví and Quintero."

In the year 2016, these women organized themselves and began a path of denunciation, but also of reflexivity and the construction of environmental and feminist knowledge, which has increasingly taken on more feminist and ecofeminist elements. Breaking the confinement of women in the area to the private sphere and dedication to the care of their children, women of different ages and backgrounds have led a territorial, social, and legal defense utilizing various strategies—among them, the judicialization of the conflict in national and international instances. This path was interrupted in August 2018 by a massive poisoning of junior and high-school students from these communes, which reached more than 1,700 cases. Women were the most affected since they were the ones expected to assume the damage to the health of their children and parents; many of them abandoning their jobs or careers or even having to move to other communes. Pregnant women had to leave the commune by medical recommendation and the chronically ill and children had to confine themselves to their homes for long periods of time. The poisonings represent a turning point in women's activism in the area that has expanded to other communities, highlighting the relationship between gender violence and extractivist violence. What is seen as an irrational form of exploitation of nature, is, at the same time, revealed as patriarchal violence aimed at perpetuating women in the private sphere and dedicated to the care of children and older adults affected by pollution. In the same way, the sexual division of labor generated by these energy and mining complexes prevents access to other sources of work and perpetuates various forms of discrimination and subordination of the generally unpaid activities of women. Extractive violence then turns into patriarchal violence.

Accompanied by various non-governmental organizations, foundations, and academic networks, these women began a process of reflection, participation in events and forums, as well as the production of texts such as articles and books that systematize their experiences as women from a damaged territory. Without excluding strategies—but rather incorporating proposals—they participated in the elaboration of the preparatory report on Human Rights through the UN Universal Periodic Review, even reaching instances such as the Inter-American Court of Human Rights in order to denounce the violation of human rights, particularly environmental, health, and children's rights. Together with local organizations and actors, they have also initiated legal actions in national instances, which concluded in a ruling of the Supreme Court in 2018 that established the responsibility of the Chilean state for the violation of environmental, health and human rights.

In closing

While political ecology highlights the inequalities of power in relation to access and diverse forms of appropriation of nature, it has been forced to territorialize and ecofeminize in the Latin American context, in particular, due to the gendered inequalities that are at the base of the economic structure and that have been underlined by the diverse feminisms and ecofeminisms of the region. Women and other diverse subjects have placed the relation between extractivism and gender at the center of the discussion, underlining how the capitalist economy is constituted on the basis of the sexual division of labor, appropriating women's reproductive work and care practices, including family care, health, community, and nature itself. In the last decades, territorial struggles have become ecofeminized as they make evident the relation between capitalist and patriarchal logics in the destructive and irrational appropriation of territory. In this sense, political ecology is not only made aware of its colonial critique and its critical-emancipatory practice, but it projects itself on the basis of a re-elaboration of feminist criticism in current territorial struggles in Latin America.

The post-neoliberal identities associated with good living and the ethics of care that have emerged in recent decades under the defense and resistance practices of socio-environmental/ territorial organizations have been configured in a political and utopian proposal of emancipation that has been enriched with increasing elements of Latin American feminisms, decolonial and ecofeminisms, and community feminisms whose critique of the extractivist model has been inseparable from their critique of the patriarchal system. Despite the fact that these processes have been slower in Southern Cone countries, where extractivist imaginaries have undergone a political trajectory that legitimized them in progressive and neoliberal visions, there too, women are embarking on a path of denaturalization of this social and sexual order. In the case of Chile, since the middle of the second decade of the new century, women's groups have initiated processes of collective mobilization and articulation in their territories and beyond, which attempt to account for the gender inequalities and gender violence involved in the zones sacrificed by the Chilean neoliberal economic model—in particular, due to the energy-mining policy that concentrates toxic and dangerous activities in coastal areas where the most affected are women.

Notes

1 Ecuador and Bolivia supported the expansion of extractivism under a plurinational discourse and a regulatory regime of recognition of the rights of nature, while others, including Chile and Peru, aligned their environmental regulations with the new international context associated with free trade agreements and, in particular, the growth in global demand for minerals. Countries including Brazil, Argentina, and Uruguay, also under politically progressive governments, have increasingly opened up to extractive activities associated with the Regional Infrastructure Project promoted by UNASUR known as IIRSA (Castro 2012, 179; Zibechi 2006, 19).

2 In this context, global geopolitics oriented toward the opening of economies within the framework of free trade agreements promoted by the World Trade Organization, as well as the Latin American agenda promoted by UNASUR, fostered regulatory processes that sought to make the extractive economy compatible with a "flexible" environmental protection system in most of the countries of the region (Tecklin, Bauer and Prieto 2011, 884–885).

References

Acosta, Alberto, and Esperanza Martínez. 2009. *El buen vivir. Una vía para el Desarrollo.* Quito: Ediciones Abya Ayala.

Acselrad, Henry. 2004. *Conflito social e meio ambiente no estado do Rio de Janeiro.* Sao Paulo: Editora Relume Dumará.

Alimonda, Hector. 2011. "'La colonialidad de la naturaleza.' Una aproximación a la Ecología Política Latinoamericana." In *La Naturaleza Colonizada*, coord. Héctor Alimonda, 21–58. Buenos Aires: Clacso.

Bolados, Paola, and Bárbara Jerez. 2019. "Genealogía de un desastre: la historia ambiental de una zona de sacrificio en la bahía de Quintero, Chile." In *Pensamento crítico latino-americano. Reflexoes sobre políticas e fronteiras*, organized by Edna Castro, 149–170. Belém: Annablume Ediciones.

Bolados, Paola, and Alejandra Sánchez. 2017. "Una ecología política feminista en construcción: El caso de las Mujeres de zonas de sacrificio en resistencia, Región de Valparaíso, Chile." *Psicoperspectivas* 16, 2: 33–42.

Bolados, Paola, Alejandra Sánchez, Katta Alonso, Carolina Orellana, Alejandra Castillo, and Maritza Damann. 2018a. "Eco-feminizar el territorio: la ética del cuidado como estrategia frente a la violencia extractivista en las mujeres de zonas de sacrificio en resistencia (zona central de Chile)." *Ecología Política* 54: 83–88.

Bolados, Paola, Fabiola Henríquez, Cristian Ceruti, and Alejandra Sánchez. 2018b. "La eco-geo-política del agua: una propuesta desde los territorios en las luchas por la recuperación del agua en la provincia de Petorca (Zona central de Chile)." *Rupturas* 8, 1: 167–199.

Bustos, Beatriz, Manuel Prieto, and John Barton. 2015. "Introduction." In *Ecología Política en Chile. Naturaleza, propiedad, conocimiento y poder*, edited by Beatriz Bustos, M. Prieto, and J. Barton, 15–59. Santiago: Editorial Universitaria.

Cabnal, Lorena. 2017. "Tzk'at, Red de Sanadoras Ancestrales del Feminismo Comunitario desde Iximulew Guatemala." *Ecología Política* 54: 100–104.

Castro, Edna. 2012. "Pan-Amazônia Refém? Expansão da fronteira, megaprojetos de infraestrutura e integração sul-americana da IIRSA." In *Megaproyectos: la amazonia en la encrucijada I*, edited by Fernando Franco, 177–216. Colombia: National University of Colombia, Campus Amazonía. Amazonian Scientific Research Institute – Imani.

Curiel, Ochy. 2009. *Descolonizando el feminismo: una perspectiva desde América Latina y el Caribe*. GLEFAS. https://repositorio.unal.edu.co/handle/unal/75231.

Dagnino, Eveligne. 2004. "Sociedade civil, participação e cidadania: de que estamos falando?" In *Políticas de ciudadanía y sociedad civil en tiempos de globalización*, coord Daniel Mato, 95–110. Caracas: FACES.

De la Cuadra, Fernando. 2015. "Buen Vivir: ¿Una auténtica alternativa post-capitalista?" *Polis* 14: 7–19.

Espinosa-Miñoso, Yuderkis. 2014. "Una crítica descolonial a la epistemología feminista crítica." *El Cotidiano* 184: 7–12.

Gudynas, Eduardo. 2015. *Extractivismos. Ecología y política de un modo de entender el desarrollo y la naturaleza*. Cochabamba: CLAES.

Guzmán, Adriana. 2019. *Descolonizar la Memoria, Descolonizar los Feminismos*. La Paz: Editorial Tarpuna Muya.

Jerez, Bárbara. 2017. "La expansión minera e hidroeléctrica a costa de la desposesión agropecuaria y turística: Conflictos ecoterritoriales extractivistas en las cuencas transfronterizas de la Patagonia argentino chilena." *RIVAR* 3, 10: 25–44.

Kruse, Thomas. 2005. "La Guerra del Agua en Cochabamba, Bolivia: terrenos complejos, convergencias nuevas." In *Sindicatos y nuevos movimientos sociales en América Latina*, compiled by Enrique De La Garza, 121–161. Buenos Aires: CLACSO.

Lerner, Steve. 2010. *Sacrifice Zones: The Front Lines of Toxic Chemical Exposure in the United States*. Cambridge, MA: MIT Press.

Losekann, Cristiana, and Roberto Vervloet. 2016. "O neoextractivismo visto através dos megaempreendimentos de infraestrutura: as zonas de sacrificios no Espíritu Santo." In *Mineração na américa do sul neoextrativismo e lutas Territoriais*, edited by Andrea Zhouri, Paola Bolados, and Edna Castro, 231–255. Sao Paulo: Annablume.

Lugones, María. 2008. "Colonialidad y género." *Tabula Rasa* 9: 73–101.

Machado, Horacio. 2015. "Ecología Política de los regímenes extractivistas. De reconfiguraciones imperiales y re-existencias decoloniales en nuestra América." *Bajo el Volcán* 15, 23: 11–51.

Morales, Manolo. 2013. "Derechos De La Naturaleza En La Constitución Ecuatoriana (Rights of Nature in Ecuador's Constitution)." *Justicia Ambiental* 5: 71–82.

Mundaca, Rodrigo. 2014. *La Privatización de las Aguas en Chile. Causas y resistencias*. Santiago: América en Movimiento.

Ojeda, Diana. 2012. "Género, naturaleza y política: Los estudios sobre género y medio ambiente." *HALAC* I 1: 55–73.

Porto-Gonçalves, Carlos and Enrique Leff. 2015. "Political Ecology in Latin America: The Social Re-appropriation of Nature, the Reinvention of Territories and the Construction of an Environmental Rationality." *Desenvolvimento e Meio Ambiente* 35: 65–88.

Santos, Boaventura. 2012. *De las dualidades a las ecologías.* La Paz: OXFAM.

Svampa, Maristella. 2015. "Feminismos del Sur y Ecofeminismo." *Nueva Sociedad* 256, 4: 127–131.

Svampa, Maristella, and Ariel Ariel Slipak. 2017. "China en América Latina: del Consenso de los Commodities al Consenso de Beijing." In *Ecología Política Latinoamericana: pensamiento crítico, diferencia latinoamericana y rearticulación epistémica*, coord. Héctor Alimonda, Catalina Toro Pérez, and Facundo Martín, 353–384. Buenos Aires: CLACSO.

Svampa, Maristella, and Enrique Viale. 2014. *Maldesarrollo. La Argentina del extractivismo y el despojo.* Buenos Aires: Katz Editors.

Tecklin, David, Carl Bauer, and Manuel Prieto. 2011. "Making Environmental Law for the Market: The Emergence, Character and Implications of Chile's Environmental Regime." *Environmental Politics* 20, 6: 879–898.

Ulloa, Astrid. 2016. "Feminismos territoriales en América Latina: defensas de la vida frente a los extractivismos." *Nómadas* 45: 123–139. Retrieved May 18, 2020, from http://www.scielo.org.co/scielo.php?script=sci_arttext&pid=S012175502016000200009&lng=en&tlng=es.

Ulloa, Astrid. 2020. "Ecología política feminista latinoamericana." In *Feminismo socioambiental. Revitalizando el debate desde América Latina*, coord. Ana De Luca Zuria, Ericka Fosado Centeno, Margarita Velázquez, and M. Gutiérrez, 75–104. Cuernavaca: National Autonomus University of México.

Zibechi, Raúl. 2006. "IIRSA: la integración regional a la medida de los mercados." *Ecología Política* 31: 19–25.

19 Environmental conflicts and violence in Latin America

Experiences from Peru

Raquel Neyra

Introduction

Latin America, like other regions of the world rich in natural resources, is the object of extractive activities, reflecting a society that puts a price on all goods in search of ever-greater profits. Corporations are deployed in the territories to be exploited, most of the time in opposition to the interests of the inhabitants of the places affected by the extractive projects. The governments of those countries that have assumed the "commodities consensus" (Svampa 2013), build an arsenal of laws, procedures, and institutions to push extractivism. Even some governments that describe themselves as "progressive" base their economies on extractivism to fund redistributive measures. These governments, like those of a neoliberal or other type, construct a specific discourse to try to convince us of the benefits of extractivism. Companies endorse this discourse by adapting their standards of Corporate Social Responsibility (CSR). Thorough analysis of CSR has revealed its contradictions (Saes, Del Bene, Neyra, Wagner, and Martínez-Alier 2021). On the other hand, we have the populations subjected to extractivism, who defend not only their territory but also the environment.

The ravages of extractivism on the environment extend to hundreds of kilometers around each project and can even be much more when we take into account winds that disperse through the atmosphere and polluted waters that seep into the water tables. The environmental degradation affects the space occupied by a society, disrupting its functioning and even its existence. The increasing deployment of extractive activities means also land grabbing and transformation of land use. Populations are displaced from the territories they have occupied for centuries, obliged to seek new economic resources and migrate to cities, where they often live in marginal neighborhoods, affecting their cultural ties with their land of origin. Socio-environmental conflict breaks out, and opposing interests emerge. Therefore, socio-environmental conflicts proliferate as extractive activities increase. To understand this complex situation, it is necessary to take an interdisciplinary approach such as that of political ecology.

In the first part of this chapter, I analyze the concept of social metabolism, and its relationship with socio-environmental conflicts. continuing in the second part with the dimension of extractivism. In the third and fourth parts I review the conflicts and the role of violence in them, finishing with an analysis of the construction of the coloniality of socio-environmental conflicts in Peru.

Changes in social metabolism and environmental conflicts

In this chapter, I employ the analytical tools provided by the study of changes in social metabolism. Social metabolism is the observation that biological and socioeconomic systems interchange a continuous throughput of energy and materials to maintain their internal

DOI: 10.4324/9780429344428-23

structure (Fischer-Kowalski 1998; Gerber and Scheidel 2018; Martínez-Alier 1990). It refers to the flows of energy and materials that enter the economy and that partly come out as waste (mine tailings, polluted water, greenhouse gases, etc.). The calculation is made in physical and tangible measures such as the ton. This allows comparisons between countries and the discovery of historical trends and gives us a more concrete vision both of what is being extracted from a country and of what is not replaced by the nature in that country. Monetary measures, by contrast, do not tell us what is withdrawn from, and transferred to the environment. The calculation methods of social metabolism are based on the system of Material and Energy Flow Accounting (MEFA), developed by Fischer-Kowalski and Haberl, among others. The MEFA indicators are commonly used by many organizations, including the UN Environmental Program (UNEP) and Eurostat. These flows also depend on the social and political relations existing at a particular moment and on a determinate local, regional, or global scale (Fischer-Kowalski and Haberl 2015; Infante-Amate, González de Molina, and Toledo 2017).

Those nation-states that depend on foreign demand and the variation in market prices experience a deterioration of the terms of trade. This exchange is ecologically unequal because the Southern countries are left with the devastation of their territory and the ravages caused by pollution, while the importing countries (the North and the East) acquire extracted materials cheaply (Martínez-Alier 1999). For a few decades, Chinese companies have predominated in extraction in Latin America, surpassing the traditionally predominant US. Chinese extractivism focuses on the strategic sectors necessary for its economy: hydrocarbons, mining, energy, and the construction of infrastructures. The presence of Canadian companies continues to be strong because Canada has become a trading platform for mining activity and facilitates the basing of firms in the country.

Those affected by extractive projects react by initially claiming a lack of consultation on the project and denouncing contamination. The mobilization of social protest takes different forms, including road blockades, mass mobilizations, the takeover of premises, roadblocks, strikes, and even hunger strikes. In other cases, national or international justice is used. According to EJAtlas (2021a), there are more than 1,000 conflicts in Latin America, most of them occurring in the mining, biomass, hydropower, and hydrocarbon sectors. Colombia has the highest number of cases with 149, followed by Peru, Argentina, and Chile with 96 cases each. Many of these conflicts result in fatalities, according to estimates by Global Witness (2021), not counting the thousands injured and ill.

The Peruvian social metabolism

Since the 1990s, changes in the Peruvian social metabolism and associated biophysical sustainability indicators show that the country has directed its entire economy toward the extraction of minerals, which is progressing at an unbridled pace (Figure 19.1). The former president Fujimori (1990–2000) opened the country to foreign investment and promulgated a strongly liberal Mining Law in 1992. The Mining Law facilitated and multiplied concessions (mining, oil, logging, etc.); increasing the extraction of metallic minerals from 100 million metric tons in 1970 to more than 400 million in 2012 (Pérez-Rincón, Vargas-Morales, and Crespo-Marin 2018). In neighboring countries, the acceleration of extraction began a few years earlier, in the 1980s, which could corroborate the assumption that the armed conflict of the Shining Path stopped extraction.

Pérez-Rincón, Vargas-Morales, and Crespo-Marin (2018) calculate that in a period of 20 years after the 1980s, Colombia doubled the extraction of materials from 200 to 450 million metric tons (metallic minerals, hydrocarbons, and biomass continue to be predominant), while

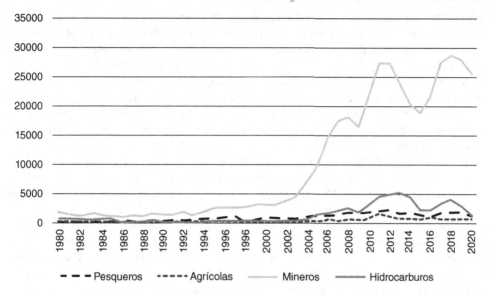

Figure 19.1 Exports of traditional products in Free On Board USD value, 1980–2020. Data Central Reserve Bank of Peru, Annual series ca_036, own elaboration.

Ecuador tripled its extraction (especially fossil fuels and construction materials), as did Bolivia (all items, including the huge areas dedicated to quinoa and soybeans).

In the case of Peru, domestic consumption (extraction plus imports minus exports, all in tons) yields a value of 15 tons per inhabitant, a very high value (in the European Union it is 13, OECD 2021). As trade deficits lead to deficits in the current account balance, there will be increased need for external financing or the use—while they last—of the international reserves (foreign currency deposits) accumulated during the boom period between 2003 and 2012. In either situation, there will be a new need for additional exports of raw materials—producing a physical trade imbalance—to pay off the debt or strengthen the country's external position, depleting resources, polluting the environment, and causing ever more socio-environmental conflicts (Samaniego, Vallejo, and Martínez-Alier 2015). At the same time, the removal of tons of earth for mining (the Yanacocha mining project, for example, removes 600,000 tons every day; Seifert 2005), as well as the increasing extraction and destruction of biomass, lead to a potential loss of energy (in the form of food, biomass for photosynthesis, etc.), which can hardly be supplied. Indeed, in 2020 Peru had the worst Environmental Performance Index (90th) in South America (Wendling, Emerson, de Sherbinin, Esty, et al. 2020).[1]

"Between 1990 and 2007, Peru received US$ 12.35 billion in mining investments, helping to transform it into one of the world's most important exporters of silver, copper, zinc, lead, and gold" (Merino-Acuña 2015, 86, citing Bebbington and Bury 2009). The contribution of extractive industries to the whole economy is seen as crucial, including an average of 22% of total tax collection, and 42% of the total income tax between 2007 and 2010 (Sotelo and Francke 2011). Mining activities expanded very quickly with the division of the entire territory square concessions available to any person or company regardless of where they are located. Although the Fujimori government ratified the International Labor Organization "Convention 169 on Indigenous and Tribal Peoples in Independent Countries," a few months later he promulgated the "Law of Private Investment in the Development of Economic Activities in the Lands of the National Territory and of Peasant and Native Communities,"[2] which attacked Indigenous rights. In any case, without ratification by congress, the law remained void. It is only in 2013 that former president Ollanta Humala (2011–2016) fulfilled his electoral promise by implementing the regulation for the Convention 169, although many aspects of it were left out. Extractive policies also tried to follow the sustainability policies promoted by the World Bank in the first years of the 2000s. The Toledo government (2001–2006) followed the same extractivist policies, consolidating neoliberalism. During his government, marked by scandals of various kinds, community members were tortured in a case that reached the English Supreme Court (EJAtlas 2021b). The government of Alan García (2006–2011) will be remembered for the Baguazo in 2009, an uprising of indigenous Awajún Wampis in the northern Amazon of Peru against extractive projects repressed *manu militari* and left a balance of 33 deaths.

The government of Ollanta Humala intended to make a monetary redistribution and reduce poverty with the benefits of extractivism. A few years before, the World Bank had financed the creation of the Environment Ministry (2008), in order to implement measures to reduce the pollution caused by extractive industries (Andreucci and Kallis 2017). To further the government vision of "economic development," a discourse emerged that attempts to justify the need for extractivism with the signing of numerous free trade agreements as well as ad hoc laws that favor and facilitate investment, such as the so-called "environmental packets" (Neyra 2016). Subsequent presidents have used the same methods (ad hoc laws and repression) until today. China is today the main trading partner, representing approximately one-third of Peruvian commercial activity (Adex 2020), focusing on the extraction of copper (the Las Bambas and Río Blanco mines, as well as the planned Don Javier and Toromocho mines), and iron (the planned Pampa de Pongo and Marcona mines), as well as the construction of infrastructure to transport materials to the Pacific coast.

Extractive companies create lobbies in the government and exert influence, as demonstrated by the Wikileaks cables,[3] including by making agreements with the National Police of Peru for the protection of mining facilities. The national policy is to favor investment, hoping to benefit from royalties and so-called "trickle down." The benefits of extractive royalties attract the interest of regional governments. Royalty policies should counter the adverse effects of the centralized appropriation of resource rents (Arellano-Yanguas 2011). However, royalties are subject to the presentation and approval of regional projects by the central authority; the inability of many regional governments to deal with this issue has led to rejection and mistrust by the local population.

The extraction

In an extractivist economy, each country specializes in the extraction of its most abundant resources. In the case of Peru, we have in the first place minerals, with the extraction of copper, gold, and silver as well as many others, providing 60–70% of export earnings. Second is biomass, including fishing and agro-industrial resources, as shown in Figure 19.1. The latter monopolizes huge territories on the coast, causing a shortage of water, which is sucked from the subsoil and diverted from rivers as in the Olmos or Majes-Siguas irrigation projects. For the export of minerals, ports are built even above archaeological sites, one of the country's most precious resources. Recently, the extractive frontiers have forced the deforestation of the Amazon by building roads or waterways through virgin or protected areas, expanding illegal logging as well as the cultivation of cocoa, oil palm, and coca for drug trafficking. This is compounded by illegal mining in many areas of the country. Examples of these cases in all Latin America can be found in the EJAtlas (2021a) among many others (Figure 19.2).

Concessions are granted indiscriminately over territories. The residents affected by the extractive projects react to such attacks spontaneously, claiming, in the first place, property rights in, and the lack of consultation on, the territory, while denouncing the contamination.

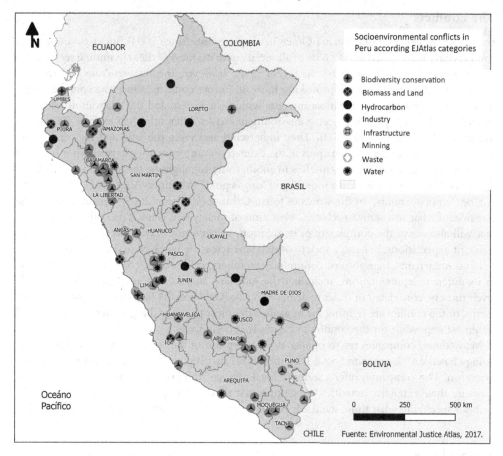

Figure 19.2 Socio-environmental conflicts in Perú according to EJAtlas categories and cases. Author: Facundo Rojas, 2017, on behalf of author.

The expansion of extractive projects leads to a process of territorial reconfiguration, where capitalist relations carry out spatial, material, and symbolic transformations that disrupt the management of space (Bebbington 2007). The interrelation between society and nature is disturbed and the use value of the territory is subordinated to the generation of profits (Luna 2019). This reconfiguration of the territory produces an ecological crisis.

Beyond claims of subsidies, Indigenous people demand participation and recognition (Rodríguez 2017; Rodríguez and Inturias 2018), the right to be heard. The resistance grows, caused by distrust toward a state and companies that marginalize them (Conde and Le Billon 2017). Companies must demonstrate ownership of the land on which the project will be carried out. They acquire the land by buying it at a low price, taking advantage of the landowners' or holders' ignorance of the monetary value of the minerals or hydrocarbons. If the owner or holder does not agree to the sale, the Mining Law grants the possibility of requesting expropriation in favor of the company. In the case of communal lands, their non-seizable (inembargable) and inalienable character was removed by the 1993 constitution of Fujimori, leaving the lands at the mercy of extractive appetites. In 1996, the new National Mining Cadaster Law[4] further weakened land tenure rights by making the possession of concessions easier. This legislation resulted in the fact that today, almost 50% of communal peasant lands are under concession (IBC 2016).

The conflicts

For as long as Peru's Ombudsman's Office has kept records (since 2004), socio-environmental conflicts have represented at least 65% of all social conflicts, most of them in mining sector (64% of socio-environmental conflicts). In recent years, however, the infrastructure construction sector in the Amazon has been provoking more and more conflicts. Once the conflicts erupt, they make evident the incompatible interests within the system and the relationship of power in society. Ordinary people perceive environmental risks differently than experts (Muradian, Martínez-Alier, and Correa 2003). They understand and value the land and the territory in ways that contrast with the evaluations of the extractive companies; they have another language of valuation, and they express themselves in another way, highlighting the unity and importance of the territory: "Water is worth more than Conga's gold" (Martínez-Alier 2005). The conflict will be "an opportunity," in the words of Johan Galtung (Calderón 2009), because it raises the question of what you want to achieve, what kind of society you want. The outbreak of a conflict will also reveal the complexity of its solution, because not only do opposing parties face different expectations, but also a variety of different actors are involved in the conflict: farmers, peasants, merchants, transporters, community members, civil society, companies, and the state in its different representations, including the judiciary and the ombudsman. These multiple levels must be considered in order to achieve an exit (Calderón 2009). However, given that the parties to the conflict are fighting for or against the most precious thing we have—life—it will be almost impossible for the conflict to be resolved definitively.

Many times, companies try to oppose this conception of the world with a technical terminology based on "Corporate Social Responsibility" that supposedly includes community participation. The companies offer a series of local projects (infrastructural, agricultural, etc.) and promote their extractive activities, suggesting that they will respect "international standards" without specifying what those standards are and whether they actually follow them (Li 2015).

The violence

The escalation in the intensity of protests also occurs because the communities have understood that it is the only way to deal on an (almost) equal footing with the government and force it to negotiate—a government that they see as captured by, or even the accomplice of extractive interests (Arellano-Yanguas 2011). In many conflicts, when the state sees that the protests get out of hand, it reacts with extreme violence, sending police and military forces as well as private security from the companies to repress. The extreme violence used against protesters often leaves some people dead and many injured, who are in turn quickly forgotten. Many institutions are dedicated to denouncing this situation and documenting this unique violence (Neyra 2019, 2020). This is a forgotten and hidden violence, as if the lives of environmental defenders and most residents of the extractive zones are not worth the same as another citizen.

Likewise, since a few years ago and during the pandemic, the deforestation of Amazonian territory for illegal activities or land trafficking has intensified, increasing violence against Indigenous peoples and environmental defenders. The Monitoring of the Andean Amazon Project (MAAP) indicates that in 2020 more hectares were deforested in the Peruvian Amazon than ever before (MAAP 2020). Extractive activities were carried out with total impunity due to the militarization and near-paralysis of the country, leaving environmental defenders at the mercy of traffickers and land invaders.

The participation of Indigenous peoples in socio-environmental conflicts occurs in approximately 90% of the cases, much more than their proportion in the population (28%, according to the censuses, INEI 2017).[5] However, they are the ones who mostly give their lives. Most of the deceased took part in some form of protest, although some of them were only spectators of the repression caused by the so-called forces of order. In almost all cases, the deceased have been murdered at point-blank range, targeted by precise bullets from the forces of order or by people engaged in extractive activities. All the passive or active protesters were brave people, young and old, who demonstrated for the defense of the territory, water, agriculture, and diversity, and took to the streets to face the bullets of repression. More than 100 people—environmental defenders—have been killed since 2002, according to the reports of Global Witness (2014, 2016) and the Peruvian Ombudsman's Office (Defensoría del Pueblo 2020) among others (see also Thomas et al., Chapter 25 in this volume; Neyra 2020).

Racism, coloniality, and violence

The violence used in socio-environmental conflicts in former colonial countries needs a historical retrospective. Latin America witnessed the largest massacre in the history of Indigenous peoples, caused by the invaders in their thirst to conquer land and material goods as well as by the diseases they brought. These (racialized) massacres laid the foundations for complete domination, motivated by the eagerness to accumulate capital (Quijano 2007). Unlike the current situation, "precious goods", or preciosities as Immanuel Wallerstein (1989) called them, were extracted from America in the beginning: silver, gold, and diamonds—high-priced and low weight. The Catholic Church played a crucial role in the colonization by forcefully catechizing people considered as irrational, wild, and sinful; this violent "crusade" or "extirpation of idolatries" and destruction of Huacas (temples), as occurred in Peru, coincided with the dispossession of Indigenous territories.

The independent Republic (1821) of white sons of Spaniards, with Western culture, reproduced the colonialism of the Spanish invasion internally, toward the Indigenous populations (González Casanova 2006a, 2006b; Svampa 2016). They took possession of the land. They reinforced the idea of race. As a result, the situation of the Indians didn't change. Drawing on historical hindsight, José Carlos Mariátegui described the plundering of natural resources, the control of the authorities exerted by the gamonal (large landowners) and recognized that the problem of land in Peru was the problem of the "Indio" (Mariátegui 1928; Scarritt 2012).

The Creoles maintained the "colonial pattern of power" with the control of the economy, state, and military authority according to a Eurocentric vision, while education and dissemination of knowledge were pursued from a Eurocentric perspective, relegating all the wisdom of Indigenous cultures. Quijano (2000) masterfully calls this "coloniality," a concept that includes the epistemic/cultural aspect to explain what he calls the "colonial pattern of power." The European linear concept of history and the rationality that separates body and mind were introduced, placing the Indigenous culture "before" the Republican era, calling it "archaic," "primitive," and "traditional" (Lander 2000). As explained by Quijano (2007), Eurocentrism spread from the center to the peripheries to "modernize" the old colonies. This same understanding of the world applies to both liberal and capitalist ideas and to the hierarchical ideas of orthodox Marxist and Left parties—that is, these two tendencies share the same Eurocentric vision.

Thus, the era of servitude and peons continues, the Indians were renamed Indigenous, while control is exercised over their imagination and their subjectivity until they are made to believe that they are inferior (Quijano 2007). Hopkins (2017) interprets this correctly as symbolic violence. In Peru, it is the cholo (half-blood), an Indigenous who lives in the cities or on the coast, already more "modern," more "integrated" in Creole society, but who is always relegated to trades such as bricklayer, pawnbroker, garbage workers, gardeners, housekeepers, guardians, low-ranking police, or military. However, what was really at stake was the possession of the products of nature and the products of labor. The use of violence found its justification when wielded against a human being considered inferior.

This brand, this wound is the one that is still carried by our country, in the study and understanding of our history and socio-environmental conflicts and continues to divide city dwellers from "provincials." The violence is practiced equally if the victim is of Amazonian, Andean, or coastal origin. It is in the first place against the non-whites who do not accept the designs of the invasive extractive activities in their territories. And in this context, it is even more important to remember that women are more discriminated against and exploited. Racism and sexual abuse have gone together in America since October 1492. In Peru, the history of extractivism goes hand in hand with violence. The current socio-environmental conflicts cannot be studied without recognizing these origins.

Violence, however, is also exercised against "whites" who take environmental stances, perhaps defending in the first place their own well-being, as with the community of Surfers defending the beach against road construction in Lima, as well as preservationists or who commit actively in favor of the affected peoples: professionals, activists, lawyers, doctors, and engineers criminalized by the state or companies. These "whites," or mestizos, are already seen with the eyes of the colonizer and repressed because they have identified with environmental struggles. Skin color becomes of secondary importance and only the political position taken by the activist or professional is considered.

Conclusion

Socio-environmental conflicts and the associated violence are closely related to the increase in, and changes to, social metabolism, while the latter finds its justification in coloniality. The extractive policies that Peruvian (like other) governments follow are pursued to the detriment of their population, of the environment, and of their own natural resources, even in the case of the new axis of conflict that is being forged with the recent discovery of the largest lithium reserves in South America for the benefit of so-called "clean" energy. These facts inextricably lead to conflict and poverty. Violence used in conflicts is first exercised by those who have power and use the coloniality of knowledge to impose their objectives. These statements can raise much debate, but it is time to face coloniality, admit that our society is racist, and observe the alienation to redirect our actions and thoughts. We have to decolonize ourselves, tear out the coloniality of knowledge from our minds.

Socio-environmental conflicts, the object of study of political ecology, have long faced, in practice, the coloniality of power and knowledge, as well as questioning it. Taken together, they will create a new socio-environmental history, analyze and subvert power relations, and contribute to awareness. In this sense, political ecology must go hand in hand with Latin American critical thought, because they share a concern for identity, seek a new historical perspective to understand and change our reality, and incorporate a critique of unequal international trade (Bravo 2017).

Notes

1 EPI ranks 180 countries on environmental health and ecosystem vitality.
2 DL No. 26505.
3 For example, cable 38881 of August 19, 2005, published by *The Guardian* (2011). (The complete bibliographical information should be placed in the bibliography.)
4 Law No. 26615.
5 210 612 indigenous Amazon, 0.9%; 5 176 809 Quechua origin, 22.3%; 548 292 Aymara, 2.4%; 49 838 other people, 0.2% (INEI 2017).

References

ADEX. 2020. *Nota semana 06l, Exportaciones peruanas. Beneficios del TLC Perú –China*. Lima: ADEX.

Andreucci, Diego and Giorgos Kallis. 2017. "Governmentality, development and the violence of natural resource extraction in Peru." *Ecological Economics* 134: 95–103.

Arellano-Yanguas, Javier. 2011. "Aggravating the resource curse: decentralization, mining and conflict in Peru." *Journal of Development Studies* 47 (4): 617–638. https://doi.org/10.1080/00220381003706478.

Bebbington, Anthony. 2007. *Minería, movimientos sociales y respuestas campesinas. Una ecología política de transformaciones territoriales*. Lima: IEP-CEPES.

Bebbington, Anthony and Jeffrey Bury. 2009. "Institutional challenges for mining and sustainability in Peru." *PNAS* 106 (41): 17296–17301. https://doi.org/10.1073/pnas.0906057106.

Bravo, Lucía. 2017. "El pensamiento crítico latinoamericano." In *La opción decolonial, América Latina ante una nueva encrucijada*, edited by Claudio Luis Tomás and Luciano Damián Bolinaga, 25–72. Buenos Aires: Teseo Press.

Calderón, Percy. 2009. "Teoría de conflictos de Johan Galtung." *Paz y Conflictos* 2: 60–81.

Conde, María and Philippe Le Billon. 2017. "Why do some communities resist mining projects while others do not?" *The Extractive Industries and Society* 4 (3): 681–697.

Defensoría Del Pueblo. 2020. *Reporte de conflictos sociales N° 202*. Lima: Defensoría del Pueblo. https://cdn.www.gob.pe/uploads/document/file/1547649/Reporte-Mensual-de-Conflictos-Sociales-N%C2%B0-202-diciembre-2020.pdf.pdf.

EJAtlas - Global Atlas of Environmental Justice. 2021a. *Proyecto EJAtlas*, Instituto de Ciencia y Tecnología Ambiental (ICTA), Universidad Autónoma de Barcelona (UAB). https://ejatlas.org/.

EJAtlas - Global Atlas of Environmental Justice. 2021b. *Río Blanco Mine Majaz/Rio Blanco Copper S.A., Peru*. Proyecto EJAtlas, Instituto de Ciencia y Tecnología Ambiental (ICTA), Universidad Autónoma de Barcelona (UAB). https://ejatlas.org/conflict/rio-blanco-mine-majaz-peru.

Fischer-Kowalski, Marina. 1998. "Society's metabolism: the intellectual history of materials, flow analysis, part I, 1860–1970." *Journal of Industrial Ecology* 2 (1): 61–78.

Fischer-Kowalski, Marina and Helmut Haberl. 2015. "Social metabolism: a metric for biophysical growth and degrowth." In *Handbook of ecological economics*, edited by Joan Martínez-Alier and Roldan Muradian, 100–138. Cheltenham: Edward Elgar Publishing.

Gerber, Julien-François and Arnim Scheidel. 2018. "In search of substantive economics: comparing today's two major socio-metabolic approaches to the economy – MEFA and MuSIASEM." *Ecological Economics* 144: 186–194.

Global Witness. 2014. *Peru's deadly environment*. Report. March. London: Global Witness, https://www.globalwitness.org/en/campaigns/environmental-activists/perus-deadly-environment/.

Global Witness. 2016. *On dangerous ground*. June. London: Global Witness https://www.globalwitness.org/en/campaigns/environmental-activists/dangerous-ground/.

Global Witness. 2021. *Last line of defence*. Report. September. London: Global Witness. https://www.globalwitness.org/en/campaigns/environmental-activists/last-line-defence/.

González Casanova, Pablo. 2006a. "El colonialismo interno." In *Sociología de la explotación*, edited by Pablo González Casanova, 185–205. Buenos Aires: CLACSO.

González Casanova, Pablo. 2006b, "Colonialismo interno (Una redefinición)." In *La teoría marxista hoy. Problemas y perspectivas*, edited by Atilio Borón, Javier Amadeo and Sabrina González, 409–434. Buenos Aires: CLACSO.

Hopkins, Seth. 2017. "Su identidad dolorosa: la colonialidad y violencia del Perú." Master of Arts diss. Bowling Green State University.

IBC, Cepes. 2016. *Directorio comunidades campesinas del Perú*. Lima: SICCAM.

INEI. 2017. *Censo Nacional XII de Población, VII de Vivienda y III de Comunidades Indígenas*. Lima: INEI.

Infante-Amate, Juan, Manuel González de Molina, and Víctor Toledo. 2017. "El metabolismo social: historia, métodos y principales aportaciones." *Revista Iberoamericana de Economía Ecológica* 27: 130–152.

Lander, Edgardo ed. 2000. "Ciencias sociales: saberes coloniales y eurocéntrico." In *La colonialidad del saber: eurocentrismo y ciencias sociales*. Perspectivas latinoamericanas. edited by Edgardo Lander, 4–23. Buenos Aires: CLACSO.

Li, Fabiana. 2015. *Unearthing conflict: corporate mining, activism, and expertise in Peru*. Durham: Duke University Press.

Luna, Josemanuel. 2019. "Reconfiguración del territorio y movimientos sociales: territorios en disputa." *Tlalli, Revista de Investigación en Geografía. UNAM* 1 (2): 55–75.

MAAP, Monitoring of the Andean Amazon Project. #136: Amazon Deforestation 2020 (FINAL). Amazon Conservation. https://maaproject.org/2021/amazon-2020/.

Mariátegui, José Carlos. 1928. *7 Ensayos de Interpretación de la Realidad Peruana, Chapter III. El problema de la tierra, El problema agrario y el problema del indio*. Lima: Amauta. https://www.lahaine.org/amauta/b2-img/Mariategui%20Siete%20Ensayos.pdf.

Martínez-Alier, Joan. 1990. *Ecological economics: energy, environment and society*. Oxford: Blackwell.

Martínez-Alier, Joan. 1999. *Introducción a la economía ecológica*. Barcelona: Editorial Rubes.

Martínez-Alier, Joan. 2005. *El ecologismo de los pobres: conflictos ambientales y lenguaje de valoración*. Barcelona: Editorial Icaria.

Merino-Acuña, Roger. 2015. "The politics of extractive governance: indigenous peoples and socio-environmental conflicts." *The Extractive Industries and Society* 2 (1): 85–92. https://doi.org/10.1016/j.exis.2014.11.007.

Muradian, Roldan, Joan Martínez-Alier, and Humberto Correa. 2003. "International capital versus local population: the environmental conflict of the Tambogrande Mining project, Peru." *Society and Natural Resource* 16: 775–792.

Neyra, Raquel. 2016. "Paquetazos ambientales o el afianzamiento del neoliberalismo en el Perú." *Ecología Política* 51: 10–14.

Neyra, Raquel. 2019. "Violencia y Extractivismo en el Perú contemporáneo." *Historia Ambiental Latinoamericana Y Caribeña (HALAC) Revista De La Solcha* 9 (2): 210–236. https://doi.org/10.32991/2237-2717.2019v9i2.

Neyra, Raquel. 2020. *Conflictos socioambientales en el Perú, Violencia y Extractivismo*. Ecuador: Editorial Abya Yala.

OECD. 2021. Material consumption (indicator). https://doi.org/10.1787/84971620-en (Accessed on 3 October 2021).

Pérez-Rincón, Mario, Julieth Vargas-Morales, and Zulma Crespo-Marin. 2018. "Trends in social metabolism and environmental conflicts in four Andean countries from 1970 to 2013." *Sustainability Science* 13: 635–648.

Quijano, Aníbal. 2000. "Colonialidad del poder, eurocentrismo y América Latina." In *La colonialidad del saber: eurocentrismo y ciencias sociales. Perspectivas latinoamericanas*, edited by Edgardo Lander, 122–151. CLACSO: Buenos Aires.

Quijano, Aníbal. 2007. "Coloniality and modernity/rationality." *Cultural Studies* 21 (2): 168–178.

Rodríguez, Iokiñe. 2017. "Linking well-being with cultural revitalization for greater cognitive justice in conservation: lessons from Venezuela in Canaima National Park." *Ecology and Society* 22 (4): Art.24.

Rodríguez, Iokiñe and Mirna Liz Inturias. 2018. "Conflict transformation in indigenous peoples' territories: doing environmental justice with a 'decolonial turn'." *Development Studies Research* 5 1: 90–105. https://doi.org/10.1080/21665095.2018.1486220.

Saes, Beatriz, Daniela Del Bene, Raquel Neyra, Lucrecia Wagner, and Joan Martínez-Alier. 2021. "Environmental justice and corporate social irresponsibility: the case of the mining company Vale S.A." *Ambiente & Sociedade* 24: 1–23. https://doi.org/10.1590/1809-4422asoc20210014vu2021L4ID.

Samaniego, Pablo, María Cristina Vallejo, and Joan Martínez-Alier. 2015. "Déficits comerciales y déficits físicos en Sudamérica." Working paper. Ecuador: FLACSO, http://www.flacsoandes.edu.ec/agora/deficits-comerciales-y-deficits-fisicos-en-sudamerica.

Seifert, Reinhard. 2005. *Yanacocha ¿El sueño dorado?* Volume 2, Cajamarca: Universidad Nacional de Cajamarca.

Scarritt, Arthur. 2012. "State of discord: the historic reproduction of racism in highland Peru." *Postcolonial Studies* 15 (1): 23–44.

Sotelo, Vicente and Pedro Francke. 2011. "¿Es económicamente viable una economía postextractivista en el Perú?" Working Paper n°5:105-125. Lima: Pontificia Universidad Católica del Perú-PUCP. http://www.redge.org.pe/sites/default/files/tema_5_Vicente%20Sotelo%20y%20Pedro%20Francke.pdf.

Svampa, Maristella. 2013. "'Consenso de los commodities' y lenguajes de valoración en América Latina." *Revista Nueva Sociedad* 244: 30–46.

Svampa, Maristella. 2016. *Debates latinoamericanos*. Buenos Aires: Editorial Edhasa.

Wallerstein, Immanuel. 1989. *The modern world-system. The second great expansion of the capitalist world-economy, 1730–1840*. Vol. III. San Diego: Academic Press.

Wendling, Z. A., J. W. Emerson, A. de Sherbinin, D. C. Esty, et al. 2020. *2020 environmental performance index*. New Haven, CT: Yale Center for Environmental Law & Policy. epi.yale.edu.

20 Quilombos and the fight against racism in the context of the COVID-19 pandemic

Givânia Maria da Silva and Bárbara Oliveira Souza

Brazilian diversity and inequality in pandemic times

The current public health emergency has presented very worrying results around the world. In Brazil, data have revealed that more Black people, who represent 56% (Alfonso 2019) of the population, die from COVID-19 (FENASPS 2021), while the group with the highest percentage of vaccinated people is white people. This evidences that the racist structure (Almeida 2018), selects its victims, even in the pandemic. In Quilombos, many of these effects have caused greater damage, due to the lack of access by Quilombolas (people who live in Quilombos) to basic health services, education, sanitation, and drinking water, among other services. In Brazil, the situation is serious, since the country is in second place in the number of deaths, on a global scale, as well as in the number of cases. As of March 23, 2021, more than 118,000 people have died as a result of COVID-19 and more than 3.7 million individuals have been infected.

The incidence of the pandemic among people in Brazil reflects the existing structural inequalities. According to the Ministry of Health, in data published in its epidemiological bulletins, there has been an increase in the percentage of black- and brown-skinned people among those either hospitalized or killed by COVID-19 (Portal Geledés 2020). In other studies, such as one conducted in São Paulo, there is an indicator that Blacks have a 62% higher probability than whites of dying from COVID-19. There are also worrisome records of the growth of the disease in Quilombos, adding up to more than 150 deaths and more than 4,200 infected (Quilombo sem Covid-19 2020). This growth is also reflected in Indigenous lands, showing the structural racism operating against Blacks and Indigenous people.

Brazil is composed of a very diverse population. According to the 2019 PNAD (Brazil National Household Sample Survey), across the country, Black and brown people are the majority. The data show that the population that declares itself Black represents 9.4%, and brown, 46.8%. Together, they equal 56.2% of the population, while whites are 42.7% (IBGE 2019). According to the IBGE (Brazilian Institute of Geography and Statistics), there are 305 Indigenous ethnic groups (IBGE 2010) and more than 800,000 gypsies (Cite organization or author here).

Quilombola communities, in turn, are present in all regions of Brazil. From north to south, from east to west, the Quilombos keep up the historical struggle to secure their rights. Today, there are more than 6,000 communities all over the country (CONAQ 2018). Of these, 3,432 Quilombola communities are certified by Palmares Cultural Foundation,[1] of which a little more than 300 have their territories titled. According to the IBGE, there are Quilombolas in 1,672 of the 5,570 Brazilian municipalities, that is, about 30% of the Brazilian municipalities have Quilombos (IBGE 2010). Of the three states with the largest number of Quilombos, two are in the Northeast (Bahia and Maranhão).

DOI: 10.4324/9780429344428-24

Brazil has at the core of its history the trafficking and trade of enslaved Africans. It was the country that imported the most slaves and the last one to legally abolish slavery. The deep Brazilian participation in the slave trade is marked by the estimate that about 40% of enslaved Africans were destined for Brazil (Gomes 1996).

In counterpoint, all this repressive apparatus existing in colonial and imperial Brazil marks the weight of Black resistance. Throughout Brazilian history, Black men and women resisted and fought against oppression and discrimination through a multiplicity of forms. One of the most significant forms of Black resistance to slavery were the Quilombola communities.

However, this rich ethno-racial diversity is anchored in structural inequality. Brazil's Gini Index is one of the worst in the world, well away from the top-ranked countries, such as Hungary (0.244), Denmark (0.247), and Japan (0.249). However, this rich ethno-racial diversity is anchored in structural inequality. Brazil The Gini Index in Brazil is one of the worst in the world, far from the top performers such as Hungary (0.244), Denmark (0.247) and Japan (0.249). Currently, Brazil has a Gini Index of 0.52, according to data from the World Bank (source: https://datos.bancomundial.org/indicator/SI.POV.GINI?locations=BR—accessed on 01/25/23). Racial and gender inequality is present in the labor market, in access to education and health, and in the policies that influence the indicators mentioned.

Structural racism is one of the pillars of the historically built inequality in Brazilian society. The deep racism that was a mark of the colonial system, present since the 16th century, was re-signified in the formation of the Brazilian state between the 19th and 20th centuries. The logic of the whitening theory permeated several sectors, such as the labor market and education. One of the most emblematic actions taken was the promotion of extensive European migration to Brazil, with various incentives such as land grants and guaranteed employment for Europeans, with a view to "whitening" the Brazilian population. The Black population, on the other hand, saw its economic and social exclusion deepened by these policies.

Studies point out that the inequalities, already strongly present among Blacks, women, Indigenous peoples, Quilombolas, and other traditional communities[2] (*comunidades tradicionais*), are expected to worsen with the pandemic caused by COVID-19. The biggest victims of the pandemic are the most vulnerable segments of our society.

The main organization of Quilombolas in the country, the National Coordination of Rural Black Quilombola Communities (CONAQ), has highlighted the structural factors that have led to the spread of the pandemic in Quilombola territories. Moreover, it has denounced this situation to public authorities, formulating proposals with the goal of reducing the damage to the communities. However, few answers have been given by the authorities.

In Brazil, in the context of COVID-19, the social movements of the countryside and in defense of forests and water are elaborating a project proposal to confront the pandemic. Even knowing that the National Congress voted for the approval of the laws PL no. 1142 and 735/2020,[3] the President Jair Bolsonaro vetoed them. This demonstrated the difficulties imposed by racism that the Quilombolas, Indigenous, family farmers, and other traditional peoples and communities (*povo e comunidades tradicionais*) have been facing in the context of the COVID-19 pandemic. The presidential vetoes are a demonstration of the lack of understanding of the vulnerabilities to which these groups are subjected. It shows that the Quilombos, in the context of the pandemic, live with and face the structural neglect of the Brazilian state. They often lack health care that meets the demands of the communities and exposes the Quilombolas' lack of access to public services.

Land conflicts, experienced in several Quilombos in the country, are another factor that has aggravated fragility in this delicate context (CONAQ 2018). In this scenario, Quilombola communities, such as those located in the Quilombola Territory of Alcântara, located in

Maranhão, experience tensions, with threats of new displacements and removals of part of the communities, due to the attempt to expand the space base located there.

Under the command of the Federal Government, a series of actions have been taken in the Alcântara territory in the middle of the pandemic—that violate the right to prior, free, and assisted consultation, and that do not ensure the right to land provided for in the Federal Constitution of 1988, in article 68 of the ADCT (Transitory Constitutional Dispositions Act).

The Quilombos have various forms of organization and location, which causes, in many cases, difficulties in accessing health policies, as well as constant complaints by CONAQ of the lack of measures by the governments.

The preventive actions recommended by the health authorities in the face of the novel coronavirus pandemic—social distancing, intensive use of masks, alcohol gel, and water-dependent care—are not effective for the Quilombolas. In addition, many Quilombola communities lack even drinking water. With the social distancing, a sanitary requirement to contain the pandemic has become a challenge, given the precarious conditions in which Quilombola communities find themselves due to poor access to public policies.

The critical situation of public policies for Quilombola communities and the Black population has other points that demand attention. A study conducted by Inesc (2020) details how recent years have worsened the situation of racial equality policies. From 2014 to 2019, there was an 80% cut in the resources allocated to racial equality policies. In 2020, the situation has worsened with the extinction of the 2034 Program, previously existing in the Pluriannual Plan, PPA 2016–2019. The PPA 2020–2023 no longer incorporates this program.

The study (Inesc 2020) also highlights the huge budget cuts for policies targeting Quilombola communities. From 2017 onward, there were no resources for land regularization of Quilombola territories, from Action 210V of the PPA (Pluriannual Plan). In 2020, of the R$3.2 million foreseen for this action, nothing was executed until August 2020. The Palmares Cultural Foundation, in turn, did not execute any resources for Quilombola communities in 2020, nor did the Ministry of Women, the Family and Human Rights.

The existing public policies for Quilombola communities, which already had a history of meeting very few of the communities' fundamental demands, are today reduced to almost nothing. At the same time, universal public policies, such as health care, have also suffered heavy blows in recent years.

In the serious pandemic context that we are experiencing, it is worth mentioning the freezing of the budget for public health and education policies for 20 years as an effect of the "public spending ceiling," as pointed out in the study by Inesc (2020). Therefore, if health services in Quilombos were already precarious, these measures have expanded even more and made access more difficult for Quilombolas, as the Brazilian Association of Collective Health (ABRASCO) points out (Silva and Silva 2020). And, in the context of the pandemic, the effects of the measures have been, and continue to be, disastrous for the lives of the Quilombolas.

Quilombola territories and identity

As we have seen in the previous reflections, the Quilombola communities have suffered a worsening in their access to fundamental public policies, including those providing land regularization of their territories. The existence of the communities is fundamentally linked to the guarantee of their territories.

The movement to fight for the guarantee of the rights of these communities is historical and political. It carries within it a secular dimension of resistance, in which men and women sought the Quilombo as a possibility to maintain themselves physically, socially, and culturally, in counterpoint to the slave logic.

Struggles for the defense of territories have occurred in different historical periods in Brazil. One of the landmarks of these struggles was the Quilombo of Palmares, a community of great proportions during the 17th century, which waged war against the colonial power for decades. It became a landmark of Black struggles in Brazil and one of its leaders, Zumbi dos Palmares, is nowadays one of the main references of the Black Brazilian resistance. Besides Palmares, Quilombos from the north to the south, and the east to the west of Brazil, bear in their history records of struggle for their lands and cultural traditions.

The processes of resistance to slavery, whether in rural or urban areas, make the Quilombos maintain deep bonds of identity with their traditional land/territory, based on their way of life. As a resistance movement, the Quilombos are located in rural and urban areas. Rurality prevails, but it is not an exclusive condition of the rural Black communities in the continent. In the case of Brazil, the Quilombola communities located in urban areas experience the advance of cities over their territories, real estate speculation, and other processes of deterritorialization:

> The logical consequence of the urban reforms followed would be the valorization of urban land and once again the segregation of the impoverished and black population [...] these territorialized other spaces and formed networks of solidarity, sociability and cultural spaces, which remain until today in the symbolic and cultural field of the city [...The notion of territory in the occupational sense notes the existence of possible racial segregation, but also highlights the appropriation that these segments perform, imprinting marks and meanings to these places [...] Thus the formation of a Black territory also goes through this notion of social exclusion, residential occupation and identity.
>
> (Anjos, Ramos, Mattos, and Marques 2008, 174)

Land and territory for Black communities, even when overlapped by the urban world, have other meanings and uses. They involve planting, production, experiences, and expressions of cultural manifestations, celebrations, sacred spaces, and the maintenance of ancestral memories linked to African heritage.

Therefore, the meanings of land and territory for rural or urbanized communities, due to deterritorialization, are not restricted to geographic, agrarian, or agricultural aspects. The relations and ways of life are associated with other characteristics of a cultural and symbolic nature, but maintain the traditionally occupied land/territory as the starting point for struggles for recognition and rights.

> The identity aspects must be taken into consideration, beyond the land issue. Land is crucial for the continuity of the group as a condition for settlement, but not as an exclusive condition for its existence[4]. And territory is not restricted only to the geographical dimension, but also encompasses broader cultural, historical, and social elements.
>
> (Souza 2016, 87)

Black communities are circumscribed and establish a vital relationship of belonging and existence/ resistance with their territories. In Brazil, the denomination recognized by the Federal Constitution of 1988 as a subject of rights was that of communities remaining from Quilombos, which means:

> (...) the ethnic-racial groups, according to criteria of self- attribution, with their own historical trajectory, endowed with specific territorial relations, with presumed black ancestry related to the resistance to the historical oppression suffered and their lands characterized as "lands occupied and used for the guarantee of their physical, social, economic, and cultural reproduction.[5]
>
> (Decree 4887/03)

In Brazil, the struggle of the Quilombola communities for their territories, now organized after the Federal Constitution of 1988, is highlighted in Article 68 of the Transitory Constitutional Dispositions Act (ADCT)[6] that assures the right to land to the quilombos in Brazil, and its articles 215[7] and 216[8] that deal with the culture and tradition of Quilombola communities and Black culture. The struggle, previously made by different strategies, gained strength after the Federal Constitution of 1988 and the recognition of rights for these social groups.

Today these movements continue to battle for the guarantee of fundamental rights, such as definitive titling of the lands traditionally occupied by Quilombola communities. An important point of the right to land titling is that the titles must be collectively assigned to the communities with mandatory insertion of inalienability, imprescriptibility, and unseizability. The relationship with the slave-owning past and the processes of resistance to the regime have become mobilizing elements to guarantee the rights to the traditionally occupied territories for legal purposes and, in them, with them, and from them, to access other public policies. The lack of autonomy over their territories has affected the Quilombolas' ways of life, and in the context of the pandemic, has left the Quilombolas even more fragile.

In the current serious public health situation that we are experiencing, most communities are not guaranteed the right to their territories, and live with conflicts, threats of expropriation, and violence (CONAQ 2018). These situations are experienced, most of the time, in conflicts with agribusiness, mining companies, and other high-impact enterprises in the regions where Quilombola communities are located. These are necessary factors for us to evaluate the current context in which the Quilombola communities are immersed in the fight against the effects of COVID-19, which is also a fight against racism that has taken on even greater proportions.

Concluding remarks

Racial discrimination and inequality, which are so present in Brazilian society, and which mark the history of the relationship of the Brazilian state with the Quilombos, are in the process of worsening in the situation of the pandemic. These are aspects that must be dealt with through measures to be taken by the local, state, and federal governments, with the necessary urgency, obeying the Federal Constitution of 1988 and the international treaties and conventions of which Brazil is a signatory.

It is up to the Brazilian state to ensure for its population, including the Quilombola communities, basic rights and health care, in order to put into effect qualified public policies for the reduction of structural inequalities that affect Blacks, women, and Indigenous people in a more determinant way, even as these have been further aggravated by the unequal effects of the pandemic.

However, what can be seen from the monitoring of public policies and government actions in this critical context of the novel coronavirus pandemic is that there has been a systematic reduction in initiatives and actions aimed at the Quilombola communities and the Black population, and an increase in the violation of the rights of these citizens. The situation of vulnerability, experienced for centuries in the Quilombos, has been worsened considerably in the current conjuncture.

However, the fight for the fundamental rights of Quilombola communities, such as the right to land and health, is still ongoing, with the incorporation of necessary strategies in this delicate context. The elaboration of a database, Quilombosemcovid19, to monitor the effects of COVID-19 in the communities and the articulation with the legislative power are some examples. With resistance and the inspiration of their ancestral roots, the strategies of the communities in the face of the unequal and racist context of Brazilian society are still in progress.

Notes

1 Certificates issued by Fundação Cultural Palmares to remaining quilombola communities. This database was updated by Administrative Rule no. 36/2020, published in the DOU on 02/21/2020. Accessed on 02/06/2020, Palmares – Fundaçao Cultural 2022.

2 The current Brazilian national policy considers the category of comunidade tradicionais or traditional communities in their National Policy for Sustainable Development of Traditional Peoples and Communities. According to this Policy, Traditional Peoples and Communities (PCTs) are defined as: "culturally differentiated groups that recognize themselves as such, that have their own forms of social organization, that occupy and use territories and natural resources as a condition for their cultural reproduction, social, religious, ancestral and economic, using knowledge, innovations and practices generated and transmitted by tradition" (Ministério da Cidadania 2022).

3 The PL 1142/2020 was about social protection measures to prevent the contagion and spread of COVID-19 in indigenous territories, and the PL 735/2020 was about emergency measures to support family farmers in Brazil to mitigate the socio-economic impacts of Covid-19.

4 Here I highlight communities that have been expropriated, such as the Amaros, and that remain with their ties of belonging as a community.

5 Brazil, Decree no. 4.887, of November 20, 2003. Regulamenta o procedimento para identificação, reconhecimento, delimitação, demarcação e titulação das terras ocupadas por remanescentes das comunidades dos quilombolas de que trata o art. 68 do Ato das Disposições Constitucionais Transitórias. Diário Oficial [da] República Federativa do Brasil, Poder Executivo, Brasília, DF, 2003a.

6 The remaining members of quilombo communities that are occupying their lands are recognized as having definitive ownership, and the State must issue them the respective titles.

7 Article 215 reads: "The State shall guarantee everyone the full exercise of cultural rights and access to the sources of national culture, and shall support and encourage the appreciation and dissemination of cultural manifestations. The State shall protect the manifestations of popular, indigenous and Afro-Brazilian cultures, and those of other groups participating in the national civilization process. The law shall provide for the establishment of commemorative dates of high significance for the different national ethnic segments. 3. The law shall establish the National Plan of Culture, of pluri-annual duration, aiming at the cultural development of the country and the integration of public power actions leading to..." (Note: this article was included by Constitutional Amendment No. 48, of 2005).

8 Article 216 reads: "Brazilian cultural heritage is constituted by goods of a tangible and intangible nature, taken individually or together, which refer to the identity, action and memory of the different groups that make up Brazilian society, including: I - the forms of expression; II - the ways of creating, doing and living; III - the scientific, artistic and technological creations; IV - the works, objects, documents, buildings and other spaces destined for artistic-cultural manifestations; V - the urban sets and sites of historical, landscape, artistic, archaeological, paleontological, ecological and scientific value. 1 - The public authorities, with the collaboration of the community, will promote and protect the Brazilian cultural heritage, by means of inventories, registers, surveillance, toppling and expropriation, and other forms of safeguarding and preservation."

References

Alfonso, Nathália. 2019. "Dia da consciência negra: números expõem desigualdade racial no Brasil". *Lupa*, November 20, 2019. https://lupa.uol.com.br/jornalismo/2019/11/20/consciencia-negra-numeros-brasil/.

Almeida, Silvio Luiz de. 2018. *What is structural racism?* Belo Horizonte, MG: Letramento.

Anjos, José Carlos Gomes dos, Ieda Ramos, Jane de Mattos, and Olavo Marques. 2008. "As condições de raridade das comunidades quilombolas urbanas". In *Diversidade e Proteção Social: estudos quanti-qualitativos das populações de Porto Alegre: afro-brasileitos; crianças, adolescentes e adultos em situação de rua; coletivos indígenas; remanescentes de quilombos,* edited by Ivaldo Gehlen, Marta Borba Silva, and Simone Ritta dos Santos, 167–178. Porto Alegre: Centhury.

Brazil: Federal Constitution of the Republic of Brazil of 1988. Brasília, 1988.

Brasil, Law Federal Decree No. 4.887, of November 11, 2003. Provides on the identification, recognition, delimitation, demarcation and titling of the lands occupied by the remaining quilombo communities.

FENASPS. 2021. "Covid-19: negros morrem mais e são vacinados do que brancos." FENASPS, March 22, 2021. https://fenasps.org.br/2021/03/22/covid-19-negros-morrem-mais-e-sao-menos-vacinados-do-que-brancos/.

IBGE - Instituto Brasileiro de Geografia e Estatística. 2010. "Gráficos e tabelas: População residente, segundo a situação do domicílio e condição de indígena." https://indigenas.ibge.gov.br/graficos-e-tabelas-2.html.

IBGE - Instituto Brasileiro de Geografia e Estatística. 2019. "Pesquisa Nacional por Amostra de Domicilios Contínua - PNAD."

Ministério da Cidadania. 2022. "Povos e Comunidades Tradicionais." *Secretaria Especial do Desenvolvimento Social.* http://mds.gov.br/assuntos/seguranca-alimentar/direito-a-alimentacao/povos-e-comunidades-tradicionais.

Palmares - Fundação Cultural. 2022. "Certificação Quilombola." https://www.palmares.gov.br/?page_id=37551.

Portal Geledés. 2020. "Em 4 semanas, mortes de pretos e pardos por Covid-19 passam de 32.8% para 54.8%." *Portal Geledés.* https://www.geledes.org.br/em-4-semanas-mortes-de-pretos-e-pardos-por-covid-19-passam-de-328-para-548/.

Quilombo sem Covid-19. 2020. "Observatório da Covid-19 nos Quilombos." Accessed August 27, 2020. https://www.quilombosemcovid19.org/.

Silva, Hilton P. and Givânia M. Silva. 2020. "A situação dos quilombos do Brasil e o enfrentamento à pandemia da Covid-19 – Artigo de Hilton P. Silva e Givânia M. Silva." *ABRASCO - Associação Brasileira de Saúde Coletiva*, September 16, 2020. https://www.abrasco.org.br/site/noticias/a-situacao-dos-quilombos-do-brasil-e-o-enfrentamento-a-pandemia-da-covid-19-artigo-de-hilton-p-silva-e-givania-m-silva/52116/.

Souza, Bárbara Oliveira. 2016. *Aquilombar-se: panorama on the Brazilian Quilombola Movement.* Curitiba: Appris Editora.

21 The "Greening" by Sustainable Development

Stretching Biopiracy

Ana Isla

Introduction

Biopiracy amounts to the appropriation of the traditional knowledge as well as biological and chemical resources of Indigenous Peoples and peasants that establishes the knowledge systems of colonialist corporations. It has evolved from no prior consent and no compensation given to communities, to consent provided by the state on behalf of any community.

This chapter argues that in the Americas, biopiracy—defined as economic activities that promote extractive recolonization processes of rights, bodies, and territories—grew exponentially when the World Bank created the concept of Global Commons to open new areas of global intervention for capital and ceded responsibility for Sustainable Development (SD), first, to governments, and later to corporations. For instance, the Convention on Biological Diversity gave countries sovereignty over their genetic resources, and left questions of ownership to national legislation. Therefore, States authorize exploration, investigation, bioprospecting, use, and exploitation of biodiversity elements. Since then, access to wildlife species has been regulated by licenses, permits, and auctions. Bioprospecting, and other genetic systems that classify and research biodiversity for commercial ends, often becomes biopiracy.

The World Bank also contributed to the increase and intensification of biopiracy since taking charge of the Global Environmental Facility to carry out the so-called "Green" Economy (GE) through non-governmental organizations (NGOs). The role of many international NGOs, such as the World Wildlife Fund (WWF), has been to act as brokers between corporations and states, and advance the language of the more ecologically friendly economic policies and programs. Within the SD paradigm, despite ongoing debates and the search for alternatives, economic growth remains a dominant objective; therefore, the destruction of subsistence economies is the central element of what today is generally understood as Sustainable Development (Isla 2015).

This chapter evaluates global capital's ecological management as discussed at the Earth Summits (1992, 2002, 2012), and extends that assessment to initiatives for "greening" the economy through SD in Loreto, Peru's Amazonia. Key concepts in SD are Natural Capital (NC) and Payment for Environmental Services (PES). Natural Capital refers to the goods and services provided by the planet's stock of water, land, air, and renewable and non-renewable resources such as plant and animal species, forests, and minerals. Payment for Environmental Services is a voluntary transaction in which a buyer from the industrial world pays a supplier for a well-defined environmental service such as a patch of forest. I draw on the work of Carolyn Merchant, Maria Mies, and Silvia Federici to critique the Global Commons perspective and evaluate three moments of biopiracy: 1) Deforestation by Forestry Concessions for wood production; 2) Forest degradation by land change use; and 3) Forest finance for REDD+. It addresses the question to what extent SD creates social equality and combats climate change.

DOI: 10.4324/9780429344428-25

The Commons Perspectives: World Bank vs Ecofeminism

Enclosures began as an English phenomenon in the 1700s. Thompson discusses the age of parliamentary enclosure (1760 and 1820) as the age of class conflict to eliminate Common Rights, at the interface between agrarian practice and political power that fenced off the common lands, putting an end to customary rights and evicting the peasantry that depended on them for their survival. Contemporary commons and enclosure debates have increased since 1968, when Garret Hardin put forward the argument that any Commons regime will result in degradation.

At the 1992 Earth Summit, economists from the World Bank proposed that the ecology must be embedded in the price system, that is, Nature requires a fully monetized world in order to be protected. They developed a key concept, Natural Capital, which refers to the goods and services provided by the planet's stock of water, land, air, and renewable and non-renewable resources Pearce and Warford 1993). At the Earth Summit, most of the countries have signed and ratified the Convention on Biological Diversity. In 1996, The Inter-American Development Bank organized a conference to discuss "mainstreaming biodiversity into development projects by promoting the creation of incentives for conservation of biodiversity and sustainable use of biological resources, such as property rights regime and markets for 'green' products" (5). Jeffrey McNeely, from the IUCN-The World Conservation Union, sustained that the main problem in financing biodiversity conservation is identifying the most suitable and equitable economic instruments that enable the full costs of exploitation to be included in prices, and proposed to charge for the use of the Global Commons, through user rights, regulations, and rents, as a way of governing and generating revenue. Since then, the local commons are being handled by global actors, while local concerns no longer matter. In effect, the Global Commons framework has extended biopiracy, granting permission to grab, particularly around Indigenous land that still conserve their local commons.

In contrast, Commons, in this paper, is defined as the natural, social, and political space that provides sustenance, security, and independence, and typically does not produce commodities for profit. To critique the Global Commons, I articulate the understanding of the Commons provided by Carolyn Merchant, who argues that the rise of modern science and technology was premised on the violent attack and rape of Mother Earth (1983), and she establishes two contradictions in the working of development. First, it arises from the assault by economic forces of production on local ecological conditions. Second, it begins from the assaults of production on biological and social reproduction (2005). Maria Mies infers that the basic precondition for economic growth continues to be former colonies. These colonies are women's housework and peasant and Indigenous Peoples' subsistence economies as well as nature. All are overexploited on what they use for their livelihood. Meanwhile, Silvia Federici draws on history, from the enclosure of the Commons and the witch trials in Europe that led to the devaluation of women, nature, bodies, and labor. She re-valorizes the work of reproduction and reconnects our relation with nature, with others, and with our bodies to regain a sense of wholeness in our lives. These authors recognize that housework as well as nature's work are productive labor and are areas of exploitation and a source of capital accumulation.

Currently, continental Amazonian women emphasize that their bodies feel the affectation of their territories. To defend the Commons of life, water, and territories, they have joined the Latin American women's movement that have developed new libertarian pedagogies with the arising of (Indigenous) community and popular feminisms—of Zapatista, Mayan, Ixhil, Quiquché, Lenca, Andean, Mapuche. Their struggles, appreciative of autonomy and self-determination, are nourished by decolonization, depatriarchalizing, socialist, anarchist, anti-dictatorship, anti-racist, and anti-neoliberal battles. For instance, for the movement *Revolucion en*

Construction Tejiendo Futuro, the imposition of 'development' projects leads us to a war declaration that intends to eradicate our resistance. Another feminist group, *Las Miradas Criticas del Territorio desde el Feminismo*, describes the body as the first territory in which the body and territory have both been historically and structurally expropriated (Cabral 2012). Consequently, they are reclaiming the Local Commons from ongoing dispossession.

In the three following examples I examine what is at stake when biological (human and biota) reproduction in Loreto's forest is disrupted by colonial and capitalist development, so-called sustainable development.

Granting Permission to Grab by:

Forestry Concessions for wood production

In this section, I argue that forestry concessions for wood production must be named as biopiracy, because they grant permission to plunder nature and its inhabitants. Following Merchant (1983), the first contradiction arising is expressed in the violence coordinated by production against the ecology. For instance, for every tree trunk mechanically extracted in a tropical forest, more than a thousand trees of all sizes are destroyed. As a result, this produces habitat alteration creating a shortage of wildlife that play an important role in the maintenance and regeneration of the forest as seed dispersers. Above all, this logging contaminates and reduces water in the ravines, rivers, and lakes, affecting all aquatic life—fish and plants. The second contradiction comes from the assaults of production on social and biological reproduction formulated by the political economy of local and global institutions. Their degradation has resulted in growing rates of chronic child malnutrition, pernicious anemia, poverty, violence against women and children.

Since the 1992 Earth Summit, a contemporary enclosure has been taking place as a result of Sustainable Development. I present the war against Loreto's forest. A conflict waged by global "good faith buyers" and local corrupt authorities through concessions. The concept of Concession was invented by European and North American imperial and colonial powers to occupy their colonies by giving complete control to private parties for their use, thereby dispossessing and disenfranchising millions of people.

In 2001, Peru introduced the concept of Sustainable Forest Development (SFD) in Forest Concessions, certified by the Forest Stewardship Council. This initiative is a failure because deforestation has been increasing (Tala de la Selva). The SFD framework established that to prevent informality and have greater control, concession areas should not be greater than 5,000 ha. One of the essential points of new forest management is the concentration of areas for forest production, 'preferably wood,' in the so-called permanent production forests. These policies were designed not to work, due to the poor budget assigned to the sector and the high level of corruption.

Between 2002 and 2004, forestry concessions were granted in Loreto (14,782,302 ha), Ucayali (4,089,926), Madre de Dios (2,522,141), and San Martin (1,501,291). These concessions covered 249,752 million hectares (SERFOR 2017). In 2005, Jose Alvarez claimed that there was ecocide in the Amazon forest as mahogany fell by forest tractors and chainsaws were placing this type of timber in serious danger of extinction. Alvarez criticized the authorities—the National Institute of Natural Resources (INRENA) and the Convention on International Trade in Endangered Species of Wild Fauna and Flora (CITES)—for resisting declaring mahogany in danger of local extinction. By 2011, social reproduction was also endangered. Alvarez (2011) denounced the invasion of the lands of uncontacted indigenous people by the loggers, and displacement from their ancestral lands to places not suitable for their way of life. At the same time, several indigenous communities in the middle basin of the Napo river were left without territory due to forest concessions. They were never informed nor consulted, in violation of ILO Convention 169, which establishes the obligation to consult any norm or project that may affect indigenous communities and their territories.

Territorial confiscation was the objective of the Fujimori government (1990–2000); however since 2008, the Concession system has been complemented by an international Free Trade Agreement with the United States (FTA) as the Amazon rainforest became the last rent frontier for economic development for the world. Former President Alan Garcia, with abusive laws unleashed horrendous tragedies like that of Bagua (Isla 2009) and prepared the ground for the forestry law that promotes Indigenous territorial confiscation through occupying large extensions of the forest under any form of concessions. Under FTA, the Forest Resources and Wild Fauna Supervision Agency (OSINFOR) was created to supervise and oversee the sustainable use and conservation of forest resources and wildlife, as well as environmental services from the forest provided by the State through various forms of use. However, to weaken OSINFOR management, in December, 2018, it was incorporated into MINAM and has decreased the collection of data, making it impossible to trace the legality of the origin of the wood (EIA 2018). After public outcry, by 2019, OSINFOR regained its independence (Global Witness 2019).

In 2017 the Regional Government of Loreto (GOREL) promoted and accepted 41 new requests for timber concessions, in four provinces (The Region 2017, 14–16). Furthermore, Servicio Nacional Forestal y de Fauna Silvestre published the map of Loreto offering 197 new "units"; of which at least 25 overlap with the areas of the Indigenous Reserves requested in 2003 in favor of Indigenous People in Isolation (IPI) in the Department of Loreto (The Region 2017, 14–16).

While it has been extremely challenging to access relevant and reliable data, for cases where we have been able to do so, we have identified several timber export shipments in which over 90% of the produce was of documented illegal origin…The Unidad de Inteligencia Financiera Report also concludes that Peru's production of illegal timber in the last 5 years (2014–2018) has been over 1,100,000 cubic meters per year – which implies the illegal harvesting of around 150,000 natural forest trees per year, degrading some 300,000 hectares of natural forests every year.

These importers are the United States, China, Mexico and 15 other countries (EIA 2018). A Global Witness publication reports on how wood exporters and buyers are complicit in the looting of the Peruvian Amazon. This maintains that "the timber industry, its main regulatory authority (SERFOR), and other government entities in Peru have denied or minimized the problem, and have tried to weaken the institutions in charge of implementing the standards."

In Amazonia, legal and illegal logging requires criminal networks, involves violations of human rights and slave labor conditions, and contributes to the degradation of biodiversity and climate change. For instance, in 2019, MAAP#123 has documented the construction of 1,500 additional linear kilometers in Loreto, San Martin, and Madre de Dios. Three types of possible illegality were detected: a) Forest Roads in Zones without Enabling Titles; b) Forest Roads in Current Forest Concessions, but in a Non-Active/Not Defined situations; and c) Forest Roads in Native Communities.

Due to the inaction of Gobierno Regional de Loreto (GOREL) and the high level of corruption in SERFOR, the Prosecutor for Environmental Matters, in Loreto, is investigating a large number of alleged crimes against forest done by logging companies, rural and native communities, local forest associations, and export companies; and it also investigates officials of government entities that intervene in the crimes against the forest (The Region, December 27, 2017).

In sum, this section has shown that Forestry Concessions for wood production, understood as Sustainable Development, assault the biological and social reproduction of Amazonia forest, and leaves behind a history of corruption, swindling, exploitation, robbery, contamination, destruction, and blood (animal and human). The sections of this chapter show how granting permission to grab is of immediate help to corporations to change the use of land, and that the same practices that finance forest can turn to in the context of large-scale expropriations The next section discusses the case of cacao production as biopiracy through land change use that produced forest degradation.

Land Change Use—The case of cacao production

The 2002 Earth Summit on Sustainable Development and the Millennium Development Goals committed rich and poor countries to a global partnership to improve the world's atmosphere, waters, and forest as well as human health. They argued that corporations were better situated to organize Sustainable Development (SD). The idea of SD is that there must be an exchange process between those with money to buy and those with natural capital to sell (Pearce and Warford 1993). Consequently, SD focuses in a narrow way on the physical nature of the forest, and evades the web of social relationships and processes in which rainforest and forest people are embedded. In the forest, the trees are interrelated with each other through the multitude of creatures that relate to them as food, shelter, and nesting; through shared access to water, air, and sunlight. The people of the rainforest are also members of this superorganism.

Maria Mies argues that basic needs are the same everywhere in both poor and rich countries and poor and rich classes. She maintains that subsistence-oriented people are overexploited by which they need for their own subsistence. Unwaged or poorly paid rural women, peasants, Indigenous peoples dependent on the commons for their subsistence, autonomy are "housewifized." This concept is applied to socially marginal and externalized economic sectors and actors as Indigenous People and peasants, when their land and products are taken from them with little or no compensation and through structural violence. By applying Mies' subsistence theory, I argue that Cacao del Norte S.A.C., has carried out biopiracy through land expropriation, forest destruction that collide with the biological and human reproduction, and unscrupulous negotiations with corrupt local authorities.

The company Cacao del Norte S.A.C. is a subsidiary of United Cacao Limited SEZC, a company incorporated and registered in Isla Caiman. The president of this company is Dennis Melka, a Czech-American national. In Peru, 17 of its companies are engaged in oil palm and 8 in cocoa production. (Dammert 2017, 17)

Land expropriation has disrupted

The reproductive labor of people

The integration of land and labor into the global economy by Cacao del Norte S.A.C, in Tamshiyacu, Loreto began in 2010 by purchasing land from "Los Bufalos Association," a group of 45 farmers and ranchers with 49.7 ha. each. The company cheapened local community land in order to expropriate it, and also bought other plots from their owners, who were told that if they did not use their land, it would revert to the state. In this way the company bought 60 plots (3,000 ha.), and paid US$1,4000 to each landowner (USD$28 per ha.) (Chirif 2016a) In addition, the company created a credit program (Programa Alianza Producción Estratégica cacao—PAPEC) aimed at other small agricultural producers to grow cacao. According to PAPEC, in 2015, it had 150 registered farmers, each with one hectare, who received cocoa seedlings and seeds, fertilizers, pesticides, plastic bags for the seedlings, and tools (fumigators etc.). The amount of the equipment received totaled US$600. By accepting the allocations, the farmers signed contracts which indebted them. None of them were able to keep a copy of the contract, which raises doubts about the company's good intentions. (Chirif 2016b)

For the surrounding communities, the destruction of habitats for endemic species (of flora and fauna) has created a shortage of raw materials, proteins, water sources, biodiversity, as well as displacement of agriculture and fisheries, losing food sovereignty. In addition, the change produced confrontation between members of the community. Those who lived in subsistence feel dispossessed of their land and their environment. Consequently, they are denouncing the company for land theft and exposing them to food insecurity (Chirif 2016ab). Since then, civil society has mobilized against the enterprise, now called Tamshi SAC, while some corrupt politicians favor it.

The reproductive labor of nature

The impacts on the forest and the local population were hidden by the corrupt government officials as the corporation began the felling of the primary forest without having the Environmental Impact Study or the Soil Classification Study. In 2015, Cacao del Norte S.A.C cut 2,380 ha. of trees for cocoa production in Loreto and 13,000 ha. in Ucayali. (The Region, September 2, 2013). In total, the corporation is responsible for over 15,000 hectares of illegal deforestation, mainly of primary tropical rainforest (EIA 2016).

As a result, biological reproduction of some forest species has been broken. For each hectare of forest that is cut and burned, there are two or three additional hectares that degrade in their ecosystem (Chirif 2016a). The Tamshiyacu area has the highest concentration of rosewood, with an estimate of 18 trees per hectare, which were already declared an endangered species by the Ministry of Agriculture. By 2015, the company is estimated to have felled around 4,284 rosewood trees. Even more serious, deforestation has been carried out in a varillal forest, on white sand soils. They are very special and fragile ecosystems that have high soil heterogeneity and different drainage conditions, which has given rise to a unique flora and fauna adapted to very poor environmental conditions. Once logged, it is difficult to regenerate, making the land unsuitable for agriculture (Chirif 2016b). Therefore, Cacao del Norte will not produce first-quality cacao, because the soils where it produces are poor (white sand forests) that require a large amount of chemical fertilizers. In addition, the falling and burning of the trees boost the emission of CO_2, increasing the temperature, decreasing rainfall by convection (convectional rainfall is very common in areas where the ground is heated by the hot sun, such as the Tropics), and breaking the balance between the absorption and the emission of carbon.

Culture of subsistence but expanded corruption

In 2016, the Public Prosecutor's Office filed an appeal in cassation before the Supreme Court to review the procedural flaws regarding the judgments that the lower courts had issued about the forest destruction (Press Release No. 001-2016). The company used Decree Law No. 838 of August 18, 1996, issued during the Fujimori dictatorship, which establishes that "the titling of plots in the depressed economy areas of Sierra, Ceja de Selva, and Selva until 31 December 1998 will be free of charge."

After many years of judicial encirclement, Cacao del Peru Norte SAC began to be investigated for the deforestation of 13,800 hectares of forest in the Loreto and Ucayali regions. In August 2018, Cacao del Norte became Tamshi SAC, and in September the oral trial began. On July 25, 2019, a victory against Cacao del Peru Norte SAC—called Cacao I or Tamshi

Figure 21.1 Deforestation in Loreto. (Picture provided by Lucila Pautrat, Kene Instituto de Estudios Forestales y Ambientales, Tamshiyacu, 2014).

SAC— materialized. A judge found it guilty of illegal logging in the district of Tamshiyacu, Loreto. The Public Prosecutor's Office requested approximately US$4,223,605 of civil reparations and 15 years in jail for managers and field staff. This case continues. In addition, there is another pending case in the Court of Loreto, Cacao II, for the deforestation of another 500 hectares of forest (Poder Judicial 2019). Furthermore, on January 4, 2020, the Office of the Comptroller General of the Republic determined that government officials, Loreto's Governor and the Regional Environmental Authority (ARA), failed to fulfill their functions of environmental oversight "to agricultural operations of high intensity." This inaction led to the deforestation of primary forests and the loss of the Forest Heritage of the Nation (Dammert (2017). In 2020, Organismo de Evaluacion y Fiscalizacion Ambiental (OEFA) resolved to impose a fine of US$35,763,792 to the company Tamshi S.A.C. for not having approved environmental management instruments. However, as corruption goes to the very top, the Ministry of Agriculture has prepared a proposal to modify the Regulation of Environmental Management of the Agriculture and Irrigation Sector by opening a two-year period so that companies with operations in the Amazon—such as Cacao l or Tamshi SAC prosecuted for illegal logging—can adapt their environmental certifications without stopping their commercial activities. This scheme opens another light of legitimation of illegal acts (Castro 2020).

In brief, capitalism causes poverty by disintegrating the culture of subsistence. This section has exhibited biopiracy in the assault of social and biological reproduction of communities by cacao production in land use changes, and connected impoverished nature and labor into international power relations. The next section discusses forest finance. The Kyoto Protocol (1997) positioned the idea that corporations and industrial states can continue polluting and achieve emissions reductions through buying forestry certifications that attest to the claim that carbon has been absorbed in the forest in question. Kyoto has not reduced greenhouse gas emissions, instead it granted capital total victory in advancing a market-based approach to climate change.

Forest Finance Capitalism—Carbon Stock: The Norway–Germany REDD+

A key concept in the carbon credit market is the Payment for Environmental Services (PES) which is a voluntary transaction where a buyer from the industrial world pays a supplier for a well-defined environmental service, such as a patch of forest or a form of land use (Fatheuer 2014).

By 2007 PES evolved into the Reduction of Emissions from Deforestation and Forest Degradation (REDD). Between the 2012 Earth Summit and the 2015 United Nations Framework Convention on Climate Change sanctioned a global market in carbon dioxide and supported the creation of new financial support programs to manage the forest as an environmental service, such as REDD and REDD+ (focused on the forest of indebted countries), and the European Emissions Trading Systems (ETS).

In Loreto, one of the several REDD+ programs/agreements is the Norway and Germany Joint Declaration of Intent to reduce deforestation and greenhouse gas emissions. The Norwegian government committed to contribute NOK 300 million (2015–2017), and NOK1,500 million for the reduction of emissions verified during the period 2017–2020. (Espinoza and Feather 2018)

The Peruvian national and regional governments have committed to the following:

- Phase I: privatizing Indigenous Peoples' territories for the implementation of REDD+;
- Phase II: communities' transformation for the implementation of activities such as microenterprises, based on biodiversity, in order to receive payment; and,
- Phase III: decentralizing forest and wildlife offices for the reception of donation or payment for verified reduction of emissions.

The Norway and Germany program is administered by the WWF Inc., which supports the Programa Nacional de Conservación de Bosques (the Forest Program) of Peru's Ministerio del Ambiente (MINAM) to finance the individual titling of Indigenous lands. For this purpose, Dirección de Saneamiento Físico Legal de la Propiedad Agraria (DISAFILPA) was created to title 43 indigenous communities. This organism intended to privatize one million hectares by December 2016 (La Region, June 8, 2016). However, due to conflicts, by 2018 only one community has been titled and another 17 are in process.

I argue that REDD+ is a new biopiracy program organizing economic growth in Loreto. Federici observes that with the advent of capitalism, reproductive labor is placed at the service of an international system of accumulation. In this sexual division of labor, women and those bodies that are feminized (nature, peasants and Indigenous) were forced into a state of reproductive labor—used as a resource, devalued, dependent, and deviant.

The reproductive labor of nature is used as a resource

The United Nations Conference on Climate Change developed numerous Payment for Environmental Services (PES) for carbon capture. Countries or industries that manage to reduce carbon emissions to levels below their designated amount would be able to sell their credits to other countries or industries that exceed their emission levels. To achieve this goal, the UN scheme offers financial incentives to "developing" countries to protect their tropical forests. ONU-REDD+ started in 2013.

The NGO participants argue that getting money now is a matter that cannot be postponed because a decent life is not possible without having access to this medium of exchange. To legalize publicly the Indigenous compromise with capital, in 2018, an Indigenous Economy Forum (2018) was organized by National Council of Protected Areas (CONAP) and Asociacion Interetnica de la Selva Peruana (AIDESEP), a Peruvian Indigenous Confederation that represents important sectors of the Indigenous People in Amazonia in association with the implementers of the Dedicated Specific Mechanism for Indigenous Peoples (MDE SAWETO Peru, initiative of the Forest Investment Fund—FIF) financed by the World Bank with the support of the World Wildlife Fund-Peru (WWF-P). Consequently, by 2018, an evaluation of the REDD

+ program in Peru suggests that some of the difficulties that REDD+ encounter in the Amazon are decreasing, and maintain that negotiated agreements have been reached (Espinoza and Feather 2018).

However, the Indigenous Environmental Network (2016) argues that REDD+ is the largest land grab in history and uses Nature as a sponge for greenhouse gas pollution instead of cutting emissions at source. Therefore, those who are truly responsible for the crisis remain untouched.

Further, it is problematic because:

> the users of the land (Indigenous people) have to describe their activities as a threat to the forest… Without such a story—that the forest would have been destroyed—there is no carbon to be saved, and no carbon credits to be sold…
>
> (Kill 2014, 10)

The reproductive labor of Indigenous People is devalued

As global capital and governments agreed to privatize communities' land, disputes became more common among Indigenous communities. In June 2016, the Organization of Indigenous Peoples of the East (ORPIO) denounced that conservation categories, brought by the agreement, reduce the collective rights of Indigenous Peoples. On the same day, the opposition to individual titling, from Indigenous communities such as Ticunas and Yaguas communities of the lower Amazon (FECONATIYA) began (La Region 2016).

Land privatization comes with a conservation incentive mechanism aimed at Indigenous communities. To this end, the Regional Environmental Authority (ARA) has been organized to promote micro-enterprise projects in exchange for 'caring for the forests.'

The Ministry of Economy and Finance (office in Lima) gives the ARA USD $ 7.16 for each hectare that enters the program, of which $ 4.29 will be spent on administration and $ 2.86 per hectare will go to the coffers of the communities for projects of micro-companies. At that time, micro-companies will open an account at Banco de la Nación and register with SUNAT to pay taxes.

In 2020, USAID, WWF, AIDESEP, and NESsT launched the Indigenous Amazon Call: Rights and Resources, offering between US$10,000 and US$40,000 to communities interested in starting an indigenous enterprise.

The reproductive labor of nature and Indigenous People drops to dependence

In March 2017, AIDESEP sent a letter to the Representatives in Assembly of the Forest Carbon Partnership Facility (FCPF), stating its disagreement with the REDD + program and demanding changes (AIDESEP 2017). However, by 2018, AIDESEP had rewritten its demands by creating a perspective that's been called REDD+ Indígena Amazónico (RIA) de Vida Plena as a socially acceptable alternative. It aims to satisfy needs by REDD+ that translates into a higher quality of life for indigenous peoples. (Espinoza 2018) For instance, in Loreto, AIDESEP accepted microenterprises, and solicited recognition and title for 844 native communities. Further, the AIDESEP pact to consolidate 12 territories and communities land with four million hectares and the collective rights of Indigenous Peoples (Espinoza & Feather 2018).

The reproductive labor of nature and Indigenous people is reduced to deviance

AIDESEP and organizations involved in REDD+, as agents of capital, have created a concept called *Articulation with Identity*, meaning that some principles of Indigenous culture are articulated in the agreement (Espinoza 2018). However, this perspective conflicts on two fronts.

On one hand, in a capitalist society there is a legitimate desire to have a better income to improve the quality of life; on the other hand, it means reviewing, modifying, or even eliminating some of the characteristics of Indigenous Peoples, culture, and livelihood. Then, if these organizations want to articulate Indigenous Peoples to the global economy, it would have to transform biodiversity through agro-exportation because that is what brings profit. But this scheme collides with the recognition that forest loss is a key factor in climate change. Furthermore, the NGOs' ideas that to better articulate Indigenous People to market the "softness" of the concept of territory was needed, infer the possibility that other figures can be incorporated into the forest, such as private property, modalities of rental of communal lands, or invitation to third parties for use or exploitation of resources (Arce 2018). Since the territory constitutes the most important social and cultural asset for Indigenous Peoples, if territories are rented, REDD+ become REDD+-based offsets, as such, they do not represent emission reduction; instead, they represent another, ostensibly lower-cost means by which firms can meet their emissions quota. An example already in Amazonia is regenera.pe, which compensates for the footprint of individuals and corporations.

As a result, REDD+ projects and programs may lead to more forests and territories of Indigenous Amazonia communities being subjected to control and monitoring of a community's land use by outsiders. Therefore, this "greening" of Indigenous territories understood as a process of capital accumulation is forming a class of land proprietors turning some Indigenous People into a commercial venture and, at the same time, with the fencing of the communal land, the formation of a population of beggars and vagabonds in the rainforest will begin.

In this chapter, I have exposed REDD+ programs as biopiracy that has placed the hands of financial capital into Indigenous culture, biodiversity, territories, and sold out the fate of Peruvian Amazon forest and its inhabitants.

Conclusion

By creating the conditions for biopiracy, global capital has launched an international organized crime where communities and nature are assaulted and diminished. The first and second sections of this chapter have exposed processes of patriarchy and colonialism by which Sustainable Development is involved in social and biological assaults, allowing the destruction of thousands of km^2 of primary forest by wood and cacao productions. While the third section on REDD+ has exposed the damage of capitalism, by land privatization, the wrecking of sustainable and livable economies, and the suppression of Indigenous rights and the rights of nature. REDD+ may increase the monetary transactions while at the same time perpetuating historical injustices, violence, and discrimination, while they dispossess hundreds of Indigenous life systems. However, it is also relevant to emphasize that measuring the emission-absorption of carbon gasses is not really possible, since forests are living organisms that breathe, are dynamic and complex systems, so their measurements are always estimates.

References

AIDESEP - "Carta No. 086-2017." Accessed on June 15, 2020 http://www.forestpeoples.org/sites/default/files/documents/Carta%20N°%20086-2017.pdf.

Alvarez, Jose. 2005. "Adios a la Caoba." *Revista Kanatari*, Iquitos, April 17, 2005.

Alvarez, Jose. 2011. "Perú: Comunidades, territorio y concesiones forestales." Accessed on November 12, 2011. https://www.servindi.org/actualidad/51918.

Arce, Rodrigo. 2018. "Reflexiones a partir del Foro de Economía indígena (FEI)." Accessed on December 15, 2018. https://www.servindi.org/actualidad-informe-especial/20/11/2018/aportes-la-conceptualización-sobre-la-economia-indigena.

Cabral, Lorena. 2012. "Agenda Feminista y Agenda Indígena: Puentes y Desafíos." In *Mujeres en Diálogo: Avanzando hacia la Despatriarcalización en Bolivia* by Coordinadora de la Mujer, 53–61. La Paz: Editora Presencia SRL.

Castro, Aramis. 2020. "Propuesta del Minagri permitira que empresas investigadas por tala obtengan permisos ambientales." Accessed on October 31, 2020. https://ojo-publico.com/2208/propuesta-del-minagri-beneficia-empresas-investigadas-por-tala?fbclid=IwAR1wv1mnfPfPkbwWXmEiPEEDserGi_VxUQ1fLXg0DA0F13qeHg2-8MD1UZ4.

Chirif, Alberto. 2016a. "Deforestación en Tamshiyacu: En qué país vivimos." Accessed on April 23, 2016. https://www.servindi.org/actualidad-noticias/23/04/2016/deforestacion-en-tamshiyacu-en-que-pais-vivimos.

Chirif, Alberto. 2016b. "Caso de Tamshiyacu: El Estado como subsidiario de la empresa." Accessed on May 27, 2016. https://www.servindi.org/actualidad-noticias/27/05/2016/caso-de-tamshiyacu-el-estado-como-subsidiario-de-la-empresa.

Dammert, Juan Luis. 2017. "Acaparamiento de Tierras en la Amazonia Peruana. El caso de Tamshiyacu." *Wildlife Conservation Society* (2017): 17.

EIA. 2016. "United-cacao linked companies ordered to stop operations by Peruvian authorities." Accessed on October 16, 2016. https://eia-global.org/blog-posts/united-cacao-linked-companies-ordered-to-stop-operations-by-peruvian-author.

EIA. 2018. "Moment of Truth: Promise or Peril for the Amazon as Peru Confronts its Illegal Timber Trade." Accessed on February 7, 2020. https://eia-global.org/reports/momentoftruth.

Espinoza, Roberto. "REDD+ Indígena Amazónico (RIA / Indigenous REDD+) Progress and challenges". "UNREDDY: A critical look at REDD+ and indigenous strategies REDDY for comprehensive forest protection". Accessed on November 18, 2018. https://www.climatealliance.org/fileadmin/Inhalte/7_Downloads/Unreddy_EN_2016-02.pdf.

Espinoza, Roberto, and Conrad Feather. 2018. "*Carrera de Resistencia, No de VelocidadEl papel de fondos climáticos internacionales en la resolución de derechos territoriales de los pueblos indígenas de Perú; Avances, retrocesos y desafíos.* ." Forest Peoples Programme and AIDESEP.

Fatheuer, Thomas. 2014. *Nueva Economía de la Naturaleza. Una Introducción Crítica.* Berlin: Heinrich Böll Foundation. Print.

Federici, Silvia. 2004. *Caliban and the Witch: Women, the Body and Primitive Accumulation.* Brooklyn: Automedia.

Global Witness. 2019. "Exportadores peruanos al descubierto. La evidencia en video." Accessed on November 19, 2019. https://www.globalwitness.org/en/campaigns/forests/exportadores-peruanos-al-descubierto-la-evidencia-en-vídeo.

Hardin, Garret. 1968. "The Tragedy of the Commons." *New Series* 162(3859), 1243–1248 (6 pages).

Indigenous Environmental Network. "Carbon Offsets cause Conflict and Colonialism." Accessed on December 15, 2016. https://www.ienearth.org/carbon-offsets-cause-conflict-and-colonialism.

Isla, Ana. 2015. *The "Greening" of Costa Rica: Women, Peasants, Indigenous People and the Remaking of Nature.* Toronto: University of Toronto Press.

Isla, Ana. 2009. "The eco-class-race struggles in the Peruvian Amazon basin: An ecofeminist perspective". *Capitalism Nature Socialism* 20(3), 21–48.

Kill, Juta. 2014. *REDD: Una coleccion de conflictos, contradicciones y mentiras.* Montevideo, Uruguay: World Rainforest Movement. Available at: https://www.wrm.org.uy/es/publicaciones/redd-una-coleccion-de-conflictos-contradicciones-y-mentiras#:~:text=REDD%3A%20una%20colecci%C3%B3n%20de%20conflictos%2C%20contradicciones%20y%20mentiras,los%20bosques%20o%20no%20abordan%20verdaderamente%20la%20deforestaci%C3%B3n.

Kill, Jutta. 2019. Bulletin 245 "REDD+: A Scheme Rotten at the Core". Accessed on January 15, 2020. https://wrm.org.uy/articles-from-the-wrm-bulletin/section1/redd-a-scheme-rotten-at-the-core/.

Kyoto Protocol. 1997. https://unfccc.int/kyoto_protocol.

MAAP#123. "Identificando Tala Ilegal en la Amazonia Peruana." Accessed on August 17, 2020. https://maaproject.org/2020/tala-ilegal/.

McNeely, Jeffrey. 1997. "Achieving Financial Sustainability in Biodiversity Conservation Programs." In *Investing in Biodiversity Conservation. Proceedings of a Workshop. 1997*, Washington, DC No. ENV-111.

Merchant, Carolyn. 1983. *The death of nature: Women, ecology and the scientific revolution.* Harper and Row: San Francisco.

Merchant, Carolyn. 2005. *Radical ecology.* Routledge: New York and London.

Mies, Maria. 1986. *Patriarchy and capital accumulation on a world scale.* London and New York: Zed Books.

Pearce, David W., and Jeremy J. Warford. 1993. *World without end: Economics, environment, and sustainable development.* New York: Oxford University Press.

Poder Judicial. Sentencia+Integral+Tamshiyacu. Accessed on December 16, 2019. (www.pj.gob.pe/wps/wcm/connect/1d9f9d004aefb0ada20ae69507b119bf/SENTENCIA+INTEGRAL+TAMSHYYACU.pdf.

SERFOR. Unidades Ofertables para Otorgamiento de Concesiones Forestales Maderables Mediante Procedimiento Abreviado Departamento de Loreto. Accessed on December 17, 2017. https://www.serfor.gob.pe/wp-content/uploads/2017/06/UNIDADES-OFERTABLES-LORETO.pdf.

Thompson, E. P. 1991. *The Making of the English Working Class.* Penguin, UK.

Tala de la selva - ¿Corrupción en los sellos ambientales? | DW Documental (n.d.) https://www.youtube.com/watch?v=6_qV9Mnz83s.

Urunaga, Julia. 2019. "Fighting illegal logging in Peru: The Government Steps Forward and the Industry Pushback." In *Deforestation in times of climate change,* edited by Alberto Chirif, 71–90. New York.

22 Territorialization through the Milpa

Zapatismo and Indigenous autonomy

Mariana Mora

Figure 22.1 Harvesting corn in the milpa. Zapatista autonomous municipality. November 17 2006. Photo credit: Mariana Mora.

Introduction

This chapter focuses on territorial struggles as part of Indigenous mobilizations. Rather than an inherent relationship to place, reconfigurations of territory emerge dynamically against broader colonial structures of power, expressed through cycles of dispossession, militarized occupation, sexual violence, and forced labor. I focus on such territorial reconfigurations as part of the daily practices of Tseltal, Tojolabal, Tsotsil, and Ch´ol Mayan communities that since 1996 form part of autonomous municipalities, under the protection of the Zapatista National Liberation Army (or EZLN, to use its Spanish acronym) in the state of Chiapas, Mexico. An analysis of the processes of territorial claims is central for understanding the struggles of many native peoples throughout Abya Yala,[1] who, as in the case of Zapatista Mayan communities, are the survivors of attempts to displace their ancestors from their lands and undermine native knowledges, ways of being, and ways of governing. It also enables methodological insights and critical reflections for politically-committed social science research aligned with these struggles. In this chapter, I address both elements.

DOI: 10.4324/9780429344428-26

Almost three decades have passed since Tseltal, Tsotsil, Tojolabal, and Ch'ol Indigenous Mayan peasants declared war against the Mexican government on January 1, 1994 and announced that the point of departure for profound anti-capitalist social transformation is redefining the terms of engagement between Indigenous peoples and the state, along with *mestizo* or culturally-mixed sectors of society. In that sense, and in contrast to other Latin American political-military organizations of decades past, who prioritized first and foremost economic redistribution as part of a narrow definition of class struggle (Saldaña-Portillo 2003), the first round of peace talks with the Mexican state in 1996 centered on Indigenous rights and culture. Implicit in the rebel army's demands was an understanding of the Mexican state as an entity founded on colonial structures of power that operate through what Patrick Wolfe refers to as a logic of elimination directed at Native populations (Wolfe 2006). Wolfe defines settler colonial states as those that continue to operate through racialized technologies and whose effects involve the systematic erasure of Indigenous peoples, their memories and histories, ontologies, and territories. Racialized technologies are social constructions that mark cultural attributes as inherent to specific populations, as if these were biological traits. Such technologies of power can involve apparently "positive" policies, such as those of assimilation and integration, as well as those founded on violence, including genocide (Wolfe 2006, 388).

The first round of Peace Accords of San Andrés Sakamchen de los Pobres, signed on February 16, 1996, recognized the rights to territory, self-determination, and autonomy as a central axis through which to begin to unravel colonial structures of power in Mexico. At the same time, as a rebel army wary of state-led reforms and suspicious of relying exclusively on state recognition, the EZLN published a communiqué at the end of December 1994 that declared their territory would be politically administered through self-government councils, as part of 38 Indigenous Zapatista autonomous municipalities (EZLN 1994) that are separate from state institutions.[2] This priority would unfold regardless of state legal reforms. The rebel army's suspicions proved well founded, as the subsequent Constitutional Reforms of 2001 recognized only highly diluted versions of Indigenous peoples' central demands regarding collective rights to autonomy, territory, and self-determination from that first round of peace accords (Gómez 2004).

The daily exercise of autonomy at the margins of the state thus became the means through which to reconstitute Indigenous Mayan territories and engender processes of re/territorialization. By territorialization, I refer to dynamic and historically-grounded collective daily practices that establish dense webs of socio-natural relationships between humans and non-human entities across particular geographies. In this sense, re/territorialization emerges from collective claims against the forces of dispossession or displacement that form part of broader structures of power. Territorial claims are thus embedded within re/territorializing struggles.

The first part of this chapter seeks to historically situate the structures of power against which Zapatista communities claim autonomy. The second section focuses on the *milpa* or cornfield, as a central locus through which broader autonomous processes of re/territorialization emanate. The third section delves into the methodological challenges of such an analysis, particularly when constricted by Indigenous rights juridical frameworks. Lastly, in the conclusions, I provide reflections for social science research supporting such claims.

Zapatista Mayan reterritorializing struggles against colonial structures of power

The practices of Zapatista Indigenous autonomy complicate legal definitions of Indigenous collective rights, as established through international treaties. The International Labor Organization (ILO) Article 169 states that the recognition of collective rights are associated with the definition of Indigenous peoples as, "peoples in independent countries who are regarded as Indigenous on account of their descent from the populations which inhabited the country, or a geographical region to which the country belongs, at the time of conquest or colonisation or the establishment of present state boundaries and who, irrespective of their legal status, retain some or all of their own social, economic, cultural and political institutions" (ILO 1989). The recognition of collective rights hinges in large part on demonstrating an inherent and culturally specific relationship to land since times before the conquest and on proving such ties in part through the continual permanence of native socio-political and spiritual forms of organization. However, only a minority of Indigenous peoples in Abya Yala can effectively demonstrate this permanence in the strict terms defined by legal frameworks (Kirsch 2018; Loperena, Hernández Castillo, and Mora 2018). This is in large part the result of centuries of forced displacement, land dispossession, and the imposition of education, land, and agrarian policies that reconfigure cultural practices, as well as political administrative reforms that undermine pluri-legal regimes. The resulting challenges lead us to critically examine claims to autonomy and struggles for territory by those peoples whose histories as survivors of conquest place them on the fringes of international legal definitions.

Such is the case of many Tseltal, Tojolabal, Tsotsil, and Ch'ol Mayan peoples in the state of Chiapas, whose territories underwent profound transformations since the beginning of the Spanish colonial era, through the *encomienda* system, the legal and illegal purchasing of their lands, and the enslavement of community members. The beginning of conquest in what is now the state of Chiapas happened early in the 16th century, starting in 1528. By 1542, Fray Bartolomé de las Casas denounced the enslavement of more than half a million Indigenous peoples (Bonaccorssi 1990). Within the colonial administrative units referred to as the Indian Republics, the Catholic Church-controlled *encomiendas* not only imposed the well-documented tributary systems but also began cultivating new crops, such as sugar and wheat.

The taxes that native communities were required to transfer to the Spanish Crown far exceeded the excess crops harvested, resulting in a profound transformation of community life so as to meet these obligations (Bonaccorssi 1990). The new crops required lands that were "voluntarily sold" by native populations to the Spanish. According to historical archives, these "parcels fractured ecosystems and transformed the limits of native communities" and many would later become the base for agricultural estates (ibid., 28). Thus, from the beginning of colonization, native Mayan communities in Chiapas underwent profound changes that directly impacted their socio-political institutions and the organization of daily life. In response, these communities staged various rebellions—one of the most well-documented being the "Indian rebellion" of Cancuc in 1712, led by María de la Candelaria, a Tseltal woman from the Zendal Province (Viqueira and Ruz 2004).

These rebellions against tributary obligations, territorial dispossession, and forced labor, as well as subsequent violent repression from the Spanish colonizers, continued through different historical periods, including during the time of the *fincas* (estates), a key chapter in Chiapas history. Beginning in the mid-1800s until their decline in the 1970s, the estate economies, based on the production of coffee, sugar, cacao, and other commodities, flourished throughout the state because of world-colonial market interests and conditions that facilitated the dispossession of Indigenous lands and the imposition of indentured servitude (Rus 2012; Gómez and Ruz 1992).

In other regions of the country, the 1910 Mexican Revolution's agrarian reforms—that centered on distributing the Indigenous and *mestizo* peasants *ejidos* or communal land holdings—dismantled the political economic control of the estates. Yet a series of loopholes in Chiapas' state juridical frameworks effectively prevented substantial land distribution (Reyes Ramos 1992). A significant percentage of estate owners relinquished only the poorest of their lands – those not suited for agriculture, such as areas on the top of hillsides and on steep mountainsides – while maintaining control of the richest lands, oftentimes surrounding river valleys. The local Tseltal, Tojolabal, Ch'ol, and Tsotsil populations no longer had to endure forced labor but, given that oftentimes their *ejidos* were founded on the edges of what remained of the estates and given that the poor soil did not allow for basic food security, most families had little choice but to continue working for estate owners (Bobrow-Strain 2007).

Despite this severe limitation, during most of the 20th century the *ejido* was, in essence, the only legal recourse that Tseltal, Tsotsil, Tojolabal, and Ch'ol communities had to reclaim pockets of their ancestral territories in Chiapas. Social memories that circulate in Zapatista communities attest to the relevance of this partial disruption of territorial dispossession. The testimonies of those elders who traveled for days on foot to the state capital of Tuxtla Gutierrez to successfully negotiate *ejido* lands by way of the Agrarian Reform in the 1930s and 1940s continue to be passed down to new generations and, as we shall see further below, oftentimes served as central elements to substantiate claims to particular communal lands. In 1992, the Constitution's Article 27, which recognized these *ejido* lands, underwent reforms that resulted in the privatization of communal land holdings and neoliberalization of agrarian policies.

In the valleys of the Lacandon jungle, as well as in the northern zone, the most immediate local result of the 1994 Zapatista uprising was the takeover of the remaining estate lands. In total, it is estimated that up to 55,000 hectares were "recovered" by EZLN sympathizers (Villafuerte Solís 1999). Such land recuperations disrupted local racialized geographies (Mora 2017) and opened the possibility for the reconstitution of Indigenous territory, beyond the patchwork of *ejidos* and networks of *ejidos* that emerged during the 1980s (Harvey 2000). The central task of Zapatista Indigenous autonomy thus became to re-establish a collective sense of belonging, within and across Indigenous communities, including redefining inter-ethnic relationships between, for example, Tseltal and Tojolabal communities or between Ch'ol and Tseltal populations, who inhabit the same regions. This task was undertaken through the founding of multi-scale autonomous governing bodies that since 1996 connect community-level decision-making with inter-community assemblies at the level of the Zapatista autonomous municipalities and, lastly, the inter-municipal level, known as *Caracol* regions. *Caracol* or snail was the symbol used in glyphs of the classic Mayan period to refer to the act of speaking (Aubry 2003). Thus, the *Caracoles* embody geographies of dialogue and exchanges that not only occur in a specific territory but also redefine sets of human and non–human relations that *create* territory.

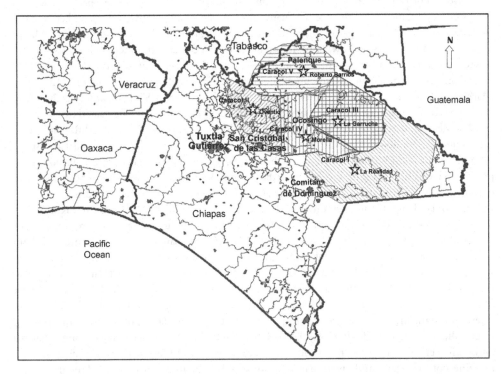

Figure 22.2 Map of Chiapas with the location of the five Zapatista regions, known as Caracoles. In 2019, eleven additional Zapatista municipalities were added to the 38 original municipalities that in 2003 were divided into these 5 Caracoles. Map designed by Edgars Martínez.

As a testament to the political imaginary emerging through the reterritorialization of Mayan ancestral lands through autonomy, Zapatista communities refer to the rebel political-administrative region now known as *Caracol IV* as the Tzotz Choj region, which joins parts of the official municipalities of Altamirano, Huixtán, Oxchuc, Cuxlujá, and Ocosingo. *Tzotz Choj*, which means the jaguar bat in Tseltal, is a figure that represents the dualities of the cosmos and the four cardinal directions.[3] Tzotz Choj was also the name given to the last governor of Toniná, one of the most important Mayan cities during the Classic Mayan era (200–900 AD). As an expression of what it means to recover the ancestry of the Toniná region against the most visible and visceral manifestations of the occupation of Mayan lands, several thousand Tseltal and Tojolabal Zapatista community members staged a protest in front of the 39th Mexican military zone and then symbolically reclaimed the nearby archaeological site of Toniná, as part of their history and territory (Balboa 1998). It is in this sense that I refer to re/territorialization as that which is both an expression and an effect of Indigenous struggles, in this case against the militarized, settler-colonial occupation of lands.

Re/territorialization through the milpa

The embodied daily practices that engender territory take place within colonial structures of power and thus tend to emerge against material manifestations of oppression. In the Wallmapu in Mapuche territory, for example, the struggle for *Weichan*—that is to say, the struggle over territory as described in Mapudungun, the Mapuche language—oftentimes is against the German descendants and other Chileans of European heritage and the agroforestry industry (Marimán Quemenado 2017; Pairican 2011). In other countries, such as Guatemala and Ecuador, struggles that in the twentieth century were largely against agricultural estates, such as coffee plantations, have now shifted against the incursion of mining and other extractivist economies (Cervone 2012; Hooker 2020; Macleod 2017). Yet, in other regions, Indigenous struggles for territory and against dispossession are situated against (il)legal markets and para/state armed forces, including regions of Colombia (Weitzner 2021).

In the case of Zapatista communities, the practices of autonomy emerge in antagonism to what in Ch'ol is known as the times of the *mosojatel*, the times of the agricultural estates. While these estates no longer exist as an economic institution, their political effects and racialized cultural projects persist as central elements that continue to act as the motor behind current local forms of Mexican state formation (Mora 2017). In that sense, the times of the *mosojatel* provide an analysis of current structures of power that condense past and current colonial forces of rule. By colonial forces of rule, I refer to the occupation of a people's land by external forces, the extraction of the life force of that land and its peoples for the benefit of others, the systematic negation of a people's ways of being, and racialized labor exploitation. In Zapatista public events, EZLN communiqués (such as the one published in July 2018, following the presidential elections), and during familial or community conversations, the times of the estates is evoked repeatedly, as a moment that reflects the logics of elimination of the Mexican state. Such an analysis is encapsulated, for instance, through such classifications of the Mexican state as the *gobierno capataz* (the "foreman government"), in reference to those who supervised workers on the estates (Desinformémonos 2018; EZLN 2018). The concept is also expressed in public events, such as the International Women's Gathering that was held in Caracol IV in March of 2018, where Zapatista Tseltal women referred to patriarchal structures that operate through the inferiorization of Indigenous men and women in relation to *mestizos* and white Europeans. They argued that Indigenous women's subordination gravitates around the figure of the *patron-marido*—that is, the *mestizo* estate owner-Indigenous husband (Mora 2021).

Against colonial structures of rule, daily practices of Zapatista autonomy anchor re/territorialization struggles that render possible conditions for *lekil kuxlejal*, what in Tseltal is referred to as a dignified collective life-existence amongst humans and between humans and non-human entities. *Lekil kuxlejal* emerges from *p'ijil o'tanil or* the wisdom-heart (feeling-knowledge) that all beings possess in order to exist in the world (Pérez Moreno 2021). *P'ijil o'tanil* is a Tseltal concept that is similar to other Indigenous concepts in various regions of Latin America that refer to *sentipensar* or feel-thinking (in contrast to the analytical versus emotional divisions of Westernized knowledges). While re/territorialization extends into multiple realms of community life, one of its central anchors is the socio-natural relationship to *maíz* (corn).

The central role of *maíz* or *ixim*, as it is named in Tseltal, as part of political struggle and as that which grants meaning to Indigenous autonomy is reflected in the embroidered cloths made by Zapatista men and women in the *Caracoles* for regional artisan markets. In these clothes, one often finds in a top corner the red star that symbolizes the EZLN. In the center of the fabric are corn stalks sprouting from the soil, the leaves partially stripped from the husks so as to reveal dark brown faces hidden behind ski masks. Other similar cloths and painted murals also depict corn husks, whose kernels are in fact the masked faces of Zapatista community members. Through such images, Tseltal, Tsotsil, Tojolabal, and Ch'ol Zapatista men and women affirm that they are people of the corn, in reference to the Popul Vuh, the sacred book of the Mayans that describes how the first four real humans (*Balam Quitzé, Balam Akab, Mahucutah,* and *Iqui Balam*) emerged from *maíz* after failed attempts to create humans with clay and wood. While Indigenous and *mestizo* peasant communities in Chiapas and throughout much of Mesoamerica depend on corn for subsistence and derive a sense of territory through the milpa, the Zapatistas have highlighted this relationship by insisting on its central role as part of practices of autonomy that maintain colonial structures of power at bay. The embroidered clothes are thus an artistic expression of a broader political affirmation—Zapatista local rebellion grows from the corn and the corn is sown in rebel territory.

REIVINDICAN TZELTALES Y TOJOLABALES LAS RUINAS DE TONINA ❂

Unos 2 mil manifestantes declararon a este sitio arqueológico
Patrimonio de los Pueblos Indígenas. Más tarde ocuparon durante 53
minutos el nuevo cuartel de la 39 Zona Militar, en demanda de la
salida del Ejército de sus comunidades Î Foto: **Guillermo Sologuren** ❂

Figure 22.3 Photo taken of the printed image published in the Mexican newspaper, La Jornada, of Tseltal Zapatista community members claiming the archeological site of Toniná as part of Mayan territory. 1998. Photo credit: La Jornada.

Autonomy, as part of Mayan Indigenous liberation struggles, is inextricably linked to *ixim*, not just as a basic staple required as part of local food security, but as that which has *ch'ulel*, the spirit energy that in English would very loosely translate into soul or heart. *Ixim* is that which forms part of the *p'ijil o'tanil* that allows all beings to exist in the world. For that reason, offerings are provided to the earth before planting. On the third day of May every year, the day of the *Ch'an Santa Cruz*, Mayan community members enact ceremonies at key water sources to

Figure 22.4 Embroidered cloth made by Zapatista artisans. Autonomous municipality. November 17, 2008. Photo credit: Mariana Mora.

ask for a healthy rain season, which will allow for a good harvest, collective health, and wellbeing for the community, the Earth, and the spirits. Offerings are similarly provided around the milpa to ask permission when harvest time arrives. *Ixim* is the primary source of nourishment for community members, which thus requires a particular type of relationship to the earth and to soil nutrients. It is important to recall that *maíz* as a wild staple was domesticated through the course of thousands of years by Mesoamerican peoples to the point that corn is no longer able to reproduce on its own, but rather depends on human agricultural practices. The cultivation of land through the milpa or cornfield emerges in a complicated ecosystem that combines numerous edible plants, such as beans, corn, and squash. The resulting symbiotic and mutually constitutive relationship results from historical interactions and, in the case of Zapatista Indigenous autonomy and other Indigenous struggles in Mesoamerica, this relationship has become a subaltern expression of the political.

Re/territorialization thus passes through the milpa. Given the *locus* of *ixim* in community life and as part of political struggle, it comes as no surprise that the first tasks of the autonomous government entailed defining the terms for the implementation of a Zapatista agrarian reform, including defining land tenure and agreements as to how to cultivate the land. More than a century of the estate economy, which, in the case of the *Tzotz Choj* region, depended primarily on cattle grazing, left the soil severely depleted and virtually barren. In fact, it is estimated that

73% of the Lacandon has been deforested, as many of these lands had been turned in to cattle pastures. For example, in the autonomous municipality of *17 de noviembre,* in the *Tzotz Choj* region, lands "recovered" after 1994 were so eroded that in the first years it was impossible to grow anything. Zapatista community members recalled that not only did the land not produce, but it was full of ticks that had previously lived among the cattle and now fed on anyone who walked by.

Redefining a connection to the earth first and foremost involved replenishing the soil of its nutrients, a mutual healing of land and community that required various agricultural cycles. At the same time, it required reinventing agricultural techniques, not only based on what the elders narrated through memories passed down through generations, but also what was to be learned from agroecological exchanges with peasant organizations, such as the *Vía Campesina,* an international network of farmers founded in 1993. Some of the first decisions of the autonomous municipal assemblies thus entailed prohibiting chemical pesticides and fertilizers and the cutting down of healthy trees, allowing the forest to rejuvenate. In addition, they created collective milpas, whose harvest would be of communal benefit.

Such practices were implemented along with the definition of autonomous land tenure. Rather than defining the terms for communal property, whereby the sum of families would be recognized as acquiring rights to land, the relationship to land was established through the acts of working the land, taking care of the soil, and participating in the cultivation of family milpas, community cornfields, and other communal agricultural spheres, such as collective coffee and banana fields. The relationship between community and nature thus does not pass-through property titles (whether in formal state terms or a title recognized by the autonomous government), but rather involves a reinscription of "the commons" (Federici 2018). Feminist Marxist Silvia Federici refers to the commons not as a past that can be excavated underneath the estates and other colonial projects, but as the reconfiguration of social memories woven into well-established communal practices and deep relations between human and non-human others.

It is in these autonomous agrarian practices and land tenure decisions centered on the milpa where the rest of the spheres of autonomy take root. The autonomous health commissions, for example, are inextricably linked to meanings of health symbolized by a diverse and nutrient-rich cornfield. The curriculum of the autonomous primary schools begins with the knowledge that is found in the milpa—what anthropologist María Bertely referred to as cornfield pedagogies (Bertely Busquets 2019). Rather than an exception, such processes of re/territorialization find echo and expressions in other claims to autonomy and self-determination throughout Latin America. In Mexico, we find various examples. In the Purépecha community of Cherán, in the state of Michoacán, the 2011 uprising against the destruction of pine forests and sacred springs by narcotraffickers, who were protected by local political parties, resulted in practices engendering territorialization that involved reforestry initiatives alongside the elaboration of local educational curriculum based on knowledges found in the forest. In the Mixe and Chatino regions of the state of Oaxaca, Yásnaya Aguilar and Emiliana Cruz engage in collaborative research in their communities that facilitates reclaiming territory, using novel pedagogical practices centered on native languages and struggle against linguistic racism (Aguilar Gil 2020; Cruz 2018). And, as we shall see in the next section, other Indigenous communities partake in similar struggles by taking their claims to state legal tribunals.

The challenges of establishing claims through legal means

Up to this point in the chapter, I have focused on the everyday practices of autonomy that form part of re/territorialization struggles at the margins of the state. Rather than demonstrating permanent native socio-political institutions and historical ties to specific lands, the emphasis has been on struggle through territorialization that re-establishes a sense of territory, with the family and community milpas acting as central anchor points. These types of daily practices and political claims, as powerful as they may be, come up against a series of challenges when they enter the judicial terrain that, as we saw at the beginning of the first section, rest upon much more narrow definitions of indigeneity, that substantiate collective rights. What happens when such calls to territorialization require verdicts in state tribunals? With what methodological challenges are anthropologists that participate as legal experts in such tribunals confronted?

In this section, I reflect on these questions through my participation as an expert witness in a land dispute case involving the Ch'ol Indigenous *ejido* of Tila, situated in the northern region of the state of Chiapas. The case reached the Supreme Court in 2010, with a verdict emitted in 2019. It involved over half a century of disputes over what seems to be a rather insignificant amount of land—i.e. 130 hectares of the more than 5,400 hectares that make up the *ejido*. The dispute began in 1966, with the expropriation of those 130 hectares to establish the political-administrative center of the local municipal government. Justifications of expropriation have included both legal and illegal means, including the alteration of the *ejido* communal property title (*plano ejidal*) in 1971 by a lawyer who argued that these lands never belonged to the *ejido*.

The disputed lands are located at the heart of the *ejido*, where the town of Tila is located, a settlement that dates back to the Mayan Post-Classic Era (900–1200 AD). This location houses the church of the Señor de Tila, a black Christ figure revered by Indigenous and *mestizo* peasant populations throughout southeast Mexico. Within those 130 hectares is where the municipal government offices are located. The local political administrative power has been largely under the rule of a small group of landowning *mestizo* families in alliance with a sector of Ch´ol and Tseltal *ejidatarios*, that is to say, members of the *ejido* Tila, and members from nearby communities. It is this same inter-ethnic political alliance that sustained and financed counterinsurgency paramilitary groups during the height of Zapatista struggle in the late 1990s, resulting in the assassination and forced displacement of many communities that sympathized with the EZLN. The Tila *ejido* and its town reflect local political divisions, with those aligned with the *mestizo* elite largely in favor of private property, while those who sympathize with the EZLN advocate for collective land use and communal property titles. In essence, what is at stake is the potential weakening of local *mestizo* privilege in the region (Mora 2020) as part of the re/territorialization of ancestral lands by a sector of the local Ch'ol and Tseltal population.

The case had been initially classified as an agrarian case, which meant that during the four decades that it passed through the lower courts, the primary identity of the affected party was construed as peasant, rather than Indigenous. However, after exhausting all other resources, the case of the *ejido* Tila reached the Supreme Court in 2010, where the human rights organization that represents the *ejidatarios* argued that Indigenous collective rights to territory were at stake. Most of the Supreme Court judges resisted such an argument, which was presented in April 2013 by then Judge Olga Sánchez Cordero, as the central argument in favor of the restitution of the 130 hectares. During that plenary session, in which the Supreme Court judges failed to reach a verdict, the majority of the judges argued that interpreting the case as one involving collective Indigenous rights amounted to processual fraud, given that it reflected the use of a particular identity at the end of the legal battle so as to favor the outcome (Mora 2020). This line of argument would have resulted in a verdict that would require that the Tila *ejido* either receive economic compensation or be granted property of equal value elsewhere. In order to circumvent this type of decision, Sánchez Cordero requested four expert witness reports, including two anthropological cultural reports, that could provide the judges with complimentary information to emit a verdict.

My colleague, Rodrigo Gutiérrez, of the Autonomous National University of Mexico (UNAM, by its Spanish acronym), and I conducted fieldwork in early 2015 that was necessary to produce one of the cultural affidavits. We traveled to Tila and conducted focus groups with diverse members of the local population. In order to search for the empirical evidence necessary to answer questions framed within legal structures, we mapped converging and diverging connections to the land, documented the land-use patterns of family members, registered the central socio-political institutions (and their decision-making practices in the *ejido*), and identified religious-cultural practices that constituted a collective identity, as Indigenous Ch'ol peoples of that locality. We interviewed not only *ejidatarios* that were in favor and against the restitution of the 130 hectares, but also Indigenous community members who were not *ejidatarios* but had moved to Tila for employment or because they had been forcibly displaced from their villages during the 1990s counterinsurgency war against the EZLN. In addition, we interviewed members of the local *mestizo* landowning elite.

In the interviews, *ejidatarios* and other community members in favor of land restitution established strong territorial claims. These were not necessarily presented in ways that would easily fit within established Indigenous rights frameworks, such as ILO 169, or within parameters of indigeneity that Supreme Court judges described in the first plenary session held in 2013. Legal imaginaries tended to reproduce essentialized definitions that did not necessarily square with how Ch'ol community members interviewed defined the relationship between communal identity and land/nature. Rather than emphasizing ties to the land since time immemorial, those interviewed were more interested in highlighting that their grandparents had struggled for that land in the way elders of past generations had struggled against the ladinos or the *kaxlanes*, those non-Indigenous people external to the region. When the *ejido* titles were finally granted in 1934 on lands that had, for more than a century, formed part of coffee plantations on which many of their relatives had been forced to work, these titles represented a triumph against *mestizo* and German landowners in the region. Time and time again, their testimonies stressed that accepting the extraction of the 130 hectares implied betraying their ancestors. And that was something that physically and spiritually they could not allow to happen because, in the words of one *ejidatario*, their responsibility was to "follow the path of their elders and continue what they had fought for."

Such explanations do not readily translate into evidence of continual ties between a peoples and a territory or establish an inherent socio-spiritual connection to the earth. This placed us as expert witnesses in a conundrum. We wondered if we should continue to seek in the interviews such narratives of ancestral memories and the role of sacred sites located within the *ejido*. We felt an ethical dilemma in insisting upon these elements given that when we had initially asked about sacred sites, specifically the caves and the rituals within these spaces, we were met with resistance. In that sense, we recognized in fieldwork the anxiety expressed by Indigenous peoples making territorial legal claims in juridical venues. In order to prove their particular socio-natural connections to a specific territory, they are expected to disclose sacred spaces and ceremonies that are used as evidence of those ties to land (Povinelli 2002). Though we gently inquired a bit further into this realm, we eventually chose not to place community members in this predicament. Rather, we opted to explain claims to territory through the processual narratives that community members insisted upon. In the expert witness report, we ended up concluding that territorial rights are founded on the struggles against what we described as cycles of land dispossession—i.e. community struggles followed by violent repression. That is, claims to Ch'ol territory are established and reinforced through generational struggles against the appropriation of lands by local *mestizo* elite in alliance with other local Indigenous populations, beginning when the Catholic Church founded the center of the local *encomienda* in the town of Tila in the 17th century.

The expropriation of the 130 hectares would fracture not only the connections established between current generations and those of the past, but also those established between various socio-natural spheres of community life. Ensuring communal lands meant that family homes would be connected to the surrounding cornfields, to the sacred caves where the elders pray and make offerings to the spirits (which is required to maintain healthy equilibriums between people, the earth, and other beings), as well as to the Black Christ of the church of Tila, to the sacred springs that also ensure sufficient water for the local population, and to the forested areas that protect and replenish the soil and bring rain. Recognizing the integrity of communal lands, we argued, ensures the reproduction of such connections and recognizes that Tila, as an Indigenous *ejido*, forms part of Indigenous Mayan struggles of *longue durée*.

Final critical reflections in the support of territorial claims

In this chapter, I have focused on Zapatista Indigenous Mayan communities in Chiapas to reflect on the ways that struggles over territory similarly engender reconfigurations of such territories. In this sense, territory is both a point of departure and a result of social struggle. For that same reason, I have insisted upon the importance of a processual analysis, rather than on essentialized and ahistorical claims. In the case of Zapatista communities, this processual analysis involved identifying first and foremost what historical conditions of exploitation and oppression do autonomous communities struggle against and what is the history of the agricultural estates that condense colonial structures of rule. Secondly, it involved tracing the production of territory as part of the everyday practices of autonomy. I did so through the political significance attached to the milpa and by describing how the daily practices surrounding the cornfields serve as grounding points for other spheres of autonomy to take root, in both a literal and a metaphorical sense.

Lastly, I have argued that such processual frameworks serve to push open limited definitions of indigeneity that are often found in juridical frameworks, making for a more inclusive terrain that better responds to the historical conditions lived by a significant number of Indigenous peoples in Abya Yala. At the same time, however, this creates a series of challenges that have methodological implications in legal cases. How can anthropologists and community members establish processual claims in ways that are anchored in the concepts, descriptions, and meanings of a people with particular epistemologies, while at the same time ensuring that these are rendered legible to those judges emitting verdicts? How can scholars and Indigenous organizations emphasize territorial claims in ways that avoid insisting upon revealing to the judges certain hidden transcripts and a people's sacred knowledge, even though in many legal proceedings, spiritual realms and a people's particular cosmovision are the most legitimate registers of alterity? And, lastly, how can anthropologists engage in legal proceedings as expert witnesses in ways that do not undermine a people's right to self-determination, including the right to represent themselves. I end this essay with these critical questions, as they are part of broader conversations regarding the political alliances through which territorial struggles take place.

Notes

1   *Abya Yala*, signifies 'Flowering Earth' in the native language of the Kuna peoples who live in what is now Colombia and Panamá. They use the term to refer to the entire American continent.
2   Glen Clouthard, in his book, *Red Skin, White Masks: Rejecting the Colonial Politics of Recognition*, would later describe the traps associated with multicultural reforms as the pitfalls of a politics of recognition. He argues that recognition exists only in the gaze of the colonizer and this maintains the underlying colonial structures intact (2014).

3 A stone sculpture of the Tzotz Choj ruler was found in the archeological site of Toniná, near the city of Ocosingo. He is represented by bats and in his belt, he holds the head of a jaguar as a trophy. According to Mayan cosmology the bats, owls and jaguars represent nocturnal deities. The sculpture represents the duality of creation and destruction.

References

Aguilar Gil, Yásnaya Elena. 2020. *ÄÄ: Manifiestos sobre la diversidad lingüística*. Ciudad de México: Almadía.

Aubry, Andrés. 2003. "Los caracoles zapatistas (Tema y variaciones)." *La Jornada, Ojarasca*. https://www.jornada.com.mx/2003/11/24/oja-caracoles.html.

Balboa, Juan. 1998. "Unos 2 mil manifestantes declararon a este sitio arqueológico Patrimonio de los Pueblos Indígenas. Más tarde ocuparon durante 53 minutos el nuevo cuartel de la 39 Zona Militar, en demanda de la salida del Ejército de sus comunidades." *La Jornada*, January 19, 1998. https://www.jornada.com.mx/1998/01/19/ocuparon.html.

Bertely Busquets, María. 2019. "Nuestro trabajo en las Milpas Educativas." *Dossiê Práticas de bem viver: diálogos possíveis entre o Núcleo Takinahakÿ e Milpas Educativas*.

Bonaccorssi, Nélida. 1990. *El trabajo obligatorio indígena en Chiapas, siglo XVI*. México: UNAM.

Bobrow-Strain, Aaron. 2007. *Intimate enemies: Landowners, power and violence in Chiapas*. Durham: Duke University Press.

Cervone, Emma. 2012. *Long live atahualpa: Indigenous politics, justice, and democracy in the Northern Andes*. Durham: Duke University Press.

Cruz, Emiliana. 2018. "Documenting landscape knowledge in Eastern Chatino: Narratives of fieldwork in San Juan Quiahije." *Anthropological Linguistics* 2(59): 205–23.

Coulthard, Glen Sean. 2014. *Red Skin, White Masks: Rejecting the Colonial Politics of Recognition*. Minneapolis: University of Minnesota Press.

Desinformémonos, Redacción. 2018. "Podrán cambiar el capataz, pero el finquero sigue siendo el mismo: EZLN." *Desinformémonos*, Julio 6, 2018. https://desinformemonos.org/podran-cambiar-capataz-finquero-sigue-siendo-ezln/.

EZLN. 1994. "Nuevos municipios autónomos". December 19, 1994.

———. 2018. "Convocatoria a un encuentro de redes de apoyo al CIG, al COMPARTE 2018: "Por la vida y la libertad"; y al 15° aniversario de los Caracoles Zapatistas: "Píntale caracolitos a los malos gobiernos pasados, presentes y futuros". July 4, 2018.

Federici, Silvia. 2018. *Re-enchanting the World: Feminism and the Politics of the Commons*. Oakland, CA: PM Press.

Gómez, Antonio, and Mario Humberto Ruz. 1992. *Memoria Baldía: Los tojolabales y las fincas. Testimonios*. Ciudad de México: UNAM.

Gómez, Magdalena. 2004. "La constitucionalidad pendiente. La hora indígena en la Corte." In *El Estado y los indígenas en los tiempos del PAN*, edited by Rosalva Aída Hernández Castillo, Sarela Paz and María Teresa Sierra, 175–206. Mexico City: CIESAS.

Harvey, Neil. 2000. *La rebelión de Chiapas: La lucha por la tierra y la democracia*. Ciudad de México: ERA.

Hooker, Juliet. 2020. *Black and Indigenous Resistance in the Americas, From Multiculturalism to Racist Backlash*. New York: Lexington Books.

Kirsch, Stuart. 2018. *Engaged anthropology: Politics beyond the text*. Oakland, CA: University of California Press.

International Labour Organization. 1989. *Indigenous and Tribal Peoples Convention (No. 169)*. https://www.ilo.org/dyn/normlex/en/f?p=NORMLEXPUB:12100:0::NO::P12100_ILO_CODE:C169.

Loperena, Christopher, Rosalva Aída Hernández Castillo, and Mariana Mora. 2018. "Los retos del peritaje cultural. El antropólogo como perito en la defensa de los derechos indígenas." *Desacatos* 57: 8–19.

Macleod, Morna. 2017. "Grievances and crevices of resistance: Maya Women Defy Goldcorp." In *Demanding justice and security, Indigenous women and legal pluralities in Latin America*, edited by Rachel Sieder: 220–241. New Brunswick, NJ: Rutgers University Press.

Marimán Quemenado, José Alejandro. 2017. *Awkan tañi müleam Mapun kimüm. Mañke ñi pu kintun*. Santiago, Chile: Centro de Estudios Rümtun – Fundación Heinrich Böll Stiftung.

Mora, Mariana. 2017. *Kuxlejal politics, Indigenous autonomy, race and decolonizing research in Zapatista communities*. Austin: University of Texas Press.

———. 2020. "(Dis)placement of anthropological legal activism, racial justice and the Ejido Tila, Mexico." *American Anthropologist* 23(12): 606–617.

———. 2021. "Reflexiones para tejer la vida-existencia." *Revista de la Universidad de México* UNAM. 871: 81–88.

Pairican, Fernando. 2011. "La Nueva Guerra de Arauco: la Coordinadora Arauco-Malleco y los nuevos movimientos de resistencia mapuche en el Chile de la Concertación (1997–2009)." *Revista www.izquierdas.cl*: 66–84.

Pérez Moreno, María Patricia. 2021. El Bats'il K'op Tseltal Frente Al Proceso Colonial. *Revista de la Universidad de México*, April: 73–79.

Povinelli, Elizabeth A. 2002. *The Cunning of Recognition: Indigenous Alterities and the Making of Australian Multiculturalism*. Durham: Duke University Press.

Reyes Ramos, María Eugenia. 1992. *El reparto de tierras y la política agraria en Chiapas. 1914–1988*. Ciudad de México: UNAM.

Rus, Jan. 2012. *El ocaso de las fincas y las transformación de la sociedad indígenas de los Altos de Chiapas*. Tuxtla Gutierrez: UNICACH.

Saldaña-Portillo, María Josefina. 2003. *The revolutionary imagination in the Americas and the age of development*. Durham: Duke University Press.

Villafuerte Solís, Daniel. 1999. *La Tierra en Chiapas: Viejos problemas nuevos*. Ciudad de México: México: Plaza y Valdés, Universidad de Ciencias y Artes del Estado de Chiapas.

Viqueira, Juan Pedro, and Mario Humberto Ruz. 2004. *Chiapas, los rumbos de otra historia*. Ciudad de México: UNAM, CIESAS.

Weitzner, Viviane. 2021. "¡Guardia, Guardia!": autonomías y defensa territorial en el contexto pos-Acuerdo colombiano. In *Autonomías y autogobierno en América diversa*, edited by M. González, A. Burguete Cal y Mayor, J. Marimán, P. Ortiz-T, and F. Ritsuko: 591–626. Quito: Ediciones Abya Yala.

Wolfe, Patrick. 2006. "Settler colonialism and the elimination of the native." *Journal of Genocide Research* 8(4): 387–409.

23 Indigenous autonomies as alternative horizons in Latin America

Societal movements and other territorialities in Bolivia and Mexico

Pabel Camilo López-Flores

Introduction

From the end of the 20th century, intense waves of social mobilization in Latin America have reshaped this part of the global South. These intense processes of social mobilization, which, in some cases, led to state transformation, have been characterized by the protagonist of Indigenous peoples placing historic claims on the national and regional stage and challenging systemic relations of exclusion, discrimination, and domination. These social movements, with their cultural, communitarian, and ancestral matrices and territorial anchors, shed light on the persistence of practices and schemas of "internal colonialism" in Latin American societies (González Casanova 1969; Rivera 2010). In some countries, these movements have challenged the anthropocentric and capitalist vision of the relationship between society and nature (Svampa 2016). They have also advanced proposals for recognition that disrupts the character of the nation-state in the region. This contestation from movements grounded in specific territories has manifested in many places across the region (Esteva 2016). This has led to the construction of Indigenous autonomies, sometimes through socio-territorial resistances, and at other times through Indigenous self-government. In that sense, it is about autonomic processes that are manifested as imaginaries and proposals for desirable futures, as well as concrete societal experiences.

The idea of autonomy radically questions the dominant state-centric conceptions of politics and liberal democracy (Zibechi 2008), acquiring central importance in ongoing discussions about Indigenous people and the reconfiguration of the state in Latin America (Burguete 2010). Autonomy, as a profoundly political notion, simultaneously updates and advances old theoretical political discussions on self-determination and decolonization, as well as other concepts such as self-management, emancipation, democracy, and liberty inside our societies within the global South. Likewise, the struggles for self-determination and autonomous processes in Latin America constitute not only resistance, but become movements for re-existences or alternatives to the capitalist extractive development models and its (neo)colonial relations of domination, dispossession, and negation of non-capitalist ways of life. Confronted with this scenario, we find the potential of diverse territorial resistances by social subjects who see their ancestral and communitarian spaces threatened (Gutiérrez 2015), and defend them by recreating, reconstructing, and reaffirming their collective ways of life and identities (Leff and Porto-Gonçalves 2015).

This chapter discusses the resurgence of territorialized movements, particularly Indigenous communitarian movements, as political subjects with a societal character, not only as social movements, but as "societal movements" (Tapia 2008), in the current Latin American context. It centers on the analysis of some of the experiences of Indigenous autonomies that are currently disputing possibilities of transformation from within or outside of the state, with

DOI: 10.4324/9780429344428-27

horizons of social emancipation or anti-capitalist social alternatives and post-extractive imaginaries in the region, which reveal new repertoires of recognition, as well as other notions of territory (Porto-Gonçalves 2010) and of relations with nature. One of the central aims of this chapter is to locate and think about autonomies in Latin America, particularly the experiences of Indigenous autonomy, either as concrete socio-territorial processes or as possible societal proposals with the potential to configure alternatives in the face of the multiple current crisis in this region and, in particular, their socio-ecological dimension.[1]

I briefly address two experiences of Indigenous autonomy in Latin America, both of which have a political-territorial character, and provoked societal disruption with de-colonial significance: the experience of Indigenous autonomy by the Guaraní people in the Chaco region of Bolivia, and the case of neo-Zapatista autonomy in Chiapas, Mexico. Both are experiences of actually-existing Indigenous autonomy in Latin America. However, the two processes are very different in terms of their relationship with the state, their positions on development and politics, their forms of administration, democratization, emancipation, and their horizons of autonomy. Thus, part of the chapter considers various interrelated but analytically distinct aspects of these movements, such as territory, the defense of self-determination, the environment and the political, and connects them to the tension dichotomy of inclusion versus subordination to the state, as well as to the real possibilities for autonomy.

Self-determination and Indigenous autonomy

In Latin America, the debate around autonomy has opened fertile fields for theoretical and political discussions about social and political territoriality and alternatives to capitalist production, drawing on the innumerable local experiences that emerge from the *"subsuelo politico"* ("political underground") (Tapia 2008). Large parts of these autonomous processes are presented as not only political-territorial dynamics but also as societal constructions, where horizontal, cooperative, and reciprocal relations come together, as well as spaces to construct prefigurative societal processes through an enormous plurality of alternative political, territorial, productive and democratic forms (Escobar 2018). Therefore, the notion of autonomy presents a multi-dimensional conception in both the theoretical and practical planes, which is complex and plural in its significance and meaning. Rather than define "autonomy" in the singular, it is better to think in terms of "autonomies" (Burguete 2010). Thus, on occasions, those who struggle, organize themselves, resist, create, and construct these social reorganization experiments—above all the subalterns—allude to the words "autonomy" to name these practices (Zibechi 2008). In this way, autonomy is expressed as a place where "community frameworks" are woven (Gutiérrez 2015) and languages, imaginaries, and autonomous practices are combined, that is, an autonomic praxis and a contra-hegemonic common sense. The notion of autonomy appears, moreover, closely associated with the territorial and cultural problematic of local communities, and to the recognition of Indigenous peoples. In this sense, the expansion and consolidation of autonomous approaches is a constitutive part of most Indigenous struggles of Latin America, because of the articulatory role that it acquires in its two manifestations: autonomy as a horizon or end, and autonomy as a process in contemporary struggles.

Thus, for at least the last 30 years, Indigenous autonomies have occupied a central place in the international debates about the collective rights of peoples. This is especially true in Latin America because of certain historical moments and events, such as the massive Indigenous mobilization in 1992 across the continent that rejected and resisted the "celebration" of 500 years since the colonial conquest; the Zapatista rebellion in January 1994 and the subsequent process of creating autonomous territories in Chiapas; the United Nations Declaration of the of the Rights of Indigenous peoples in 2007; the Indigenous mobilizations and constituent processes and assemblies in Bolivia and Ecuador that achieved the incorporation of Indigenous autonomy into the constitutions of both countries—to mention but a few of the enormous number of events, processes, and forms in which Indigenous autonomy has come to be recognized, demanded, disputed, constructed and defended across Latin America (Díaz and Sánchez 2002; López 2008; Burguete 2010; Composto and Navarro 2014; Esteva 2016; Ulloa 2017; Escobar 2018; Rousseau and Morales 2018). In this period, the theme of Indigenous autonomies has already assumed relevance at the level of general recognition, from the concrete demands and proposals of Indigenous movements and organizations, to the aforementioned debates and processes of reform and/or re-foundation of states in Latin America (Santos 2010).

Likewise, although previously the idea of autonomy was considered an isolated proposal exclusive to some groups, today it has been renovated and become not only a discourse but also an everyday practice and construct of political territorial organization and action, of social and cultural identity and of critical meaning, of potential societal alternatives or, simply, as a modality of territorial resistance and defense of collective rights. Thus, currently the actors of Indigenous movements have not only acquired experience of, but have also changed the initial conditions for, demanding and struggling for autonomies. Additionally, the demand of Indigenous autonomy has currently placed the right to land and territory as an axis or nucleus of discursive diffusion of the demand for self-determination. Thus, the demand for self-determination of Indigenous peoples can be found contained in demands for the reconstitution of land, perceived as a base for communitarian self-determination, as well as in demands for the recognition of the territory, as a foundation for the self-determination of a people.

Following this line, Zibechi (2008) affirms that the socially-controlled territory is the primary basis for the construction of autonomy. Thus, autonomy is linked with difference; Indigenous peoples need territorial autonomy to protect their culture, their cosmovision, and their world as something distinctly different from the hegemonic world (Zibechi 2008, 137). In this perspective, autonomy is the result of a kind of trilogy of inseparable aspects: territory-self-government-self-determination. Following this approach, López y Rivas (2011) propose reflecting from the "reconstitution of Indigenous peoples," understanding this concept not in its nativist sense of reconstituting an idealized past, but rather in it connotations toward and future of unification of Indigenous peoples, of intra-communitarian, regional and macro-regional articulation, of the strengthening of their autonomous conscience, of the construction or reconstruction of other broader forms of political, territorial, and cultural organization in the face of nation-states, society, and the dominant socio-economic system. Likewise, Esteva (2016) contends that the condensed notion of Indigenous autonomy should be understood as a political project that gives historical continuity to the long-standing resistance of Indigenous peoples, transforming it into a proposal for liberation and emancipation shared by many other groups and social subjects.

In this sense, is necessary also to examine the relational and contested nature of territories and territorial praxis, focusing on the centrality of power to territory and counter-hegemonic territorial politics (Clare and Liz 2018), or in communitarian counter-hegemony processes (López Flores 2019). While such an understanding of power frames the complex production of territories, as pointed out by Clare and Liz (2018), it is important to also reflect on how movements intervene in producing their own territories. In these contexts, "the demands of environmental self-determination of Indigenous peoples enter into a discussion with extractivisms and all the forms of appropriation of their territories" (Ulloa 2017, 177). In this way, as Composto and Navarro (2014) point out, it is here where the political, the community, and the territory weave powerful capacities against the enclosure of the common and for the asymmetric battle against capital. Likewise, as recalled by Escobar (2018), in the context of many grassroots communities, autonomous design takes place under conditions of ontological occupation: "The concept of autonomy should thus be seen in terms of ontological struggles for the defense of people's territories and life worlds. It is precisely in those cases that the idea of autonomy is flourishing and the hypothesis of design for autonomy is taking on the cimcticst meaning (Escobar 2018, 167).

In parallel, these contemporary struggles for Indigenous autonomy would reflect, among other aspects, unresolved and ongoing conflicts and tensions between the nation-state—in its postcolonial and neocolonial condition—with the Indigenous peoples or the original nations, where dispute also the contents, consolidation, and respect for collective and historical political-territorial rights around, precisely, the demand for Indigenous self-government. Likewise, these processes would be subject to contentious processes of negotiation, pressure, and/or harassment, with the presence of external actors and factors including governments, political parties, military forces, extractive companies, NGOs, non-Indigenous social groups and other actors—but also by endogenous phenomena that sometimes represent contradictions and limits within the Indigenous movements themselves. In this sense, the struggle of the peoples for Indigenous political-territorial autonomy faces multiple conflicts, obstacles, limits and challenges (Mora 2017). However, Indigenous movements that claim and exercise autonomy have achieved in these years both recognition by the state, as in the case of Bolivia, and the construction of de facto autonomy "without permission" from the State (Burguete 2010), as in the case of Zapatismo in Mexico.

I focus on exploring some processes and experiences of socio-territorial resistance and the construction of political-territorial autonomy led by Indigenous social movements in Latin America, particularly in Bolivia and Mexico, which, as political subjects in some cases, become societal movements (Tapia 2008), with the capacity not only to question some forms and policies of nation-states, but also to challenge their assumptions and homogenizing schemas (one state=one nation) and their inheritance built on the 'coloniality of power' (Quijano 2000) and forms of 'internal colonialism' (González Casanova 1969; Rivera 2010).

The autonomy of Guaraní people in Bolivia: between subordination and self-determination

In Bolivia, the perspective of state transformation through the construction of a plurinational state had as one of its central nuclei the recognition of the Indigenous Autonomies, as a concrete proposal to move toward democratization and, in turn, the decolonization of the State, as a condition of making plurinationality possible. Both plurinationality and Indigenous autonomy would then be the result of collective imaginations, historical demands, and political struggles of social movements of an Indigenous nature, which would lead to the constitutional mandate of construction of a "plurinational community and autonomous state," which in turn implies breaking with the monopoly of politics—that is, it would entail the establishment and development of a plurality of spaces and forms of Indigenous self-government.

In 2009, with the approval of the current Political Constitution of the State (CPE), after a Constituent Process (2006–2008) with a new autonomous regime and with the entry into force of the *Ley Marco de Autonomías y Descentralización* (Framework Law of Autonomies and Decentralization LMAD) in 2010, Indigenous autonomies passed from a sphere of tension and conflict to a more "administrative" and "legal" dynamic at the territorial level, focused on the statutory process. That is, in the elaboration of autonomous statutes and the follow-up of the procedure stipulated by the respective normative framework, after December 2009 eleven

Figure 23.1 Some varieties of corn from the Gurani people of the Bolivian Chaco, 2020.

municipalities expressed, via referendum, their will to become *Autonomias Indigenas Originarias Campesinas* (Original Peasant Autonomies, AIOC). This historical claim had as one of the main protagonists and promoters the Guaraní people of the Bolivian Chaco, self-identified as the "Guaraní Nation," politically and territorially organized through their communities, their traditional authorities, and their parent organization, the Assembly of the Guaraní People (APG), which for decades has been claiming the community reconstitution of their ancestral territory (see Figure 23.1), managing to incorporate and dispute their demand for autonomy in the Bolivian constituent process of the past decade. Once the new Political Constitution of the State was approved in Bolivia, and with the opening of the post-constituent period from 2009 onwards, growing tensions emerged between the *Movimiento al Socialismo* (Movement to Socialism, MAS), Evo Morales's governing party, and Indigenous movements. This was due to disputes over territory and natural resources.

The question of Indigenous autonomy appears here as a concrete horizon of the Guaraní people, of its political organization and all its communitarian basis in its ancestral territories, as also evidenced in their demands on the state for cultural and territorial recognition, the claims around natural resources, and the defense of the "Plurinational State." With these demands, the Guaraní Nation not only proclaimed territorial autonomy but also that their vision has horizons of recognition that are linked, in a certain dimension, to the reconstitution of their ancestral territory, which in turn entails, in a certain sense, communitarian control over the natural resources that are found in the subsoil.

With the new political and juridical scenario configured in Bolivia through the current Political Constitution (2009), particularly in relation to the theme of territorial organization and the new autonomy regime established in the LMAD in 2010, previously in December

2009 eleven municipalities initiated the process of becoming autonomous Indigenous communities. In 2017 the municipality of Charagua (region of Santa Cruz) in the Chaco Region became the first autonomous Indigenous community in Bolivia. This process was considered by its protagonists as a central part of the struggle of the Guaraní people for autonomy and a fundamental historical moment within the horizon of territorial reconstitution as a people. However, this Indigenous autonomous process, despite having completed all of the prerequisites demanded by the constitution and the corresponding law of implementation of self-government, clashed with the obstacles and bureaucratic state instances which subordinated the character and reach of Indigenous autonomy to a municipal system under bureaucratic control of the state (López Flores 2017).

Since the promulgation of the current Bolivian Constitution, especially in the last two MAS governments (2010–2014 and 2015–2019), it is possible to account for the deployment of a state project that, in reality, both in its norms and in its concrete policies, has not only represented a setback for the most "transformative" principles of the constitution but also a denial of the real possibilities of state decolonization—thus undermining the conditions for the construction of a plurinational state. It is undeniable and more than paradoxical the way in state institutionality promoted a new cycle of capitalist expansion over community ancestral territories, for the most part recognized as collective (Makaran and López 2018). In that sense, specifically with regard to the AIOC, it is evident that the MAS government opted to limit, obstruct, protect, and subordinate them, which would correspond to a monopolization ambition in politics and a hyper-extractive turn in terms of the economic model. This trend was manifested in the case of the first AIOC, the Guaraní Autonomy in Charagua (see Figure 23.2), both in its long and conditioned process of transformation, and especially in the governmental and partisan subjection of this new "autonomous government."

In this sense, a crucial issue, among others, to understand the contradictions and limitations of a possible Indigenous autonomy is the extractivist agenda of the state. The leaders conceive autonomy primarily as a means to protect their territory and natural resources and

Figure 23.2 Possession of the authorities of the Autonomous Indigenous Government of Charagua Iyambae, January, 2017.

ensure autonomous and sovereign decision over them. However, these ambitions are far from the reality in which the interest of the state prevails, which is presented in official speeches as "general/national interest," above Indigenous rights and plurinational principles. The conflict over natural resources, the main factor in the paralysis of Indigenous autonomist processes by the central government, not only consists of legal obstacles, fraudulent practices in terms of Prior Consultation or attempts at co-optation, but also entails an important dose of direct violence exercised by the state apparatus against the communities and their leaders, who oppose the exploration and exploitation of hydrocarbons without prior consultation.

The rebel Zapatista Indigenous autonomy in Chiapas: the construction of self-government from below and without permission from the state

The demand for autonomy has a long tradition in the Indigenous peoples in Mexico, and from the first day of the Zapatista uprising on January 1, 1994 in Chiapas it was present in the identity of the movement, the result of a process of accumulation and evolution in the Indigenous claims of the Mayan people of south-eastern Mexico (Burguete 2004). A great majority of the peasant-Indigenous organizations, of community *ejidos*, strengthened their struggles around the demand for land, among these several identified with the principles of the original Zapatismo (of the Mexican Revolution) and there will be various dynamics of "land recuperation" (Díaz and Sánchez 2002). Thus, the option to the "de facto" route" almost always involved confrontation with the State and violence. On the other hand, given the weak presence of the state, especially the social state, its functions were partially covered by other actors, mainly the Church and NGOs. With the breach of the San Andrés Accords of 1996 by successive Mexican governments, the autonomies in Chiapas were built de facto, and the EZLN then decided to apply on its own what was agreed in San Andrés and proceeded to establish autonomous Zapatista jurisdictions as a strategy to establish its own path of compliance with the agreements (Burguete 2004). Thus, the Zapatista uprising had put autonomy, together with the demands of democracy, freedom, justice, etc., at the center of the demands and demands of the Indigenous peoples.

After the process of land recuperation, the other fundamental aspect of territorial expansion was the constitution of the *Municipios Autónomos Rebeldes Zapatistas* (Rebel Zapatista Autonomous Municipalities, MAREZ). In the MAREZ, respective authorities govern through the principle of *mandar obedeciendo* (leading by obeying), giving them, from their very first day, a functioning distinct from that of the formal municipalities recognized by the Mexican state. Since the foundation of the MAREZ after 1994, the construction of material autonomy has been a central aspect of the Zapatista struggle, something that represents the "material conditions for resistance" (Zibechi 2008). They form its basis and allow the articulation of the distinct actions of the Zapatista movement. They are where the Indigenous communities denominated as the "*bases de apoyo*" ("bases of support") constitute a vertebral column and fundamental movement, and in a sense represent the autonomous process as a whole. From this point onward, de facto autonomy has been the principal Zapatista strategy and that which has sustained their rebellion as a form of socio-territorial resistance, something that has permitted the Zapatistas to question and interpellate the Mexican state itself.

As a result of this process, in 2003 the *Caracoles Zapatistas* (Zapatista Snails) and the *Juntas de Buen Gobierno* (JBG, Good Government Boards) were created as new instances of regional coordination, strengthening the Indigenous autonomist impulse. Since then, they have acquired centrality in the Zapatista structure, suggesting that the socio-territorial project has displaced the military project as the principal priority of the Zapatistas. This civil Zapatismo has since become the face of the movement both nationally and internationally (Burguete 2004). Together with the construction of forms of self-government, this has advanced the implementation of justice and the creation of health, education, and production programs, increasing the direct participation of community members in the decision-making processes (Zibechi 2008). For that purpose, the control of territory is the primary basis on which Zapatista autonomy is built. However, this autonomy does not consist of a statement or represent an ideological goal. Indigenous peoples need territorial autonomy to protect their culture, their worldview, and their world as something different from the hegemonic world, in a kind of trilogy: territory-self-government-autonomy (Zibechi 2008). Through this political and territorial reconfiguration by the Zapatista movement, new socio-territorial imaginaries have been, and continue to be, produced over the last decade and a half in the territory of Chiapas, in tension with and disputing the territorial and political imaginaries of the nation-state and neoliberal globalization in this country. In this sense, through the construction of de facto autonomies "without the permission of the state" (Burguete 2010), the Zapatista movement re-signified their own understanding of recognition and autonomy, which, for the Zapatistas, had already been reclaimed from the state and become practices for the exercise of the right to self-determination and territorial self-government, in resistance to the Mexican state and with a rebel character. In this perspective, these self-controlled spaces are defined as a laboratory for the transformation of social relations.

In this way, the Zapatistas are perhaps promoting a true socio-political change and societal disruption that runs counter to the exclusionary practices and logics of capital, with said socio-political transformation producing a socio-territorial transformation. In this sense, this display of radical democracy centered upon *mandar obedeciendo* is organized from the multiscalar into the multidimensional; from the local, regional and zonal, as much in political as in social-cultural and economic aspects, that which develops the meaning of social political practice and of the socio-territorial advances which configures the Zapatista autonomy project in all its density as societal movement (Tapia 2008). It was created as a new form of organization and a new level and way of governing in the Zapatista autonomous communities. Thus, the Caracoles were born as a territorial configuration, where their heart is the JBG. Five JBGs were created to function in each Caracol, one for each rebel region and where one or two delegates from the MAREZ of the region were sent to be part of the JBG. The JBGs coordinate a certain number of MAREZ, but they do not replace their functions. As a result, in each Zapatista Rebel Autonomous Region, there are three levels of civil government: the regional, with a JBG; the municipal, made up of an Autonomous Council of a MAREZ; and the community, with representatives of the Communities in Resistance.

Thus, the creation of the Caracoles and the JBG sought to overcome some problems or tensions that had been generated in the process of building autonomies. In this sense, the *neo-zapatismo* has had a political impact in the Latin American Indigenous movements, including the struggle for new models of post-liberal citizenship, defining the limitations of neoliberal multiculturalism, and encouraging the adoption of de-colonial forms of resistance and liberation, in which autonomy is constructed through the recreation of social ties in local, national, and international spaces (Harvey 2016). Likewise, in August 2019, the Zapatista Army of National Liberation (EZLN) reported that it had created seven new Caracoles or "Centros de Resistencia Autónoma y Rebeldía Zapatista" (Centers of Autonomous Resistance and Zapatista Rebellion, CRAREZ). The majority will be the headquarters of the JBG, in addition to the five that had already been functioning for 15 years and four new MAREZ in Chiapas. In total, the Zapatistas will have established 43 instances of self-government, unrelated to the official government institutions.

Since the uprising, the Zapatista support bases, their political-territorial organizational bodies, and the EZLN military command that protects them, have opted for the possibilities of social change outside the state. In this scenario, it is possible to reflect on the senses in which Zapatismo has taken root and marked differences in the daily life of the communities and localities in the State of Chiapas; how it is changing around the resolution of central problems like the agrarian question; in what way it makes a difference in basic social issues, such as health, education and collective production; and how it has transformed the frameworks of politics and democracy through a new organization for self-government. Mora (2017) points to the central spheres of Zapatista Indigenous autonomy, particularly governing practices, agrarian reform, women's collective work, and the implementation of justice, as well as health and education projects. These findings allow us to critically analyze the deeply complex and often contradictory ways in which the Zapatistas have re-conceptualized the political and contested the ordering of Mexican society along lines of gender, race, ethnicity, and class (Mora 2017). In this way, from the margins and rejecting official institutions, the Zapatistas have been building a constitutive political power "from below", which generates cultural practices and produces alternative knowledge to the socio-spatial logics of capital and the nation state.

Territorialities in dispute, autonomous and counterhegemonic re-existences

In the case of Bolivia, the constitutional acknowledgement of the plurinational character of the state has recognition as one of its central axis. This, in turn, has given it the mandate to implement Indigenous autonomies as a concrete proposal and modality for the plural transformation of the state structure, assuming a strongly communitarian social matrix. Bolivia's multisocietal condition (Tapia 2008) implies that there is a need to undo historic logics, practices and schemata of coloniality and of internal colonialism. However, the governmental structures of the current "state," have paradoxically followed productivist visions in recent years, and policies based predominantly on models of extractivist development have prevailed, marking a contradiction with the meaning and reach of the political-territorial autonomies of Indigenous peoples established in the constitutional text and various international treaties. This has affected Indigenous territories, peoples and communities, generating and intensifying socio-territorial and socio-environmental conflicts that have put into question and disputed the conditions and real possibilities of Indigenous autonomies, distancing the horizon of democratization, decolonization, and state transformation in this Andean-Amazonian country.

In this sense, struggles for Indigenous autonomy have reconfigured State society. The autonomy project of the Guaraní, originally constructed from below, in their territory, was ultimately subordinated to the state (see Figure 23.3). At the same time, however, through these demands, Indigenous organizations and peoples, in particular of the lowlands as is the case with the Guaraní people, have become constituted as contra-hegemonic nuclei in the confrontation with the capitalist policies and the power of the state, despite its rhetoric definition as "plurinational."

For its part, in the Mexican case, the neo-Zapatista uprising of 1994 by the Mayan Indigenous people in south-eastern Mexico became an essential reference in the imaginary of the historic struggles of the Indigenous peoples in Latin America for self-determination. Thus, more than 25 years after the Indigenous uprising in Chiapas, and more than 15 years after the consolidation of its political territorial forms of de facto autonomy, through the Caracoles, the MAREZ, and the *Juntas de Buen Gobierno*, a process of Indigenous self-government has been configured with a communitarian base and "insurgent territoriality." This is an autonomous and decolonizing societal process that has developed as much in material as in symbolic dimensions.

Figure 23.3 Map of the Indigenous Autonomous Municipality of Charagua Iyambae, January 2017.

Conclusion

The goal of this short text has been to explore the experiences and processes derived from some movements struggling for Indigenous autonomy, which, from their everyday resistance and construction, can produce and re-signify socio-territorial and communitarian spaces (López et al. 2017), with a counter-hegemonic character. In this way, this text recognizes how in some countries in Latin America, faced with possible tendencies toward and processes of reduction, restriction, and/or negation of the recognition or exercising of the right to self-determination of Indigenous peoples, some processes of resistance, socio-territorial mobilization, and experiences of Indigenous autonomies could represent spaces and practices for communitarian re-existence and political-territorial construction.

At the same time, it focuses on identifying and problematizing how some socio-territorial processes and Indigenous movements are occupying spaces and re-signifying collective imaginaries of the political and of democracy within the scope of "other territorialities," away from the institutional space of the state. These socio-territorial subjects become societal movements, in particular through the communitarian organizational forms with cultural bases, through the practice of government and, in turn, the reclamation, defense, and dispute of Indigenous autonomy. This implies a scenario of conflict and tension in relation to the debates over the character and nature of the state, of its visions and politics of development, concurrently questioning and problematizing its historic presuppositions, its current forms, ends and narratives, such as the modality in which it changes and is reconfigured, its relations within society and, in particular, with the Indigenous peoples and movements, who are continually pushed the margins through processes that have been called "internal colonialism" in Latin America.

In this sense, from collective organization and everyday resistance, Indigenous autonomous processes seem to represent strategies to confront and develop alternatives to the hegemonic ideologies and dominant paradigms that legitimize and perpetuate relations of domination, subordination, and exclusion experienced of Indigenous peoples. These organizational forms that claim Indigenous communitarian cosmovisions and practices can open spaces for the construction of innovative alternatives at all levels, from the communitarian, to the regional, and national structures, a shift that reflects scenarios that perhaps makes possible horizons of improved democratization, decolonization, emancipation and social dignity.

Note

1 It is also about processes of re-emergence of "societal movements," as configurations of protest and collective action that express a more dense character than a classical social movement and emerge with a prefigurative mode and other organizing principles (Tapia 2008).

References

Burguete, Araceli. 2004. "Una década de autonomías de facto en Chiapas (1994–2004)": los límites. *Pueblos Indígenas, Estado y Democracia*. CLACSO, 239–278.

Burguete, Araceli. 2010. "Autonomía: la emergencia de un nuevo paradigma en las luchas por la descolonización en América Latina". *La autonomía a debate: autogobierno indígena y Estado plurinacional en América Latina*. FLACSO-Ecuador/GTZ/IWGIA/CIESAS–UNICH, 63–94.

Clare, V. Habermehl, and Mason-Deese Liz. 2018. "Territories in contestation: relational power in Latin America". *Territory, Politics, Governance* 6(3): 302–321.

Composto, Claudia, and Mina L. Navarro. 2014. "Claves de lectura para comprender el despojo y las luchas por los Bienes Comunes Naturales en América Latina". In *Territorios En Disputa. Despojo Capitalista, Luchas En Defensa de Los Bienes Comunes Naturales y Alternativas Emancipatorias Para América Latina*, eds. Mina Navarro y Claudia Composto, 33–75. México, DF: Bajo Tierra Ediciones.

Díaz, P. Hector, and Consuelo Sánchez. 2002. *México diverso. El debate por la autonomía*. México: Siglo XXI.

Escobar, Arturo. 2018. *Designs for the pluriverse: radical interdependence, autonomy, and the making of worlds*. Durham, NC: Duke University Press.

Esteva, Gustavo. 2016. "La hora de la autonomía". en *Pueblos originarios en lucha por las autonomías: experiencias y desafíos en América Latina*, eds. P. Lopez and L. Garcia, 29–58. Buenos Aires: CLACSO.

Gonzáles Casanova, Pablo. 1969. *Sociología de la explotación*. Siglo XXI: México.

Gutiérrez, A. Raquel. 2015. *Horizonte comunitario-popular: antagonismo y producción de lo común en América Latina*. Cochabamba: SOCEE/Autodeterminación.

Harvey, Neil. 2016. "Practicing autonomy: zapatismo and decolonial liberation". *Latin American and Caribbean Ethnic Studies* 11(1): 1–24.

Leff, E., and C.W. Porto-Gonçalves. 2015. "Political Ecology in Latin America: the social reappropriation of nature, the reinvention of territories and the construction of an environmental rationality". *Revista Desenvolvimento e Meio Ambiente* 35: 65–88.

López, Bárcenas Francisco. 2008. *Autonomías Indígenas en América Latina*. México: COAPI/AC/MC.

López y Rivas, Gilberto. 2011. Autonomías indígenas, poder y transformaciones sociales en México. In *Pensar las autonomías. Alternativas de emancipación al capital y el Estado*, Jóvenes en Resistencia Alternativa (JCR editors) 103–115. México: Bajo Tierra.

López Flores, Pabel C. 2017. "¿Un proceso de descolonización o un periodo de recolonización en Bolivia? Las autonomías indígenas en tierras bajas durante el gobierno del MAS". *RELIGACIÓN. Revista de Ciencias Sociales y Humanidades* II(6): 48–66.

López Flores, Pabel Camilo. 2019. *Contrahegemonía comunitaria: las experiencias autonómicas del pueblo Guaraní en Bolivia y del Zapatismo en México*. Buenos Aires: Editorial El Colectivo.

López, M.F. Sandoval, A. Robertsdotter, and M. Paredes. 2017. "Space, power, and locality: the contemporary use of territorio in Latin American geography". *Journal of Latin American Geography* 16(1): 43–67.

Makaran, Gaya and Pabel López. 2018. *Recolonización en Bolivia. Neonacionalismo extractivista y resistencia comunitaria*. México: Bajo Tierra.

Mora, Mariana, 2017 *Kuxlejal politics indigenous autonomy, race, and decolonizing research in Zapatista communities*. Austin: University of Texas Press.

Porto-Gonçalves, Carlos Walter. 2010. *Territorialidades y lucha por el territorio en América Latina: geografía de los movimientos sociales en América Latina*. Caracas: IVIC.

Quijano, Anibal. 2000. "Colonialidad del poder y clasificación social". *Journal of World-Systems Research*, Immanuel Wallerstein (Coord.) XI(2): 342–386.

Rivera, C. Silvia. 2010. *Ch'ixinakax utxiwa: reflexión sobre prácticas y discursos descolonizadores*. Buenos Aires: Tinta Limón.

Rousseau, Stéphanie y Anahi Morales. 2018. *Movimientos de mujeres indígenas en Latinoamérica: género y etnicidad en el Perú, México y Bolivia*. Lima: Fondo Editorial PUCP.

Santos, De Sousa Boaventura. 2010. *Refundación del Estado en América Latina, Perspectivas desde una epistemología del Sur*. Lima: IIDS/PDGT.

Svampa, Maristella. 2016. *Debates latinoamericanos. Indianismo, Desarrollo, Dependencia, Populismo*. Buenos Aires: Edhasa.

Tapia, Luis. 2008. *Política Salvaje*. Buenos Aires: Muela del Diablo, CLACSO.

Ulloa, Astrid. 2017. "Perspectives of environmental justice from Indigenous peoples of Latin America: a relational Indigenous environmental justice". *Environmental Justice* 10(6): 175–180.

Zibechi, Raul. 2008. *Autonomías y emancipaciones. América Latina en movimiento*. México: Bajo Tierra.

24 Land occupations and land reform in Brazil

Nashieli Rangel Loera

Occupations, landless encampments, and agrarian reform movements

At the end of the 1990s, some specialists, fellow anthropologists, and sociologists with a long experience of fieldwork and research among trade unions and rural workers' organizations in northeastern Brazil, were surprised by a novelty in the rural landscape of the region: a series of encampments, a consequence of land occupations carried out within the properties of large landowners, formerly known in the region as owners of the (sugarcane) mills. The camps were occupied by families of rural workers, many of them former workers of those same properties and sugarcane fields, and were made up of rows of small huts of wood and straw covered with black plastic sheeting, locally called *tarpaulin*, to which were added flags indicating the organization, or, rather, the *movement* that had organized the occupation and was in charge of the camp.[1] According to Sigaud et al. (2006, 29–30), for scholars of the social relations of the large sugarcane plantations of the *zona da mata* of Pernambuco or for those familiar with the literature on the subject, what was seen was surprising: we were confronted with a set of signs indicating a major discontinuity in the social order [own translation].

The land occupations and agrarian reform camps, as described by the authors, were the work either of the Landless Rural Workers Movement (MST) or of the rural workers' unions, which, with such installations, demanded the expropriation of those properties. Such an undertaking, for scholars familiar with this social world, could not go unnoticed in a region historically dominated by the authority of the sugarcane industry owners and bosses. In Pernambuco, as in other states in the country with a strong concentration of land and presence of rural unions, the demands and struggles for land through occupations and the setting up of camps were not, until the 1980s, part of the horizon of possibilities for these organizations. As shown by some studies (Sigaud 1979; Sigaud 1980; Rosa 2011a; Heredia 1989; Menezes 1997), strikes were, *par excellence*, the forms of mobilization present among rural workers to demand mainly wage rights. The demand for land at various times throughout the 20th century was a central guideline among various categories of social subjects and in several regions of the country; however, this demand, gathered more broadly as *agrarian reform*, is updated as a banner of struggle or main guideline of the *landless movements*, in the late 1980s and early 1990s, in a moment of opening The landless movements were able to take up their struggle in the late 1980s and early 1990s, at a time of democracy post-military dictatorship, and of the process of elaboration and implementation of the 1988 Constitution, a context favorable to listening to the State.[2] At this time, the conditions of possibility were created for the *movements* and participants of landless occupations and camps to be recognized as fundamental interlocutors in the struggles for land (Servolo de Medeiros 2002; Sigaud, Ernandez, and Rosa 2010). In this context, we also witnessed a national expansion of *landless movements*, mainly of the MST which, in this period, also gained international projection.[3]

DOI: 10.4324/9780429344428-28

As Servolo de Medeiros (2002, 9) mentions, "the issue of agrarian reform has been present in the Brazilian public debate, more or less intensely, at least since the first decades of the twentieth century, assuming, however, over time, differentiated forms and meanings" [own translation]. As this Brazilian sociologist summarizes well:

> Throughout the years, the updating of the keyword "agrarian reform" was given mainly by the actions of important contingents of workers who locate themselves or make themselves recognized in public spaces through the struggle for land. "squatters," "tenants," "land renters" in the 1960s; "squatters" in the 1970s; "rubber tappers," "landless peasants," "dam-affected people" "riverside dwellers" in the 1980s; "landless peasants" which includes wage earners of large productive units in crisis, as in the case of sugarcane fields, up to inhabitants of urban peripheries, in the 1990s: socially differentiated characters, diverse geographical axes, different social and political identities, which indicate permanence of struggle for land but also show that its terms change and that the meaning of the agrarian question is transformed.
>
> (Servolo de Medeiros 2002, 10–11)

For Mattei (2005) and Almeida (2007), the MST played a central role in the 1990s, repositioning in the public debate the peasantry's problems, the importance of agrarian reform, and the problem of land concentration in Brazil.[4] According to Mattei (2005), it is also in this context that agrarian reform reappears linked to the idea of social transformation and the reduction of inequalities in the country. However, the novelty is not in this conjunction, nor in the action of rural workers' organizations that open the possibility of the production of a state policy of land distribution. As mentioned above, in several regions and moments of Brazilian history, organized rural workers played a central role in the struggles for land and the production of social rights. The novelty lies in the implementation of the *conquest* of unproductive lands through massive land occupations and encampments (Stédile and Fernandes 1999).

> The occupations appeared as an alternative to pressure the State, creating political facts and attracting the attention of public opinion. Initiated in Rio Grande do Sul, at the end of the 1970s, others quickly followed in Santa Catarina, Paraná, Sao Paulo, Mato Grosso do Sul [...].
>
> (Servolo de Medeiros, 2002, 49 [own translation])

Anthropologist Lygia Sigaud (2000), in an indispensable text for understanding the consolidation of landless encampments as a modality of agrarian reform demands in contemporary Brazil, proposes the existence of a "camp form," that is, a symbolic language of social demand whose central elements are the black canvas huts, known as *barracos*, and the flags of the *movements*. The agrarian reform encampments would be, according to this author, "a way of saying," i.e. those who wanted certain lands to be expropriated and distributed, should occupy them and set up an encampment and from there formulate their demands to the State.[5] The banners indicated which movement was organizing such a demand: MST, trade unions, or one of the more than 100 rural workers' *movements in existence* throughout the national territory until the beginning of the 2000s. By the end of the 1990s and the beginning of the first decade of the 2000s, landless encampments could already be found throughout the country, becoming part of the Brazilian rural landscape in practically all states and regions. As Fernandes (2005) demonstrates, in this period practically 77% of the land expropriations present in 30 states of the country were the result of, or originated from, massive land occupations organized by *movements*.[6]

Macedo (2005) points out a kind of common sense or naturalization existing in some analyses that attribute a 'spontaneous' character to the land mobilizations in Brazil, as if they had arisen almost suddenly, and likewise to the movements and the subjects that participate in them. These analyses do not take into consideration a detailed process of organization and transfer of knowledge and the implementation of a whole social mechanism that made their existence possible. In the second part of this text, I will try to recover this social dynamic in the context of São Paulo, but not without first making it clear, as Macedo mentions, "that this language or social form has its own processes of constitution, with a technology of mobilization of families and realization of occupations that contains variations according to the specific conjunctures" (Macedo 2005, 474). In a recent work (Rangel Loera 2015a), I ethnographically explored some of these variations, which are due to different regional conditions, from the existing heterogeneity of the families participating in the encampments and landless movements, to the political configurations of municipalities and states, and to the dominion status of the disputed lands. Recently, with the ascendance to the presidency of Brazil a president who openly opposes land expropriations and sees all those who claim them as enemies (Rangel Loera 2019), we can also mention the importance of considering the national political conjuncture and configuration of the federal government.

We can affirm that the link between occupations and landless encampments and agrarian reform was produced jointly, and literally, in the movement. In this process, MST *militants*[7] from the south of the country circulated and settled in various states of Brazil, bringing the *know-how* of the mobilizations and a whole social technology mobilized in the occupations initially carried out in the south of the country, thus having a fundamental performance to consolidate this language of social demand nationally, but also internationally, mainly through their performance and the networks formed in the Via Campesina space. As demonstrated by the work of anthropologist Maite Yie (2018) for the case of peasant mobilizations in Colombia, such as those in Nariño, for example, in the organization of the Mingas for sovereignty and for territory, part of the ritual repertoire of the MST, such as the *mystique* and land takeovers was incorporated and reworked as in a kind of bricolage, to which elements belonging to the history and struggle of Colombian peasant communities were added. It is necessary to point out that the transfer of knowledge and mobilization technology has been a back-and-forth movement, as MST militants have also circulated throughout Latin America and other parts of the world learning from the experiences, history and concrete practices of other struggles.

However, occupations and encampments as a symbolic language of claim is not necessarily generalizable to any context of struggle for land, as Rosa (2011b, 2012) demonstrates from the South African case of the Landless People Movement, where the claims are inscribed in a history of relationships with the territory in which graves, ancestors, marriage and funeral rituals, the dead and ritual objects, determine the relationship with the place and with the land they wish to recover. This is the case in other Latin American contexts with a strong presence of indigenous peasantry, such as Bolivia, Guatemala, Mexico, and Peru. It was also this difference in the meaning of land claims and the history imbued in them, the argument mobilized by Guarani Ñandeva Indians from the south of the state of Sao Paulo for not accepting the invitation made by MST militants to join the mobilizations and occupations of unproductive lands organized in the region.[8] As we were told by Guarani leaders, it was not a question of fighting for "a" piece of land, nor was their claim based on the principle of unproductive land. In the case of these Guarani Ñandeva, it was a question of returning to "their land," now owned by a congregation of monks, and which had historically been occupied by their *relatives*.[9]

The conditions for the possibility of agrarian reform settlements[10]

In the Brazilian context, the mobilizations for land and agrarian reform promoted by rural workers' organizations, organized under the slogan "Occupy, resist, produce," opened up the possibility not only of democratizing mass access to land in practically all regions of the country with the creation of rural settlements, but also the possibility of producing social effects and changes in the relations of domination in the Brazilian countryside and, more broadly, of diversifying agricultural production by reconfiguring large areas of monoculture into small and medium rural properties dedicated to the production of diverse foods, with palpable effects on the environment as well.[11] In the northeast region, as Sigaud and de L'Estoile (2006) mention in the case of Pernambuco, the land occupations that led to expropriations by the State allowed rural workers, who depended on the region's estate owners, to acquire the status of small producers installed in the new rural settlements created by INCRA. In the case of the state of Sao Paulo, where I conducted most of my fieldwork and research on the landless peasants, the rural settlements, the result of the mobilizations of the last four decades, occupy important extensions of land that used to belong to large landowners and ranchers.[12] In the far west of the state, for example, several properties that were known as "grileiros" lands[13] are now occupied by families of settlers, small rural producers whose main means of subsistence is the production of milk and vegetables. In the last of the settlements created in the region, 235 families have settled on the lands of a former unproductive 4,640-hectare farm, owned by a former mayor of one of the largest cities in the Pontal region. This hacienda, known as Nazareth, was disputed and re-occupied several times by landless families who suffered numerous dispossessions for more than 20 years. In 2017, the hacienda was finally expropriated for land reform purposes.

In just one decade, from the creation of the first rural settlements, according to data from the first report of the Data Luta database (1998), a total of 79 settlements were created as a result of massive land occupations in the state of Sao Paulo. And, according to data from the Land Institute of Sao Paulo (ITESP), up to 2019, this body counts 47,027 land titles issued, and a total of 140 rural settlements. The Pontal de Paranapanema region concentrates the largest number of settlements in the state, with a total of 98.

The first land occupations that culminated in the creation of the first rural settlements in the 1980s were promoted by members of the Catholic Church, specifically the Basic Ecclesial Communities (CEB) linked to the Pastoral Land Commission (CPT), by leaders linked to the PT (Workers' Party) and the MST (Fernandes 1999). Other similar alliances between rural unions, the Catholic Church, political parties, and other rural workers' organizations made possible the first mobilizations for land and the creation of agrarian reform settlements in other regions of the country. However, what seems to be a key point in all the experiences of land mobilization in the last 40 years is the participation of those who have already conquered a plot of land, the settlers, as shown by several case studies carried out in various regions of the country. In the case of the state of Sao Paulo, the settlers have been a central element in the social machinery of the mobilizations, the subsequent land expropriations and the maintenance of the agrarian reform territories. In an ethnography carried out in two landless camps and in one of the first settlements created in the state, I demonstrate the existence of a wide network of relatives, friends, and acquaintances of the settlers, who are mobilized or encouraged to participate in the occupations or to join an already constituted landless camp, forming what I called the "spiral of land occupations" (Rangel Loera 2006). These pre-existing relationships of kinship, friendship, and neighborhood are fundamental to bringing together new families willing to spend years under a black tarp hut waiting to be compensated with a plot of land. This 'spiral' connecting people and relationships extends temporally to the earliest settlements, giving us a diachronic view of at least 40 years of the history of land struggle in the state, and connects spatially as well, forming a broad network of settlements and encampments spaced over a wide territory of more than 40 municipalities in the state.

This 'spiral,' however, is not restricted to Sao Paulo; it extends to other states as well. Following the trajectory of a camp, I was able to map a network of relatives settled in the two states, Bahia and Sao Paulo (Rangel Loera 2010). Edismaria had even begun to participate in camps and landless movements in her home state, Bahia. Due to a series of circumstances and family problems and encouraged by her sister, settled in the municipality of Promissão in Sao Paulo, she decided to go to a landless camp in that state, where she would eventually settle. For Fernandes (1999), this social technology of mobilization, and the distribution of land with the creation of settlements, is the result of what this geographer calls the spatialization and territorialization of the struggle for land in Brazil.

> The grassroots efforts may be the result of the 'spatialization' or 'spatiality' of the struggle for land. Spatialization is a process of concrete movement of the action in its reproduction in space and territory. In this manner, the grassroots efforts may be organized by people who came from elsewhere, where they constituted their experiences.
>
> (Fernandes 2005, 320)

I will not go into the ethnographic details, but I think it is important to mention that in this mobilization process, especially when a large portion of land is in dispute, as was the case with the Nazareth hacienda mentioned above, the participation of settlers in the so-called *grassroots* work and in the *correrias* is central. The first term refers to the invitation made door to door in rural neighborhoods, towns, and cities neighboring the property, for people to participate in the occupation or in the newly formed camp. As part of these visits, informative meetings are also organized with local leaders who encourage the participation of their neighbors or acquaintances. The term "*correrias*" refers to the various negotiations that are carried out with local authorities, or with representatives of government institutions, or even with rural unions or *movement* leaders to resolve all kinds of practical issues related to the organization of mobilizations and the daily life of the campers. Thus, settlers who have already been provided with a plot of land and already know the region where a particular landless camp is located and maintain relations of proximity or even friendship with local leaders or authorities, are fundamental pieces in the implementation of this "technology of mobilization of families" (Macedo 2005, 473). They are also key to the maintenance, over the course of years, of landless camps located in the vicinity and to establishing the conditions for the possibility of the existence of newly formed settlements.

Data collected during my fieldwork conducted in the Pontal region in the last three years indicate that settlers from the settlement called Irmã Dorothy, one of the most recent in the region (created in 2016), maintain daily exchanges with settlers, relatives, and acquaintances from at least six other neighboring settlements. There is an intense circulation of domestic animals such as chickens, pigs, and even puppies, who are very often seen as guardians of the houses and plots. These animals are being added to the newly constituted family lots and are forming, in addition to the settlement's own families, extensive genealogies of relatives scattered throughout various settlements. A great variety of plants and seeds also circulate and are replanted and cultivated in the new plots of land, thus reproducing true gardens in the newly conquered territories. These exchanges are part of a wide network of humans and non-humans whose existence depends on the attention and continuous effort of the other, and this includes land, plots, pastures, all kinds of domestic animals, and neighbors. José, a former settler of the Pontal, summarized this assemblage of exchanges as being "*a whole system*," a form of interaction that constitutes a process of mutual creation and growth, of life support and cohabitation.

Relationships of livelihood and dwelling that Ingold (2000) considers central and constitutive processes of the *environment*.

Anthropologist Ana Luisa Micaelo (2016), in her ethnography conducted in the Araupema settlement in the Mata Pernambucana region, demonstrates the importance of fruit trees for the settlers. Fruit trees are considered perennial crops and become material evidence of the time and history of the settlers in the place. They are also, therefore, a way of legitimizing and affirming their belonging to this conquered land.

Pedro, an interlocutor, settled in Irmã Dorothy, showed me a variety of fruit trees and bushes in his orchard: papaya, acerola, avocado, mango, coffee, strawberry, passion fruit and others. Some of them he had collected from his parents' lot, who have lived for twenty years in another rural settlement in the same region. Pedro's plot is one of the 35 that make up the settlement and contrasts in size and diversity with other properties of large landowners in the region, where sugarcane monoculture prevails. Pedro, as well as other neighbors of the settlement, has been producing his vegetable garden, his plot of land and his house, thanks to the network of relationships he maintains with other settlers, relatives, and neighbors. Silvana, Pedro's sister-in-law, who is also settled in Irmã Dorothy, has daily recourse to her ex-sister-in-law, known as Preta, and her husband, José, who live in the neighboring settlement, which was established 25 years ago. Irineu, Silvana's husband goes daily to Preta's lot to fill bottles with whey from the milk left over from the morning. He uses it to feed his pigs daily.[14] It is also from the well of the neighboring settlement that Irineu and Silvana, in exchange for a small contribution, have access to water on their lot. As is common in the region, it is the older settlements that provide basic services and inputs to the newer ones, at least until state arrive. However, exchanges are maintained and intensified in moments when "*a precisão aperta [necessity imposes itself]*" as it is common to hear among the settlers. Daily, through small actions, which often go unnoticed even by the technicians or representatives of the State who circulate among the settlements, a whole network of aid and care is activated in this extended territory of the settlements, and it is precisely these daily nourished relationships which are part of the social machinery of the agrarian reform that make possible and create conditions of possibility for the democratization and diversification of land in Brazil.

Conclusions

In August 2020, bill 529 was sent for approval in the legislative assembly of the state of Sao Paulo, which imposes some measures for the auditing of public accounts and opens the possibility of demission, privatization, and extinction of public bodies. The Land Institute Foundation of the State of Sao Paulo (ITESP), which mediates the distribution of land that has been subject to expropriation, and also provides technical advice to rural settlements and communities in the state, is one of the bodies targeted by the new law. The project was approved in October of the same year, giving the endorsement for the extinction of six public entities of the state. This institute was maintained thanks to a strong reaction and mobilization by universities and various public institutions, organized civil society groups, among them the Landless Rural Workers Movement (MST) and non-governmental organizations, such as the renowned Socio-environmental Institute (ISA), whose main focus is the defense of social and collective rights related to the environment, cultural heritage, and human rights of the peoples.[15]

However, the continuity of ITESP's activities continues to be under threat. With few resources and no new technical staff hired for several years now, this foundation has been gradually dismantled, and several of its municipal offices have already been closed. The ITESP, although with a limited and often precarious performance, is a public organ of the state that plays an important role in the maintenance, advising, and diversification of production in the settlements, especially among those farmers who lack close support networks. Its performance, although often little noticed, is fundamental when it has been necessary to mediate conflicts and balance the scales of power in relation to large landowners in the region. This is the case, for example, of some of the settlements in two municipalities of the Pontal that had entire crops, fruit trees, and their silkworm production[16] completely destroyed due to the spraying of sugarcane plantations, close to the settlements, by a large sugar and agrofuel plant in the region. ITESP followed this case closely, even intervening in the mediation for the repair of the crops and the production of the settlers. As one of the ITESP technicians mentioned to me during an interview, "[they] are vigilant of the behavior of the large landowners in relation to the settlements, and the landowners know it."

The dismantling of this foundation and other public bodies related to land expropriations or to the guarantee of social rights is not only at the state level, but also a project of the national government that bets on the "every man for himself" mentality, when it is not about the government's own allies. The federal government has given support and publicly declared in favor of large landowners and agribusiness entrepreneurs, and, in this context, landless movements and camps, seen as enemies, are threatened and criminalized. As I mention in Rangel Loera (2019, 2020), the last three years have been the most violent for the landless since 2003. We witnessed at the national level the murders of peasants and movement leaders, violent dispossessions, evictions and dismantling of camps, but also of families of settlers, who, in the absence of a nearby support network, leave or abandon their lots as they can hardly maintain themselves without the support of the institutions and social programs of the State, currently in the process of being dismantled. In this context, the very existence of the settlements and of lands already expropriated and distributed, as a result of mobilizations of rural workers' organizations such as the MST, is also threatened. Families in this situation seek daily and creatively to intensify and strengthen ties and modes of existence in a conquered land. Once they have been provided with an agrarian reform lot, they begin fencing, planting, building their houses, having *criação*, producing links with their neighbors. Thus, *taking care of their land* become central concerns for my interlocutors, as they are ways of legitimizing their presence in that place.

It is undeniable that the existence of rural settlements in the Brazilian countryside, of interconnected, extensible and prolongable communities, sets in motion a material and symbolic logic that goes against the grain of the neoliberal rationality that predominates today. As Fjeld and Quintana (2019, 3) would say, "a neoliberal rationality as a de-democratizing logic that produces a tendential privatization of the world." This form, following these authors, "implies the reduction of spaces and forms of use and work in common." And although in the current situation, at least on the part of the Brazilian government, there seems to be a determination to destroy any other alternative to this dynamic, the daily life and continuity of the agrarian reform settlements show us that, in spite of the attacks and difficulties, the production of life in common in these spaces is persistent. Pedro, one of my interlocutors, explained to me, with regard to his lot, that some fruit trees had taken a little longer than others to grow and bear fruit, and he compared them with his life in the settlement: *trees are like us, their roots have to get used to the land, to their new place, they look for their little place in the ground, and when they settle down and get used to it, then they grow upwards and give us fruit that later give us other trees.*

Notes

1 Local/native terms will be highlighted in italics.
2 As Servolo de Medeiros (2002:36) explains; it is in the 1988 constitution that agrarian reform appears for the first time in the constitutional text. "The 1988 Constitution has agrarian reform inscribed in its text as a theme of the chapter on 'economic and social order'. The text states that property should fulfill its social function (art. 5, XXIII)."

3 As I show in another work (Rangel Loera, 2015b), it is common to relate the Landless Movement, almost metonymically, with the MST. However, until the late 1990s there were, at the national level, more than 100 peasant organizations that acted in mobilizations related to the demand for land, including rural unions or other entities such as the Pastoral Land Commission (Comisión Pastoral de la Tierra, CPT). The MST undoubtedly became the largest peasant movement at the national level during this period. However, the Landless Movement is made up of this group of rural workers' organizations that have occupations and encampments as an instrument of struggle for land.

4 It is important to mention that, in Brazil, the land issue linked to the indigenous question has also been a central and disputed issue throughout the 20th century, but has gained greater visibility in the public and international debate since the 1988 Constitution. However, the indigenous issue is a separate debate among scholars of the agrarian question and the peasantry in the Brazilian context. Even institutionally, for the Brazilian State, the land management policy in relation to these two categories of subjects (indigenous and peasants) is carried out, separately, by two different government bodies, FUNAI (National Indian Foundation) and INCRA (Institute of Colonization and Agrarian Reform).

5 Sigaud (2000) in an innovative analytical proposal based on the encampments, the object of his ethnographic research in the Mata Pernambucana region of northeastern Brazil, formulates that the 'encampment form' became a model of social demand legitimized by the Brazilian State, once the occupation of land and the setting up of an encampment, rural workers' organizations 'tell' the State which properties should be expropriated and the State 'responds' by opening a process of technical review of these lands to determine whether or not they are unproductive and whether or not they can be used for agrarian reform, i.e., whether they can be distributed among land claimants. Sigaud's formulation has been used in other ethnographic contexts to analyze similar processes of social demand. This is the case, for example, of the 'piquete form' in Buenos Aires, Argentina (Manzano, 2009) and of the 'retaken form' among the indigenous Tupinambá (Alarcón, 2013) and the Guaraní and Kaiowá, in Mato Grosso do Sul (Corrado, 2013).

6 As shown by studies on the sociogenesis of land mobilizations in the country, occupations and encampments of rural farmers are commonly associated with the origin of the MST "as if there were a symbiosis between the act of occupying and organizing" (Sigaud, Ernandez and Carvalho Rosa, 2010: 14), and although indeed, according to the cited authors, some founders of the MST were present in 1979 in the occupation of the Macali farm, considered by the MST itself the zero frame of the movement in the state of Rio Grande do Sul, there are precedents of these repertoires of action in the states of Rio Grande do Sul and Rio de Janeiro in the period prior to 1964, the year of the military coup in Brazil.

7 Land conquest is the name given by MST militants to the action of occupying new unproductive lands and/or rural settlements: communities or rural productive units, as a result of the processes of expropriation and distribution of land by the Brazilian State. Already, the militant category is how those who have an organic link with the MST and are part of the leading cadres of this movement call themselves.

8 Data from the research project "As formas de acampamento" (2010/02331-6) funded by Fapesp.

9 *Pariente* is a generic term present in several ethnic groups in Brazil to designate those who are indigenous but it can also designate the link to a specific ethnic group, or even the link to an extended indigenous kin. In this context the term also referred to their Guarani Ñandeva ancestors. For more details on this case see the work of Maiane Fortes (2016).

10 Rural settlements are made up of a group of plots of land that have been divided and granted by the State (through INCRA) to landless families as a concession of use. Their size, format and organization can vary greatly from region to region. They are also made up of a heterogeneity of agricultural and livestock farming activities and a diversity of crops, but also of families, which can sometimes accommodate two or three generations on the same plot. They are also spaces that, over time, form configurations of plots, to which relatives and non-human animals are added. Rural workers participating in massive land occupations are known as *campers* (acampados). When those workers finally get a plot of land, they are locally known as *settlers* (assentados).

11 As shown in the "Atlas of the Agrarian Question" available at http://www2.fct.unesp.br/nera/atlas/luta_pela_terra.htm, the areas occupied by landless movements and considered as "reform settlements" are mainly large estates and rural properties where labor and environmental laws were not respected.

12 According to Mattei (2005: 354) based on 1998 IBGE data, Brazil, until the beginning of the 1990s, was one of the countries with the highest land concentration in the world. Landowners, with more than 2000 hectares of land, owned 43% of the land in the country. Small landowners, with less than 10 hectares of land, occupied 2% of the national territory.

13 Fernandes (1999) takes up in his study several land registries in this region whose titles were falsified and were the scene, since the 1960s, of numerous conflicts between the occupants who lived and worked there, the so-called 'posseiros' (squatters) and those who were known as 'grileiros'. The latter appropriated the land by presenting forged ownership papers. To give them the appearance of being old, they were meticulously written in handwriting no longer in use, old imperial seals were affixed to them and they were made to 'yellow' by keeping them in drawers full of crickets (grilos), whose urine made them yellow. Hence the word 'grileiros'.

14 In the Brazilian rural world, the set of farmed animals in small lots are generically called *criação*, the same term that is used to refer, for example, to foster children [filhos de criação], or those children that have been affectively and socially added to a family unit.

15 For more information on the approval of this law and the importance of the performance of this public body for rural communities see ISA's report published on October 15, 2020:
https://www.socioambiental.org/pt-br/noticias-socioambientais/instituto-que-reconhece-quilombos-de-sp-sobrevive-ao-pl-529-mas-ha-risco-de-sucateamento.

16 In the Pontal region there is a group of 20 settlers who are dedicated to silkworm production. In their plots they mainly plant blackberry, and the leaves of this plant are used to feed the silkworms. The production is marketed to a company established in the neighboring state of Paraná, and exports the production of silk threads. These settlers, distributed in 4 settlements in the region, also have kinship ties among themselves and have inherited not only the knowledge of this type of production to their children and grandchildren, but also the trade relations.

References

Alarcon, Daniela. 2013. "A forma retomada: contribuições para o estudo das retomadas de terra, a partir do caso Tupinambá da serra do Padero". *Revista Ruris*: 7 (1): 99–126.

Almeida, Barbosa Mauro. 2007. "Narrativas agrárias e a morte do campesinato." *Revista Ruris* 1 (2): 157–186.

Corrado, Elis. 2013. "Acampamentos kaiowá. Variações da 'forma acampamento'." *Ruris* 7 (1): 127–151.

Dataluta. 1998. Banco de dados da luta pela terra: relatório das ocupacões de terra. *Unesp*. Available at http://www2.fct.unesp.br/nera/projetos.php.

Fatos da Terra, N. 27. 2019. Fundação Instituto de Terras do estado de São Paulo (ITESP).

Fernandes, Mançano Bernardo. 1999. *MST. Formação e territorialização*. São Paulo: Hucitec.

Fernandes, Mançano Bernardo. 2005. "The occupation as a form of access to land in Brazil: a theoretical and methodological contribution." In *Reclaiming the land. The resurgence of rural movements in Africa, Asia and Latin America*, edited by Sam Moyo and Paris Yeros. London/New York: Zed Books.

Fjeld, Anders and Laura Quintana. 2019. "Reinstitucionalización, formas de vida y acciones igualitarias: reinvenciones de lo común hoy contra el capitalismo neoliberal." ["Reinstitutionalization, life forms and egalitarian actions: reinventions of the common today against neoliberal capitalism."] *Revista de Estudios Sociales* 70: 2–9. https://doi.org/10.7440/res70.2019.01.

Fortes, Maiane. 2016. "As retomadas Guarani Ñandeva de Barão de Antonina e Itaporanga-SP: Etnicidade como linguagem de demanda e (re) apropriação territorial." Bachelor's thesis. Universidade Estadual Paulista Jullho de Mesquita Filho (UNESP).

Heredia, Beatriz. 1989. *Formas de dominação e espaço social: a modernização da agroindustria canavieira em Alagoas*. São Paulo: Marco Zero.

Ingold, Tim. 2000. *The perception of the environment. Essays on livelihood, dwelling and skills*. London and New York: Routledge.

Macedo, Marcelo Ernandez. 2005. Entre a "violência" e a "espontaneidade" reflexões sobre os processos de mobilização para ocupacões de terra no Rio de Janeiro. *Mana* 11(2): 473–497. https://dx.doi.org/10.1590/S0104-93132005000200006.

Manzano, Virginia. 2009. "Piquetes y acción estatal en Argentina: Un análisis etnográfico de la configuración de procesos políticos. [Piquetes and state action in Argentina. An ethnographic analysis of the configuration of political processes]." In *Estado y movimientos sociales: estudios etnográficos en Argentina y Brasil. [State and social movements: ethnographic studies in Argentina and Brazil]*, edited by Mabel Grimberg, Maria Ines Fernández A. and Marcelo Rosa, 15–36. Buenos Aires: Antropofagia.

Mattei, Lauro. 2005. "Agrarian reform in Brazil under Neoliberalism: evaluation and perspectives." In *Reclaiming the land. The resurgence of rural movements in Africa, Asia and Latin America*, edited by Sam Moyo and Paris Yeros. London/New York: Zed Books.

Menezes, Marilda. 1997. *"Peasant-migrant workers: social networks and practices of resistance."* PhD diss., University of Manchester.

Micaelo, Ana Luísa "Essa terra que tomei de conta". *Pertencimento e territorialidade na zona da Mata de Pernambuco*. Lisboa: ICS, 2016.

Rangel Loera, Nashieli. 2006. *A espiral das ocupações de terra*. São Paulo: CERES/POLIS.

Rangel Loera, Nashieli. 2010. "'Encampment time': an anthropological analysis of the land occupations in Brazil." *The Journal of peasant studies* 37 (2): 285–318. https://doi.org/10.1080/03066151003594930.

Rangel Loera, Nashieli. 2015a. "Mecanismos sociais da reforma agrária em São Paulo pelo viés etnográfico." *Lua Nova: Revista de Cultura e Política* 95: 27–56. https://doi.org/10.1590/0102-6445027-056/95.

Rangel Loera, Nashieli 2015b. *Tempo de acampamento*. São Paulo: Unesp.

Rangel Loera, Nashieli. 2019. "Liberté et assujettissement comme conditions de possibilité au Brésil: notes sur la production quotidienne 'd'ennemis'." In *Liberté de la recherche. Conflits, pratiques, horizons*, edited by Duclos Mélanie and Anders Fjeld, 107–118. Paris: Kimé.

Rangel Loera, Nashieli. 2020. "Uno jala al outro": barracos and movements as social gears in the world of land occupations in the Brazilian context. *Revista de Antropología Social* 29 (2): 167–184. https://dx.doi.org/10.5209/raso.71679.

Rosa, Marcelo. 2012. "A Terra e seus Vários Sentidos: por uma Sociologia e Etnologia dos moradores de fazenda na África do Sul contemporânea." *Revista Sociedade e Estado* 27 (2): 361–385. https://doi.org/10.1590/S0102-69922012000200008.

Rosa, Marcelo. 2011a. *O engenho dos movimentos. Reforma agrária e significação social na zona canavieira de Pernambuco*. Rio de Janeiro: Garamond.

Rosa, Marcelo. 2011b. "Mas eu fui uma estrela de futebol! As incoerencias sociológicas e as controvérsias sociais de um militante sem-terra sul-africana." *Mana* 17 (2): 364–394. https://doi.org/10.1590/S0104-93132011000200005.

Servolo de Medeiros, Leonilde. 2002. *Movimentos Sociais, Disputas Políticas e Reforma agrária de Mercado no Brasil*. Rio de Janeiro: CPDA/UFRRJ/UNRISD.

Sigaud, Lygia. 1979. *Os clandestinos e os direitos: estudos sobre trabalhadores da Cana de açúcar de Pernambuco*. São Paulo: Duas cidades.

Sigaud, Lygia. 1980. *Greve nos engenhos*. Rio de Janeiro: Paz e Terra.

Sigaud, Lygia. 2000. "A forma acampamento. Notas a partir da versão pernambucana." *Revista Novos Estudos Cebrap* 58: 73–92.

Sigaud, Lygia, Gautié, David Fajolles, et al. 2006. "Os acampamentos da reforma agrária: história de uma surpresa." In *Ocupações de terra e transformações sociais*, edited by Benoît de L'Estoile and Lygia Sigaud, 7–18. Rio de Janeiro: FGV Editora.

Sigaud, Lygia, and de L'Estoile, Benoit. 2006. Introdução. Uma etnografia coletiva em terras pernambucanas. In *Ocupações de terra e transformações sociais*, edited by Benoît de L'Estoile and Lygia Sigaud, 29–63. Rio de Janeiro: FGV Editora.

Sigaud, Lygia, Ernandez, Marcelo, and Rosa, M. 2010. *Ocupações e acampamentos. Sociogênese das mobilizações por reforma agrária no Brasil*. Rio de Janeiro: Garamond Universitaria/FAPERJ.

Stédile, João Pedro and Fernandes, Bernardo Mançano. 1999. *Brava gente. A trajetória do MST e a luta pela terra no Brasil [Brave people: the trajectory of the MST and the struggle for land in Brazil]* São Paulo: Editora Fundação Perseu Abramo.

Yie, Garzón Maite. 2018. "¡Olha só, os camponeses aqui estamos: etnografia do (re)aparecimento do campesinato como sujeito político nos Andes nariñenses colombianos" ["See, we the peasants are here! Ethnography of the re (emergence) of the peasantry as a political subject in the Colombian Nariñense Andes."] PhD thesis in Social Sciences, Unicamp.

25 From Chico Mendes to Berta Cáceres

Responses to the murders of environmental defenders

Diana Jiménez Thomas Rodríguez, Grettel Navas, and Arnim Scheidel

Introduction

'Environmental defenders' (hereafter defenders) are participants in collective struggles to protect nature against socially and ecologically destructive development projects (UNEP 2018). In response to their activism, defenders face criminalization, threats, aggression, and assassination. The violence that defenders face has led to global declarations and the creation of various international frameworks for their protection (ibid.; UN 2018). Despite these efforts, defenders have suffered increasing levels of threats and violence over the past decade (Butt et al. 2019; Middeldorp and Le Billon 2019). Latin America is one region where this trend is most acute, as 60% of the murders of defenders have occurred in Latin America. Violence against defenders is most extensive in Colombia, Brazil, Guatemala, Mexico, and Honduras (Global Witness 2019).

As a response to increasing violence, the United Nations Environment Program (UNEP) launched two protection initiatives in 2018: the Defenders' Policy and the Environmental Rights Initiative. There is also a growing body of literature regarding defenders' protection. Knox (2015) and Tanner (2011), for example, argue that the effectiveness of institutionalized instruments is limited by a lack of specificity in their content and by the enforcement capacity of states. Moreover, Esguerra-Muelle et al. (2019) and IM–Defensoras (2013) argue that such mechanisms fail to address the particular forms of gendered and racialized expressions of the violence that female defenders face, such as smear campaigns and sexual harassment (Deonandan and Bell 2019; Tapias Torrado 2019; Sieder 2017).

In this chapter, we aim to contribute to the understanding of the ways in which the Latin American and international community have responded to the murder of environmental defenders, as well as the limitations of such responses in preventing future violence against defenders. We take as examples the responses to the high-profile murders of two defenders: Chico Mendes (murdered in Brazil in 1988) and Berta Cáceres (murdered in Honduras in 2016). Chico Mendes (1980–1988), the leader of Acre's Union of Rubber Tappers, was committed to contesting land grabs in the Brazilian Amazon, and Berta Cáceres, an indigenous Lenca leader in Honduras, spent most of her life fighting for the rights of indigenous peoples, in particular indigenous women.

Both assassinations drew a range of responses. We propose a schema which maps these responses onto two axes: 1. their institutional character, whether or not they come from within state or inter-governmental institutions, and 2. their transformative capacity, the extent to which they challenge the dominant extractivist paradigm. We argue that conventional institutional and non-institutional responses are crucial to address the violence perpetuated against defenders. However, as these responses are inherently limited by their individualized character and their focus on retributive justice, we argue that the more transformational responses aiming

DOI: 10.4324/9780429344428-29

to recast the logic of extractivism are those that protect environmental defenders best in the long run. If these more transformational responses are to increase their effectiveness, they need to expand beyond a narrow understanding of what violence is. Doing so would expand responses beyond demands for retributive justice and toward demands for the protection of the territories and ways of life protected by defenders. On this basis, the chapter will illustrate how responses to violence against defenders can be understood and it will establish why a change in the way we approach activist protection is vital.

The legacy of Chico Mendes

During the mid-1960s, the Brazilian military government aimed to economically integrate the Amazonia into the country by opening the region to the transnational private sector and promoting new migration patterns (Hecht and Cockburn 2010). This was done through *Operação Amazônia* (Operation Amazonia) and its discourse of national development, productivity, and sovereignty, combined with an imaginary of the Amazon as *terra nullius*. The government backed investments and fiscal incentives for the expansion of cattle and industrialized agriculture, and began to develop infrastructure for new industries, such as hydroelectric plants and transregional highways. It promoted the 'occupation' of the Amazon through ambitious migration programs, relocating its vast landless population of approximately 30 million people from the Northeast to the Amazon region. However, as most migrants were not able to secure the promised land rights, a majority of them were pushed into the rubber trade as *seringueiros*, people who collect and sell latex (or *seringa*), despite exploitative labor conditions (Shoumatoff 1991; Hecht and Cockburn 2010). As cattle ranching expanded, land disputes arose. Previous rubber estates and 'new' land were sold and cleared for cattle ranching and industrialized agriculture, destroying the rainforest-based economies and livelihoods of seringueiros, peasants, brazil-nut gatherers, and indigenous peoples (Mendes 1989).

Francisco Alves Mendes Filho, better known as Chico Mendes, became a key figure in this context. Like his father and grandfather, Mendes was a rubber tapper. He received his education from Euclides Fernandes Távora, a communist revolutionary and trade unionist. His contact with Tavora prompted him to join the Rural Workers' Union (*Sindicato dos Trabalhadores Rurais* or STR) in 1976, and to create the local chapter of the Union in Xapuri in 1977. Founded in 1975 by Wilson Pinheiro, who led the union until his assassination in 1980. The STR sought to liberate rubber tappers in Acre from peonage and prompted seringueiros to collectively sell to middlemen and to refuse to pay rent to *seringalistas*—rubber estates landowners. As government policies began to shift the dominant economic activity in Acre from rubber to cattle-ranching, the Union switched to protesting against massive land grabs by a new elite. It was in this context that Pinheiro devised the 'standoff' or *empates*, a strategy in which human blockades prevent the entry of bulldozers and other machinery into the rainforest (ibid.).[1]

After Pinheiro's assassination in 1980, Chico assumed leadership of the STR. Under his leadership, the movement gained national and international attention and advanced an alternative economic model for the Amazon. In fact, the international recognition that anthropologists Mary Allegretti and Steve Schwartzman sought to instill was, in part, intended to protect Chico, as he had already received death threats and had survived several assassination attempts (Shoumatoff 1991).

By 1987, his international recognition was at its peak. Chico received the Global 500 Roll of Honor Award bestowed by UNEP, as well as an award from the Better World Society, which reframed the STR's struggle from one for workers' rights to one for the environment. In that same year, Mendes flew to the United States to meet with the Inter-American Development Bank (IDB) and the US Senate Appropriations Committee, the body responsible for releasing funds to multilateral financial institutions, to dissuade both from funding the extension of the BR-364 highway[2] in Brazil—a key project for the expansion of cattle ranching in the country (ibid.).

While meeting with the IDB, Mendes proposed an alternative model for the Amazon—'extractive reserve' or *resex*. Resexs are environmental conservation zones under public owner-ship which continue to be inhabited by local communities with life-long usufruct rights to the use and management of rainforest resources. For *seringueiros*, extractive reserves institutionalized their access to land and its collective management (Allegretti 1998, 2008; Barbosa de Almeida et al. 2018). By simultaneously protecting rainforest areas from deforestation while sustaining local livelihoods, the model of resexs showed how ecological conservation and social justice intertwine into what is known as 'environmental justice'. A year after Chico's visit, the IDB and the Senate stopped all funding for BR-364. The World Bank declared cattle ranching a destruc-tive form of colonization and the extractive reserve a promising alternative (Shoumatoff 1991).

Chico Mendes was murdered in his home on December 22, 1988 by two cattle ranchers. Chico had been vocal about the death threats he had received earlier that year. He appealed to authorities to intervene, and although he was granted police protection, police were woefully unprepared. Chico's murder was the 90th assassination of a rural worker in Brazil in 1988 (Gross 1989), but, unlike prior murders, his did not go unnoticed. A few hours later, it made headlines in national and international newspapers, inspiring campaigns in solidarity with the STR's struggle and its defense of the Amazon.

Thanks to international pressure, his murderers did not escape. Both culprits, Darly and Darcy Alves, were sentenced to 19 years' jail. However, charges against them for other assassi-nations were never investigated. The role of *União Democrática Rural* (Rural Democratic Union or UDR), a right-wing vigilante organization of cattle ranchers and landowners which was suspected to be the intellectual authors of the murder, remained unprobed.

Chico's murder had concrete consequences for the Amazon. In 1990, the first *resex*, Alto Juruá, was created, and the presidential decree 98.987-01/30/1990 which established it became the first legal instrument to recognize extractive reserves in Brazil. Ten years later, *resexs* were part of the newly-created National System of Protected Areas (Law 9.985), which formalized a collection of conservation models at different governance levels. By 2018, 76 *resexs* had been created in the Amazon rainforest, covering 14 million hectares—today the most used model of people-based conservation in the region (Aguiar Gomes et al. 2018). Furthermore, in 1996 the Chico Mendes Memorial was established as an advisory group to support local communities on technical matters, networking, and the promotion of sustainable extraction of natural resources.

Chico Mendes has been widely celebrated for his work. The arts, especially music and film, pay him homage. Songs, such as "Ao Chico" by Tião Natureza, and the documentary *Chico Mendes, Eu quero viver* by Adrian Cowell (1989), stand out. Belatedly, the Brazilian government has recognized Chico by creating the Chico Mendes Institute of Biodiversity Conservation (ICMBio) in 2007 and naming him Patron of the Brazilian Environment[3] (Watts 2013).

The seeds of Berta Cáceres

In 2009, the coup d'état organized against Manuel Zelaya reversed a democratization process underway since the 1980s, and halted significant agrarian reforms designed to support peasant organizations. As foreign private investment was promoted, monocultures expanded, moratoria on mining concessions were withdrawn, and private sector plans for hydroelectric dams set in motion (León Araya 2015; Flores and Ardón 2017). Between 2010 and 2017, the Honduran government specifically targeted defenders who contested privatization processes via land claims (Middeldorp et al. 2018). Indigenous and peasant defenders were among those most commonly targeted. Berta Cáceres Flores (1971–2016) was one such case.

Berta, an indigenous Lenca leader of the Intibucá region, dedicated her life to fighting racial and gender discrimination and protecting indigenous territories. In 1993, Berta and Salvador Zuñiga founded the Civic Council of Popular and Indigenous Organizations of Honduras (COPINH) as a response to illegal logging and its effects on the wellbeing of indigenous communities in Honduras. Its mission is to protect the environment and promote indigenous rights and worldviews, with a particular concern for Lenca culture and beliefs, and to support indigenous women's rights and quality of life (Goldman Prize 2016; COPINH 2020a; Korol 2018).

The organization has to date confronted many regional development projects, in particular contesting 17 water privatization projects (Lewis 2016). In Lenca, cosmology, rivers and other bodies of water are agents in and of themselves as well as sacred beings (Méndez 2018). As Berta would later declare in her acceptance speech of the Goldman Environmental Prize in 2015:

> In our worldviews, we are beings that have emerged from the earth, water and maize. We, the Lenca people, are ancestral guardians of the river… The Gualcarque river has called us, as all the others who are seriously threatened all around the world…
>
> (Avila 2016)

In contesting the privatization of rivers, Berta's work in COPINH challenged the notion of development that instrumentalizes nature and made environmental wellbeing central to the program of social justice for indigenous communities.

Equipped with Convention 169 of the International Labor Organization (ILO),[4] by which signatory countries are obligated to obtain prior informed consent from indigenous peoples regarding any projects that impact their territories, COPINH challenged numerous water concessions. It confronted the state, major transnational companies like Sinohydro and Siemens, and bilateral/multilateral institutions such as the World Bank, Finnfund, and USAID (Lewis 2016; Korol 2018).

In 2015, Berta Carceres was awarded the Goldman Environmental Prize for her work, particularly for her leadership in the fight against the Agua Zarca dam project. In 2016, members of the community of Río Blanco asked COPINH for help to contest the concession, as the project would destroy the Gualcarque river and displace Lenca communities. That same year, Berta was murdered.

Her assassination was the culmination of the violence that Berta had experienced throughout her activism. She had been intimidated, threatened, criminalized, and temporarily detained since 2013, despite an Interamerican Human Rights Commission (CIDH) resolution mandating that the Honduran state guarantee her protection (Frontline Defenders 2018; CIDH 2018). In fact, Berta continually expressed fear for her life, evidenced by the complex protocols she followed to maximize her safety (Lewis 2006; Korol 2018). However, on March 2, 2016 a group of men fatally shot her and wounded Gustavo Castro, a Mexican activist.

Cáceres' death provoked a national outcry. COPINH and Berta's family demanded an investigation and the prosecution of both the material and intellectual authors of her death, highlighting that those who had perpetrated the crime were not the only ones responsible (COPINH 2020b; Lakhani 2018a). Among the people identified as the intellectual authors of her murder was David Castillo, an ex-military officer and DESA's president. The campaign demanding an investigation and trial soon acquired the label and online hashtag #JusticiaparaBerta.[5]

Berta's death resonated beyond national borders. The United States, Spain, and the European Commission all issued statements about her death. Likewise, the CIDH, a group of 50+ international organizations, and other well-known international human rights activists repudiated her murder. The responses at both national and international levels relied on the use of the web and social media. COPINH created a website dedicated specifically to Berta (see COPINH, 2022), together with social media accounts on Twitter and Facebook.

Following national and international outcry, a long judicial process convicted the seven men who perpetrated the attack. However, attempts to investigate the intellectual authors of the crime and to hold all of those involved in the attack fully accountable have been largely futile (Lakhani 2018a, 2018b; Sandoval 2019; CIDH 2018). One recent exception is David Castillo, soon to face trial in Honduras (Lakhani 2021).

People continue to protest and demand justice for Berta. Two iconic phrases have appeared in protests: "*Todas somos Berta*" (We are all Berta) and "*Berta no se murió, se multiplicó*" (Berta did not die, she multiplied). Through these phrases, Berta's murder has been re-signified by activist and defender communities. Her murder transformed her into a seed planted in the Earth and growing within every environmental activist who gathers under her image.

Berta's death is also commemorated through the arts. Street art has emerged worldwide celebrating Berta's struggle. In 2016, an artistic anthology was created under the direction of the *Biblioteca de las Grandes Naciones*. Several documentaries and songs pay tribute to her, such as Campaña Madre Tierra's film *Guardiana de los Ríos* (2016). In 2018 and 2021, COPINH organized (virtual) art festivals for Berta with the participation of a number of Latin American artists.

Berta's murder may have played a role in prompting the Escazú Agreement (2018). Signed on the second anniversary of her death, this Agreement represents the second legal regional instrument, after the Aarhus Convention, to implement Principle 10 of the 1992 Rio Declaration establishing the importance of citizen participation in environmental matters (UN 2018). The treaty is designed to function as a tool for environmental activists, recognizing their right to access relevant information regarding environmental issues and urging the state to assume responsibility for protecting the rights of defenders (Stec and Jendrośka 2019).

Lastly, as evident in her rhetoric, Berta understood that environmental injustice results from links between capitalism, racism, and patriarchy (Korol 2018; Lewis 2016). Her death has promoted a consequent dialogue about the intersection of these three systems and called attention to women environmental defenders—in particular, the contributions they make and the gender-based injustice and violence they face (Red Latinoamericana de Mujeres Defensoras et al. 2018).

Situating responses to environmental defender killings

The responses to the murders of Chico and Berta fall into eight distinct groups, as shown in Table 25.1.

Table 25.1 Types of responses of the murders of environmental defenders

1. International pronunciations
2. Retributive justice for direct perpetrators
3. Retributive justice for intellectual authors of murders (in Berta's case only, and in process)
4. Creation of national/international legislation and/or policy (ie. the presidential decree 98.987–01/30/1990, and the Escazú Agreement)
5. Financial decisions of multilateral agencies
6. Creation of artistic and media content
7. Protests and mobilization of civil society demanding retributive justice for both material and intellectual authors of the murders
8. Protests in solidarity with the struggles of Chico and Berta (pushing for the incorporation of resexs into Brazilian law and for the halt of the Agua Zarca project, respectively)

The responses noted above vary according to two factors. The first is whether they are 'institutional' responses, that is, responses that occur within organs of state (from prosecution, government declarations and policies, the creation or reform of legal frameworks) or inter-governmental organizations (for instance, declarations and new policy frameworks). 'Non-institutional' responses, by contrast, come from within civil society but fall outside the formal channels of state or inter-governmental organizations (i.e., street protests, digital activism, and artistic commemorations).

Secondly, responses vary by the extent to which they seek to challenge extractivism as a paradigm—a scale we term 'transformative potential'. By extractivism, we mean an economic and political process which rely on resource extraction along unequal international patterns, and which tend to create environmental harm and deepen socio-economic inequality (Svampa and Viale 2014; Li 2015; Gudynas 2009). More 'conventional' responses seek to address the individual incident of a particular murder, accepting as given the current economic model of the state, whereas 'transformative' responses challenge that paradigm by demanding the recognition of nature's rights. Unlike 'institutional character', transformative potential should be thought of as a continuum, allowing a nuanced analysis of responses as shown below.

Figure 25.1 depicts where the responses to Berta and Chico's murders fall within this schema. Institutional responses include (from least to most transformative): international pronunciations, retributive justice, the creation of international frameworks, financial decisions of multilateral agencies, and the creation of new national policies and legislation. Non-institutional responses include (from least to most transformative): civil society mobilization demanding retributive justice for both material and intellectual authors of the murders, the creation of artistic and media content, and protests in solidarity with the struggles of Berta and Chico.

In both cases, there are institutional responses, such as national and international pronunciations and retributive justice against the murder's perpetrators. These responses fall to the conventional side of the spectrum, as they use already available state mechanisms, are highly individualized, and do not address the root causes of the murder of either defender. The creation of new international frameworks is slightly more 'transformative'. It moves away from an individualized response, seeking to create new legal avenues and protections for environmental defenders more generally. Though these responses seek to enhance the protection of defenders, they do not address the root causes of violence against them. Furthermore, such international frameworks tend to be limited by a lack of specificity, a historical contextual base

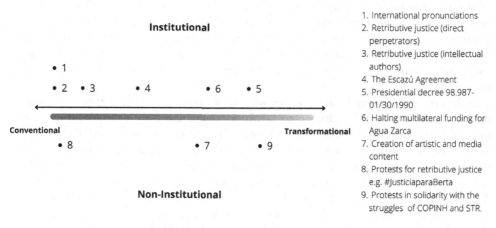

Figure 25.1 Situating the responses to the murders of Cáceres and Mendes according to the proposed schema.

(Knox 2015), scale mismatches, and tensions surrounding sovereignty this produces (Stec and Jendrośka 2019).

Two responses fall closer to the middle of the spectrum or lean toward the transformative end: the withdrawal of funding for Agua Zarca and the presidential decree which institutionalized extractive reserves. These two responses, due in large part to civil society pressure, challenge extractivism to different extents. The withdrawal of funding halted the privatization of the Gualcarque River, but no others, and, unfortunately, was likely the product of bad press rather than a reconsideration of water privatization. The institutionalization of extractive reserves was more transformative as it institutionalized the paradigm of environmental justice. However, as *resexs* are not yet the main model in the Amazon, the transformative potential of this response has been limited.

While institutional conventional actions are essential, their efficacy in eliminating violence against defenders is limited. Firstly, they fail to address the root causes of violence against environmental defenders—that is, the extractivist politics at odds with environmental justice. Secondly, these strategies tend to be highly individualized, and to focus only on violence against high-profile individuals, such as in the case of efforts for retributive justice and existing protection mechanisms. This individualized approach makes these responses incompatible with the collective form that grassroots resistance tends to follow,[6] as well as rendering them unscalable to collective protection. This leaves less well-known defenders vulnerable to aggression, as violence against them goes unpunished. This is evident in the STR's struggle, where the murders of Pinheiro and others were not treated with the same importance as that of Chico. Thirdly, as our cases show, institutional strategies are limited by the co-optation of the state by local and transnational elites. In the case of Chico, this resulted in a lack of political will and capacity to investigate the role of the UDR in his murder. Lastly, our cases show that institutional strategies can be limited by the weak institutional capacities of the state, and its lack of institutional transparency and accountability. This not only makes retributive justice harder to accomplish; it also undermines the effectiveness of international frameworks, such as the Escazú Agreement, as its implementation is in the hands of the state.

Given the barriers to effective institutional responses, non-institutional responses have been crucial. The strength of non-institutional responses lies in their capacity to exercise 'soft power,' what Nye (2009) explains as the capacity of a given actor to shape the actions and preferences of other actors without the use of coercion. Street protests, digital activism, and pronunciations of well-renowned individuals use soft power by engaging in what Keck and Sikkink (1998) call 'naming and shaming'. This strategy uses international norms to increase the political costs of a given action or inaction, and consequently to persuade the actor in question to comply with the demand. Non-institutional conventional responses attempt to circumvent, through political and social pressure, barriers to institutional responses such as political apathy, corruption, or entrenched interests.

In the cases of both Mendes and Cáceres, we find that non-institutional responses encompass the creation of artistic and media content, civil society mobilizations, demands for retributive justice for both material and intellectual authors, and protests in solidarity with their environmental struggles. Both cases show a strong and extensive use of the arts, from music and painting to film and poetry. These responses are located near the center of the spectrum toward the transformative end, as art contributes to change toward social justice (hooks 1995; Milbrandt 2010). In the case of street art, Latorre (2019) shows how murals are a way of writing an oppositional history, contesting dominant narratives, and creating and communicating alternative knowledge. Consequently, artistic responses have the potential to contest extractivism and deepen a conversation about how to move beyond it.

The public outcry at Mendes and Cáceres' deaths created significant pressure, instrumental in the prosecution of the perpetrators of the crimes. In both cases, the outcry succeeded in having the material authors convicted. In the case of Berta, the campaign #JusticiaparaBerta led to the arrest and trial of one of the suspected intellectual authors of her murder. These protests were highly individualized, demanding retributive justice for both defenders, but were mostly silent on the larger question of the extractivist model itself. Thus, they are located toward the conventional end of the spectrum. The same applies to existing initiatives and campaigns within civil society that have emerged to counteract the violence that defenders face such as: the work of Global Witness and Frontline Defenders, who produce annual reports with much-needed statistical data and up-to-date information on at-risk defenders;[7] the Environmental Defender Law Center (EDLC) and Earth Rights International, who offer direct legal support to environmental activists; and award schemes, such as the Goldman Environmental Prize and the Equator Prize which recognize and increase the visibility of defenders.

In contrast, other protests in solidarity with the struggles of Berta and Chico are located toward the transformative end of the spectrum. In the case of Brazil, Mendes' murder and the outcry that followed contributed to the increased visibility of the socio-environmental concerns that were emerging in the country, and to the creation of a legislative framework for extractive reserves. In the case of Berta, the protests following her murder pressed for the withdrawal of financial support for Agua Zarca, and increased awareness of the intersectionality of indigeneity and gender in environmental conflicts.[8] These protests in solidarity with Mendes and Cáceres' struggles, unlike campaigning for retributive justice, demanded at least some shift away from extractivism. However, they also remained limited insofar as they targeted the particular conflict that led to these two activists' deaths, and did not expand into other environmental conflicts or environmental justice more generally. Regardless, protests in solidarity remain crucial, particularly since paradigm change is not likely to come willingly and organically from within the state.

Protests in solidarity can also be transformative because of the discourses present therein. In the phrases of *"mataram nosso líder, mas não nossa luta"* [they killed our leader, but not our struggle] (Gross 1989, 79) in the case of Chico, read in a banner at his funeral, and of *"Todas somos Berta"* and *"Berta no se murió, se multiplicó,"* chanted at protests against her murder, defenders are performing acts of subversion and protection. They reject violence as a strategy to provoke fear and silence, announcing how violence against defenders multiplies resistance instead of erasing it. These phrases attempt to nullify the rationale behind violence against environmental defenders, and it is in these responses that we can find a blueprint for the types of response that can successfully prevent violence against defenders.

Physical assaults against environmental defenders have been frequently conceptualized as a result of corrupt project management and low governmental presence and capacity. Yet the scale of violent attacks against defenders occurring globally shows that these are systemic features of extractivism rather than outliers (Méndez et al. 2018). Eliminating violence against environmental defenders thus necessarily requires tackling its root cause: the inherently violent nature of extractivism. As Navas et al. (2018) and Jiménez Thomas (2018) argue, violence in environmental conflicts comprises not just physical attacks against defenders, but a wide and complex process of structural, epistemic, symbolic, and gendered violence. As such, only working toward a guarantee of environmental justice can achieve the right to a life free of violence for environmental defenders.

For this reason, protests in solidarity with environmental struggles are crucial, but they need to transcend a particular issue or cause. Broad advocacy is needed, which takes a more multifaceted aim at the violence experienced by defenders and communities, encompassing the protection, not only of individuals, but also of the territories and ways of life involved in environmental conflicts, and contesting the notions that underpin extractivism in the first place.

The best way to protect environmental defenders is by creating the circumstances under which it is unnecessary to defend nature and related ways of being. A collective shift toward environmental justice is the only truly scalable safety mechanism for the protection of collectives as a whole, and not just for high-profile individuals.

Conclusions

To bring about effective change, institutional conventional responses must be pursued at the national and international level. They may offer a sense of justice to the relatives of the activists murdered and to their communities of struggle, and they may establish institutional mechanisms that widen the rights and tools at the disposal of defenders. Given the range of vested interests and other obstacles to institutional action, it is also important that non-institutional conventional responses occur, such as protests demanding retributive justice. The pressure that protests create can help to counteract state co-optation, political apathy, or institutional weakness, which tend to limit retributive justice mechanisms. Moreover, more transformative non-institutional responses, such as protests in solidarity with environmental struggles, are crucial. These are key in pushing for the concrete advancement of the struggles to which the defenders belong—recall the *resex* in Brazil and the withdrawal of funds for Agua Zarca in Honduras.

Increased activism is key to securing the safety of defenders since it subverts the rationale of deterrence and intimidation which motivates violence in the first place. This activism ought not to be limited to a particular individual or cause, but should encourage support for and involvement in as many similar struggles as possible, over other territories and ways of life under threat. This is what the cries of *"Berta no se murió, se multiplicó"* and of *"mataram nosso líder, mas não nossa luta"* seek to obtain.

As direct grassroots actions become crucial, support from non-governmental organizations and civil society collectives is essential. Such support could take the form of direct financial and legal assistance for environmental groups at risk. Direct support for networking between environmental defenders may also be useful, so that groups can share knowledge and strategies. Academics can play an important role here by co-producing knowledge on the adverse social-environmental impacts of extractivism. Evidence is not only important to advance the understanding of environmental conflicts; it can also be used to mobilize affected groups, to leverage responsible stakeholders, and to positively influence the trajectory of concrete environmental conflicts.

Responses to violence against defenders must change from a reactive to a proactive stance. Support must be given before individuals and communities become at-risk, or, rather, so that communities never become at-risk. The possibility of moving toward such types of support depends on moving toward a broader understanding of violence. If violence only calls to mind physical aggression, protection measures will only ever address the pinnacle of the process of violence in environmental injustice. A lasting and more transformative way to protect environmental defenders requires to understand the complex processes of violence at play and to create unconventional responses that challenge the paradigms that sustain them.

Acknowledgments

While writing this chapter, Jehry Rivera Rivera, an environmental defender from the Bröran indigenous people in Costa Rica, was murdered. We dedicate this work to him, his family, and his struggle to defend indigenous land rights. We would also like to thank the editors, Diana Ojeda and Felipe Milanéz, and our colleague, Thais de Carvalho, for their generosity. Your comments made our work infinitely better.

Notes

1 While women were barred from actively participating in the STR, they played a crucial role in supporting the movement. They performed care tasks such as providing food during the *empates*, as well as occupying the frontlines as a strategy to deter violence against union members (Campbell 1996).

2 The first leg of highway BR 361 was finished by 1970. It connected Rio Branco to Porto Velho and Cuiaba in the state of Mato Grosso. Highways were central to the government plans for integrating the Amazon. In other regions, the Belem–Brasilia Highway and the Trans-Amazon Highway were crucial (Hecht and Cockburn 2010; Shoumatoff 1991).

3 These events took place under the government of Lula da Silva due to the centrality of the workers' struggle under his administration. Today, ICMBio has lost authority under Bolsonaro's presidency.

4 COPINH played a key role, along with other organizations, in Honduras's signature and ratification in 1994 (Korol 2018).

5 This hashtag has been recently accompanied by that of #CastigoALosAtala and #5AñosJuntoABerta.

6 Scheidel et al. (2018) have documented how environmental defenders mobilize most commonly as a collective struggle over land and customary resources.

7 In the case of Brazil, both organizations rely heavily on the documentation work done by the Pastoral Land Commission and the Indigenist Missionary Council (*Conselho Indigenista Missionário* or CIMI).

8 In the case of Berta, her identity as a Lenca woman may have additionally helped to render more visible her murder, as both indigenous and gender rights have consolidated as norm regimes in the international sphere.

References

Aguiar Gomes, Carlos Valerio et al. 2018. "Extractive Reserves in the Brazilian Amazon thirty years after Chico Mendes: Social Movement Achievements, Territorial Expansion and Continuing Struggles." *Desenvolvimento e Meio Ambiente* 48: 74–98.

Allegretti, Mary. 1998. "Reservas Extrativistas: Uma Proposta de Desenvolvimento da Floresta Amazônica." *Pará Desenvolvimento*, 25: 2–29.

———. 2008. "A Construção Social de Políticas Públicas. Chico Mendes e o Movimento Dos Seringueiros." *Desenvolvimento e Meio Ambiente* 18: 39–59.

Barbosa, de Almeida, Mauro, Mary Allegretti, & Augusto Postigo. 2018. "The Legacy of Chico Mendes: Successes and Obstacles in the Extractive Reserves." *Desenvolvimento e Meio Ambiente* 48: 25–55.

Butt, Nathalie, Frances Lambrick, Mary Menton, & Anna Renwick. 2019. "The Supply Chain of Violence." *Nature Sustainability* 2(8): 742–47.

Campbell, Cynthia. 1996. "Out on the Front lines but Still Struggling for Voice". In Rocheleau, D. et al. (Eds.), *Feminist Political Ecology: Global Issues and Local Experiences*, 27–61. New York; Routledge.

CIDH. 2018. Comunicado de prensa: Ante el próximo fallo en el Caso de Berta Cáceres, OACNUDH y CIDH expresan su preocupación por la exclusión de la representación de las víctimas y las demoras injustificadas en el proceso.

COPINH. 2022. *Justicia para Berta Cáceres.*

———. 2020a. *¿Quiénes somos? ¿Que es COPINH? Consejo Cívico de Organizaciones Populares e Indígenas de Honduras (COPINH)*, La Esperanza, Intibucá.

———. 2020b. Berta Cáceres: Información, noticias y análisis sobre el proceso de búsqueda de justicia de la gran causa Berta Cáceres. Consejo Cívico de Organizaciones Populares e Indígenas de Honduras (COPINH), La Esperanza, Intibucá.

Cowell, A. (Director). 1989. *Chico Mendes: Eu Quero Viver*. Verbo Filmes.

Deonandan, Kalowatie & Colleen Bell. 2019. "Discipline and Punish: Gendered Dimensions of Violence in Extractive Development." *Canadian Journal of Women and the Law* 31(1): 24–56.

Esguerra-Muelle, Camila, Diana Ojeda, Tatiana Sánchez, & Astrid Ulloa. 2019. "Introducción: Violencias Contra líderes y liderasas Defensores del Territorio y el ambiente en América Latina." *LASA Forum* 50(4): 4–6.

Flores, Daysi & Patricia Ardón. 2017. "Berta lives! COPINH Continues…." *SUR International Journal on Human Rights* 25: 09–117.

Frontline Defenders. 2018. *Case History: Berta Cáceres*. Dublin: Frontline Defenders.

Global Witness. 2019. *Enemies of the State? How Government and Business Silence Land and Environmental Defenders*. London.

Goldman Environmental Prize. 2016. *Berta Cáceres 2015 Goldman Prize Recipient South and Central America*. Goldman Environmental Foundation, San Francisco.

Gross, Tony. 1989. Epilogue. In Mendes, Francisco. 1989. *Fight for the Forest: Chico Mendes in his own words*. Adapted from O testamento do Homen da Floresta edited by Candido Grzybowski. London: Latin America Bureau.

Gudynas, Eduardo. 2009. Diez Tesis Urgentes Sobre el Nuevo Extractivismo: Contextos y Demandas bajo el Nuevo Progresismo Sudamericano Actual (187–224), In Schuldt et al. (eds.), *Extractivismo, política y sociedad*. Quito: Centro Andino de Acción Popular.

Hecht, Susanna & Alexander Cockburn. 2010. *The fate of the forest: Developers, destroyers, and defenders of the Amazon* (2nd edition). Chicago; University of Chicago Press.

hooks, bell. 1995. *Art on My Mind: Visual Politics*. New Press: W.W. Norton.

IM–Defensoras. 2013. "A Feminist Alternative for the Protection, Self-Care, and Safety of Women Human Rights Defenders in Mesoamerica." *Journal of Human Rights Practice* 5 (3): 446–459.

Jiménez Thomas, Rodríguez Diana 2018. *Soybeans, development, and violence: The environmental resistance of Mayan women and men in Hopelchen, Campeche, Mexico* (Master's thesis). University of Oxford.

Keck, Margaret & Kathryn Sikkink. 1998. *Activist Beyond Borders*. Ithaca: Cornell University Press.

Knox, John H. 2015. "Human Rights, Environmental Protection, and the Sustainable Development Goals." *Washington International Law Journal* 24(3): 517–36.

Korol, Claudia. (2018). *Las revoluciones de Berta: Conversaciones con Berta Cáceres, 'Guardiana de los ríos'*. Buenos Aires: La Fogata Editorial.

Latorre, Guisela. 2019. *Democracy on the Wall: Street art of the Post-dictatorship Era in Chile*. Ohio: Ohio State University Press.

Lakhani, Nina. 2018a, October 17. "Berta Cáceres murder trial to be monitored by international lawyers". *The Guardian*.

———. 2018b, December 1. "Berta Cáceres: conviction of killers brings some justice, but questions remain". *The Guardian*.

———. 2021, March 2. Honduras: accused mastermind of Berta Cáceres murder to go on trial next month. *The Guardian*.

León Araya, Andres. 2015. *Rebellion under the Palm Trees: Memory, Agrarian Reform and Labor in the Aguán, Honduras* (Doctoral dissertation). The City University of New York.

Lewis, Chris. 2016, March 8. "They Want to Prohibit Us From Dreaming": An Interview with Berta Cáceres. *Jacobin*.

Li, Fabiana. 2015. *Unearthing conflict: corporate mining, activism, and expertise in Peru*. Durham: Duke University Press.

Mendes, Francisco. 1989. *Fight for the Forest: Chico Mendes in his own words*. Adapted from O testamento do Homen da Floresta edited by Candido Grzybowski. London: Latin America Bureau.

Méndez, María José. 2018. "'The River Told Me': Rethinking Intersectionality from the World of Berta Cáceres". *Capitalism Nature Socialism* 29 (1): 7–24.

Middeldorp, Nick & Philippe Le Billon. 2019. "Deadly Environmental Governance: Authoritarianism, Eco-Populism, and the Repression of Environmental and Land Defenders." *Annals of the American Association of Geographers* 109(2): 324–37.

Middeldorp, Nick, Carmen Elena Villacorta, & Esteban De Gori. 2018. 'Violencia y Represión contra Defensores de Tierra y Territorio en Honduras: Desde el Golpe de Estado hasta la Actualidad' 89–96. In *Golpe Electoral y Crisis Política En Honduras*. CLASCO.

Milbrandt, Melody. 2010. Understanding the Role of Art in Social Movements and Transformation. *Journal of Art for Life*, 1(1): 7–18.

Navas, Grettel, Sara Mingorría, & Bernardo Aguilar. 2018. "Violence in Environmental Conflicts: The Need for a Multidimensional Approach." *Sustainability Science* 13(3): 649–60.

Nye, Joseph. 2009. *Soft Power: The Means to Success In World Politics*. New York: Public Affairs.

Red Latinoamericana de Mujeres Defensoras et al. 2018. Informe sobre la situación de riesgo y crimi-nalización de las defensoras del medioambiente en América Latina. Red Latinoamericana de Mujeres Defensoras de Derechos Sociales y Ambientales, Grufides, Enginyeria Sense Fronteres, CATAPA, Ghent University - Human Rights Centre, Fedepaz, Demus-Studies For Women. Cajamarca.

Sandoval, Elvin. 2019, December 2. Condenan a Siete Implicados en el Asesinato de Berta Cáceres; Pagarán Entre 30 y 50 Años de Cárcel" News article in CNN Latinoamérica.

Shoumatoff, Alex. 1991. *Murder in the rain forest: The Chico Mendes story*. London: Fourth Estate.

Sieder, Rachel. 2017. *Demanding Justice and Security: Indigenous Women and Legal Pluralities in Latin America*. New Brunswick: Rutgers University Press.

Stec, Stephen & Jerzy Jendrośka. 2019. The Escazú Agreement and the Regional Approach to Rio Principle 10: Process, Innovation, and Shortcomings. *Journal of Environmental Law* 31 (3): 533–545.

Svampa, Maristella & Enrique Viale. 2014. *Maldesarrollo: La Argentina del Extractivismo y el Despojo*. Buenos Aires: Katz.

Tanner, Lauri. 2011. "Kawas v. Honduras – Protecting Environmental Defenders." *Journal of Human Rights Practice* 3(3): 309–26.

Tapias Torrado, Nancy. 2019. *Situación de las Lideresas y Defensoras de Derechos Humanos: Análisis Desde una Perspectiva de Género e Interseccional*. Bogotá: CAPAZ.

UN. 2018. *Regional Agreement on Access to Information, Public Participation and Justice in Environmental Matters in Latin America and the Caribbean*. Santiago: United Nations.

UNEP. 2018. *Promoting Greater Protection for Environmental Defenders: Policy*. Kenya: United Nations Environment Program.

Watts, J. 2013, December 20. "Brazil salutes Chico Mendes 25 years after his murder." *The Guardian*. Last accessed 1 March 2023: https://www.theguardian.com/world/2013/dec/20/brazil-salutes-chico-mendes-25-years-after-murder

Part V

Environmental disputes and policies

26 Latin America's approach in the international environmental debate

Between "eco development" and "sustainable development"

Fernando Estenssoro Saavedra

Introduction

The environmental debate, understood as everything that involves overcoming the global environmental crisis, formally arrived on the world political agenda with the holding of the United Nations (UN) conference on the Human Environment in Stockholm in June 1972 (Stockholm 1972). The final document of that conference ordered the creation of a broad collaboration between the nations of the world and international organizations to overcome this serious threat. This effort continues until the present and everything indicates that it will continue to be a relevant issue during the first half of this century because this global problem is still not possible to overcome.

From the outset of this debate, different ways of understanding the characteristics of the environmental crisis emerged, namely: who is responsible for it and, above all, who must bear the main costs of overcoming it. Differences that persist to this day and which, to a large extent, explain the slow progress in reaching a solution. In this sense, the so-called underdeveloped countries, which in the Cold War times identified themselves as Third World and are known today as the global South, and which includes Latin America and the Caribbean (hereinafter AMLC, according to its Spanish acronym), have always maintained that this crisis was generated mainly by the world's richest, developed and industrialized countries. These rich countries were known in their time as either the First World (or the developed capitalist world), or the Second World (the highly industrialized communist countries) and which today, after the end of the Cold War, are known as the global North. For the South in general, and AMLC in particular, the global North must assume its responsibility in the generation and exacerbation of this crisis, as well as paying the ecological debt that they have acquired with the rest of humanity due to the implementation and imposition of their growth and development models, which although it took them to the pinnacle of global power, also generated the environmental crisis (Estenssoro 2020).

To get a general understanding of AMLC's historical approach in this debate, we can periodize it in four main stages: 1) Its beginnings in Stockholm 72; 2) The consolidation of AMLC's thinking and the concept of ecodevelopment; 3) from Sustainable Development to Rio 92; 4) from Rio 92 to Rio + 20.

Stockholm 72: Latin America and the first North–South confrontation in the environmental debate

The convening of the first World Conference on the Environment was a First World initiative, that is, it was convened by and for the most industrialized and developed capitalist countries. They were aware of the environmental damage they had caused through their process of economic growth and development and expressed great fear that the growth and industrialization efforts of the underdeveloped or Third World, as well as their demographic growth, could

DOI: 10.4324/9780429344428-31

further exacerbate this crisis. For this reason, the First World presented a public discourse very critical of development, seeking to curb and redirect the industrialization efforts of underdeveloped countries in works such as the first report of the Club of Rome, *The Limits of Growth*. The chosen body to implement these policies were to be the UN, since these necessarily had to be global in scope. As pointed out by Maurice Strong, Secretary General of the Stockholm Conference 72, it was in "industrialized countries where concern for pollution created the original idea of the Stockholm Conference" (quoted in Estenssoro 2020, 110).

As expected, Third World countries in general and Latin Americans in particular reacted with deep distrust to this initiative. They correctly suspected that the developed countries were directing the efforts of the international system to prioritize the resolution of those issues that had been affecting the quality of life of their already affluent societies, leaving aside efforts to overcome the underdevelopment of the majority of humanity.

Then came an attempt to boycott this conference by representatives of developing countries that forced Strong to negotiate with them. Strong invited a select group of 27 world-renowned personalities, mainly representatives and experts from the Third World, to a meeting in order to reach an agreement that would include the concerns regarding overcoming underdevelopment and connecting them to the issue of overcoming the environmental crisis. This meeting was held in the Swiss town of Founex in June 1971. From this negotiation, in which important Latin Americans like Enrique Iglesias, Felipe Herrera, and Miguel Ozorio da Almeida, among others, participated, a document known as the Founex Report emerged. For the first time, two ideas that up to that point had appeared to be contradictory were joined together: protect the environment and achieve full development:

> It can be said that, to a large extent, the current interest in environmental issues has originated from the problems experienced by the industrially advanced countries. These problems are in themselves, to a large extent, the result of a high level of economic development (…) However, the main environmental problems in developing countries are basically different from those perceived in industrialized countries. They are mainly problems that have their roots in poverty and the very lack of development of their societies. In other words, they are problems of rural and urban poverty (…) For these reasons, concern for the environment should not weaken, and need not do, the commitment of the world community—both developing countries and of the industrialized—to dedicate themselves to the most important task of developing the most backward regions of the world. On the contrary, it underlines the need not only to commit fully to achieving the goals and objectives of the second development decade, but also to redefine them in order to attack misery, which is the most important aspect of the problems that afflict the environment of the world majority of humanity.
>
> (El Informe de Founex, 1971)

The Founex Report (El Informe de Founex in Spanish)was incorporated as part of the discussion document of the Stockholm Conference, which precipitated the end of the boycott threat and allowed the conference to continue its pre-established course.

Environment and development—a single issue for Latin Americans

Since Stockholm 72, for developing countries the issues of environment and development were absolutely connected and should be understood as a single process and ideas from the developing world became very influential in international debate. In this sense, Latin Americans played a leading role with works like *Catastrophe or New Society? Latin American world model*, developed by specialists linked to development issues and international relations and supported by the Bariloche Foundation, for which it is also known as the Bariloche Group report. In the report they argued that the environmental problems were not due to the irremediable crisis that would come from the scarcity of resources in the face of accelerated growth of the world population, as proposed by neo-Malthusian theorists, as well as the attempts of the Third World to industrialize, as the representatives of the First World proposed. For them, and for Latin Americans in general, the solution to the environmental crisis was sociopolitical in nature and its solution began, first of all, with ending the unequal distribution of power and wealth in the world (Herrera et al. 1976).

The ecodevelopment concept

Also from the Economic Commission for Latin America and the Caribbean (ECLAC), important efforts were made to unite the issue of caring for the environment with efforts to achieve development, especially since this part of the world, rich in natural resources and dependent on them for its economic growth, required rational and sustainable exploitation that does not deteriorate ecosystems in order to ensure their present and future interests (Estenssoro 2020). In this sense, one way to reconcile the issue of caring for the environment with the development needs of AMLC was its "appropriation" of the concept of ecodevelopment.

This concept was developed by the Polish economist Ignacy Sachs, who was naturalized French and later Brazilian. Sachs assumed that human beings are subjects aware of their belonging to nature and their future. Therefore, the issue of development should better capture the interaction of natural and social processes, in which humans are both subjects and objects. In this sense, ecodevelopment meant adapting the production process to the natural environment, for which it was necessary to know in depth, both the specific ecosystem and the cultural solutions of each society within that ecosystem:

> Ecodevelopment is a style of development that insistently seeks in each ecoregion specific solutions to particular problems, taking into account ecological but also cultural data, as well as immediate, and long-term needs.
>
> (Sachs 1974, 363)

This concept was immediately made their own by important figures in the Latin American environmental debate of the 1970s. As Enrique Leff rightly put it:

> the first proposals on ecodevelopment found in Latin America a favorable territory for its promotion.
>
> (Leff 2009, 221)

In fact, Leff continues

> Ignacy Sachs himself considered Latin America the potentially most fertile region to receive his proposals and over the years he traveled to various countries—mainly to Mexico and Brazil, a country in which he had links of second citizenship—to promote ecodevelopment.
>
> (Leff 2009, 222)

In this way, many Latin American intellectuals and academics began to work, very early on, on environmental problems based on the concept of eco-development. Mention should be made of Iván Restrepo, Enrique Leff, Vicente Sánchez, Héctor Sejenovich, Jaime Hurtubia, Francisco Szekely, Francisco Mieres, Hilda Herzer, Margarita Merino de Botero, Raúl Brañez, and Augusto Ángel, among many others. These were mainly academics and scientists linked and related to institutions such as ECLAC, UNEP, CLACSO, as well as institutions dedicated to the study of natural resources and environmental problems, such as the Institute for the Development of Natural Resources of Colombia (INDERENA), or the very foundation of the Centro de Ecodesarrollo de México founded in 1974 by Leff himself. For example, in 1977 Leff wrote about eco-development:

> It is neither a question of cultural and ecological conservationism, nor of substituting traditional knowledge for modern science and technology, a product of capitalist rationality. Ecodevelopment is not a social project founded on the extremely efficient energy or caloric content of a new technocratic philosophy. On the contrary, the objective of eco-development can be defined as a series of actions conducive to creating the knowledge and techniques necessary for the use of the resources of each ecosystem within the ecological criteria that guarantee their reproduction. The selection of resources and the uses to which their production is assigned continue to be determined by the different social projects in which this eco-development strategy is defined.
>
> (Leff 1977, 107)

Similarly, professionals linked to the UNEP Regional Office for Latin America the Caribbean (ORPALC), as well as to ECLAC, published in 1978:

> We consider eco-development as a modality of economic development that postulates the use of resources to satisfy the needs of current and future generations of the population, by maximizing the functional efficiency of ecosystems in the long term, using adequate technology to this end and the full use of human potential, within an institutional framework that allows the participation of the population in fundamental decisions.
>
> (Sánchez and Sejenovich 1978, 154)

In summary, ecodevelopment was an avant-garde concept in AMLC when the environmental political debate was in its infancy, which united the ideas of the environment and development and which also refuted the First World theses that blamed population growth—especially in the Third World—for the global environmental crisis and that, by placing absolute physical limits on the Earth, proposed to stop all kinds of development and economic growth. For Latin Americans, these First World theses, applied in a world so unequal in terms of wealth, quality of life, opportunities for human development and political power, condemned the peoples of the Third World to remain as an eternal underdeveloped periphery.[1]

From Sustainable Development to Rio 92

Finally, and despite all the effort made by Latin Americans, ecodevelopment was not the concept that dominated the international debate when defining the unity between environment and development, but was displaced by the concept of sustainable development.

In this regard, it should be remembered that after the completion of Stockholm 72, it was agreed that humanity would meet every ten years in order to study and evaluate the sustained progress against the environmental crisis and define new tasks. To this end a second great environmental summit was to be held in Nairobi in 1982 (Nairobi 82). However, this initiative totally failed and resulted in a smaller administrative meeting. The failure was due to the conjugation of two main factors. On the one hand, the differences between the developed world and the developing world had deepened regarding how to understand the concept of global environmental crisis, and therefore where it places the emphasis on policies aimed at overcoming it. On the other hand, the international climate between the communist bloc and the capitalist bloc had worsened, after the arrival of Ronald Reagan and Margaret Thatcher to the governments of the United States and the United Kingdom, respectively.

After the failure of the Nairobi 82 meeting, the following year the UN decided to create the World Commission on Environment and Development (WCED), aimed at putting the environmental issue on the agenda of world political priorities and for which it was essential to find a consensus formula between the opposing perspectives of developed and underdeveloped countries, originating from Stockholm 72. The Norwegian, Gro Harlem Brundtland, was appointed president of this commission, which is why it was called the Brundtland Commission. This commission met for four years looking for a consensus formula, where the Latin American proposal was the concept of ecodevelopment. Finally, in 1987, this commission presented its historic report, *Our Common Future*, in which the concept that was promoted was not eco-development but Sustainable Development.

That same year, 1987, the UN General Assembly commended the work of WCED and accepted the concept of Sustainable Development as they had defined it. In 1989, based on this concept, the UN called a new major summit on the environment, to be held in Rio de Janeiro in 1992 (Rio 92). But this time titled as the United Nations Conference on Environment and Development, also known as the Earth Summit, and intended to relaunch the environmental issue as one of the most important on the world political agenda.

The Rio 92 Environmental Summit

Undoubtedly, the general spirit with which the countries of AMLC arrived at the Rio 92 Conference was very different from that of Stockholm 20 years earlier. This time, all developing countries came up with their own much more developed theses to deal with this problem. Therefore, in the call to Rio 92, it was much clearer that the environmental crisis was the main responsibility of the highly industrialized and developed countries, they had caused it and they had to assume their responsibility for it, including the global social inequality that they caused with their expansion and development model. By the way, Latin Americans were aware that this was a global problem and that they, too, had to make their contribution to overcome it, but they clearly insisted that facing the global environmental crisis there were common but differentiated responsibilities, and the most developed, wealthy, and industrialized countries had to take care of the ecological debt they had contracted with the least developed countries due to their polluting and ecologically harmful lifestyle and development.

In addition, Latin Americans had produced their own pre-call report, entitled, "Nuestra Propia Agenda" ("Our Own Agenda"). This outlined the three main points that AMLC considered essential to resolve at the Rio 92 summit: first, regional strategies for dealing with the environmental crisis, an aspect emphasized by Latin America and the Caribbean; secondly, issues that affected the world ecosystem, such as global warming, which the First World countries emphasized, but for which the responsibility of the region was much less; and, finally, world policy issues, such as the ecological debt that the North had towards the Third World (Estensoro 2020).

The Rio 92 Environmental Summit was quite successful. Among its results, there are five main documents, two global conventions ("The Framework Convention on Climate Change" and "The Convention on Biological Diversity") and three agreements ("A Declaration of Principles on Management, Conservation and Sustainable Development of All Types of Forests," the "Rio Declaration on Environment and Development" and "Agenda 21"), which were to guide the environmental discussion in the following years.

Obviously, not all the hopes and expectations going into this Summit were fulfilled. The demand that the North recognize and assume their ecological debt was not met. The representatives of the North, "proposed a program with strictly environmental issues and were reluctant to incorporate development issues in the analysis of environmental problems," and only prioritized those issues they considered global such as "Climate Change, biological diversity, oceans and the ozone layer" (Glender 1994, 269). Despite its shortcomings, Rio 92 marked an important milestone in the fight against the global environmental crisis and has been, to date, the most successful of all the environmental summits that have been held. Likewise, hundreds of NGOs from the world, and mainly from the Third World, participated in Rio 92, and held a Global Forum criticizing the traditional and dominant views of development and economic growth while advancing other forms of development with social justice, with a vision much closer to the original idea of ecodevelopment in Latin America. From these debates, the proposals of *buen vivir* or "good living" would later emerge, and, later still, would be reflected in the constitutions of Ecuador and Bolivia.[2] The truth is that, as of Rio 92, civil society would be increasingly present at these large environmental conferences.

From Rio 92 to Rio + 20, from illusion to disappointment

Despite the progress made at Rio 92, the main tension between developed and underdeveloped countries did not disappear, which after the end of the Cold War came to be known as North–South tension. On the contrary, it only showed an evolution produced by the new historical circumstances.

Ten years after Rio 92, the World Summit on Sustainable Development was held in Johannesburg in June 2002 (Johannesburg 2002). At this summit, the former First World countries now identified as global North, continued with their evasive policies regarding their responsibility in generating the environmental crisis and paying the ecological debt contracted with humanity as a whole and, above all, with the least developed countries. They avoided any legally binding obligation of financial aid to the South, except in those aspects of environmental protection that the North considered important for their interests. Likewise, the representatives of the North weakened the Principle of Common but Differentiated Responsibilities (CBDR, by its acronym in English), making the most diverse interpretations (Guimarães and Da Fontoura 2012).

Thus, when the Action Plan of this summit was discussed, the countries of the South, which were the great defenders of this principle and who had Brazil as their leader, sought to advance in concrete actions for its implementation. Although they managed to do so in principle, as stated in various parts of the document, the representatives of the developed world were adamant about avoiding any type of wording that would allow an interpretation that the North acquired, "a legally binding obligation to grant financial aid to poor countries" (Guimarães and Da Fontoura 2012, 24; Fuentes 2003, 50).

Therefore, Johannesburg 2002 represented a setback in all matters related to the North's committed aid for the development of the South, even at pre-Stockholm-72 levels. In this sense, the great historical demands of the least developed countries, including Latin American ones, were again frustrated (Estenssoro 2019).

Ten years after Johannesburg 2002, the United Nations Conference on Sustainable Development was held in 2012, again in Rio de Janeiro. Since it occurred twenty years after the Rio 92 Summit, it was also known as Rio + 20. At this conference, not only were the trends registered in Johannesburg 2002 accentuated regarding the decline of consensual

North–South agreements, but in several respects they went back to the pre-Rio-92 and even pre-Stockholm-72 reality.

Rio + 20, unlike Rio 92 and Johannesburg 2002, was not convened as a Summit meeting by the UN, which meant that the presence of Heads of State was not necessary. In fact, neither the President of the United States nor the Chancellor of Germany attended the event, and the European Commission sent a much smaller delegation of representatives than for previous summits. Again, the Principle of Common but Differentiated Responsibilities (CBDR) was quite relativized and simplified in its final document, *The Future We Want* (Wenceslau et al. 2012, 597). In the writing of this document, developed countries opposed explicit approaches from the South that had been won at previous summits, such as "the Right to Safe and Clean Water and Sanitation or the regulation of financial and commodity markets" (Guimarães and Da Fontoura 2012, 27). Likewise, the promised increase in development aid from the North to the South never materialized. The North managed to relativize any obligatory commitment to pay its ecological debt. For this reason, what remained of Rio + 20, as has been pointed out, were "empty phrases with practical content such as those of 'promoting efficiency' or perfecting access'" (Guimarães and Da Fontoura 2012, 27).

The climate change example: The Ecuadorian case of Yasuni ITT

One of the best examples of how any environmental initiative from the South that may affect its hegemonic interests is blocked by the North, was what happened with the Ecuadorian Yasuní-ITT initiative, which was carried out within the framework of the fight against climate change.

In this regard, it should be remembered that the Kyoto Protocol expired in 2012 and as this date approached the annual meetings of the Conference of the Parties to the United Nations Framework Convention on Climate Change (COP), were aimed at achieving a new agreement that would replace it and that, this time, would be binding on all the member countries of the COP. In this spirit, in 2007 the President of Ecuador, Rafael Correa, proposed in the General Assembly of the United Nations, the Yasuní-ITT Initiative, which consisted in keeping underground, that is, not exploiting, more than 840 million barrels of oil found in the Amazon subsoil of the Yasuní National Park and equivalent to 20% of Ecuador's oil production.

In this way the generation of 407 million tons of CO_2 into the atmosphere would be avoided, in addition to other harmful effects on the Amazonian ecosystem such as deforestation, the spread of methane, the migration of species, and water pollution, among others (Vásquez 2015a, 2015b). In exchange, Ecuador asked the international community for a financial contribution of US$3.6 billion according to international prices for a barrel of crude, equivalent to 50% of the income that the country would stop receiving for not exploiting this source of oil. This amount of money was to be raised in 13 years from 2007, with a determining milestone of having US$100 million by December 2011 (Vásquez 2015a, 2015b).

These funds would be gathered in an International Trust (FI), administered by the United Nations Development Program (UNDP) and governed by a Steering Committee, in which the Government of Ecuador, taxpayers, and Ecuadorian civil society would have representation. The money from the Yasuní Fund would be invested in the financing of strategic sustainable development programs defined in the National Development Plan of Ecuador, all directly related to the conservation of biodiversity and ecosystems, the promotion of renewable energy, research, science, innovation, and technology aimed at developing biological knowledge and changing the energy matrix. Likewise, the proposal stated that if the monies from the international community were not gathered within the established deadlines, the extractive plans would be carried out (Vásquez 2015a, 2015b). Finally, on August 15, 2013, the government of

Rafael Correa terminated the initiative, signaling its failure due to the lack of an adequate response from the global North, since, after six years of efforts, only US$13.3 million was contributed to the Yasuní Fund, representing only 0.37% of the expected amount (Estenssoro and Vásquez 2017).

A critical reason for this failure were the disagreements that occurred related to the administration of the funds to be collected by the so-called International Trust (FI). In July 2008, the government of Ecuador constituted the Administrative and Directive Council of the Yasuní-ITT Initiative in charge of outlining policies to ensure its success, develop financial mechanisms, and develop strategies for promotion and negotiation, in order to address these disagreements. This Council established negotiations with UNDP to finalize the signing of an IF, which was to be presented for the 2009 Copenhagen Climate Change Summit; however, the negotiations were suspended prior to the holding of this Summit by the President of Ecuador. Ecuador's objections indicated that the UNDP demanded that the country relinquish its sovereignty by proposing that the FI be administered by a majority of members representing the contributing countries from the global North, which left its strategic administration (the use of its funds and what types of projects to carry out with those same funds) outside the sovereignty of Ecuador. Obviously, this attitude of the Ecuadorian President was criticized by the governments of the United States, Great Britain, Germany, and other Northern countries, accusing him of being hyper-nationalist or an economic nationalist, as well as not granting sufficient guarantees to international taxpayers. Later, it was learned, from documents published by Wikileaks, that the US was concerned about what they called Correa's "obsession with sovereignty" when he insisted on defending Ecuador's natural resources, as well as the difficulties that the Ecuadorian government made in establishing military cooperation with the United States. This was compounded by the expulsion of two North American agents from Ecuador in February 2009. For their part, in September 2010 the German government, which in the beginning had expressed great rhetorical support for this initiative, changed its attitude and began to boycott it due to the fact that President Correa refused to leave control of the initiative in international hands, and in June 2011 Germany declared that it would no longer continue its contributions (Estenssoro and Vásquez 2017). The German Government viewed, with great concern, the strategic trajectory that Ecuador's environmental proposal implied because it threatened to deliver the international environmental initiative to the countries of the periphery. Specifically, the German Minister of Cooperation, Dirk Niebel, in a letter addressed to Green Deputy Ute Koczi, reported that Germany would not contribute to this initiative, arguing that there were no guarantees on the part of Ecuador to keep oil underground, but also stated that "the Ecuadorian proposal could form a dangerous precedent for other oil-producing countries to make similar requests for compensation" (quoted in Estenssoro and Vásquez 2017, 74).

In summary, it can be pointed out that with the Yasuní ITT initiative, Ecuador gathered the essence of the historical approaches of Latin America in the international environmental debate. As noted, for Latin America the fight against the global environmental crisis is closely related to the fight against poverty, social inequality, and unequal relations in international trade, in addition to always bearing in mind that this environmental crisis was generated by the colonialist model of development, economic growth, and lifestyle imposed by the global North. A development model that maintains Northern countries as the hegemonic powers of the international system. This is the ecological debt that the North owes to the South and that AMLC has placed in the environmental debate from the beginning. As President Correa well pointed out when he presented this proposal, "there is no worse enemy for the environment than poverty" (Correa 2008, A5). Likewise, the proposal was inspired by the highly valued principles of AMLC, such as that of Common but Differentiated Responsibilities. Finally, it included a specific idea that AMLC had been pointing out since the 1980s, referring to putting economic value on the multiple and transcendental environmental services provided by the ecosystems of the region not only for the maintenance of the balance of the global ecosystem, but also for the maintenance of the high standard of living of Northern societies (Estenssoro 2019).

Despite the expectations that this proposal originally generated, the story ended up giving reason to the worst fears that Latin Americans had since the beginning of the global environmental debate, and which was brilliantly summarized by the Brazilian, Helio Jaguaribe, when pointing out in 1971 that, it could not be ruled out that, if the cooperative, multilateral, and just solutions of the international system to face the global environmental crisis failed, the countries of the global North would finally end up carrying out ecological-imperialist actions in order to ensure the supremacy of their interests and their own survival:

> … It is particularly to be feared that, confronted with various forms of ecological scarcity, the Imperial Centers and the developed nations and groups will reserve all or almost all of the scarce facilities for their own use, while imposing on the rest of the world, restrictive policies on population expansion, economic and technological, necessary to return the world to the ecological balance.
>
> (Juagaribe 1972, 122)

In the case of the Ecuadorian proposal, the North simply could not accept it because, if they did, would it not be replicated by other peripheral countries that also depend on the export of commodities and natural resources to highly industrialized countries? What would happen in the international economic and geopolitical order if countries that export gas, uranium, copper, lithium, or other natural resource began to demand that countries from the global North pay a fair price according to what they would stand to gain by not exploiting these resources, so as not to cause damage to the environment? Wouldn't the global capitalist industrial and financial system be seriously threatened if this type of environmental justice were imposed? It was evident to the global North that this initiative of Ecuador, which brilliantly captured the historical approach of Latin America, must necessarily fail. And its result finally proves that North–South tensions in the environmental debate will not disappear, and AMLC as a member and defender of the ideas of the global South will continue to be an important region.

Notes

1 For further information on the concept of ecodevelopment, see Estenssoro (2020), *Historia del Debate Ambiental en la Política Mundial 1945–1992. La perspectiva latinoamericana (Nueva edición corregida y aumentada)*.
2 For deeper into the proposals of good living, to see Friggeri (2021), "Buen Vivir y Socialismo Indoamericano. Una búsqueda epistémico-política".

References

Correa, Rafael. 2008. *"Ecuador pedirá compensación por no explotar su petróleo"*. El Mercurio, Santiago (December 11, 2008): A5.

"El Informe de Founex". En *Ecodesarrollo. El pensamiento del decenio*, compilado por, Margarita Marino de Botero y Juan Tokatlian, 51–85. Bogotá: INDERENA/PNUMA, 1983.

Estenssoro, Fernando. 2020. *Historia del debate ambiental en la política mundial 1945-1992. La perspectiva latinoamericana (Nueva edición corregida y aumentada)*. Buenos Aires: Biblos.

Estenssoro, Fernando. 2019. *La Geopolítica Ambiental Global del Siglo XXI. Los desafíos para América Latina*. Santiago: RIL editores.

Estenssoro, Fernando and Juan Pablo Vásquez. 2017. "Las diferencias Norte-Sur en el debate ambiental global. El caso de la propuesta del Ecuador: Yasuní ITT." *Universum*, 32 (2017): 63–80.

Friggeri, Félix Pablo. 2021. "Buen Vivir y Socialismo Indoamericano. Una búsqueda epistémico-política". *Revista Brasileira de Ciências Sociais*, 105: 1–16.

Fuentes Torrijo, Ximena. 2003. "Los resultados de la Cumbre de Johannesburgo", *Revista de Estudios Internacionales de la Universidad de Chile*, 36: 32–33.

Glender Rivas, Alberto I. 1994. "Las relaciones Internacionales del desarrollo sustentable". In *La Diplomacia Ambiental. México y la Conferencia de Naciones Unidas sobre Medio Ambiente y Desarrollo, compilado por Glender, Alberto Glender y Víctore Lichtinger*, 254–282. México D.F.: Fondo de Cultura Económica.

Guimarães, Roberto P. and Yna Souza dos Reis Da Fontoura. 2012 "Rio+20 ou Rio-20? Crônica de um fracasso anunciado". *Ambiente & Sociedade*, 15: 19–39.

Herrera, Amílcar O., Hugo D. Scolnick, Graciela Chichilnisky, Gilberto C. Gallopin, Jorge E. Hardoy, Diana Mosovich, Enrique Oteiza, Gilda Romero Brest, Carlos E. Suárez, and Luis Talavera. 1976. *Catastrophe or new society: A Latin American model*. Ottawa: International Development Research Center.

Juagaribe, Helio. 1972. "El equilibrio ecológico mundial y los países subdesarrollados". *Estudios Internacionales* 17: 92–123.

Leff, Enrique. 1977. "Etnobotánica, biosociología y ecodesarrollo". *Nueva Antropología* 6: 99–109.

Leff, Enrique. 2009. "Pensamiento Ambiental Latinoamericano: Patrimonio de un Saber para la sustentabilidad". En, *VI Congreso Iberoamericano de Educación Ambiental*, 215–236 Buenos Aires: Secretaría de Ambiente y Desarrollo Sustentable.

Sachs, Ignacy. 1974. "Ambiente y Estilos de Desarrollo". *Comercio Exterior* 24: 360–368.

Sánchez, Vicente and Hécgtor Sejenovich. 1978. "Ecodesarrollo: Una estrategia para el desarrollo social y económico compatible con la conservación ambiental". *Revista Interamericana de Planificación* 47–48: 152–160.

Vásquez, Juan Pablo. 2015a. "Elementos y claves al calor de la conflictividad, para una perspectiva latinoamericana en el debate ambiental: el caso de la iniciativa Yasuni ITT". *REDESG* 4: 4–25.

Vásquez, Juan Pablo. 2015b. "La tensión histórica norte – sur global en el debate ambiental. El conflicto en torno a la iniciativa Yasuní ITT". *Revista Estudios Hemisféricos y Polares* 6: 1–28.

Wenceslau, Juliana, Natalia Latino Antezana, and Paulo du Pin Calmon. 2012. "Políticas da Terra: Existe um novo discurso ambiental pós Rio +20?". *Cuadernos EBAPE.Br*, 10: 584–604.

27 Degrowth and Buen Vivir

Perspectives for a great transformation

Alberto Acosta

Introduction

Degrowth and Buen Vivir (roughly translated as Good Living), without establishing linear parallels, may well be seen as perspectives for the great transformation towards societies characterized by equity and justice, freedom, and equality, and living in harmony with Nature. However, if they are placed in a broader context, condensing them (not homogenizing them) and, as necessary, offering other elements to guide reflection, they are connected by several elements. In fact, from this exercise, bridges will be formed and consolidated between these and other perspectives while other historical meanings are built, all within a necessarily post-capitalist horizon.

This discussion offers us a double contribution. On the one hand, it allows us to understand the world from visions other than dominant ones. On the other hand, it gives us clues as to what another world could be like; let us say it from the outset, a world where many worlds fit: *the pluriverse*.[1] All human and non-human beings can live with dignity, that is, satisfying their basic needs without endangering the lives of future generations.

In such a short text, it is difficult to give a detailed explanation of all the existing trends in such a convulsive and changing world, plagued by the multiple pandemics so characteristic of the civilization of capital. That is why what is presented here are only sketches. In the case of Sumak Kawsay,[2] we refer to the experiences emerging from the global South, and, in the case of Degrowth, we draw from a certain kind of Degrowth movement, particularly in Europe.

In this endeavor, we propose a discussion that opens with some reflections on Degrowth, understood as an option for rethinking the world, freeing it from the religion of permanent economic growth, that is, breaking with the paradigms of progress as the driving force of Modernity. This Modern era is a time of profound transformations in the history of Humanity, in which, among other fundamental issues, reason has prevailed over religion (Enlightenment), myth lost its place in the explanation of the universe, and the causes of all phenomena began to be sought through science, while the human being came to occupy the center of thought and action (anthropocentrism). Sumak Kawsay, then, will allow us to broaden the analysis by criticizing development, the favored child of progress. And in this context, let us begin by noting that development means anything, from raising skyscrapers to installing latrines, drilling oil wells as well as water wells, it is a concept of monumental emptiness… It is a testimony to the power of ideas that an empty concept has dominated discussions for half a century, in the words of Wolfgang Sachs, who as early as 1992 also anticipated that "the arrow of progress is broken and the future has lost its shine: what awaits us are more threats than promises" (Sachs 1992).

Let us look for elements of convergence between Sumak Kawsay and Degrowth.

DOI: 10.4324/9780429344428-32

Central elements of the Degrowth perspective

At present, there are increasing calls, especially in the industrialized European countries, for an economy that favors not only a more responsible (?) growth but even *Degrowth*.[3] Clearly, Degrowth is not synonymous with economic recession because Degrowth is part of a social, political, cultural, and economic process programmed to build another type of society. In contrast, a recession is a moment of crisis in the capitalist process of economic growth.

The conclusion reached is that growth cannot be the engine of the economy and even less its ultimate goal. Therefore, it is urgent to have a profound and responsible discussion on Economic *Degrowth*, initially in the global North (steady-state growth is not sufficient), which must necessarily go hand in hand with post-extractivism in the global South. It is necessary to assume *Degrowth* as an option to address economic issues and at the same time for socio-ecological transformation.

Copying, almost verbatim, the definition proposed by Serge Latouche and Federico Demaria,[4] Degrowth questions the hegemony of economic growth and calls for a democratically managed redistributive reduction of production and consumption, especially in industrialized countries, to achieve environmental sustainability, social justice, and well-being. Degrowth is often associated with the idea that small can be beautiful, but what counts here is that the emphasis should be not only on the less but also on the different. In a Degrowth society, everything will be different: activities, forms and uses of energy, relationships (including political ones), gender roles, the distribution of time between paid and unpaid work, and relationships with the non-human world. The goal of Degrowth is to break out of a society dominated by the fetishism of growth. Such a rupture is therefore related both to words and to things, to symbolic and material practices. It implies the decolonization of the imaginary and opening the door to other possible worlds (which are already underway). The Degrowth project does not aspire to another form of growth, nor to another type of development (sustainable, social, fair, etc.), but the construction of another society, a society of frugal abundance, a post-growth society (Niko Paech), or of prosperity without growth (Tim Jackson). In other words, it is not initially an economic project, not even of another economy, but a social project that implies escaping from the economy as a reality and as an imperialist discourse.

Having accepted this definition, we may well enter into a subject that a long list of prestigious thinkers has worked on. At the risk of leaving out some key figures, let us recall some critics of growth who, from different perspectives, have contributed and continue to contribute to building the thesis of Degrowth: Nicholas Georgescu-Roegen, Iván Illich, André Gorz, Kenneth Boulding, Serge Latouche, Tim Jackson, Niko Paech, Enrique Leff, José Manuel Naredo, Joan Martínez-Alier, among others. Even Amartya Sen, a winner of the Nobel Prize in Economics, broke with economic growth as a synonym of development, although he cannot really be considered a de-growthist.

Reflections on *Degrowth* somehow find other entries in works such as those of John Stuart Mill,[5] who foresaw the need to think of a stationary economy. As early as 1848 this English economist had made these prescient remarks:

> I do not know why there should be any reason to rejoice that people who are already richer than anyone need be, have doubled their means of consuming things. which produce little or no pleasure except as representative of wealth (…).
>
> (Mill 1848)

More current ideas, such as those of Herman Daly (1999), reaffirmed that the capitalist economy, the economy of growth, functions as an idiotic machine. That is to say, it is a machine that metabolizes natural resources, processes them by exhausting them, and discards them by polluting, and that it has to do more and more of the same in order to be able to function, which is the logic of accumulation under capitalism.

We have two clearly identified limits from this approach: the ecological limit and the consumption saturation point. From these limits, we can also draw fundamental conclusions that feed the debate. What counts now is that these ideas are expanding more and more.

Initially, the Degrowth perspective was generated primarily (although not exclusively) from the academic world, but different movements have also made it their own in recent years. Generally, movements are not born as actors of *Degrowth*; however, with their struggles and demands, they enter the political-conceptual level of the *Degrowth* perspective implicitly and, increasingly, explicitly. We have as examples various resistance movements against megaprojects and industrial agriculture, or transformers of cities, or promoters of energy democracy, food sovereignty, the commons, climate justice, etc. *Degrowth* questions the "green economy" since it is a tributary to the economy of growth and is nourished by the matrix idea of the permanent commodification of Nature, such as the well-known "carbon markets." Nor does it coexist amicably with eco-Keynesian responses, which consider the need for economic growth to depend at best on a "qualitative" or "selective" dimension.

Degrowth is a twofold proposal. On the one hand, by not focusing exclusively on the traditional sphere of the dominant economy, it formulates a perspective of integral social change, identifying as a fundamental problem the imperative of capitalist economic growth. On the other hand, it suggests placing the diverse and multiple concrete experiences in a broad and integral context.

Let us summarize some of the foundational elements of *Degrowth*:[6]

- What is fundamentally criticized is the almost linear fixation of economic management and the consequent policy responses to achieve "development" or maintain welfare, both of which imply that the "engine of growth" must be restarted to solve the problems.
- In times of multiple crises, and especially under the conditions of capitalism dominated by financial markets, growth becomes a destabilizing factor as financialization increasingly distorts economic structures and flows. Even the supply of natural resources, which are increasingly difficult to access, is involved in this speculative logic.
- This desperate search to reactivate the "engine" of the economy leads, as is increasingly evident, to growing environmental deterioration. Global climate collapse generates more and more destruction at the local or regional level. In turn, it limits the possibilities of supporting economic growth as a tool for solving the problems and demands of societies. A situation that is perceived with redoubled force when trying to return to the normality experienced before the coronavirus pandemic. Will we return to the (a)normality caused by the civilization of capital that led to covid and so many other pandemics?
- Similarly, the growing consumption of goods and unbridled competition, increasing waste and squandering, increases inequalities and inequities, leading to frustrations and imbalances of all kinds.
- Increasing pressure on labor and social polarization are other negative consequences of focusing economic management exclusively on growth.
- Growth impulses bring with them problems at the political level. Behind these processes of endless accumulation or fed by them, an escalation of violence emerges in multiple forms (the disappearance of peoples and communities, the assassinations of activists, the criminalization and persecution of rights defenders, and the destruction of territories and subjectivities) encouraged, for example, by the unbridled extractivism that seeks to secure primary resources for the "idiot machine," causing conflicts and even wars over these raw materials.
- Another cause of the problems linked to the logic of growth is the predominant forms of subjectivization. The consumerist and professionalizing approaches are deeply rooted, and

social changes also have a dimension of psychological-social, and cultural customs. Consumerism and extreme professionalization signify status, and offer purpose; however, happiness does not follow. On the contrary, there is suffering: stress due to permanent competition, stress due to consumption, stress in free time, stress due to lack of time. Even the increasing quantities of products manufactured and sold end up becoming a burden, more destruction, more frustration.

- As Mill anticipated, there is no doubt that there are already many people, especially in the global North, who have saturated their capacity to satisfy their needs with more and more material goods. These are fundamental issues. Recall another English economist, John Maynard Keynes (1933), who also raised the issue of the absolute limit of saturation in terms of consumption.

For the time being, there are relatively few economic policy proposals promoted from governmental spaces. Many more are conceived both at the level of grassroots initiatives and the level of transforming institutions.[7]

At this point, it is worth mentioning some possible ambivalences and complications of the *Degrowth* perspective: the conflict between concrete projects and a more comprehensive social vision, the lack of power and class stratification issues, and questions of the organization of paid work. Knowing the origins of this discussion centered mainly in Europe, one cannot demand of it a reading from the decoloniality that characterizes many debates in the global South, but this is a question that will have to be addressed at some point.

For this very reason, it is worth recovering the contributions of one of the most lucid Latin American thinkers, Enrique Leff. Leff (2008) recommends a transition towards a different form of organization of production and society itself, taking on the challenges involved. To achieve this, he asks and proposes:

> How to deactivate the growth of a process that has in its original structure and its genetic code an engine that drives it to grow or die? How to carry out such a purpose without generating an economic recession with socio-environmental impacts of global and planetary scope? [...] this leads to a strategy of deconstruction and reconstruction, not to explode the system, but to re-organize production, to disengage from the gears of the market mechanisms, to restore the shredded matter in order to recycle it and rearrange it in new ecological cycles. In this sense, the construction of environmental rationality capable of deconstructing economic rationality implies processes of reappropriation of nature and reterritorialization of cultures.
>
> (Leff 2008)

The capacity of institutional policy to face these challenges and direct these processes is minimal. Let us insist; *Degrowth* is not synonymous with recession; neither is it an issue that is exhausted in the economic field. It requires a transition towards another form of organization of production, consumption, distribution, and society itself, assuming these challenges as an increasingly present issue in industrialized countries, the most responsible for the global environmental debacle and many of the serious conflicts facing Humanity. Let us not forget that these countries are, in many ways, "underdeveloped," not only[8] because of their excessive accumulation of material wealth, which has caused and continues to cause great environmental destruction; they have not even managed to eradicate poverty, while the cracks of inequality grow unstoppably in areas where unhappiness is increasingly present.

A crucial point: Degrowth in the global North in no way implies that impoverished countries maintain their situation of poverty and misery so that ecological balances are not affected.

It is urgent to dismantle the "imperial way of life" (Brand and Wissen 2017) (existing above all in the rich countries) which threatens the world's ecological balance, exacerbates social inequalities, and deepens political conflicts. In this endeavor, these countries will have to retrace much of the road travelled, reversing this form of growth, which is unrepeatable at the global level.

In a context of *Degrowth*, economic growth could be limited to the expansion of specific products that should be made when there is a need to overcome specific shortages or when societies face some contingency; outside these cases, the basic principle of *Degrowth* (ensuring decent living conditions for the entire population) can be understood as the trend towards less consumption and a longer duration of the objects that are produced, seeking to reduce destructive economic production in a controlled manner, but simultaneously enhancing higher levels of happiness.

At the same time, countries that have become rich through the impoverishment of many others must assume their responsibility to make way for a global restoration of the socio-environmental damage caused. In other words, these countries must pay their ecological debt and even their colonial debt. Therefore, what should be a matter of attention in the South is not to try to repeat socially and ecologically unsustainable lifestyles.

Buen Vivir, an alternative to development

In a context of criticism and alternative constructions (during the turn of the century) the contributions of native peoples have gained prominence, within which the visions of Afrodescendant peoples should be incorporated. Their values, their experiences, their practices, in short, their worldviews, are not new. They were present (in one way or another) since before the arrival of the European conquerors, and many survived in the endless colonial republican period but were made invisible, marginalized or openly fought against. These worldviews include various challenges to "development" and the economics of growth, both practically and conceptually.

These proposals emerged at a time of a generalized crisis of the oligarchic and colonialism-rooted nation-state, thanks to the growing organizational and programmatic strength of indigenous and popular movements. Their eruption in the recent turn of the century (as vigorous political subjects) explains the emergence of Sumak Kawsay at that juncture. In this period, in various parts of the planet, ecological questions and alternatives have also started to consolidate. Many aligned with the vision of harmony with Nature, so characteristic of Good Living.

It is essential to insist that these ideas come from the indigenous world. A world where written culture does not prevail, which limits, but does not prevent, the recovery of their visions.[9] A world that cannot be romanticized and whose history is not exempt from various forms of violence.

Indigenous people are not premodern, nor are they backwards. Their values, experiences, and practices synthesize a living civilization capable of confronting a colonial Modernity. With their proposals, they imagine a different future, which is already nourishing global debates. Buen Vivir, then, seeks to gather the main values, some experiences and above all practices existing especially in the Andes and the Amazon.

When indigenous peoples speak of Good Living or Sumak Kawsay,[10] they propose, first and foremost, a reconstruction based on the utopian vision of the future of several indigenous peoples and nationalities of Abya-Yala (or Our America).

The principle that inspires the Good Life and defines it is harmony or, if we prefer, balance. Balance and harmony in the lives of human beings with themselves, in the lives of individuals in the community, in life between communities, and between peoples and nations. And, finally, all individuals and communities, from a rich cultural diversity, living in balance with Nature. Therefore, it is convenient to speak of a good life, not better, nor better than that of others, nor in a continuous process of living life in order to uselessly try to improve it.

To begin with, we cannot confuse the concepts of Good Living with that of "better living" since the latter assumes unlimited material progress. "Better living" encourages us to compete permanently with others with whom we compare our lives, in order to produce more and more, in the process of endless material accumulation. This "better living," then, encourages competition, not harmony. Let us remember that, for some to "live better," millions of people have had and still have to "live badly." Good Living does not imply repeating such a process of exponential and permanent material accumulation.

Political responses are needed to enable an evolution driven by "culture being in harmony" and not by a "civilization of better living" (Freire 2011). It is a matter of building a society of solidarity and sustainability within the framework of institutions that ensure life. The Good Life, let us repeat, aims at an ethic of sufficiency for the whole community, and not only for the individual.

Hence, its use as a simplistic notion, devoid of meaning, constitutes one of the greatest threats as a political device. The self-interested and accommodating discursive definitions in its formulation ignore its emergence from traditional cultures. This generalized tendency in various governmental spheres in Ecuador (and Bolivia) could even lead to a "new age" Sumak Kawsay, which would transform it into mere fashion. Along this path, Good Living could become a new name for "development": the "development of good living." Such a version of Good Living has been transformed into a simple device of power and a tool for propaganda, which serves to control and domesticate societies, as has already happened with the progressive governments of Ecuador and Bolivia.

Atawallpa Oviedo Freire, an outstanding scholar of the subject, goes even further, proposing not to translate Sumak Kawsay into any language because it would deform its spirit, and its transformative potential would be lost. This is not a minor issue. On the one hand, there is the risk of creating a new dogmatism and purism; on the other, of falling into new fashions with simple bureaucratic and bureaucratizing actions from government agencies or simple academic readings that would end up vampirizing the essence of Good Living.

Here, without going into more details due to space limitations, what stands out is the possibility of assuming Good Living as an open concept, recognizing its deep indigenous roots, from which we can build other worlds—all this without closing the possibility of a broad and enriching debate and dialogue with other knowledge and knowledges. At this point, post-developmentalist debates can be inserted, as well as others, such as the de-growthist, committed to overcoming modernity.

At this point, other contributions from different parts of the planet cannot be minimized. Good Living, as a culture of life, with diverse names and varieties, finds parallels in different periods in the different regions of Mother Earth, each with its specificities, such as Ubuntu in Africa or Swaraj in India, from spaces that we could assume as the world of indigeneity, as Aníbal Quijano said. We could also incorporate the powerful reflections of the Svadeshi, which includes a large part of Gandhi's thought. Ivan Illich's conviviality proposals could also be mentioned. The Good Life, whose discussion expands throughout the world like concentric circles caused by a stone falling into a lake,[11] also integrates (or at least it should integrate) different humanist and anti-utilitarian visions coming from The North.

Moreover, in order to prevent a singular and indisputable concept, it would also be better to speak of "buenos vivires" or "good livings;" that is to say, buenos convivires (or good coexistences) of human beings in community, good convivires of communities with others, good convivires of individuals and communities in and with Nature.

What matters is that in recent decades, Latin America has seen the emergence of profound proposals for change that are emerging as possible paths to civilizational transformation. Popular mobilizations and rebellions, especially from the indigenous world, appear as the force of long-standing historical, cultural, and social processes. These struggles of resistance and construction of alternatives are the basis of what we could understand as Good Living

Good Living constitutes a critical qualitative step forward by overcoming the concept of "development" and all its multiple synonyms, introducing a different vision, much richer in content and more complex. Let us assume Buen Vivir as an opportunity to collectively build new ways of life. It is not a model, and it is definitely not a new development regime. Good Living, in essence, is a process of life that comes from the communitarian matrix of peoples living in balance with Nature.

We cannot expect to overcome capitalism first in order to make Buen Vivir a reality. As has been demonstrated throughout the centuries, in the midst of permanent colonization, the values, experiences, and multiple practices of Sumak Kawsay are present. And it is precisely from these spaces, with diverse experiences, that the indispensable alternatives to "civilization" are constructed.

Without assuming the State as the only (or the most critical) sphere of strategic action, it is crucial to rethink it from a plurinational and intercultural perspective: these are dimensions to be built from the community. This is a historical commitment. It is not a question of modernizing the current State by bureaucratically incorporating the indigenous and afro-descendent, or by favoring spaces, such as intercultural bilingual education, only for the indigenous. Another State[12] requires adopting and processing the cultural codes of the indigenous peoples and other nationalities, as well as those of the traditionally marginalized groups. That is to say; a broad debate must be opened in this regard in order to move towards a State free from Eurocentric ties. From this process, which requires rethinking existing structures and institutions to make way for the payment of historical debts due to processes of conquest and colonization still present in Our America, there must emerge institutions that make the horizontalist exercise of power a reality.

This possibility also depends on how well we can understand and confront the interests that seek to maintain the capitalist *status quo* to preserve their power, interests that are precisely opposed to Good Living, as an alternative to development. Thus, it is clear that it is not a matter of doing better than what has been done so far and waiting for society to transform. What is sought is to collectively build new covenants of social and environmental coexistence, which requires creating new spaces of freedom and breaking all the fences that impede its validity. Such a process undoubtedly implies confronting a myriad of currently dominant interests.

Overcoming all types of inequalities and inequities is unavoidable. The Good Life cannot admit a society divided into social classes. Decolonization and overcoming profoundly entrenched racism, as well as depatriarchalization, are also fundamental for its construction. Likewise, cultural and territorial issues require urgent attention.

In short, the struggle is of a civilizational nature, implying that there are dozens, perhaps hundreds of dimensions, that must be addressed with equal care. Thus, today more than ever, in the midst of the serious and multiple global difficulties we face (just a few facets of the civilizational crisis that looms over Humanity) it is essential to build other forms of life that are not regulated by the accumulation of capital. Without being a unique and indisputable proposal, the Good Life is useful for that, including for its transforming and mobilizing of political value.

The search for new forms of living implies revitalizing the political discussion, obfuscated by the economistic and technical vision of ends and means. By deifying economic activity, particularly the market and productivism and consumerism, including individualism, many non-economic instruments, indispensable for improving living conditions, have been abandoned. These transitions will undoubtedly have to involve other economic logics and the convivial use of technological advances.[13]

Another economy (why not a post-economy[14]) should be wrought from the search for and construction of alternatives applied with a holistic and systemic vision, based on the actual validity of Human Rights and the Rights of Nature, with all the potentialities that these rights entail. This should be seen as a starting point and not as a point of arrival of public policies, which should be designed in such a way that their application implies, to cite just two aspects, the elimination of poverty and of any process that leads to the disappearance of species.

In the context of a new economy, the need to strengthen and dignify labor, outlawing any form of labor precariousness, is evident. However, this is incomplete. Buen Vivir provides a space in which people must organize themselves to recover and take control of their own lives. But we must go further. It is no longer just a matter of defending the labor force and recovering surplus labor time for workers, that is, opposing labor exploitation. It is time to rethink the organization of society based on leisure as a right, which even implies the need to redistribute the division of labor along with wealth and income (Acosta 2020).

Another element arises here: it is indispensable to open the door to the decommodification of Nature. To begin with, we must overcome the privatization of water, which must always be subordinated to the demands of human beings living in harmony with Nature. At stake, then, is the defense of all species' lives against anthropocentric schemes of socioeconomic organization, destruction of the planet through environmental depredation and degradation. It follows from the above that it is urgent to overcome the divorce between Nature and human beings. To write this historical change is the greatest challenge of Humanity if we do not want to risk the very existence of human beings. This is what the Rights of Nature, included in the Constitution of Ecuador (2008), is all about, which does not necessarily imply its enforcement. Recognizing Nature as a subject of rights assumes a biocentric position, which could be extended to a position devoid of any center, based on an alternative ethic by accepting intrinsic values in the environment. All beings, even if they are not identical, have an ontological value, even if they are of no use to humans.

As a prologue to a permanent transformation

The challenge is there. There is an urgent need to stop the vortex of economic growth and even decline, especially in the global North. Meanwhile, the South must move towards post-development options since doubts increasingly plague even those who enthusiastically promoted development. It is enough to consider that the Economic Commission for Latin America (ECLAC) itself says that development is exhausted, according to the executive secretary of ECLAC, who even pointed out that extractivism is also exhausted.[15] As Eduardo Gudynas rightly summarizes when analyzing this statement, there is *"a remarkable opportunity to address other types of alternatives that are located beyond development"* (Gudynas 2020), although he himself doubts that those in power understand what this possibility means.

On this point, we can find convergences between Degrowth and post-development, with Buen Vivir as one of its referents. What is important is to be clear that neither in the North nor the Global South is there room for opulent forms of life, the "imperial way of life," which always occurred at the cost of the vital stagnation of others and the destruction of Nature. This even leads to a rethinking of the issue of economic growth in the global South.

Such *Degrowth* implies not only physically reducing "economic metabolism." The economy must subordinate itself to the mandates of Mother Earth and the demands of Humanity. Other options of life and social organization outside the utilitarianism and anthropocentrism of

Modernity, i.e. capitalism, are indispensable. This requires another rationality that deconstructs the current logic of production, distribution, circulation and consumption. We must begin to disengage from the perversity of global capitalism, especially speculative capitalism.

In synthesis, to achieve this, it is necessary to overcome the fetish of economic growth, escape from the useless race in pursuit of a phantom that is development, decommodify Nature and strengthen the space of common goods, introduce interrelated and community criteria to value goods and services, decentralize and deconcentrate production, change consumption patterns, and significantly redistribute wealth and power. These are some of the bases for building another civilization collectively.

Notes

1 See Ashish Kothari, Ariel Salleh, Arturo Escobar, Federico Demaria, and Alberto Acosta, *Pluriverso - Diccionario del Postdesarrollo* (Barcelona: ICARIA, 2019).
2 Many reflections on this topic were collected in the book by Alberto Acosta, *El Buen Vivir Sumak Kawsay, una oportunidad para imaginar otros mundos* (Barcelona: ICARIA, 2013).
3 For further details on Degrowth, one can review Serge Latouche, *Degrowth and post-development: creative thinking against the economy of the absurd* (Barcelona: ICARIA, 2009); Tim Jackson, *Prosperity without growth: Foundations for the Economy of Tomorrow* (London: Routledge, 2017); Giacomo D'Alisa, Federico Demaria, and Giorgios Kallis, *Degrowth. Vocabulary for a new era* (Barcelona: ICARIA, 2015).
4 See the chapter on "Decrecimiento" in Ashish Kothari, Ariel Salleh, Arturo Escobar, Federico Demaria, and Alberto Acosta, *Pluriverso—Diccionario del Postdesarrollo* (Barcelona: ICARIA, 2019); also edited at Abya-Yala/ICARIA in Quito; also available in English in India.
5 John Stuart Mill, *Principles of Political Economy with Some of Their Applications to Social Philosophy*. (Toronto: University of Toronto Press, 1848).
6 Some of these reflections are part of the work of Alberto Acosta and Ulrich Brand, *Salidas del laberinto capitalista - Decrecimiento y Postextractivismo* (Barcelona: ICARIA, 2017).
7 See article "Si podemos decrecer," December 14, 2014. Available at https://www.eldiario.es/ultima-llamada/decrecimiento-programa-economico-podemos_132_4466451.html.
8 José María Tortosa, "Maldesarrollo y mal vivir - Pobreza y violencia escala mundial." In *Debate Constituyente series, Abya-Yala*, edited by Alberto Acosta and Esperanza Martínez. (Quito, 2011).
9 However, there are fundamental documents from the indigenous movement itself, especially that of the Confederation of Indigenous Nationalities of Ecuador: CONAIE, *Political project for the construction of the Plurinational and Intercultural State - Proposal from the vision of CONAIE* (Quito, 2013). Other indigenous contributions have been important for the dissemination of these ideas, in Ecuador, Carlos Viteri Gualinga, *Visión indígena del desarrollo en la Amazonía* (Quito, 2000) (mimeo); in Bolivia, Fernando Huanacuni Mamani, *Vivir Bien / Buen Vivir Filosofía, políticas, estrategias y experiencias regionales* (La Paz: Convenio Andrés Bello, Instituto Internacional de Investigación y CAOI, 2010). A contribution where an interesting compilation of indigenous thought on the subject is made is that of Antonio Luis Hidalgo-Capitán, Alejandro Guillén García - Nancy Deleg Guazha, *Antología del Pensamiento Indigenista Ecuatoriano sobre Sumak Kawsay*, (Universidad de Cuenca and Universidad de Huelva, 2014), in which important texts of representatives of the indigenous intelligentsia such as Luis Macas, Nina Pacri, Blanca Chancoso, Arirura Kowii, Luis Maldonado, among others, are collected.
10 The best-known expressions of Good Living or Living Well refer to concepts existing in indigenous languages of Latin America, traditionally marginalized but not disappeared: *sumak kawsay* or *allí kawsay* (in Kichwa), *suma qamaña* (in Aymara), *ñande reko* or *tekó porã* (in Guaraní), *pénker pujústin* (Shuar), *shiir waras* (Ashuar), among others. Similar notions exist in other indigenous peoples, such as Mapuches in Chile, *kyme mogen*; Kunas in Panama, *balu wala*; Miskitus in Nicaragua, *laman laka*; and other similar concepts in the Mayan tradition of Guatemala and Chiapas in Mexico.
11 There are, by the way, many contributions on Buen Vivir from those who do not necessarily come from the indigenous world, we recall the works of Atawallpa Oviedo Freire, *Qué es el sumakawsay - Más allá del socialismo y capitalismo*, (Quito, 2011); Xavier Albo, "Suma qamaña = el buen convivir", *Obets*, Alicante (2009); Josef Estermann, *Más allá de Occidente - Apuntes filosóficos sobre interculturalidad, descolonización y el Vivir Bien andino*, Abya-Yala (Quito, 2015); Eduardo Gudynas, "Buen Vivir: Sobre secuestros, domesticaciones, rescates y alternativas," in *Bifurcación del Buen Vivir y el sumak kawsay*, (Quito: Ediciones SUMAK, 2014); Francois Houtart, "El concepto del sumak kawsay (Buen Vivir)

y su correspondencia con el bien común de la humanidad," Revista Ecuador Debate N° 84 (2011); Omar Felipe Giraldo, *Utopías en la era de la supervivencia - Una interpretación del Buen Vivir*, (México: Editorial ITACA, 2014); Ramiro Avila Santamaria, *Los derechos de la naturaleza y el buen vivir en el pensamiento crítico, el derecho y la literatura*, (Akal - Interpares, 2019). Also, it is necessary to highlight the important research on the origin of the concept of Good Living by David Cortez; one of his most outstanding and recent contributions is "Sumak Kawsay, Buen Vivir y Cambio Climático- Genealogías," in *Cambio climático. Lessons from and for Latin American Cities*, edited by Sylvie Nail, 143–173 (Bogotá: Universidad Externado de Colombia, 2016).

12 See the reflections on this topic in the article by Alberto Acosta, "Rethinking the State anew: Rebuilding it or forgetting it?" In *Repensando nuevamente el estado: reconstruirlo u olvidarlo*, edited by Pascual García, Jessica Ordoñez, Ronaldo Munck (Dublin: Glasnevi Publishing, 2019), also available at https://ecuadortoday.media/2019/01/04/repensando-nuevamente-el-estado-reconstruirlo-u-olvidarlo/.

13 Without denying the possible importance of the rapid technological advances achieved, particularly in recent decades, and which will continue to surprise us day by day, we must bear in mind that not all of humanity benefits from these achievements. There are, for example, vast segments of the world's population that do not have equal access to the world of information technology. On the other hand, technology can be a tool for domination, as is the case with the growing expansion of mechanisms for controlling societies and individuals.

14 A task that demands, for example, to put economics in its true place within the social sciences by removing it from their "imperial" pedestal, sciences (also of European heritage) that must also be questioned. See, for example, Alberto Acosta (2015); "Las ciencias sociales en el laberinto de la economía", *Revista Latinoamericana POLIS* 41, available at https://journals.openedition.org/polis/10917; Alberto Acosta, John Cajas-Guijarro (2018), "De las "ciencias económicas" a la posteconomía - Reflexiones sobre el sin-rumbo de la economía", *Revista Ecuador Debate* 103, CAAO, Quito, available at rebelion.org/docs/242595.pdf.

15 See the interview with Alicia Barcena, Executive Secretary of ECLAC, who assumes that the development model of the subcontinent has been exhausted, in Diaro el País of Spain, February 7, 2020, available at https://elpais.com/economia/2020/02/05/actualidad/1580921046_527634.html.

References

Acosta, Alberto. 2020. "Hoy más que nunca - El derecho al Ocio", available at https://ecuadornoticias.org/hoy-mas-que-nunca-por-el-derecho-al-ocio-no-al-trabajo/.

Brand, Ulrich and Markus Wissen. 2017. *Imperiale Lebensweise - Zur Ausbeutung von Mensch und Natur in Zeiten des globalen Kapitalismus*. München: Oekom Verlag.

Freire, Atawallpa Oviedo. 2011. *Qué es el Sumakawsay - Más allá del Socialismo y Capitalismo*. Quito: Ediciones Sumak.

Gudynas, Eduardo. 2020. "El agotamiendo del desarrollo: La confesión de la CEPAL", available at http://economiasur.com/2020/02/el-agotamiento-del-desarrollo-la-confesion-de-la-cepal/.

Keynes, John Maynard. 1933. "Autosuficiencia Nacional." Revista Ecuador Debate No. 60, CAAP, Quito December 2003.

Leff, Enrique. 2008. "Decrecimiento o deconstrucción de la economía". Peripecias No. 117.

Mill, John Stuart. 1848. *Principles of Political Economy with Some of Their Applications to Social Philosophy*. Toronto: University of Toronto Press.

Sachs, Wolfgang. 1992. *Dictionary of Development - A Guide to Knowledge as Power*. London: Zed Books.

28 Social cartographies in Latin America

Gerónimo Barrera de la Torre

Introduction

Social cartographies have become a common tool for communities, activists, state agencies, and researchers in Latin America involved in the rise of environmental conflicts and territorial struggles.[1] The involvement of communities in the map-making process has introduced a different approach to the production, sharing, and representation of knowledge about territory and landscapes, and has become a tool for marginalized communities (Indigenous, Afrodescendant, *campesino*, mestizo) to contest environmental exploitation. Maps are political artifacts, always produced in a multiplicity of power relations; participatory perspectives bring another layer to the production process (Torres, Gaona, and Corredor 2012; Barragán-León 2019; Ospina Mesa 2018). These "political acts" have become central to communities' territorial claims: as map making "has always been a state enterprise in Latin America" (Herlihy and Knapp 2003, 310), they represent a device to convey different understandings and valuations of the environment. As evidenced by the literature and projects discussed in this chapter, social cartographies are evolving, bringing renewed interest in, and reflection on, the production of maps, renovating perspectives on territorial and environmental struggles (Sletto 2020).

The colonial legacies in Latin America evidenced by the extractivist gaze (Gómez-Barris 2017; Svampa, 2019), marginalizing environmental conservation schemes (e.g., De Matheus e Silva, Zunino, and Huilñir Curío 2018), and land dispossession processes (e.g., Hernández Reyes 2019) have shaped the appropriation of resources and given rise to an understanding of nature that collides, merges, and interweaves with communities' usages and ways of seeing the world. These legacies pervade maps as a device of state control, where the coloniality of knowledge, being, and nature remains critical (Melin, Mansilla and Royo 2019). However, social cartographic approaches have brought a concomitant openness and decentralization in cartographic and territorial knowledge production, making maps at once contentious tools for development schemes, scientific research, and emancipation, revealing the diversity of cartographic approaches in Latin America. This diversity expands the landscape of social cartographic practices and experiences, bringing a plurality of worldviews to the reworking of maps while navigating the challenges of representation.

This chapter engages with these debates, revealing the plurality of social cartographies that, on the one hand, have allowed for emancipatory struggles to achieve territorial recognition and self-determination, and, on the other, impose representational and onto–epistemic models to render territories legible or extractable. The knowledges produced through social cartography are equally convoluted by a myriad of approaches, where epistemologies and ontologies render the environment differently. The validation and translation of knowledges remains a key issue in terms of how maps gain recognition by communities, scientists, and the state. I suggest that such diversity points toward pluriversal cartographies that engage with the inherent

DOI: 10.4324/9780429344428-33

contradictions of map making, and open possibilities of representing multiple, becoming worlds, highlighting maps as processes to think and use the environment otherwise.

I begin this chapter situating social cartographies in the region of Latin America through a brief historical reflection to contextualize their inception. Next, I discuss the conflicted development of this mapping approach, arguing for an ongoing differentiation; that promotes a wide range of cartographic possibilities (Barragán-León 2019; Bryan 2020; Cockayne, Ruez, and Secor 2017). Building on this discussion, I turn to examine the articulations of land titling, environmental conservation, and (neoliberal) sustainability schemes applied in and by communities in different parts of Latin America, giving a sense of the various perspectives, possibilities, contradictions, and unexpected consequences of social cartography. Lastly, I reflect on current trends toward decoloniality and performativity in mapping, which pose new critical stances from which to re-think representation and knowledge production, and to advance praxes for more just social cartographies.

Social cartographies in Latin America

The first participatory mapping projects arrived in the region in the 1980s in a context of legislative and structural changes as well as social unrest that put pressure on states to regularize collective property rights, making marginalized populations' territories legible for development and environmental management (Bryan 2011, 2015; Offen 2009; Sletto et al. 2013). Indigenous social movements had already fought important struggles for the recognition of territories and control over the use of resources in their lands. The approval of the ILO Convention 169 in 1989 was crucial for changing the political geography of Indigenous people and other marginalized communities (Herlihy and Knapp 2003; Offen 2009; OIT 2014). Most of the first efforts were directed toward land rights and titling that helped numerous communities gain control over ancestral lands—a "quiet cartographic revolution"—that had Central American communities as some of its earliest examples (Herlihy and Knapp 2003): for instance, the Miskito and Garifuna countermapping initiative in Honduras that satisfied the state's desire for legibility and countered settler and state intromission in their territory (Mollett 2013).

Moreover, participatory mapping began in the context of the decentralization of environmental policies, concern over the consequences of capitalist depletion that expanded conservation practices (Anthias and Radcliffe 2015), which, together with the process of map making and knowledge production itself, generated spaces for dialogue between communities, activists, academics, and the state. Regarding the first, the World Bank and other international entities have had a significant role in pressing states across the region to make legislative changes recognizing communities' ancestral lands to create certainty over ownership and boundaries (Bryan 2015). Regarding the second, participatory research methods gained traction among researchers introducing approaches that shifted roles in knowledge production and spatial representations to participants (Herlihy and Knapp 2003). These cartographic approaches served to shape negotiations between communities and the state, adding to land recognition efforts to gain control over environment elements and for the implementation of biodiversity conservation and development schemes. The case of Indigenous peoples' territories in Bolivia, which incorporate participatory mapping for the definition of Native Community Lands (*Tierras Comunitarias de Origen*) (Anthias 2019), is an instance of the use of such maps in the context of negotiation with the state. Thus, facing the expansion of (neo)extractivism over communities' territories (Svampa 2019), as well as the environmental crisis heralded by climate change, mapping has become a useful tool for supporting territorial struggles, strengthening communities, providing evidence of Indigenous and peasants sustainable use of the environment, and communicating their diverse ways of understanding and relating to their landscapes.

Cartographies' becoming through difference

Social cartographies are considered to name an "array of community-based research and development approaches" (Herlihy and Knapp 2003, 306), but from their inception in Latin American communities, cartographies have passed through a process of differentiation due to the variety of approaches to representation, the diverse goals of the mapping process, and, of course, the plurality of worldviews through which maps have been re-shaped. It has been recognized that these mapping projects are potentially more inclusive regarding conceptions of, and relations with, the environment (Teague 2011, 100) and diversify the approaches to spatial representation (Sletto, 2020). The diversification process has brought to the fore other mapping projects that challenge universalistic representations reflecting statist and capitalist views, seeking to question the coloniality of nature/power/being/knowing enmeshed in cartographic endeavors. Moreover, this differentiation process, as well as demonstrating potential to propose different territorialities, is instrumental in unveiling the violent work of representations and fixed identities.

I suggest here, following Bryan (2020) and Cockayne, Ruez and Secor (2017), that difference is critical in social cartographies, as a means of confronting to face the coloniality of statist-capitalist-patriarchal homogenization of space (Mansilla, Quintero and Moreira-Muñoz 2019). Difference, through the plurality of understanding of territories and the environment, brings new territories into existence (Bryan 2020, 214) and makes communities visible while challenging hegemonic representations. Still, differentiation may also be "taken up by representation and shackled to identity" (Cockayne, Ruez, and Secor 2017, 11), resulting in instantiations of boundaries that further the frontierization process, rendering places into land (Cons and Eilenberg 2019) and the environment as extractible through the colonial gaze (Gómez-Barris 2017), while imposing representational boundaries fixing identities rendering them legible to state/capital logics.

Furthermore, the naturalization of differences that veils hierarchies within communities and in their relationship with external actors (Mansilla, Quintero and Moreira-Muñoz 2019) signals the mapping process as interlocking relations of force that assembles practices of inclusion and erasure (Sletto 2015). This dynamic in the production of social cartographies remains poorly documented (Barragán-León 2019; Sletto 2020), particularly its implication for communities in the long term, and how, internally, intersections of social differentiation also affect the production of knowledge and the power relations through which certain understandings of the environment are represented. Sletto (2020, 1) proposed that the new phase of mapping projects with and by communities is characterized by a plurality of approaches, proposes, and techniques, which represent the "radical edge of new social cartography." These "radical social cartographies" open up possibilities to assert the different understandings and knowledges of the environment that have been systematically invisibilized and marginalized, as the complex and problematic setting in which this mapping takes place is acknowledged (Bryan 2011).

Difference also names the excess—of knowledge, of life, of sacredness—that transcends the limitations of representing or apprehending spaces through conventional cartographic approaches. Excess in social cartographies may also make it possible to surpass hierarchies and forms of violence that such technologies of power inherited through forms of discipline, surveillance, and legibility (Scott 1998). Difference as generative entails recognizing the multiplicity of life against the universalist, monocultural approaches established by coloniality and statism providing for rampant capital accumulation (Bryan 2020; Cockayne, Ruez, and Secor 2017).

There are two implications of social cartographies' treatment of marginalized communities' territorialities and identities that I want to highlight, because they raise important issues about how the environment is known and represented.

As for the first, social cartographies remain contentious, embodying different forms of understanding space, territories, and the environment under uneven relations of power (Sletto 2015). The problem of "scientific rigor" and legibility remains a thorny issue, related to the translation of knowledges and experiences of the world that have systematically invisibilized the epistemic diversity of communities through cartographic convention. Facing such epistemic violence, maps are regarded as windows to visualize other territorial knowledges and dialogically include other dimensions of life, such as sacredness (Leal-Landeros and Rodríguez-Valdivia 2018). Erasures of epistemic diversity are unveiled through the "geography of absences" (Mansilla, Quintero and Moreira-Muñoz 2019, 153) which, drawing on Boaventura de Sousa Santos's *Epistemologies of the South* (2016), "highlights those blind points in the epistemic map of geography" that have advanced the colonial project and negated the existence of alternative territorialities. The collective, dialogic production of territorial knowledges (Ospina 2018) asserts the difference of epistemes as maps are "exceeded by the dialogue that people generate while mapping" (Leal-Landeros and Rodríguez-Valdivia 2018, 108).

The second implication is that through mapping, representational categories become fixed, imposing new hierarchies in the differentiation process that are functional for the colonial project and statist legibility. Difference, paradoxically, may serve to capture and accommodate the "Other" into accessible, racialized, gendered, and disciplined alternatives, where "placed-based epistemologies" are translated into the language of conventional cartography (Preci 2020). The spatialization of difference through social cartographies has transformed identities and territories in many communities (Herrera 2016), while the complex Indigenous, Black, and *campesino* understandings and forms of representation of the world—as well as communality— are subsumed. In that sense, social cartographies risk establishing univocal representations of identities and ontologies that crystallized fluid, transgressive life projects (Bessire 2014; Vindal and Rivera 2019). While these mapping practices assert differentiation between communities' epistemes and state/capital logics (Oslender 2017), the dispute is circumscribed by reinstating a dichotomic approach that renews the linear, universal purview, as well as dichotomic notions such as nature-culture and traditional-modern. Moreover, conventional cartography, based in obtrusive and dominant objectivity, figures maps as part of a stable ontology that captures/ erases, "fixing culture in place" (Sletto 2015, 926)—a crucial issue considering today's concerns over the lasting implications of social cartographies.

The myriad landscapes of social cartographies

In this section, I explore, through examples from communities' experiences and collectives, a critical issue throughout the implementation of social cartographies: the contentious articulations of land, autonomy, and environmental struggles. Even though these articulations were signaled long ago (Peluso 1995), they are currently relevant, given that the multicultural assertion of collective rights is diminishing (Hale 2020). However, it is also important to consider that, while it is true that social cartographies remain an important means for claiming land rights and advancing struggles against expropriation and exploitation, it does not necessarily follow that community participation and the co-production of knowledges ensures that maps escape conventional and state territorial logics. As many studies carried out in the preceding decades have evidenced, social cartography has had paradoxical implications for communities and their environment (Anthias 2019; Anthias and Radcliffe 2015; Mollett 2013; Oslender 2017; Preci 2020; Wainwright and Bryan 2009). Many of the cartographic endeavors have been assimilated into a "territorial order whose lines [are] codified by the law," and the shift form territorial rights to property has replaced the emphasis on autonomy with that of ownership and to the facilitation of state legibility (Bryan 2015, 97–98). The first element of this articulation, land, remains central to social cartographies' support in the struggles over collective rights and the commons.

These struggles over land rights, particularly communal and customary land tenure, are inextricably linked to environmental struggles and the environmental crisis that extractivist logics have imposed over communities' territories. The discourses of environmental conservation are underpinned by the relation between Indigenous communities and biodiversity hotspots, as well as by the naturalization of collective land rights for such populations (Li 2010), defining them as natural guardians of "natural resources" (Anthias and Radcliffe 2015). This assumed relation between Indigenous peoples and biodiversity is associated with conservation schemes rooted in the participation of inhabitants, and has been decisive in establishing collective land rights as necessary to protect populations and the environment (Anthias and Radcliffe 2015). With the experience gained over the years, however, it is becoming evident that the adjudication of land rights guarantees neither sustainability nor the maintenance of the community (Teague 2011; Anthias 2018), and that the "solution to dispossession is not possession" (Bryan 2020, 208), as land titling reinserts territories in the state territorial logic and the capitalist process of frontierization.

Thus, securing ownership and (private) property is based on a particular relation with the environment and among humans, which poses a contradiction for the advancement of a territorial epistemic plurality, and is also based "in the conflation of racialized bodies" (Mollett 2013, 1234). Whereas the validation of collective rights to territory and resources under the state logic of multicultural governmentality (Hale 2020, 219) remains critical for communities' struggle, formal state recognition entails "the duty to act as any other property owner with regard to the claims of another" (Bryan 2020, 209). Even as collective rights are grounded in cultural difference (Hale 2020, 219), such difference relies on representational categories that fix marginalized communities. This has significant effects on communities as maps convey "the idea of an immemorial and immutable Indigenous territory" (Preci 2020, 29), foreclosing internal differences, inequalities, and peoples' futures. Moreover, as Cruz (2020, 28) explains for the context of Zapotec communities in Mexico, as collective rights are granted, maps flatten space, turning "the commons inside out, transforming it into resources that the state controls as property and transforming knowledge into data." Maps may portray Indigenous through essentialist representation of identity, binding people to certain aspects of their environment (e.g. forest, water, fire), practices (e.g. agriculture, conservation, commonality), and ontologies that serve to categorize, regulate, and naturalize, hindering "future opportunities of self-determination" (Li 2010; Preci 2020, 30). As Wainwright and Bryan explain, "the legal and cartographic strategy thus confronts a racist and exclusionary colonial past, yet reinforces differences and inequalities in the colonial present" (2009, 154).

Invariably, "contradictory lines" (Sletto 2009, 254) are drawn across communities' territories. Frontiers remain a point of inflection in cartographic practices toward more equitable and just access and use of the environment. Frontiers are necessary to produce "the bounded spatial units for environmental management" (2009, 256), held up by "abstract knowledges of postcolonial state [and state-led decolonial] bureaucracy" (Anthias 2019, 20) that deterritorialize features of the environment for its conservation or commercialization. Meanwhile, their definition represents "a symbolic rift in space" (Sletto 2009, 273): ultimately, what is at stake here "is that violence in abstraction which hides the negotiation of uses inherent in commoning" (Linebaugh 2011, 24). But instead of confounding frontiers to an inexistent univocal state or to an essentialist notion of communities' territorialities, Sletto's examination of how the map-making process sheds light on the production of frontiers is useful for understanding the paradoxes of social cartographies as "boundaries assume different meanings in different social and historical contexts" (2000, 253).

In Latin America, numerous groups have engaged in the defense of communities' territories, using diverse technologies to document as well as to rethink cartographic languages. These cartographic endeavors, or "*maptivismo*," interrogate the territorial logics that define access to land, conservation, and resource management (Barragán-León 2019). These groups integrate social organizations, communities, and academia, revealing the ample range of approaches as well as emerging epistemologies. An example of this variety of approaches is *GeoComunes* in México, a collective that has develop a platform to visualize conflicts emerging from degradation, privatization, and dispossession of the commons, which, for instance, has generated a cartographic analysis of the Yucatan peninsula's capitalist expansion and its pressure on social property ("GeoComunes," 2021). Also, from a different epistemic approach, *GeoBrujas* and *Colectivo Miradas Críticas al Territorio desde el Feminismo* ("Mapeo del Cuerpo," 2021) draw on feminist perspectives and corporeal techniques to engage with the violence and extractivism that women face, emphasizing the corporeal-territorial connection in the experience of different forms of oppression.

Whereas Anthias (2018) shows that the participatory cartographies implemented for land titling in the Chaco region by the Bolivian state have had a limited capacity to counteract dispossession processes, increasing fragmentation of indigenous territories and even increasing resource conflict, Melin, Mansilla, and Royo (2019), and Mansilla and Melin (2020) present an critical view into the decolonization of cartographic practices, in which maps are used as tools to transform understandings of territory through Mapuche worldviews. In their effort to stop hydroelectric projects and privatization of their lands, such mapping practices opened spaces for territorializing difference through Mapuche concepts that developed specific cartographic languages (orientation, symbology, toponomy) and made visible sites valued in Mapuche culture for their medicinal plants and spiritual entities, portraying the Mapuche's communities' relation with the environment. In this manner, such examples of decolonial social cartographies play a key role in denouncing and visibilizing environmental conflicts and the depredation underpinned by the racist and gendered extractivist gaze (Gómez-Barris 2017). Adding to these efforts from Chile, *Territorio Alternativos* as "popular territorial consulting" articulates with experiences of autonomous movements to promote alternative approaches to territory and nature through participatory mapping and decolonial methodologies. This consulting is formed by a research group integrated to the Institute of Geography at the Pontificia Universidad Católica de Valparaíso ("Cartografías radicales," 2021).

Mapping, as a political vehicle, holds possibilities for elaborating communities' own territorial narratives (Barragán-León 2019) and supporting the process of differentiated territorialization. For example, projects as the Nova Cartografia Social in Brazil are focused on the practices of auto-cartographies by marginalized communities in the Amazon ("Mapas," 2021), strengthening social movements facing expropriation and providing a defensive response for territorial resistance and environmental management (Malagodi 2012). Another instance of collectives working with communities in developing cartographic alternatives is Iconoclastas, which has developed a series of collective mapping workshops influenced by different epistemic and methodological approaches, and cartographies integrating artistic, political, and academic dimensions ("Cartografías," 2021). Their maps represent a range of conflicts across Latin America, for example, the main socioenvironmental problems and alternatives facing the consequences of soy plantation expansion.

One of the important features of social cartography is the variety of representations that range from drawings in the soil or on pieces of paper, to the use of Participatory Geographic Information Systems (PGIS). Such variety also serves different purposes, for example, in the case of the socioenvironmental struggles in Quilombolas, Brazil, PGIS maps were part of communities' training to face dispossession, showing communities and the areas in dispute (Figure 28.1).

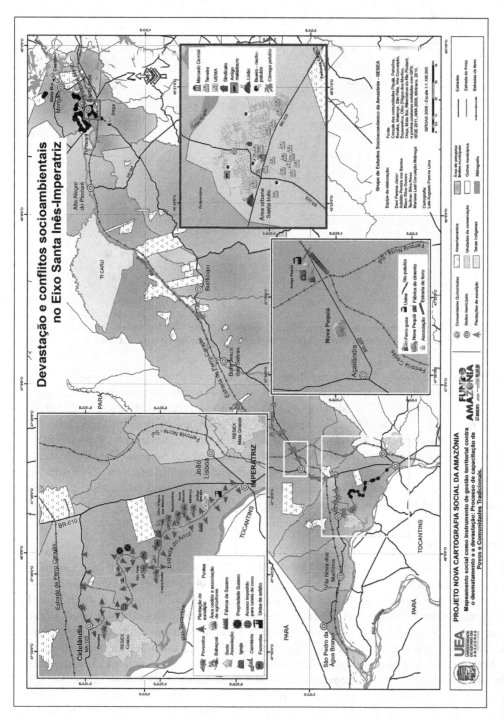

Figure 28.1 Such variety also serves different purposes, for example, in the case of the socioenvironmental struggles in Quilombolas, Brazil, PGIS maps were part of communities' training to face dispossession, showing communities and the areas in dispute.

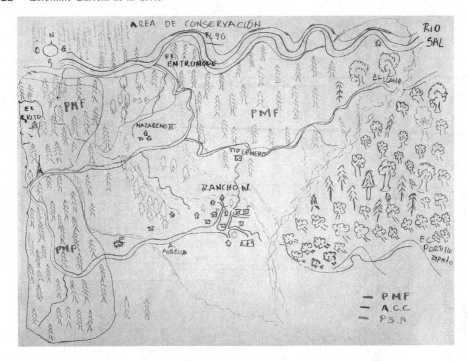

Figure 28.2 Another example is a map drawn by a community in the Chatino region, Mexico, showing the conservation areas that established new borders within their common lands.

Another example (Figure 28.2) is a map drawn by a community in the Chatino region, Mexico, showing the conservation areas that established new borders within their common lands.

Thus, the new radical cartographies emphasize "long-standing and emergent forms of self-determination" (Hale 2020, 219) which draw on "other" geographies emerging from the communitarian organization and defense of territories and environments (Ramos 2018). However, such cartographies need to shift away from spreading the concept of territorial exclusivity (Preci 2020) and bend toward reflexive, dialogic, and decolonial openness. Even with the paradoxical outcomes referred to, communities negotiate, resist, and take advantage of the cartographic tool to re-think autonomy and the use of the environment.

Final thoughts

The 2019 *Enemies of the State?* report (Global Witness 2019) shows the trend of criminalization and death that continues to affect Latin America, as half of the murders of environmental and land defenders around the world took place in that region, especially targeting Indigenous and other marginalized communities. This is a critical, dreadful, and urgent issue that points to the structural state violence to which marginalized communities are subjected. As Leanne Betasamosake Simpson asserts (2020, 43), dispossession is what defines the relationship of Indigenous, and other marginalized communities, with the state. As I intended to demonstrate through this chapter, social cartographies are becoming through their diversification, through difference that questions capture and contains the multiplicity of life. I suggest that these pluriversal cartographies, as new radical approaches, encounter paradoxical legacies through processual and decolonial perspectives. Several examples discussed here problematize the shift to the processual character of maps and their ontological instability, leading toward new ways of mapping (Bryan 2020; Salamanca and Astudillo 2018). In addition, social cartographies may support reflection and analysis of the "legal-bureaucratic knowledges of the state 'from the margins'" (Anthias 2019, 225) and the construction of territorial narratives beyond the constrained boundaries of the state's territorial logics.

One controversial issue has been the lack of reflection about the cartographic legacies which confront notions of tradition and identities, as well as the assumption of participatory methodologies, as better suited to confronting environmental conflicts. These pluriversal cartographies are turning to the "messiness" (Bessire 2014) of marginalized communities' realities, acknowledging that "maps and subjects that conduct mapping are never pure" (Ospina 2018, 12). In weighing the problematic legacies of cartographies in terms of relations of power and the context in which environments are known and represented through maps, the interweaving of structural and local dynamics and inequalities become crucial. Centering on these issues, the mapping process may be instrumental in generating profound reflections about what relations with the environment, and among community members, are being confronted and devised.

Rather than regarding maps only as a result, it should be recognized that their richness comes from the possibility to "incite dialogue" (Barragán-León 2019) and share experiences, practices, and knowledge about a territory. The focus on the dialogic and performative subverts exploitative relations, as well as the locus of enunciation of cartographers (locals and outsiders) (Schulz 2017), opening spaces for other cartographic languages (Cruz 2020). However, the focus on processual and decolonial approaches does not exclude struggles over land and against environmental depletion—on the contrary, I find such arguments have been and will be fruitful in expanding the scope of social cartographies to assume and confront its inherent contradictions. As we reveal ourselves through the map—and while we become different—the horizons of social cartographies expand towards more just and equitable landscapes, something that, as Betancourt (2015) asserts, is not possible "without global epistemic and territorial justice."

Note

1 I use "social cartographies" to denominate the great variety of cartographic projects in the region such as participatory mapping, counter-mapping, *mapas parlantes*, etc. Social cartographies have been used widely in Latin America, signaling new trends toward critical approaches to socio-environmental conflicts and epistemic justice.

References

Anthias, Penelope. 2018. *Limits to Decolonization: Indigeneity, Territory, and Hydrocarbon Politics in the Bolivian Chaco.* Ithaca: Cornell University Press.
———. 2019. "Ambivalent Cartographies: Exploring the Legacies of Indigenous Land Titling through Participatory Mapping." *Critique of Anthropology* 39 (2): 222–42.
Anthias, Penelope, and Sarah A. Radcliffe. 2015. "The Ethno-Environmental Fix and Its Limits: Indigenous Land Titling and the Production of Not-Quite-Neoliberal Natures in Bolivia." *Geoforum* 64: 257–69.
Barragán-León, Andrea Natalia. 2019. "Cartografía Social: Lenguaje Creativo Para La Investigación Cualitativa." *Social Cartography: Creative Language for Qualitative Research* 36: 139–59.
Bessire, Lucas. 2014. *Behold the Black Caiman: A Chronicle of Life among the Ayoreo.* Chicago: University of Chicago Press.
Betancourt Santiago, Milson. 2015. "Giro Descolonial y Nuevas geocartografías." *Ecología Política* 48: 109–11.
Bryan, Joe. 2011. "Walking the Line: Participatory Mapping, Indigenous Rights, and Neoliberalism." *Geoforum* 42 (1): 40–50.

———. 2015. *Weaponizing Maps: Indigenous Peoples and Counterinsurgency in the Americas*. New York, NY: The Guilford Press.

———. 2020. "Commentary: What sort of Territory? What Sort of Map?" In *Radical Cartographies. Participatory Mapmaking from Latin America*. Edited by Bjorn Sletto, 203–216. Austin: University of Texas at Austin.

"Cartografías," Iconoclastas, accessed October 7, 2021. [https://iconoclasistas.net/cartografias/].

"Cartografías radicales," Territorios alternativos, accessed October 7, 2021. [http://territoriosalternativos.cl/cartografias-radicales/].

Cockayne, Daniel G., Derek Ruez, and Anna Secor. 2017. "Between Ontology and Representation: Locating Gilles Deleuze's 'Difference-in-Itself' in and for Geographical Thought." *Progress in Human Geography* 41 (5): 580–99.

Cons, Jason, and Michael Eilenberg. 2019. *Frontier Assemblages*. Oxford: John Wiley & Sons.

Cruz, Melquiades (Kiado). 2020. "Oral Narratives in the Rincón Zapoteco: A Cartography of Processes." In *Radical Cartographies. Participatory Mapmaking from Latin America*. Edited by Bjorn Sletto, 19–24. Austin: University of Texas at Austin.

De Matheus e Silva, Luis Fernando, Hugo Marcelo Zunino, and Viviana Huiliñir Curío. 2018. "El negocio de la Conservación Ambiental: Cómo la Naturaleza se ha Convertido en una Nueva Estrategia de Acumulación Capitalista en la Zona Andino-lacustre de Los Ríos, sur de Chile." *Scripta Nova* 22 (February). Available at https://revistes.ub.edu/index.php/ScriptaNova/article/view/19021/28907.

"GeoComunes: Cartografía colaborativa en defensa de los bienes comunes," GeoComunes, accessed October 7, 2021. [http://geocomunes.org/presentaci%C3%B3n/quienes.html]

Global Witness. 2019. Enemies of the State? How governments and businesses silence land and environmental defenders *Global Witness*, Available at: https://www.globalwitness.org/en/campaigns/environmental-activists/enemies-state/.

Gómez-Barris, Macarena. 2017. *The Extractive Zone: Social Ecologies and Decolonial Perspectives*. Durham: Duke University Press.

Hale, Charles R. 2020. "Afterword." In *Radical Cartographies. Participatory Mapmaking from Latin America*. Edited by Bjorn Sletto, 217–220. Austin: University of Texas at Austin.

Herlihy, Peter H., and Gregory Knapp. 2003. "Maps of, by, and for the Peoples of Latin America." *Human Organization; Oklahoma City* 62 (4): 303–14.

Hernández Reyes, Castriela Esther. 2019. "Black Women's Struggles against Extractivism, Land Dispossession, and Marginalization in Colombia." *Latin American Perspectives* 46 (2): 217–34.

Herrera, Johana. 2016. *Sujetos a Mapas: Etnización y Luchas Por La Tierra En El Caribe Colombiano*. Bogotá: Pontificia Universidad Javeriana.

Leal-Landeros, Joselin, and Alan Rodríguez-Valdivia. 2018. "Cartografía Social de Chapiquiña: Reivindicando os Direitos Territoriais Indígenas nos Altos de Arica, Chile." *Íconos. Revista de Ciencias Sociales*, no. 61 (August): 91–114. http://scielo.senescyt.gob.ec/scielo.php?script=sci_abstract&pid=S1390-12492018000200091&lng=pt&nrm=iso.

Li, Tania Murray. 2010. "Indigeneity, Capitalism, and the Management of Dispossession." *Current Anthropology* 51 (3): 385–414.

Malagodi, Marco Antonio Sampaio. 2012. "Geografias do Dissenso: Sobre Conflitos, Justiça Ambiental e Cartografia Social no Brasil." *Espaço e Economia. Revista brasileira de geografia econômica*, no. 1. Available at: http://journals.openedition.org/espacoeconomia/136.

Mansilla Quiñones, Pablo, Quintero Weir, José, and Moreira-Muñoz, Andrés. 2019. "Geography of Absences, Coloniality of the Being and the Territory as a Critical Substantive in the South Epistemologies" *Utopía y Praxis Latinoamericana* 24 (86): 148–161.

Mansilla Quiñones, Pablo and Melin Pehuen, Miguel. 2020. "Mapuche Cartography: Defending *Ixofillmogen*." In *Radical Cartographies. Participatory Mapmaking from Latin America*. Edited by Bjorn Sletto, 81–96. Austin: University of Texas at Austin.

"Mapas," Nova Cartografia Social da Amazonia, accessed October 7, 2021 [http://novacartografiasocial.com.br/mapas/]

"Mapeo del Cuerpo como Territorio," Miradas críticas del territorio desde el feminismo, accessed October 7, 2021. [https://territorioyfeminismos.org/metodologias/mapear-el-cuerpo-como-territorio/]

Melin, Miguel, Pablo Mansilla Quiñones, and Manuela Royo. 2019. *Cartografía Cultural Del Wallmapu: Elementos Para Descolonizar El Mapa En Territorio Mapuche*. Temuco: Pu Lof Editores.

Mollett, Sharlene. 2013. "Mapping Deception: The Politics of Mapping Miskito and Garifuna Space in Honduras." *Annals of the Association of American Geographers* 103 (5): 1227–41.

Offen, Karl. 2009. "O mapeas o te Mapean: Mapeo Indígena y Negro en América Latina." *Tabula Rasa, 10 (enero junio): 163 89.*

Oslender, Ulrich. 2017. "Ontología Relacional y Cartografía Social: ¿hacia Un Contra-Mapeo Emancipador, o Ilusión Contra-Hegemónica?" *Tabula Rasa* 26 (June): 247–62.

Ospina Mesa, César Andrés. 2018. "Tensiones y potencias de la Cartografía Social en la producción colaborativa de conocimientos territoriales," VI Encuentro Latinoamericano de Metodología de las Ciencias Sociales, 7 al 9 de noviembre de 2018, Cuencua, Ecuador.

Peluso, Nancy Lee. 1995. "Whose Woods Are These?: Counter-Mapping Forest Territories in Kalimantan, Indonesia." *Antipode* 27 (4): 383–406.

Preci, Alberto. 2020. "Fixing the Territory, a Turning Point: The Paradoxes of the Wichí Maps of the Argentine Chaco." *The Canadian Geographer/Le Géographe Canadien* 64 (1): 20–31.

Ramos, David Jiménez. 2018. "Gestión Social y Defensa de los Territorios en México: Aproximaciones Teórico- metodológicas Desde la Cartografía Social para el Estudio de Territorios Bioculturales." *Desenvolvimento, fronteiras e cidadania* 2 (1): 51–55.

Salamanca Villamizar, Carlos, and Francisco Astudillo Pizarro. 2018. "Justicia Ambiental, Metodologías Participativas y Extractivismo en Latino América." *Justice spatiale | Spatial Justice* 12: 16.

Santos, Boaventura de Sousa. 2016. *Epistemologies of the South: Justice against Epistemicide / Boaventura De Sousa Santos*. London: Routledge.

Schulz, Kastern A. 2017. "Decolonizing Political Ecology: Ontology, Technology and 'critical' Enchantment." *Journal of Political Ecology* 24 (1): 125–43.

Scott, James C. 1998. *Seeing like a State: How Certain Schemes to Improve the Human Condition Have Failed*. New Haven: Yale University Press.

Simpson, Leanne Betasamosake. 2020. *As We Have Always Done: Indigenous Freedom through Radical Resistance*. Minneapolis: University of Minnesota Press.

Sletto, Bjørn. 2009. "'Indigenous People Don't Have Boundaries': Reborderings, Fire Management, and Productions of Authenticities in Indigenous Landscapes." *Cultural Geographies* 16 (2): 253–77.

———. 2015. "Inclusions, Erasures and Emergences in an Indigenous Landscape: Participatory Cartographies and the Makings of Affective Place in the Sierra de Perijá, Venezuela." *Environment and Planning D: Society and Space* 33 (5): 925–44.

———. 2020. "Introduction: Racial Social Cartographies." In *Radical Cartographies. Participatory Mapmaking from Latin America*. Edited by Bjorn Sletto, 1–18. Austin: University of Texas at Austin.

Sletto, Bjørn, Joe Bryan, Marla Torrado, Charles Hale, and Deborah Barry. 2013. "Territorialidad, Mapeo Participativo y Política Sobre Los Recursos Naturales: La Experiencia de América Latina." *Cuadernos de Geografía: Revista Colombiana de Geografía* 22 (2): 193–209.

Svampa, Maristella. 2019. *Las Fronteras Del Neoextractivismo En América Latina*. Costa Rica: CALAS.

Torres, Irene Velez, Sandra Rátiva Gaona, and Daniel Varela Corredor. 2012. "Cartografía Social como Metodología Participativa y Colaborativa de Investigación en el Territorio Afrodescendiente de la Cuenca alta del río Cauca." *Cuadernos de Geografía: Revista Colombiana de Geografía* 21 (2): 59–73.

Vindal Ødegaard, Cecilie, and Juan Javier Rivera Andía, eds. 2019. *Indigenous Life Projects and Extractivism: Ethnographies from South America*. Gewerbestrasse: Palgrave Macmillan.

Wainwright, Joel, and Joe Bryan. 2009. "Cartography, Territory, Property: Postcolonial Reflections on Indigenous Counter-Mapping in Nicaragua and Belize." *Cultural Geographies* 16 (2): 153–78.

29 Rights of nature in the courts of Latin America

Moving forward to better protect our environment?

Ximena Insunza Corvalán

Nature as a subject of law in Latin American constitutions

Since the 2000s, certain Latin American countries have included the legal recognition of nature as a subject of rights in their constitutional texts. The enshrinement of the rights of nature in the constitution of a country has been conceived as a way of recognizing the interdependent relationship between nature and human beings, and of providing essential tools for more effective advocacy and defense. As Zaffarroni (2011) points out

> [...] the most important thing is that, by recognizing nature as a subject of rights, it acquires the status of an abused third party when it is illegitimately attacked and, therefore, enables the exercise of legitimate defense in its favor (the legitimate defense of third parties).
>
> [Own translation]

In order to understand the constitutionalization of nature as a subject of rights, it is fundamentally important to recognize the existence and, consequently, the particular attributes of at least four legal families : (1) the Roman-Germanic legal family, "which is characterized by the presence of codes and written rules of law, drafted prior to the problems that practice presents" ; (2) the common law family, in which precedents are the main source of law; (3) the socialist legal family, in which law had to be applied in a manner conforming to a party's ideology; and (4) the religious legal family, in which law is created through "the interpretation made of the sacred and revealed books, and religious traditions"(Ayala 2013, 238–239). Latin American countries are mostly framed within the Roman-Germanic tradition, commonly known as continental law, which is a relevant antecedent in terms of understanding the constitutionalization of the rights of nature.

This paradigm shift began in Ecuador with the inclusion of the following phrase in its 2008 Constitution: "Nature or Pacha Mama, where life is reproduced and realized, has the right to the full respect of its existence and for the maintenance and regeneration of its vital cycles, structure, functions and evolutionary processes" (Article 7). Shortly thereafter, Bolivia enshrined the following in Article 8 of its 2009 Constitution's enlisting several principles that citizens and society must follow:

> I. The State assumes and promotes the following as ethical, moral principles of the plural society: ama qhilla, ama llulla, ama suwa (do not be lazy, do not be a liar or a thief), suma qamaña (live well), ñandereko (live harmoniously), teko kavi (good life), ivi maraei (land without evil), and qhapaj ñan (noble path or life). II. The State is based on the values of

DOI: 10.4324/9780429344428-34

unity, equality, inclusion, dignity, freedom, solidarity, reciprocity, respect, complementarity, harmony, transparency, balance, equal opportunities, social and gender equity in participation, common welfare, responsibility, social justice, distribution and redistribution of social wealth and assets for wellbeing.

In Article 135, the same Constitution grants the right to a Popular Action (*Acción Popular*), which can be exercised by anyone,

> against any act or omission by the authorities or of individuals or collectives that violates or threatens to violate rights and collective interests related to public property, space, safety and health, the environment and other rights of a similar nature recognized by this Constitution.

Colombia, although following a different path, is a remarkable case study since its courts, particularly the constitutional court, have in several cases recognized various constituent elements of the environment as legal subjects entitled to protection. Clear examples of this trend are: the Atrato River (Chocó),[1] Cauca, Magdalena,[2] Quindío Pance (Valle del Cauca),[3] La Plata (Huila),[4] Otún (Risaralda),[5] the Pisba moors (Boyacá),[6] among others. On the first three cases, the court recognized the rights of rivers to be protected from pollution and other forms of degradation, imposing on the state the responsibility to implement measures to repair and prevent new damages. On the la Plata case and Pisba Moors cases, the courts acknowledged the need to protect future generations from the effects of climate change; thus the ruling mandated the state to implement mitigation measures considering the rights of future inhabitants of the forests. Similarly, the Colombian Supreme Court of Justice accepted a *habeas corpus* writ recognizing the legal personality of the spectacled bear called "Chucho,"[7] and more recently a judge in Cartagena issued a judgment granting constitutional protection in which it compelled the state to create a policy for the preservation of bees, although this decision was eventually overturned.

Among the aforementioned cases, the Rio Atrato judgment issued in 2016 is noteworthy. In its recitals it states that:

> In this order of ideas, the greatest challenge of contemporary constitutionalism in environmental matters is to achieve the safeguarding and effective protection of nature, cultures and forms of life associated therewith, as well as of biodiversity, not because of the simple material, genetic or productive utility that these may represent for human beings, but because, being a living entity composed of other multiple forms of life and cultural representations, they are subjects of individualizable rights, which renders their integral protection and respect as a new imperative for States and societies. In sum, only from a stance of profound respect and humility toward nature, its members and its culture, is it possible to relate to them in fair and equitable terms, leaving aside any concept that is limited to the simply utilitarian, economic or efficient.[8]

The decision then holds the meaning of the new conception which implies several relationships that will inform the new paradigm that Colombia Constitutional Court will follow in the future:

> In short, it can be concluded that the central premise on which the conception of bioculturality and biocultural rights is based is the relationship of profound unity between nature and the human species. This relationship is expressed in other complementary elements such as: (i) the multiple ways of life expressed as cultural diversity are intimately linked to the diversity of ecosystems and territories; (ii) the richness expressed in the diversity of

cultures, practices, beliefs and languages is the product of the co-evolutionary interrelationship of human communities with their environments and constitutes an adaptive
response to environmental changes; (iii) the relationships of different ancestral cultures
with plants, animals, microorganisms and the environment actively contribute to biodiversity; (iv) the spiritual and cultural meanings nature for indigenous peoples and local communities are an integral part of biocultural diversity; and (v) the conservation of cultural
diversity leads to the conservation of biological diversity, so that the design of policy, legislation and case law should focus on the conservation of bioculturality.

[my translation]

The recognition of the Atrato river as a legal subject was extremely relevant because it recognized the interdependence between humans and nature, under the term "bioculturality."
Although this concept had already been used in the literature, it is now beginning to be widely
disseminated and used by stakeholders in countries comprising Latin America, not only at the
academic level, but also across civil society and, most importantly, by decision-makers.
Consequently, a transition toward integrating this new paradigm has begun.

However, nowadays, there is much debate about the effectiveness of the enshrinement of nature
or its elements as a subject of law, independent of the rights of individuals. Generally speaking, it
is possible to classify these criticisms into two categories: Firstly, there are theoretical critiques,
concerning the design of laws and the question of who represents nature. Secondly, there are
criticisms of a practical nature, related to the enforcement of decisions issued by the courts or
tribunals—whether these be general or specialized—which order compliance with specific protection regulations or the implementation of reparation measures, but whose enforcement depends
on governmental agencies or private organizations, and which may lack effective monitoring or
follow-up and, in some cases, even the financial resources for ensuring compliance.

The creation of special courts and tribunals as a guarantee of access to environmental justice

On the other hand, Latin America has also experienced changes in the institutional framework
for resolving socio-environmental conflicts. The proliferation of this type of dispute in the
region in recent decades has been permeated by a greater awareness of, and adhesion to, the values of protection of our environment by the general public, and broader coverage from the
media, especially social networks, which have brought to the forefront the depth of the crisis
that the region's current development model is going through, beset by an inadequate use and
exploitation of its natural resources (Gligo et al. 2020). The way in which these types of disputes are resolved is constantly evolving and, in our opinion, there are currently no adequate
mechanisms for their proper resolution. The interests at stake, often conflicting and difficult to
harmonize, have caused such tensions that tribunals or courts specializing in environmental
matters have been created both worldwide and in Latin America.

Specialization as a phenomenon has encompassed various areas of knowledge, also reaching
the legal sciences (Baum 1997; Legomsky 1990). In general terms, it can be said that the literature has pointed out the following as advantages of the creation of specialized courts and tribunals (Legomsky 1990; Baum 1997; Zimmer 2009): (1) efficiency, since conflicts are resolved
in a shorter amount of time and are limited to a problem of which all actors have a basic
knowledge; (2) quality, since the adjudicators' expert knowledge allows for a better application
of the law; (3) uniformity and consistency of decisions throughout a jurisdiction; (4) better case
management, through the construction of mechanisms within judiciaries that more quickly
relate the issues and allow for a more expeditious processing of cases; (5) the elimination of
jurisdictional disputes and forum-shopping; and (6) better response in reviewing the actions of

Table 29.1 Environmental Courts & Tribunals in South and Central America

Country	Tribunals/Courts/Other	Year	Competences and jurisdiction
Bolivia	Agrarian Environmental Court	2012	Imparts justice in agrarian, livestock, environmental, water, and biodiversity related matters that do not fall under the jurisdiction of administrative authorities.
Brazil	Specialized Court of the Environment and Agrarian Affairs of the State of Amazonas	2011	No information.
Chile	Environmental Courts	2012	Heras recourses against environment-related administrative decisions that are judicially reviewable and actions seeking reparation for environmental damages.
Costa Rica	Environmental Administrative Court	1995	Protection and prevention of environmental damage.
Peru	Environmental Compliance Enforcement and Inspection Court Forestry and Wildlife Court	2011 2017	Second and final administrative instance of the OEFA. Second and final administrative instance of Line Directors.
Trinidad & Tobago	Environmental Commission	2000	Environmental Management Law.

Source: Own elaboration in base a (CEPAL 2018).

administrative agencies, since specialized courts and tribunals are better positioned than the general courts.

Internationally, 'the explosion' in the number of environmental courts or tribunals since 2000 is astounding. Today, there are over 1,200 environmental courts or tribunals in 44 countries at the national or state/provincial level, with some 20 additional countries discussing or planning environmental courts or tribunals. (Pring and Pring 2016)

Most relevant, and surely a direct cause of the creation of these institutions, has been the development of international law, namely through international conferences and agreements such as Stockholm 1972; Earth Charter 1982; Rio 1992; Johannesburg 2002; Rio +20 2012; the United Nations Framework Convention on Climate Change and Convention on Biological Diversity; the Aarhus Convention and Escazú Convention.

In the case of Latin America, the following courts or tribunals have been recognized by international organizations (Table 29.1); however, only in the case of Chile are they judicial bodies as such—the rest, while having some degree of autonomy, belong to the executive branch.

We will now focus on two countries. The first, Costa Rica, was chosen for being one of the countries with the best performance indexes in environmental matters and for being one of the first to establish this type of specialization (at the administrative level). The second, Chile, was chosen as its model is the only one in the region in which specialized agencies can exercise jurisdiction as such, even though they are not part of the Chilean judiciary.

The cases of Costa Rica and Chile

Costa Rica

Costa Rica's Environmental Administrative Tribunal (TAA), created in 1995, was a pioneering initiative in the region. This tribunal is a "decentralized body of the Ministry of the Environment and Energy, with exclusive jurisdiction and functional independence in the performance of its duties. Its rulings exhaust the administrative channel, and strict and compulsory compliance shall be had with its resolutions."⁹ It is made up of three regular members and three alternates, all appointed by the National Environmental Council for a six-year term. Members are required to be professionals with experience in environmental matters; only one regular member and one alternate member must be lawyers. The TAA has jurisdiction to, among other things, as it is established in article 103 Organic Environmental Law 7554

> [h]ear and resolve, in administrative proceedings, claims lodged against all persons, either public or private, for violations of the environmental and natural resources protection laws, as well as to hear, process and resolve, *ex officio* or at the request of a party, claims referring to active and omissive conducts that violate or threaten to violate the regulations of the environmental and natural resources protection legislation.

It is a hybrid body; however, it has received support from the Constitutional Chamber, which has afforded it the powers and responsibilities related to the prevention, prosecution, and penalization of environmental damages as a direct component of the constitutional right to a healthy environment, enshrined in Article 50 of the Costa Rican Constitution.

The problem in the case of the Costa Rica Environmental Administrative Tribunal is related to its dependence on the political power, which sometimes empowers and gives it support. On other occasions, however, it is not part of the political agenda, so the TAA is unable to provide adequate protection. It is thus highly recommended its independence of the Executive power.

Chile

The creation of the Environmental Courts in Chile took place in the midst of the legislative process entailing the redesign of the environmental institutional framework. In 2008, in order to render feasible the approval of Law No. 20,417—which created the Ministry of the Environment, the Environmental Evaluation Service, and the Superintendency of the Environment—the executive branch had to compromise by sending a bill that created, at that time, a single environmental court, which would become a "life raft" for the completion of the processing of the aforementioned law. Although the bill began its legislative process in 2008, the enactment of Law No. 20,600 only took place four years later.

There are currently three environmental courts in Chile, which are divided between the territorial jurisdictions of the Northern Zone (Antofagasta), the Central Zone (Santiago), and the Southern Zone (Valdivia).

The environmental courts are composed of three members, two of whom are lawyers specializing in environmental or administrative matters, while the third is a graduate in the sciences with a specialization in environmental matters. All members of the court must have at least ten years of professional practice. There are also two alternate members, one a lawyer and the other a graduate in the sciences with a specialization in environmental matters. With regard to the appointment system, it is worth noting that Article 2 of Law 20,600 establishes a very specific appointment system in which the Public Senior Management Post Appointment System (the "SADP") intervenes first, followed by the Supreme Court, and finally the presidency, which proposes a candidate to the Senate. The candidate must be ratified by three-fifths of the Senate's members in office.

The environmental courts have various competences and jurisdiction to hear certain proceedings, which can be classified into four types: (1) Disputes regarding environmental administrative decisions; (2) suits seeking reparation for environmental damages; (3) authorizations of certain provisional measures ordered by the Superintendency of the Environment; and

(4) consultations regarding the most severe sanctions imposed by the Superintendency of the Environment.[10] Only final judgments issued in the proceedings set out in paragraphs a) and b) may be challenged through the filing of appeals by way of cassation (on the merits and/or on procedural grounds). In general, jurisdiction is associated with the environmental management instruments and not with the elements that make up the environment, or with the environment as a legally protected interest or right in itself.

Law 20,600 provides for two types of proceedings, one related to disputes regarding administrative decisions that, broadly speaking, involve the filing of a written claim alleging unlawful decisions or actions taken by the administrative agencies of the state; i.e., a claim against an administrative act. The agency that issued such an administrative act must subsequently inform the court of the basis or reasons for its issuance. Later, the parties make their pleadings based on the aforementioned briefs, so that the court may subsequently issue its judgment. The other type of proceeding relates to suits seeking reparation for environmental damages, i.e. a more limited version of an ordinary tort trial governed by common law, albeit with some modifications.

Effectiveness of environmental protection

For some, the deficiencies of the Chilean environmental courts are related to their design. A first criticism relates to the lack of independence of the members of environmental courts. As explained above, the three branches of government intervene in the appointment of all members of the environmental courts, which must take place every six years. As such, the way in which court members rule in each case has an impact on the possibility of their being reappointed, even more so if their ratification by the Senate has a quorum of 60%, which implies the need to engage in negotiations at the time the candidate is proposed.

A second criticism is linked to the jurisdiction and competences of the environmental courts and the legal standing to resort to them. In terms of these courts' competences, in disputes related to administrative decision-making, what is challenged is the legality of the administrative acts issued by the respective ministry or services of the environmental sector; however, the claims lodged are not general claims. Rather, they must comply with a number of requirements including, among others, first exhausting the administrative channel. As for legal standing, generally speaking, an injury or damage must be proven (a concept that has been expanding through case law); however, the party bringing the claim must also have been a party to the sanctioning administrative procedures, or have been directly impacted, among other requirements. In other words, the conjugation of Articles 17 and 18 of Law No. 20,600 restricts access to environmental justice, given the absence of a single claim of illegality in broad terms.

A third criticism is associated with procedural matters. The two types of proceedings mentioned above have very regulated stages, which are neither necessarily expeditious nor timely in terms of resolving conflicts. The current processes are not flexible enough to handle some of the simpler environmental legal conflicts that are not necessarily associated with environmental management instruments.

These deficiencies have translated into the ordinary justice system being in permanent tension when it comes to providing environmental protection, so much so that cases involving, for example, the need to enter into the Environmental Impact Assessment System are better handled by the ordinary justice system than by the specialized judiciary. The same has occurred in the case of citizen participation in environmental impact assessments, proceedings in which the ordinary justice system opens up stages for citizen observations without these cases even being heard by the specialized courts.

Conclusions

While the enshrinement of the rights of nature in the constitutions of some Latin American countries has not been as effective as desired—as a result of the problems of justiciability and the normative incoherence of the regulation of natural resources mentioned above—there is no doubt that it has permeated these countries' internal regulations, above all with respect to the understanding of nature as a decisive element when thinking about development models and, consequently, about public policies. Likewise, and with greater repercussions, the emergence of nature as an object of protection in itself (and not necessarily as a subject of right) has had an undeniable impact on the behavior of certain judges (constitutional, general, or specialized) who have gone beyond the most conservative conceptions of law to guarantee the conservation of ecosystems and ways of life.

At the same time, although the creation of specialized judiciaries has improved access to environmental justice, it has not yet been developed in such a way that one can categorically affirm that their creation implies a better or greater protection of the environment as a legally-protected interest or right. The main reason for this is that their designs have not brought about the necessary independence or tools, either procedural or substantive, to institute a profound change in the asymmetry between the parties, namely, public administration versus private parties, and private parties versus private parties who caused environmental damage. Nevertheless, if some reforms were to be made, better standards could be achieved and the purpose for which the specialized judiciaries were created could be fulfilled.

Notes

1 T-622 of November 10, 2016, issued by the Sixth Review Chamber of the Constitutional Court.
    As claimed by the action, intensive mining destroys the river bed, produces the dumping of highly polluting substances such as mercury and cyanide, and disperses mercury vapors as waste from waste treatment, therefore "the pollution of the Atrato River is threatening the survival of the population, the fish and the development of agriculture, which are indispensable and essential elements of food in the region, which is the place where communities have built their territory, their lives and recreate their culture" [Own translation].
2 STC 3872-2020 Luís Miguel Llorente (*Grupo de Litigio de Interés Público de la Universidad del Norte Vs. Ministerio de Ambiente y Desarrollo Sostenible (MADS), a la Unidad Administrativa Especial de Parques Nacionales Naturales, a la Procuraduría General de la Nación, a la Fiscalía General de la Nación, a la Policía Nacional de Colombia y otros* (2020), Supreme Court of Justice, Civil Cassation Chamber.
    In such cases, the Supreme Court of Justice declared the Salamanca Island Parkway as a subject of law and ordered the drawing up of a strategic and efficient five-month action plan to reduce its deforestation and degradation levels to zero (0). "Such planning must contain commitments, responsible authorities, lines of action and specific dates for the promotion of actions for the prevention and restoration of the VPIS, as well as the consequences in case of non-compliance, in accordance with the legal provisions on environmental matters," held the judgment [Own translation].
    The novelty of this new ruling lies in understanding man, flora and fauna on an "equal and interdependent plane, in which they cohere to render the lives of all on Earth bearable, from which it follows that they must work harmoniously to avoid environmental degradation, air pollution, the extinction of animal species, drought of water basins, collective diseases (pandemics) and all negative impacts produced by the excessive, uncontrolled, abusive and inadequate use of the so-called natural resources" [Own translation].
3 *Juan Luis Castro Córdoba y Diego Hernán David Ochoa Vs. Ministerio de Ambiente y Desarrollo Sostenible, EPM, Hidroeléctrica Ituango S.A. E.S.P. y otros.* Superior Court of Medellín, Fourth Civil Decision Chamber (2019-076). The Court concluded, "(i) that future generations are subjects of rights of special protection, (ii) that they have fundamental rights to dignity, water, food security and a healthy environment, and (iii) that the Cauca River is a subject of rights, which implies, as was done with the Atrato River, its protection, conservation, maintenance and restoration, under the responsibility of

the Municipal Public Entity and the State" [own translation], ordering the National Government to exercise the guardianship and legal representation of the river's rights, through the Institution to be appointed by the President.

4 Sole Municipal Civil Court of La Plata. March nineteenth (19), two thousand nineteen (2019). Docket No.: 41-396-40-03-001-2019-00114-00. the judge decided to assume an eco-centric approach, noting that the La Plata River was being polluted and recognizing it as a subject of rights.

   In the decision, which has an "eco-centric and anthropogenic" approach, this tributary was declared as a subject of rights. "Thus, for this strict case, this court, with deep respect for nature and following the environmental case law, will recognize the "La Plata River" as a subject before the law, will evaluate the facts reported that impacted this water resource due to this condition, and will adopt the protection measures it considers necessary, once the foregoing has been examined vis-à-vis the rights of the plaintiffs," held the judge in the ruling. [Own translation]

5 Fourth Court of Enforcement of Judgments and Security Measures of Pereira. September eleventh (11) of two thousand nineteen (2019). Protection Ruling 036/2019.

   The action for protection was filed against the Ministry of the Environment, the Department of Risaralda, the Municipalities of Pereira and Dosquebradas, as well as the Risaralda Regional Autonomous Corporation (Carder) and the company Aguas de Pereira for failing to stop the pollution of the river. In the action, citizens requested protection of their collective rights to a healthy environment, ecological balance and access to drinking water.

   In the protection ruling, the judge ordered the authorities of the region to allocate the necessary budget to clean up the basin and build an aqueduct and sewerage plan in villages such as La Florida, La Bananera, La Bella, and other areas located upstream of the intake and that did not have this service.

   The judge also ordered the development of educational activities in the neighborhoods near the Otún River in order to raise awareness about waste management.

6 The Ministry of the Environment and Sustainable Development of Colombia filed an appeal against the judgment of June 29, 2018 issued by the 2nd Oral Administrative Court of the Circuit of Duitama that accepted the constitutional protection action seeking the protection of the fundamental rights of citizen participation and due process. In said case, the claimants reproached the Ministry of the Environment and Sustainable Development for having failed to socialize the delimitation process of the Pisba Moor, thus violating their right to due process, which precluded them from evaluating the economic and social conflict that could arise from the revocation of the mining permit and, with it, the workers' labor contracts. Consequently, the claimants requested that the delimitation process of the Pisba Moor be stopped. The Colombian environmental authority's appeal is based on the need to prioritize the general interest over the private interest and argued that, in its opinion, said authority complied with the participation process regulated in the Colombian legal framework. In the second instance, the Administrative Court of Boyacá issued a judgment dated August 9, 2018, upholding the first instance decision and additionally declared that the constitutional precedent established by judgment T-361 of 2017 was fully applicable to the delimitation process of the Pisba Moor.

7 *Luis Domingo Gómez Maldonado Vs. Fundazoo, a Corpocaldas, a Aguas de Manizales, a la Unidad Administrativa Especial del Sistema de Parques Nacionales Naturales y al Ministerio del Medio Ambiente* (2020) Action for protection, Judgment SU-016, Constitutional Court.

   In the Court's opinion, while in principle *habeas corpus* was designed to guarantee the freedom of individuals, this does not exclude its use to demand the protection of animals as sentient beings and as subjects before the law. The Court's central argument embodied a shift from a merely anthropocentric vision to an eco-centric and anthropogenic one, in which man is primarily responsible for the conservation of the environment, within the framework of a "universal and biotic citizenship", which cannot ignore "our condition as living beings and animals." The losing parties in these proceedings filed an action for constitutional protection with the Labor Chamber and the Criminal Chamber of the Supreme Court of Justice, which ultimately granted the constitutional action, leaving the *habeas corpus* without effect. In other words, this mechanism vacated the decision that granted the remedy for the protection of the bear "Chucho" and maintained the conditions of captivity in the zoo of the city of Barranquilla.

8 Recital 5.1.

9 Article 103 Organic Environmental Law 7554.

10 Law 20,600, Article 17.

References

Ayala, Alfonso. 2013. "Una nueva teoría integral de comportamiento judicial. Entendiendo las verdaderas motivaciones de los jueces." *Revista del Tribunal Electoral del Poder Judicial de la Federación* 1 (Number 11): 235–264.

Baum, Lawrence. 1997. *The Puzzle of Judicial Behaviour*. University of Michigan Press.

CEPAL. 2018. Acceso a la información, la participación y la justicia en asuntos ambientales en América Latina y el Caribe: hacia el logro de la Agenda 2030 para el Desarrollo Sostenible. Pag 152.

Gligo, Nicolo, Gisela Alonso, David Barkin, Antonio Brailovsky, Francisco Brzovic, Julio Carrizosa, Hernán Durán, Patricio Fernández, Gilberto C. Gallopín, José Leal, Margarita Marino de Botero, César Morales, Fernando Ortiz Monasterio, Daniel Panario, Walter Pengue, Manuel Rodríguez Becerra, Alejandro B. Rofman, René Saa, Héctor Sejenovich, Osvaldo Sunkel, and José J. Villamil. 2020. *La tragedia ambiental de América Latina y el Caribe, Libros de la CEPAL, N° 161 (LC/PUB.2020/11-P), Santiago, Comisión Económica para América Latina y el Caribe (CEPAL)*. Edited by CEPAL. Vol. 161, *Libros de la CEPAL*. Santiago: CEPAL.

Legomsky, Stephen H. 1990. *Specialized Justice: Courts, Administrative Tribunals, and a Cross-National Theory of Specialization*. Oxford, UK: Clarendon Press.

Pring, G., and C. Pring. 2016. Environmental Courts & Tribunals: A Guide for Policy Makers. UNEP.

Zaffarroni, Eugenio. 2011. "La Naturaleza Como Persona: de la Pachamama a la Gaia." In *Los Derechos de la Naturaleza y la Naturaleza de sus Derechos*, edited by Carlos Espinosa Gallegos-Anda and Camilo Pérez Fernández, 3–33. FLACSO: Ecuador.

Zimmer, Markus B. 2009. "Overview of specialized courts." *IJCA*.

30 How can tenure reform processes lead to community-based resource management?

Experiences from Latin America

Iliana Monterroso

Introduction

Commons play a key role in sustaining the livelihoods of forest-dependent communities; in particular, vulnerable groups such as Indigenous peoples and poor women (Agrawal 2014). Globally, communities and Indigenous peoples are estimated to hold as much as 65% of the available land under customary tenure but only 18% of this land has been formally recognized either as owned or designated for the use for Indigenous peoples (RRI 2015; RRI 2020a). In sub-Saharan Africa, as much as 70% of the land can be categorized as customary common property; nonetheless, only 3% of this land is formally recognized in law (Alden-Wily 2018). In contrast, for decades Latin America has witnessed widespread policy reforms that redefined tenure rights over natural resources as the result of changes in regulations over who governs the appropriation and use of forests[1] and lands.

Despite this important progress, increasing conflicts surface as overlaps and pressures endanger the livelihoods and cultures of Indigenous peoples due to ongoing contestations over lands targeted for development and/or extractive investment by external interests (Gligo 2020). Furthermore, despite efforts to recognize rights, formalization does not always resolve internal disputes that put pressure on resource control and extraction of resources (Finley-Brook 2007; Sikor and Lund 2009). Nor does it stop external pressures, such as land invasions or pressure from the private sector and the government for resource access, including the rollback of rights previously granted (Monterroso et al. 2019a).

This chapter discusses two reform processes that shaped environmental governance of protected areas and forestlands in Latin America in different ways. The first, in Guatemala, analyzes the emergence of community forest concession contracts as co-management schemes in protected areas. The second, in Peru, discusses the recognition of native communities and the demarcation and titling of their communal lands in the Amazon. Based on extensive research on forest tenure reforms conducted by the Center for International Forestry Research, this chapter draws on research that analyzed reforms around recognition of collective rights and the extent to which these processes have led to community-based resource management[2]. We note that despite neither of these reforms having the recognition of collective rights as their main goal, they provided opportunities for the recognition of rights, opening spaces for broadening the participation of local communities in natural resource management and governance.

The chapter is organized in three sections. The first provides an overview of recent tenure reforms taking place in Latin America, introducing our two case studies. The second section discusses the content and the outcomes of the reforms analyzed. In the discussion section, we argue that rights devolution processes are an enabling condition for long-term community stewardship leading to the emergence of more equitable models of resource management. To do so, however, reforms need to recognize the emergence of collective political subjects and

DOI: 10.4324/9780429344428-35

guarantee those subjects opportunities to benefit from the rights acquired. This will be key to address ongoing governance challenges.

Tenure reforms in Latin America: implications over natural resources governance

Tenure reforms stem from changes in institutions, often statutory regulations, that (re)define the bundle of rights and responsibilities over who uses, manages, and controls resources and how (Larson et al. 2010). Although progress across countries is uneven, Latin America has led tenure reform processes; especially around two resource systems: land and forests (RRI 2015). These reforms are the result of social struggles, many of which were led by Indigenous peoples' movements to recognize their rights, a process that has had significant influence on environmental governance (Roldan 2004). In fact, in the region large portions of land have been formally recognized as either designated for or owned by indigenous groups and local communities, including Mexico (52%), Peru (35%), Colombia (34%), Brazil (23%), and Nicaragua (28%) (IBC 2016; Larson et al. 2016; Herrera 2017).

As of 2020, at least 31% of land area (571 Mha) was under some type of collective tenure regime, most of which are lands owned by Indigenous Peoples, local communities, and Afrodescendants either owned and customarily managed or designated for their use, while 137 Mha remain legally unrecognized (RRI 2020). In terms of forestlands, as of 2017[3] nearly 29.9% of forests in Latin America were under some type of collective tenure regime owned by communities, and another 6.3% of the area had been designated for their use (RRI 2018). The remaining 59% is owned or administered by the state[4] (RRI 2016). Table 30.1 describes statutory tenure regimes that recognize tenure rights to Indigenous Peoples and local communities in studied communities.

The legal entry point and institutional mechanisms to recognize different sets of rights vary from collective land and territorial titling to co-management schemes and concession contracts that recognize lesser or temporary rights (Larson et al. 2010). Early reforms, like those implemented in Mexico and Peru, were the outcome of major agrarian changes during the 1900s. In Mexico the revolutionary government recognized indigenous *ejidos* and agrarian communities in 1910 (Bray 2013, 2020), while a similar process led to the recognition of native communities

Table 30.1 Formal recognition of Indigenous Peoples' and local communities' tenure rights in studied communities

Country	Designated of Indigenous Peoples and Local Communities		Owned by Indigenous Peoples and Local Communities	
	Area (Mha)	*Percent of country area*	*Area (Mha)*	*Percent of country area*
Guatemala**	0.38	3.55	1.40	13.04
Peru*	9.27	7.24	35.29	27.57
Latin America	99.80	5.32	335.4	17.87

Source: RRI, 2015: 6–7.

* Data for areas designated for Indigenous Peoples and local communities include communal reserves (co-management protected area mechanism) and Indigenous reserves (recognizing isolated and initial contact) indigenous populations). Areas owned by Indigenous Peoples and local communities include native (Amazon indigenous) community lands and peasant (Highlands and Coast indigenous) community lands.

** Data for areas designated for Indigenous Peoples and local communities include community concession areas while the area owned by Indigenous Peoples and local communities refers to communal lands.

in the Peruvian Amazon in the 1970s (Monterroso et al. 2017). Others were more directly linked to the struggle for recognition of Indigenous territorial claims over land, such as those implemented in Panama, which resulted in the first *comarca*[5] in 1938, following social mobilization (Runk 2012). A similar process occurred in Colombia, when constitutional reforms in early 1990s followed the endorsement of the International Labor Organization ILO Convention 169, propelled a series of reforms that protected Indigenous Peoples' collective rights, and for the first time recognized Afro-descendants as subjects of collective tenure rights (Ortiz-Guerrero et al. 2018; Bolaños et al. 2021).

More recent reforms are linked to emerging interests in conservation and processes of decentralization. For instance, Brazil accounts for the largest portion of the region recognizing collective rights to communities such as through extractive reserves and the recognition of Indigenous Peoples. However, as of 2012, 47% of the Brazilian Amazon overlapped with protected areas and/or indigenous lands (RRI and ISA 2015). Other examples include the communal reserve[6] protected area category in Peru and the community forest concessions[7] in Guatemala. These processes of the formalization of rights are also linked to efforts to decentralize protected areas and introduce participatory conservation co-management schemes.

The analysis of historical trajectories in countries that have undergone reforms in the region demonstrate that while occasional, individual, dramatic events propelled progress toward meaningful rights devolution, multiple changes and feedback loops evolved over a period of years until the context was notably different (Gnych et al. 2020). Reforms in favor of communities often emerged from social struggles, sometimes as a part of broader national processes that follow constitutional reforms and supported by wider networks and alliances of civil society and indigenous organizations that push for reform implementation (Monterroso et al. 2019a).

Although the region is considered to be at the forefront of the recognition of tenure rights, the implementation of reforms in the region reveals gaps in progress and many community and indigenous groups are still pending legal recognition. For example, in Brazil, it is estimated that more than 100 million hectares reserved as indigenous lands are still pending formalization, while more recently the country experiences major rollback of rights (O'Callaghan et al. 2021). In Peru, the national indigenous organization AIDESEP argues that over 20 million hectares claimed by indigenous people remain unrecognized by the state (AIDESEP 2013). In Colombia, a recent analysis has found that 271 afro-Colombian councils are pending recognition and formalization of their collective lands (Guerrero et al. 2017). Furthermore, while formalization has progressed, challenges remain, including the consolidation of institutional arrangements, authority to reach consensus on territorial boundaries, and internal rules that strengthen governance[8] over territories and resources (Monterroso et al. 2019a).

Guatemala: Community Forest Concessions (CFC)

The Maya Biosphere Reserve (MBR) in Petén, Guatemala, is a prominent example of successful reforms promoting the devolution of forest rights to local communities (See Figure 30.1). Under the oversight of the National Council for Protected Areas (CONAP), the MBR constitutes the largest protected area in the country (2.1 million hectares) and the largest tropical forests in Central America. In 1994, the Government of Guatemala promoted an innovative approach for managing the reserve; community forest concessions were granted under contract to community-based organizations for a 25-year cycle, with the possibility of extension. Concession contracts were issued, contingent upon obtaining forest certification by the Forest Stewardship Council (FSC) within three years of being granted the concession. Between 1994 and 2002, 12 community forest concessions were granted to resident and non-resident communities in the Multiple Use Zone (MUZ) of the reserve on an area of close to 400,000 ha

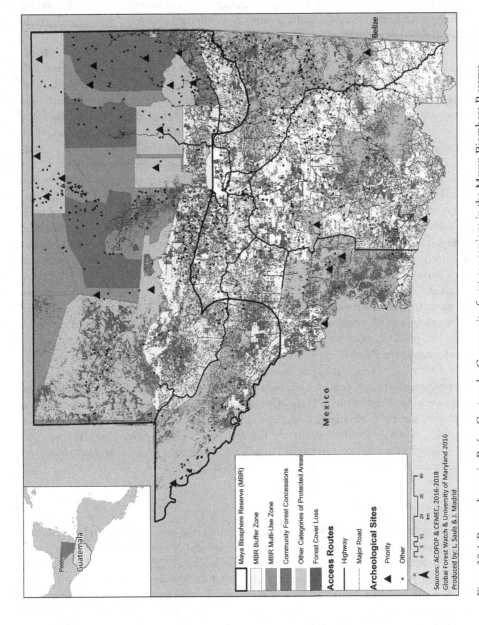

Figure 30.1 Research area in Petén, Guatemala. Community forest concessions in the Mayan Biosphere Reserve.

Source: ACOFOP y CEMEC 2016, 2018 in Monterroso et al., 2019b.

(Radachowsky et al. 2012). Each concession is operated by a community forest enterprise (CFE) responsible for sustainably managing the forest and sharing the resulting benefits among members (Butler 2021).

The devolution of rights provided incentives for long-term community forest stewardship, generating incomes, employment, and new forms of investment both at the level of community enterprise as well as at the household level. The community forest concessions have shown to be a successful model for combining sustainability objectives with mechanisms that improve the quality of life and promote local development, notably contributing to attaining diverse Sustainable Development Goals (SDGs) (Stoian et al. 2019).

Sustainable forest management in the MUZ, as shown by independent third-party certification, has been able to strongly reduce deforestation rates in the MUZ (0.4% per annum over the period 2000–2013) (Hodgdon et al. 2015). Deforestation in active concessions has been registered at close to zero, compared with in the core and buffer zones of the MBR where deforestation rates over the same period were 1.0% and 5.5% per annum, respectively. Additionally, CFEs invest about US$500,000 each year to organize patrols against forest fires and protect forest boundaries from illicit activities (Monterroso et al. 2018a). Beyond forest conservation, there is also strong evidence that CFEs drive socioeconomic progress in the communities managing the forests, reflected in increased income, investments, savings, capitalization of community enterprises, and asset building at both household and enterprise levels (Bocci et al. 2018). CFEs forest value chains generate over US$5 million and represent an average of 40% of family income, allowing them to invest in health and education, keeping many of their members out of poverty (Stoian et al. 2018). Community forest enterprises' operating concessions helped to solve social, economic, and governance challenges that the state proved incapable of addressing alone. They succeeded in strengthening governance against illicit activities, providing income and employment to keep people from being displaced (Devine et al. 2020). Additionally, through the creation of secondary-level organizations, such as the Community Forest Association of Peten (ACOFOP), they have been able to strengthen collective action strategies to resist ongoing pressures and build governance over time (Millner et al. 2020). Despite meeting goals of conservation and livelihood development, uncertainties about the renewal of concession contracts have emerged as divergent proposals to put part of the MBR under strict protection contrast with interests to expand tourism development in El Mirador (Devine 2018).

Peru: Indigenous native community titling

Peru has been at the forefront of Latin American land struggles with a large number of Indigenous Peoples' lands titled as native communities in the Amazon. Efforts to assert state's control over the Amazon started with the agrarian reform in the 1970s and progressively moved forward due to the mobilization of Indigenous organizations (Monterroso et al. 2017). Legal recognition of Indigenous communities promoted demarcation of their territorial claims which were formalized through communal land titling. Since 1974, the Peruvian government has formalized collective property rights by titling more than 1,300 native communities on over 12 million hectares (IBC 2016). This recognition is important for the peoples who live in the Peruvian Amazon and directly depend on these forests. It also has implications for the conditions of the forests they live in, which represent 17% of the national forest area (MINAM 2016). Despite this progress, there is still a considerable gap in the process of formalizing the lands claimed by the indigenous peoples. According to SICNA-IBC (2016), over 600 native communities (some 5.5 million hectares) are still pending titling.

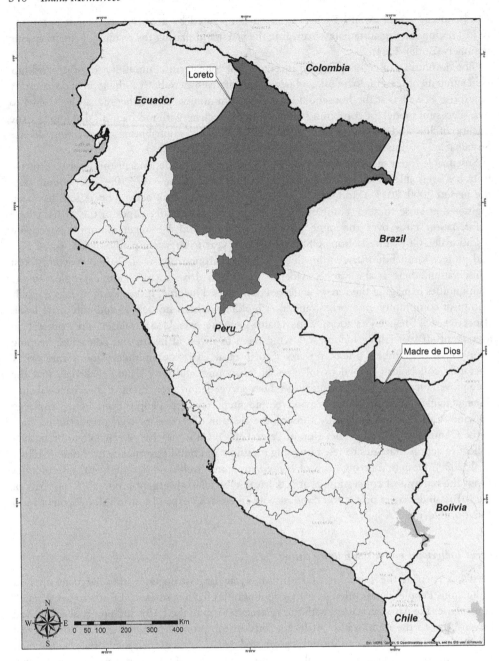

Figure 30.2 Research area in Peru. Loreto and Madre de Dios.

Source: Monterroso et al. 2019b.

Our analysis in Peru focused on the Madre de Dios and Loreto regions (See Figure 30.2) where government policies promoted investment and colonization of the Amazon since the early 1900s (Monterroso et al. 2017). Loreto holds 50% of the country's forests, and the region has remained sparsely populated. While smaller, Madre de Dios is considered a biodiversity-rich region, with 60% of the land under different categories of protected area.

While Peru recognized indigenous collective rights to land in 1974, in 1975 forests were declared public property and forest lands claimed by indigenous communities have been allocated under long-term usufruct contracts since then. Although the Law of Native Community recognizes the right of communities to manage the forests within their communal lands, communities are required to follow provisions in the Forest Law to apply for an authorization for subsistence uses and for a management plan for commercial purposes. Communities on the borders of protected areas must comply with existing regulations and develop rules for managing forest and fauna resources.

Local communities rely on forest resources for livelihoods. Timber, palm fruits, and bushmeat are traded locally in the Loreto region while in Madre de Dios extraction of brazil nuts and timber are the most important income-generating activities. In both, agriculture and fishing are the most significant subsistence strategies. In Madre de Dios, while most of the communities have been formalized, increasing pressure emerging from investment interests in mining, petroleum, and infrastructure initiatives threaten the ability to benefit from acquired rights; conflicts are also visible in petroleum extraction areas in Loreto (Zamora and Monterroso 2019).

Research analyzing the impact of the formalization of collective rights to land and forests in Peru argues that outcomes are linked to the perception of tenure security. In 24 native communities analyzed, 80% of those surveyed consider their rights to be stronger since titling, although this percentage is higher in men (85%) than women (75%) (Cruz-Burga et al. 2019). In terms of livelihoods, 83% of the members of communities analyzed in Madre de Dios consider their livelihoods to have become more constrained since titling took place. Despite progress, only 35% of those surveyed in Madre de Dios and Loreto say they participate in forest management. In fact, less than 40% had forest management plans registered (ibid.).

Discussion

These case studies show variations in reforms promoted in Latin America, in the content of rights granted, the emergence of new forms of community organization, the consideration of livelihoods both for subsistence and commercial purposes, and the conditionality on forest resources access and use as well as the external pressures and conflict. Table 30.2 synthesizes these differences.

These two cases showcase how reforms have the potential to enhance community-based resource management, triggering the emergence of new collective political subjects. This process nonetheless requires the assurance that local communities and indigenous peoples can benefit from rights acquired and are able to address existing socio-environmental challenges.

Intersection between the recognition for rights and the ability to realize rights and benefit from reform processes

Both reforms recognized decision-making rights over land and, to a certain degree, over forests. In both cases, the state retains rights over subsoil resources, although in the case of native communities in Peru initiatives are subject to the implementation of consultation processes following regulations. In the case of Guatemala, decisions over subsoil and definition of rights derived from benefits from ecosystem service provision, as in the case of carbon sequestration, are also retained by the state. Another difference, in terms of the content of rights between the two processes, is related to the duration of rights. According to the existing regulations, rights over communal lands titled in Peru do not expire; in Guatemala, however, usufruct rights are subject to renewal after 25 years. Another important conditionality in concession contracts linked to

Table 30.2 Comparison of reforms analyzed

	Guatemala: Community forest concessions	Peru: Titling of Native Communities
Content (bundle) of rights granted by reform	Use and decision-making rights, State retains alienation rights and rights over sub-soil (petroleum) and other ecosystem services (carbon).	Use and decision-making rights (over lands classified as agriculture) including alienation rights, State retains rights over sub-soil (petroleum and mining).
Forms of community organization	Community forest enterprises, organized through secondary level organizations (ACOFOP).	Legal recognition and formalization of native communities, organized through Indigenous Federations.
Consideration of livelihoods	Communities are able to benefit from timber and non-timber forest management activities both for subsistence and commercial purposes.	Restricted to subsistence, management and commercialization require following different legal procedures
Conditionality on forest resources access and use	Access to forest resources is subject to third party certification (FSC) and land and forests are subject to 25-year contract renewal	Collective titling rights do not prescribe, however rights to lands classified as forest are only granted under usufruct.
External pressures and conflict over resources	Conflicts around tourism and petroleum interests, expansion of illegal activities, infrastructure initiatives	Conflicts due to overlap with subsoil rights, expansion of illegal activities, infrastructure initiatives

Source: based on Monterroso et al. 2017; Cruz-Burga et al. 2019; Stoian et al. 2018; Gnych et al. 2020.

the renewal of contracts is sustained FSC certification, verified through annual evaluations (Carrera et al. 2006). After more than 20 years of certification, positive environmental impacts are multiple and have been widely documented (Radachowsky et al. 2012). The role of community concessions in the stabilization of land use changes, especially fire control management, has also been documented as a model of sustainable forest management in the tropics (Grogan et al. 2016; Blackman 2015). Furthermore, concessions have played an important role in maintaining populations of large and medium-sized mammals, including jaguars (Tobler et al. 2018).

In the case of Peru, while collective rights to land are recognized through titling, formalization of rights over forestlands since 1975 are granted as usufruct contracts—all forestlands are considered public lands. Therefore, demarcation processes require an agroecological evaluation that determines which portion of land is suitable for titling—determined by soil classifications which differentiate agrarian from forest lands. In practice, while the titling process has progressed, most native communities are waiting for usufruct contracts (Monterroso and Larson 2018). Even for those few that have been able to formalize forestlands, further procedures are needed for approval of forest management plans, permits, and authorizations. Cumbersome procedures for formalization and costs linked to their implementation, coupled with additional requirements, limit communities' ability to benefit forests in their lands (Notess et al. 2020). These are important challenges in the realization of rights for indigenous communities, hindering the ability to improve livelihoods (Larson et al. 2019). While information on the impact of formalization processes, and particularly titling in environmental conditions, is scarce, available information suggests that titling has the potential to reduce land clearing and forest disturbance, at least in the short term (Blackman et al. 2017). These results are in line with other studies in Amazon regions where titling interventions have been promoted, including Bolivia, Ecuador, Brazil, and Colombia, where titling has shown to have a significant effect in controlling deforestation (Blackman and Veit 2018). However, critics also highlight the limits of titling to slow forest change, especially when communities lack the ability to make land use choices and the legitimacy to hold these claims against external pressures, especially in the Amazon (Robinson et al. 2017). Despite differences in tenure regimes, community forestry has been widely promoted as an intervention. After four decades of study, FAO defines community-based forestry as "initiatives, sciences, policies, institutions and processes that are intended to increase the role of local people in governing and managing forest resources" (Gilmour 2016). Community forestry approaches are premised on secure rights to land and forest resources improving resource management, contributing to the achievement of development goals, and supporting self-determination of indigenous and customary groups (Coleman and Mwangi 2013). Despite larger investments, the outcomes of community forestry initiatives depend on a series of interconnecting factors, including existing socioeconomic dynamics, secure tenure rights, internal community governance, government support, and material benefits (Baynes et al. 2015; Hajjar et al. 2020). The analysis of these case studies show that despite differences, both processes depart from the recognition of rights and the ability to organize which have led to the emergence of forest-based organizations playing important roles in monitoring implementation and expanding the set of rights for both indigenous and forest-dependent communities.

Emergence of community-based organizations: making up new political subjects

In Peten, changes in government policy regarding land use and tenure, moving from a focus on agriculture, opened opportunities for community involvement in co-management of protected areas (Monterroso et al. 2018). These efforts put conservation outcomes at the center of processes, providing incentives for long-term community forest stewardship, incomes, employment, and new forms of investment, leading to one of the largest certified forests managed collaboratively by communities. The process of recognition of rights provided an important entry point for broadening spaces for participation and organization of local communities around natural resource management and governance (Taylor 2012; Millner et al. 2020). In the case of Peten, between 1994 and 2002 local communities organized to form community-based forest enterprises which led to the formation of a second level organization, the Community Association of Peten (ACOFOP), in 1995 (Taylor 2010). ACOFOP played a critical role, both in enhancing collective action at the local level and in claiming the recognition of forest communities as political subjects in the negotiation of management activities in the protected area (Millner et al. 2020; Paudel et al. 2012). Their role in negotiating conditions for communities managing protected areas has been key in ensuring continued support against external pressures (Devine et al. 2018).

In Peru, rights recognition of indigenous communities in the Amazon has been possible because of social movements that pushed forward changes in regulations and mobilized political and financial support to implement reforms (Monterroso et al. 2017). This progress was largely possible due to the emergence of national indigenous organizations, such as the Interethnic Association for the Development of the Peruvian Amazon (AIDESEP), in parallel with continued mobilization by several subnational indigenous federations in the Amazon. Further, in 1984, AIDESEP catalyzed the emergence of the Coordination Body of Indigenous Organizations of the Amazonian Basin (COICA), bringing together national indigenous organizations from nine countries (Smith et al. 2003). AIDESEP and COICA pressured to promote advances in the recognition of collective rights in the Amazon. They also played key roles in monitoring and lobbying government agencies and internationally funded projects to ensure the specifically targeted recognition and titling of native communal lands (Monterroso and Larson 2018). They were part of a broader regional movement that mobilized around claims for the recognition of their customary systems, set of norms, customs, traditions and practices developed locally and accepted, reinterpreted and enforced by indigenous groups to govern local resource access and use (Larson et al. 2019). In Peru, crisis situations such as the violent confrontations of Bagua in 2009 and the deaths of indigenous people in Saweto in 2014 were key in promoting shifts in the trajectory of policy processes. They repositioned indigenous organizations in the political arena and mobilized public opinion in favor of indigenous collective rights, prompting discussion of the challenges to implementing existing reforms. In the Peruvian Amazon, titling has been promoted as a critical enabling condition for national REDD+ initiatives (Blackman and Veit 2018; Monterroso and Sills 2022). While titling started in the late 1970s, it stalled for decades due to lack of political support, changes in the institutional framework, and cumbersome procedures. In 2014, during COP 21, international supporters called for action to overcome challenges, including lack of financial support, to complete regularization of communities in target areas (Monterroso and Larson 2018b). Since 2015, international funding has flowed into multiple environmental projects that support the recognition, demarcation, and titling of the communal land holdings of native communities (Monterroso and Larson 2018). However, as formalization remains highly conditioned by the control of deforestation, land use change, and colonization pressures, most communities benefiting from these programs argue that most of their lands classified as forests remain under usufruct contracts while lands classified as agricultural lands subject to title are proportionally smaller. As has been documented elsewhere, these different approaches in formalization mechanisms may result in increased tenure insecurity if they do not address underlying challenges from increasing pressures (Sunderlin et al. 2014; Buntaine et al. 2015).

Ongoing governance challenges

Community forest enterprises in the Mayan Biosphere Reserve have played a critical role in territorial stability for over 20 years, supporting government and non-state actors in strengthening governance against illicit activities and providing income and employment to keep people from being displaced (Devine et al. 2018, 2020). However, despite meeting goals of conservation and livelihood development, uncertainties about the renewal of concession contracts emerge with divergent proposals to put part of the Mayan Biosphere Reserve under strict protection (Devine 2018). Since the early 2000s, new initiatives have emerged with renewed interest in the forests as spaces for economic development based on agribusiness (e.g., palm oil, rubber), expansion of cattle ranching, and extensive agriculture. In stark contrast to these proposals for economic development, and in opposition to the community forest concessions, influential actors have been campaigning for expanding the area under full protection and to allow only tourism development, undermining existing governance arrangements in the protected area and promoting green land grabs (Devine 2018).

Remaining challenges in the governance of native communities in the Peruvian Amazon are manifold (Monterroso and Larson 2018b). Restrictions on the use of forest resources, often including additional cumbersome requirements to prove eligibility for these rights, prevent communities from asserting them in practice (Notess et al. 2020). Moreover, beyond the scope of the rights granted during formalization, another problem is the extent to which formalized rights have been compartmentalized in areas claimed as integral of indigenous territories. This compartmentalization is further evident in regulations that recognize rights to overlapping claims to different resources in the same area. For example, national regulations formalize rights to lands but are not clear regarding other resources in the territories, such as water, forest, petroleum, gas, or the benefits these generate. As regulatory frameworks recognize rights to the same land for different stakeholders, additional types of legal overlaps affect indigenous territories, undermining their legitimacy over existing claims. This compartmentalization of rights is contradictory to the notion of territorial rights inherent to the claims of indigenous peoples, hindering the ability of indigenous groups to be able to exercise the rights gained. Moreover, overlaps may bring out differing views around resource management, sometimes exacerbating conflict. At the local level, ongoing resource conflicts frame the configuration of territorial practices and institutions, requiring improved understanding of the external pressures driving conflict over resources. In the Amazon, as well as in Peten, increasing pressures exist to expand development, infrastructure and/or extractive investment in collectively managed and owned areas (Runk 2012). Despite these ongoing threats, the formalization of indigenous rights to land and resources is still perceived as the most important mechanism to secure indigenous rights and livelihoods (Seymour et al. 2014).

Tenure interventions should be tailored to particular tenure challenges, i.e., whether rights are unclear, insecure, or in conflict (Monterroso and Sills 2022). Addressing these tenure challenges is a highly contested and political process (Naughton-Treves and Wendland 2014). Strengthening tenure arrangements requires constant negotiation between the State, indigenous groups, and local community groups not only to gain access to resources but also to increase local decision-making around forests and resources (Lemos and Agrawal 2006). While land titling is widely considered to provide the greatest tenure security, some interventions have recognized different subsets of the full bundle of rights (access, management, exclusion) to different subsets of the resources or services associated with land (e.g., wood, non-wood forest products, carbon rights, water provision) (Naughton-Treves and Wendland 2014). Furthermore, while tenure interventions are promoted to clarify rights, formalization does not entirely guarantee tenure security or conservation outcomes (Holland et al. 2017; Robinson et al. 2014).

Concluding remarks

This chapter contributes to the analysis of ongoing socio-environmental transformations by analyzing two reform processes in Peru and Guatemala. Results highlight how these processes have shaped environmental governance of protected areas and forestlands in the region in different ways. Furthermore, these cases show underlying contestation processes around the recognition and exercise of rights in Latin American forest areas, as forest-dependent and Indigenous communities face increasing external pressures for interests over their resources.

Results presented here suggest that in the case of reforms that favor the recognition of collective tenure rights, what happens on the ground depends on the existence of a legal framework that guarantees the collective right to land and forest, and on how reforms are implemented, including the role of government officials and social mobilization. In both cases, secure tenure is considered as an important element to sustain livelihoods and strengthen governance, but differences between processes highlight that they are highly contested as Indigenous Peoples and local communities try to realize their rights and benefit from reform processes.

There are different factors that affect the ability of indigenous groups and forest-dependent communities to gain tangible benefits from the recognized rights; including those related to how formalization processes are designed and implemented. While legal reforms provide the basis for accessing forest and land resources, too often regulatory frameworks do not facilitate the conditions needed to exercise these newly acquired rights. Rights granted may be conditional and non-permanent, imposing further obligations and limitations to sustain them on the long-term basis. This is particularly the case with Peten community forest concession contracts, where rights are granted conditionally on demonstrated compliance with forest management regulations. The resulting co-management regimes could potentially allow government oversight and support to local forest managers, but when poorly designed or implemented they impose costs and other burdens. In the case of Peru, it appears state actors consider the work to be done once they provide the title, failing to provide technical assistance or support of any kind to the titled communities to ensure the ability to benefit from rights acquired (Monterroso and Larson 2018b).

Both cases highlight the emergence of community-based forest organizations as new political subjects that have been key in both sustaining ongoing reforms and mobilizing against existing pressures as infrastructure development and extractive interests try to undermine communities' livelihoods and forest sustainable conditions. In the case of Guatemala, community forest enterprises have been able to position themselves as key stakeholders in strengthening environmental governance in Peten, in the context of increasing pressures from renewed interest in the forests as space for economic development. In Peru, indigenous organizations were able to position formalization as a precondition for meeting existing commitments around climate change mitigation. Both cases highlight the interlinked challenges required to address issues of rights and governance in the context of socio-environmental transformations.

Notes

1 Forests refer to "all kinds of forests, ranging from relatively untouched "natural" ones to those with high levels of intervention and management" (Sunderlin 2005, 1386).
2 This chapter draws from ongoing work by the CGIAR Policy Institutions and Markets Program conducted by CIFOR in Peru and Guatemala. In the case of Peru, the chapter draws on findings from the Global Comparative Study on Forest Tenure Reform https://www2.cifor.org/gcs-tenure/. In the case of Guatemala, it draws from ongoing research conducted jointly with ICRAF https://pim.cgiar.org/2019/05/15/in-peten-guatemala-community-forestry-is-viable-business/.
3 The analysis is based on nine complete cases including Bolivia, Brazil, Colombia, Costa Rica, Guyana, Honduras, Mexico, and Suriname.
4 Compared to approximately 92% in Africa and 62% in Asia.
5 *Comarca* is an indigenous region recognized by Panamanian laws as customary territories with self-government and autonomy. In total the government has recognized five comarcas https://www.iwgia.org/en/panama.html.
6 The communal reserves are a protected area category that recognizes the communities' management rights (Special Regime for the Management of Communal Reserves, Resolution 019-2005-INRENA-IANP).
7 Community forest concessions are 25-year contracts between a legally recognized community-based organization with the Guatemalan State that grants management rights over forest management units.
8 We define environmental governance as the set of institutions guiding how decisions around natural resources are made and by whom, influencing processes, policies, and their implementation at multiple scales and levels (Lemos and Agrawal 2006).

References

Alden-Wily, L. A. 2018. "Risks to the Sanctity of Community Lands in Kenya. A Critical Assessment of New Legislation with Reference to Forestlands." *Land Use Policy* 75: 661–72.

Agrawal, A. 2014. "Studying the Commons, Governing Common-Pool Resource Outcomes: Some Concluding Thoughts." *Environmental Science & Policy* 36: 86–91.

Asociación Interétnica de Desarrollo de la Selva Peruana (AIDESEP). 2013. *Territorialidad y Titularidad en la Amazonía Norte del Perú, Alto Amazonas y Datem del Marañón – Pueblos.* Lima, Peru: Achuar, Kukama, Shapra, Kandozi, Shiwilo, AIDESEP/CORPI.

Baynes, J., J. Herbohn, C. Smith, Rober Fisher, and David Bray. 2015. "Key Factors Which Influence the Success of Community Forestry in Developing Countries." *Global Environmental Change* 35: 226–38.

Bocci, Corinne, Lea Fortmann, Brent Sohngen, and Bayron Milian. 2018. "The Impact of Community Forest Concessions on Income: An Analysis of Communities in the Maya Biosphere Reserve." *World Development* (107): 10–21.

Bolaños, O. C., J. Herrera-Arango, C. Guerrero-Lovera, and H. Molina. 2021. Bridging Research and Practice to Influence National Policy: Afro-Colombians Territorial Rights, from Stagnation to Implementation. *Bulletin of Latin American Research* 41(3): 387–403.

Blackman, A. 2015. "Strict Versus Mixed-use Protected Areas: Guatemala's Maya Biosphere Reserve." *Ecological Economics* 112: 14–24.

Blackman, A., L. Corral, E. S. Lima, and G. P. Asner 2017. "Titling Indigenous Communities Protects Forests in the Peruvian Amazon." *Proceedings of the National Academy of Sciences of the United States of America* 114(16): 4123–4128.

Blackman, A., and P. Veit 2018. "Titled Amazon Indigenous Communities Cut Forest Carbon Emissions." *Ecological Economics* 153: 56–67.

Bray, D. 2013. "When the State Supplies the Commons: Origins, Changes, and Design of Mexico's Common Property Regime." *Journal of Latin American Geography* 12(1): 33–55.

———. 2020. *Mexico's Community Forest Enterprises: Success on the commons and the seeds of a good Anthropocene.* Tucson: The University of Arizona Press.

Buntaine, M. T., S. E. Hamilton, and M. Millones 2015. "Titling Community Land to Prevent Deforestation: An Evaluation of a Best-case Program in Morona-Santiago, Ecuador." *Global Environmental Change* 33: 32–43.

Butler, M. 2021. "Analyzing Community Forest Enterprises in the Maya Biosphere Reserve Using a Modified Capitals Framework." *World Development* 140: 105284.

Carrera, Fernando, Dietmar Stoian, José Joaquín Campos, Julio Morales, and Gustavo Pinelo. 2006. "Forest Certification in Guatemala." In *Confronting Sustainability: Forest Certification in Developing and Transitioning Countries.* Edited by Benjamin Cashore, Fred Gale, Errol Meidinger, and Deanna Newsom, 363–406. Yale Forestry & Environmental Studies Publications Series. 28. New Haven.

Coleman, E., and E. Mwangi. 2013. "Women's Participation in Forest Management: A Cross-Country Analysis." *Global Environmental Change* 23(1): 193–205.

Cruz-Burga, Z., I. Monterroso, A. M. Larson, F. Valencia, and J. S. Saldaña. 2019. *The Impact of Formalizing Rights to Land and Forest: Indigenous Community Perspectives in Madre de Dios and Loreto.* Vol. 242. Bogor, Indonesia: CIFOR.

Devine, J., N. Currit, Y. Reygadas, L. Liller, and G. Allen. 2020. "Drug Trafficking, Cattle Ranching and Land Use and Land Cover Change in Guatemala's Maya Biosphere Reserve." *Land Use Policy* 95: 104578.

Devine, J., D. Wrathall, N. Currit, B. Tellman, and Y. Langarica. 2018. "Narco-cattle Ranching in Political Forests." *Antipode* 52(4): 1018–38.

Devine, Jennifer A. 2018. "Community Forest Concessionaires: Resisting Green Grabs and Producing Political Subjects in Guatemala." *The Journal of Peasant Studies* 45(3): 565–84.

Finley-Brook, M. 2007. "Indigenous Land Tenure Insecurity Fosters Illegal Logging in Nicaragua." *International Forestry Review* 9(4): 850–64.

Gilmour, Don. 2016. *Forty Years of Community-Based Forestry: A Review of Its Extent and Effectiveness.* Rome: FAO.

Gligo, N. (Ed.). 2020. *La tragedia ambiental de América Latina y el Caribe.* Libros de la CEPAL, N° 161 (LC/PUB.2020/11-P). Santiago: Comisión Económica para América Latina y el Caribe (CEPAL).

Gnych, S., L. Steven, R. McLain, I. Monterroso, and A. Adhikary. 2020. "Is Community Tenure Facilitating Investment in the Commons for Inclusive and Sustainable Development?" *Forest Policy and Economics* 111: 102088.

Grogan, J., C. Free, G. Pinelo, A. Johnson, and R. Alegría. 2016. *Estado de Conservación de las Poblaciones de Cinco Especies Maderables en Concesiones Forestales de la Reserva de la Biósfera Maya.* Ciudad de Guatemala: CATIE.

Guerrero, L. C., A. J. Herrera, M. E. Helo, A. Beltran-Ruiz, A. Aramburo, S. Zapata, and M. Arrieta (2017). *Derechos de las Comunidades Negras: Conceptualización y Sistema de Información Sobre la Vulnerabilidad de los Territorios sin Titulación Colectiva.* Bogota, Colombia: Observatorio de Territorios Interetnicos.

Hajjar, Reem, Johan A. Oldekop, Peter Cronkleton, Peter Newton, Aaron J. M. Russell, and Wen Zhou. 2020. *A Global Analysis of the Social and Environmental Outcomes of Community Forests. Nature Sustainability* 4: 216–24.

Herrera, J. 2017. *La Tenencia de Tierras Colectivas en Colombia: Datos y Tendencias.* CIFOR InfoBrief No. 203. Bogor, Indonesia: CIFOR.

Hodgdon, B., D. Hughell, V. H. Ramos, and R. McNab. 2015. *Deforestation Trends in the Maya Biosphere Reserve, Guatemala.* New York, NY: Rainforest Alliance.

Holland, M. B., K. W. Jones, L. Naughton-Treves, J.-L. Freire, M. Morales, and L. Suárez 2017. "Titling land to conserve forests: The case of Cuyabeno Reserve in Ecuador." *Global Environmental Change*, 44: 27–38.

Instituto del Bien Común (IBC). 2016. *Tierras comunales: más que preservar el pasado es asegurar el futuro. El Estado de las comunidades indígenas en el Perú*. Informe 2016. Lima: Instituto del Bien Común.

Larson, A. M., Deborah Barry, and Ganga Ram Dahal. 2010. "New Rights for Forest-Based Communities? Understanding Processes of Forest Tenure Reform." *International Forestry Review* 12(1): 78–96.

Larson, A. M., I. Monterroso, N. Liswanti, T. Herawati, A. Y. Banana, P. Canturias, K. Rivera, and E. Mwangi. 2019. *Models for Formalizing Customary and Community Forest Lands: The Need to Integrate Livelihoods into Rights and Forest Conservation Goals*. Vol. 253. Bogor, Indonesia: CIFOR.

Larson, A. M., F. Soto, D. Mairena, E. Moreno, E. Mairena, and J. Mendoza-Lewis 2016. "The challenge of 'Territory': Weaving the Social Fabric of Indigenous communities in Nicaragua's Northern Caribbean Autonomous Region." *Bulletin of Latin American Research* 35(3): 322–37.

Lemos, M. C., and A. Agrawal. 2006. "Environmental Governance." *Annual Review of Environment and Resources* 31: 297–325.

Millner, Naomi, Irune Peñagaricano, Maria Fernandez, and Laura K. Snook. 2020. "The Politics of Participation: Negotiating Relationships through Community Forestry in the Maya Biosphere Reserve, Guatemala." *World Development* 127: 104743.

Monterroso, Iliana, P. Cronkleton, D. Pinedo, and A. M. Larson. 2017. *Reclaiming Collective Rights: Land and Forest Tenure Reforms in Peru (1960–2016)*. Vol. 224. Indonesia: CIFOR.

Monterroso, Iliana, P. Cronkleton, and A. M. Larson 2019a. "Commons, Indigenous Rights, and Governance." In *Routledge Handbook of the Study of the Commons*. Edited by Blake Hudson, Jonathan Rosenbloom, and Dan Cole, 376–91. London: Routledge.

Monterroso, Iliana, and A. M. Larson 2018a. *Progress in Formalizing "native community" Rights in the Peruvian Amazon (2014–2018)*. Indonesia: CIFOR Infobrief (233).

———. 2018b. *Challenges in the Process of Formalization of Rights of Native Communities in Peru*. Indonesia: CIFOR Infobrief (220).

Monterroso, Iliana, A. M. Larson, E. Mwangi, N. Liswanti, and Z. Cruz-Burga. 2019b. *Mobilizing Change for Women Within Collective Tenure Regimes*. USA: Resource Equity Research Consortium.

Monterroso, Iliana, and E. Sills 2022. "Interaction of Conditional Incentives for Ecosystem Conservation with Tenure Security: Multiple Roles for Tenure Interventions." In *Land Tenure Security and Sustainable Development*. Edited by Margaret Holland, Yuta Masuda, and Brian Robinson, 201–23. Cham, Switzerland: Palgrave Macmillan.

Monterroso, Iliana, D. Stoian, S. Lawry, and A. Rodas 2018. *Investigación y Política Sobre Concesiones Forestales Comunitarias en Petén, Guatemala: Lecciones Aprendidas y Desafíos Pendientes*. Indonesia: CIFOR Infobrief (236).

Naughton-Treves, L., and K. Wendland 2014. "Land Tenure and Tropical Forest Carbon Management." *World Development* 55: 1–6.

Notess, L., P. Veit, I. Monterroso, E. Sulle, Anne M. Larson, Anne-Sophie Gindroz, Julia Quaedvlieg, and Andrew Williams. 2020. "Community Land Formalization and Company Land Acquisition Procedures: A Review of 33 Procedures in 15 Countries." *Land Use Policy* 110: 104461.

O'Callaghan, M., S. Dil, C. Ewell, and A. Wherry. 2021. *Rolling Back Social and Environmental Safeguards in the Time of COVID-19: The Dangers for Indigenous Peoples and for Tropical Forests*. London: Middlesex University, Lowenstein Human Rights Clinic Yale Law School, Forest People Programme.

Ortiz-Guerrero, Cesar, J. Herrera Arango, V. Guaqueta Solórzano, and P. Ramos Barón. 2018. *Historical Trajectories and Prospective Scenarios for Collective Land Tenure Reforms in Community Forest Areas in Colombia*. CIFOR InfoBrief (237).

Paudel, N., I. Monterroso, and P. Cronkleton. 2012. "Secondary Level Organisations and the Democratisation of Forest Governance: Case Studies from Nepal and Guatemala." *Conservation and Society* 10(2): 124–35.

Radachowsky, J., V. H. Ramos, R. McNab, E. H. Baur, and N. Kazakov 2012. "Forest Concessions in the Mayan Biosphere Reserve: A decade later." *Forestry Ecology and Management* 268: 18–28.

Robinson, B., M. Holland, and L. Naughton-Treves 2014. "Does Secure Land Tenure Save Forests? A Meta-Analysis of the Relationship between Land Tenure and Tropical Deforestation." *Global Environmental Change* (29): 281–293.

Robinson, B. E., M. B. Holland, and L. Naughton-Treves 2017. "Community Land Titles Alone will not Protect Forests." *Proceedings of the National Academy of Sciences* 114(29): E5764 LP-E5764.

Roldan, R. 2004 *Models for Recognizing indigenous Land Rights in Latin America.* Biodiversity Series No. 99. Washington, DC: The World Bank.

RRI. 2018. *At a crossroads: Consequential trends in recognition of community based forest tenure from 2002 and 2017.*

———. 2015. *Who owns the world's land? A global baseline of formally recognized indigenous and community land rights.*

Rights and Resources Initiative (RRI). 2016. *Closing the Gap: Strategies and Scale Needed to Secure Rights and Save Forests.* Washington, DC: Rights and Resources Initiative.

———. 2020a. *Estimate of the area of land and territories of Indigenous Peoples, local communities, and Afro-descendants where their rights have not been recognized.* Technical Report.

———. 2020b. *Urgency and Opportunity: Addressing Global Health Climate and Biodiversity Crises by Scaling-Up the Recognition and Protection of Indigenous and Community Land Rights and Livelihoods.* Washington, DC: Rights and Resources Initiative.

Rights and Resources Initiative (RRI) and ISA. 2015. *Advances and Setbacks in Territorial Rights in Brazil.* Washington, DC: Rights and Resources Initiative, Instituto Socioambiental.

Runk, J. V. 2012. "Indigenous Land and Environmental Conflicts in Panama: Neoliberal Multiculturalism, Changing Legislation, and Human Rights." *Journal of Latin American Geography* 11 (2): 21–47.

Seymour, F., T. La Vina, and K. Hite. 2014. *Evidence Linking Community-Level Tenure and Forest Condition: An Annotated Bibliography.* San Francisco, CA, USA: Climate and Land Use Alliance, 61.

Sikor, T. and C. Lund. 2009. "Access and Property Regarding Natural Resources: A Question of Power and Authority." *Development and Change* 40(1): 1–22.

Smith, R. C., M. Benavides, M. Pariona, and E. Tuesta 2003. "Mapping the Past and the Future: Geomatics and Indigenous Territories in the Peruvian Amazon." *Human Organization* 62: 357–68.

Stoian, D., I. Monterroso, and D. Current 2019. "SDG 8: Decent work and economic growth–potential impacts on forests and forest-dependent livelihoods." In *Sustainable Development Goals: Their Impact on Forests and People,* 237–78. London: Cambridge University Press.

Sunderlin, W. D., A. M. Larson, A. E. Duchelle, I. A. P. Resosudarmo, T. B. Huynh, A. Awono, and T. Dokken 2014. "How are REDD+ Proponents Addressing Tenure Problems? Evidence from Brazil, Cameroon, Tanzania, Indonesia, and Vietnam." *World Development,* 55: 37–52.

Stoian, D., A. Rodas, M. Butler, I. Monterroso, and B. Hodgdon. 2018. *Forest Concessions in Petén, Guatemala: A Systematic Analysis of the Socioeconomic Performance of Community Enterprises in the Maya Biosphere Reserve.* Indonesia: CIFOR.

Taylor, P. 2010. "Conservation, Community, and Culture? New Organizational Challenges of Community Forest Concessions in the Maya Biosphere Reserve of Guatemala." *Journal of Rural Studies* 26 (2): 173–84.

———. 2012. "Multiple Forest Activities, Multiple Purpose Organizations: Organizing for Complexity in a Grassroots Movement in Guatemala's Petén." *Forest Ecology and Management* 268: 29–38.

Tobler, M., R. Garcia Anleu, S. Carrillo-Percastegui, G. Ponce, J. Polisar, A. Zuñiga, and I. Goldstein. 2018. "Do Responsibly Managed Logging Concessions Adequately Protect Jaguars and other Large and Medium-sized Mammals? Two Case Studies from Guatemala and Peru." *Biological Conservation* 220: 245–253.

Zamora, Alejandra, and Iliana Monterroso. 2019. *Regional and Local Perspectives on Tenure Insecurity in the Loreto and Madre de Dios Regions of Peru.* CIFOR InfoBrief 246.

31 Environmental policy and institutional change

The consequences of mobilization

Ezra Spira-Cohen and Eduardo Silva

Introduction

The decision to hold the United Nations' 1992 Conference on Environment and Development (UNCED) in Rio de Janeiro signaled a new place for Latin America in global policy and potential environmental leadership. The Conference highlighted the close relationships (and tensions) between environmental issues and economic development—in Latin America, the former had historically been sacrificed to advance the latter. The Rio Declaration on Environment and Development advanced the concept of sustainable development (Principle 1), highlighted the role of poverty eradication for sustainable development (Principle 5), and enshrined the concept of common but differentiated responsibilities (Principle 7). This outlined a promising path for states in the region to grow their economies and globally compete. Other key outcomes of the UNCED were of considerable interest to Latin American countries: namely, the signing of a forest conservation agreement and the advancement of biodiversity protections as an issue.

Although Latin American countries have become key players in global environmental politics, there is considerable variation in environmental policy and institutions across the region. Environmental governance is multifaceted and has included the creation of government ministries, green parties, and the adoption of policies at multiple levels of government that adhere to international frameworks and respond to internal pressures (de Castro et al. 2016). The IDB and initiatives like REDD+ have shaped rigorous rules about forest management. Institutions focused on environmental governance such as ministries for the environment and green parties have expanded since the 1980s. More Latin American states than anywhere else have ratified the ILO Convention 169 on indigenous rights, which advances prior consultation as a tool for governance and has impacted policy on land and natural resource use (Table 31.1).

Policies and institutions for natural resource management (such as forests and water) and activities that impact the environment vary significantly in their mandates. Furthermore, rigorous rules often exist and forms of environmental rights have been enshrined in constitutions, but there is a gap between laws and policies on paper and their implementation and enforcement. Often environmental governance is subordinated to economic development goals and forms of development that rely on extractive activities that are detrimental to the environment. As a result, social movement pressure that can have indirect or mediated impacts on the policy process is key for the adoption of natural resource protections and the implementation and enforcement of existing environmental policies.

In the context of democratization and the increased international salience of environmental issues since the 1970s, social mobilization against environmental degradation has become a hallmark of Latin American societies and politics. The expansion of extractivism has spurred strong resistance across the region. The national importance of extractive activities means that local community resistance is relevant at the national scale. Social movement action is often necessary

DOI: 10.4324/9780429344428-36

Table 31.1 Environmental Governance in Latin America★[1]

Environment Ministries (Year created)★★	Green/Ecologist Parties (Years active)★★★	Green/Ecologist Parties (Years active)★★★★
Argentina – Ministry of Environment and Sustainable Development (2015)[†]	Argentina (2006–2010)	Argentina (2000)
Bolivia – Ministry of Environment and Water (no data) Brazil – Ministry of Environment (1985)	Bolivia (2007–) Brazil (1986–) Chile (1987–2001 and	Bolivia (1991) Brazil (2002) Chile (2008) Colombia (1991)
Chile – Ministry of Environment (2010)[†]	2008–) Colombia (1998–2005 and	Costa Rica (1993) Ecuador (1998)
Colombia – Ministry of Environment and Sustainable Development (1993)[†]	2005–) Costa Rica (2004–)	Guatemala (1996) Honduras (1995
Costa Rica – Ministry of Environment and Energy (1980)	Mexico (1993–) Nicaragua (2003–)	Mexico (1990) Nicaragua (2010)
Ecuador – Ministry of Environment (1996)	Peru (no data)	Paraguay (1993)
El Salvador – Ministry of Environment and Natural Resources (1997)	Uruguay (1987–2009 and 2013–)	Peru (1994) Venezuela (2002)
Guatemala – Ministry of Environment and Natural Resources (2000)	Venezuela (2010–)	
Honduras – Secretary of Natural Resources Environment and Mines (no data)		
Mexico – Secretary of Environment and Natural Resources (1994)		
Nicaragua – Ministry of Environment and Natural Resources (1979)		
Panamá – Ministry of Environment (2015) [†] Paraguay – Secretary of Environment (2018) Perú – Ministry of Environment (2008) Uruguay – Ministry of Housing, Territorial Administration, and Environment (1990) Venezuela – Ministry of Popular Power for Ecosocialism and Water (1979)		
Source: UNEP. Data on years was collected by authors from multiple sources.	Source: Global Green Membership and Federation of Green Parties of the Americas	Source: ILO

★ Only Spanish speaking countries and Brazil, excluding Caribbean.

★★ Current Environment Ministry. Environmental ministries and related agencies often change names and mandates. Dates refer to earliest known year of an existing ministry for the environment. Different governments, at different times have raised environmental governance to cabinet level ministry, lowered it, or placed it within other agencies.

★★★ Green or Ecologist parties.

★★★★ Countries that have ratified ILO Convention 169. Not all listed have incorporated this into constitution/legislation/jurisprudence.

[†] Another environmental agency is known to have existed prior to the year the ministry was created.

to prevent environmentally destructive development projects, but it is generally insufficient on its own for broader policy and institutional change. Thus, this chapter highlights important international and domestic political and economic dynamics that mediate the impact of contentious action on environmental governance in relation to extractive development projects. On a more general plane, the examination of the consequences of mobilization in specific cases offers insights into the complex array of factors that shape environmental policy and related institutional change.[2]

Because this process is complex and occurs at multiple levels of analysis, the cases discussed in the chapter were chosen with a focus on extractivism as an arena for mobilization and environmental policy-making that is relevant to the entire region and elicits future research. These cases highlight dynamic interactions between the international norms, institutions for citizen participation, political opportunities, and political economy that shape environmental governance. To frame this discussion and in advance of any conclusions, some theoretical considerations are helpful to understand the role of social mobilization in broad environmental policy and institutional change.

Theoretical considerations

In addition to external pressures, literature on environmental governance highlights the role of social mobilization and internal political and economic factors on policy and institutional change (Hochstetler and Keck 2007). Mobilization can have an influence at many stages in the policy process, such as agenda setting and policy initiation, formulation, adoption, implementation, evaluation, and feedback stages (Amenta and Young 1999). Environmental policy outcomes can take the form of legislation, decrees, and regulations. These can include bans, moratoria, community consulting mechanisms, and oversight. Additionally, regulatory mechanisms include environmental impact reporting, support for eco-friendly development, and community compensation, as well as the decentralization of revenue and environmental protection. Institutional change can take the form of increased state capacity including budgets, personnel, and expertise, services offered, pay, new bureaus. Change can occur in different types of institution, including ministries, coordinating boards, agencies that handle environmental impact reporting, ombudsman offices, boards, and commissions, arbitration or mediation agencies, and environmental tribunals (Silva 2017).

In this chapter we analyze the factors that explain social mobilizations' impact on a variety of policy and institutional outcomes. Chile replaced its decentralized inter-ministerial environmental agency with a centralized Environment Ministry in 2010 after sustained social pressures and in the context of Chile's process of ascension into the OECD. Bolivia and Ecuador adopted prior consultation as a mechanism for channeling community demands and social participation in decision-making around environmental protection, but guidelines to mitigate negative effects of proposed development projects have become a measure to control protest and ensure the continuation of extractive activities. In the Amazon region, mobilization (which began in the 1970s) has advanced forest conservation efforts, influencing the creation of parks, ecological reserves, or national monuments, but the economic drive to expand agricultural frontiers and increase extractive activities mitigates their impact on conservation outcomes.

Assessing the dynamics between existing environmental governance, resistance to extractive activities, market forces, and international factors requires careful attention to variation in cases as well as attention to abrupt changes over longer periods of time. Overall, policy and institutional changes that result from resistance to extractivism are not unidirectional and can move toward more sustainability, preservation of the status quo, or even regression towards further environmental degradation.

Movement impacts on environmental governance can be direct, but are more often indirect, joint, or mediated by political actors. Direct effects occur in the form of immediate short-term consequences that reflect explicit movement demands. Indirect effects are more common, taking the form of setting agendas and influencing allies or public opinion, which can have an impact but lack activist participation in the policy process once politicians begin to address those issues. However, identifying direct or indirect policy effects can be difficult because movements often have multiple goals that can develop over short, medium, and long time horizons. Similarly, causality and the attribution of impacts to protest or mobilization are difficult to establish since social movements have multiple demands and strategies which are shaped by existing political opportunities, making it difficult to differentiate between cause and effect (della Porta and Diani 1999; Amenta et al. 2010).

Movement organizations often have mediated effects. This occurs when they act in conjunction with political or institutional allies and benefit from support in public opinion (Giugni 2004), which increases the likelihood that movements will achieve their goals (Amenta et al. 2010). Social movement leaders are tasked to change the perceptions of the public and of institutional and political actors (McAdam and Schaffer Boudet 2012), which is done through a variety of strategies. Issue framing is essential as well as insider and outsider tactics that combine negotiations with politicians or bureaucrats and street protest (Paredes and de la Puente 2017). Moreover, structural and economic factors determine if and how social movements affect environmental policy and institutional change. The nature of demands shapes political mediation and the impact movements have on political outcomes; so more politically salient issues will likely encounter more resistance (Giugni 2004). Thus, the broader context of political opportunities and of political economy varies for different environmental policy areas. Specific issues impact different sectors, which have different places in the structure of the economy; so interests and stakes that emerge in the process of social mobilization can change the balance of power between related social, political, and economic forces. Compared to some environmental issues that have low political salience, which suggest movements are more able to impact politics, the economic weight of extractive activities may increase the political stakes. The remainder of this chapter examines cases of environmental policy and related institutional change that highlight the effects of social mobilization and important political and economic factors that shape their impacts.

Institutions and policy change in Chile

Since the 1990s, socio-territorial conflicts have increased, as has resistance to industrial mining and related hydropower projects in Chile. Despite their territorial component, mobilization around extractive activities has had national-level policy impacts (Delamaza et al. 2017). A campaign of resistance to hydroelectric dams impacted energy policy, resulting in the creation of Chile's 20/25 Law in 2013 which mandates achieving 20% renewable energy production by 2025. The campaign leveraged political opportunities such as Chile's ascension into the OECD, with environmentalists contributing to the legislative process. By contrast, when Chile's centralized Environment Ministry was created in 2010, also as a response to pressure in the context of OECD ascension, movement effects were indirect as activists were excluded from participating in the policy process and the institution was given weak regulatory authority (Silva 2018).

The Patagonia Without Dams campaign in southern Chile successfully challenged the construction of a series of mega-dams designed to power the mining industries in northern Chile. After emerging in 2006, the campaign grew and advanced an alternative development model for the region based on tourism and alternative energy, which imbued the local anti-dam struggle with an important national concern. Bringing local demands to the national scale linked the campaign to national environmental politics and garnered widespread public support, which increased political opportunities for the movement and produced broad consequences for energy policy.[3] After an eight-year campaign, which articulated common interests across a multi-class coalition, linking international organizations with local, rural, and national urban actors in Chile, the result was the government's decision not to permit construction in 2014.

In addition to challenging and stopping specific dams, key social movement organizations involved in the campaign had mediated effects on energy policy and the historic energy bill.

This involved shifting tactics toward insider politics and close interaction with national legislators. Mediated influence on energy policy from the movement increased when parliamentary elections in 2009 shifted the balance of power in the legislature, favoring more environmentally friendly policy and renewable energy. As a result, in an unprecedented move, progressive legislators in direct consultation with environmentalists authored an energy policy bill requiring 30% energy production from renewables. After three years of negotiation with the executive branch, the result was the 20/25 Law (Silva 2016, 2018).

Since the 1990s, powerful economic interests enshrined in Chile's political pact established during the transition from dictatorship blocked social pressure to change environmental governance, which was the case when a decentralized inter-ministerial environmental agency was replaced with the centralized Environment Ministry in 2010 (Silva 2018). Ever since Chile's transition to democracy in 1991, activists have pushed for the establishment of a ministry for the environment. However, the balance of power within the government favored a loose ministerial coordinating agency, which was established in the mid-1990s. Here, environmental activists from major social movement organizations managed to put the issue on the political agenda, which resurfaced with every presidential campaign cycle. It was not until the mid-2000s, however, that an Environment Ministry was established. This was an indirect effect of the environmental social movements who kept the issue alive during electoral presidential campaigns. Presidential candidates generally entertained the issue to court electoral support. A change in the political balance of power and international pressure turned proposals into reality. In the mid-2000s, a shift to the left by an outlier within the ruling center-left coalition—Michele Bachelet—and OECD insistence on stronger environmental institutions as a condition for Chile's accession[4] induced Bachelet's center-left government to create a ministry for the environment. Contrasting the mediated impacts on renewable energy policy, where opposition to movement organizations was fragmented and declined over time, impacts on this institutional reform were indirect. Activists framed the issue and kept it alive on the policy agenda, but politicians took up the issue without activist involvement in policy formulation (Silva 2018).

Beyond energy policy, this campaign was a formative component of a larger cycle of contention in Chile that has generated significant and ongoing impacts on institutions and national politics beyond environmental governance. In May 2011, the anti-dam campaign mounted the largest environmental protest ever in the capital, and, coincidentally, the largest protest in re-democratized Chile since the student mobilization of 2006. That protest helped set off a new wave of massive student mobilizations as well as a cycle of regular protest marches bringing together environmentalists, students, labor and other sectors. This multi-sectoral movement has called for the transformation of Chile's economic model that had been imposed during the military dictatorship (Silva 2016). They put a constitutional convention on the agenda in Chile, and after the momentous 2019 protest cycle a successful plebiscite was held in October 2020, installing a constitutional convention in 2021.

The political consequences of resistance to extractivism in Chile introduced here highlight the multitude of factors that shape policy and institutional change (Giugni 2004). While the anti-dam campaign transformed national energy policy debates, political, economic, and international factors determined if and how outcomes materialized. The indirect and mediated effects of protest are made possible through interaction with political and institutional actors, but these are constrained by the broader political and economic context. This explains why opposition to mining activities have had virtually no impact on mining policy and activities in Chile. Mining is historically central to Chile's political economy. Minerals, specifically copper, generate the lion's share of Chile's exports. Hence, they contribute a large share of government revenues and to Chile's GDP. This ensures strong political clout for the sector (de Castro et al.

2016). Elsewhere in the region, however, in different political and economic settings, anti-mining efforts have had more progressive policy consequences.

Mining policy in Central America

Although mining activities intensified in Latin America during the recent global commodity boom, there is significant variation in policies and regulations. Central America is illustrative, where Costa Rica and El Salvador have imposed bans, Guatemala and Honduras have suspended mining pending court decisions, and Nicaragua has uninterruptedly advanced gold mining (Spalding 2019). Metal mining is not a dominant economic sector in Nicaragua (accounting for only 4% of GDP), which suggests that the weight of the mining sector alone may not be enough to account for divergent policy outcomes across Central America. A comparison of the moratorium implemented in El Salvador in 2017 with the persistence of mining in Nicaragua provides insight into important political factors that explain different consequences of anti-mining mobilization and related policy outcomes.

El Salvador's national anti-mining movement developed over a long timeframe, shaping the mining ban through important alliances across local and national scales and across social and economic sectors. Sustained mobilization produced broader awareness about the movement, which shifted public opinion to oppose mining and support a moratorium. Ten organizations came together in 2005 to form the National Roundtable Against Mining (La Mesa), connecting activists locally and nationally and networking across sectors. This was done largely by highlighting the issue of water rights, a message that resonates with communities across the country as well as with the then opposition party, the Farabundo Martí National Liberation Front (FMLN), which has had close links to social movements since the civil war. But the movement's impact on mining policy was only possible through direct engagement with executive and legislative officials since business sectors remained interested in mining and El Salvador does not have formal mechanisms for direct citizen participation in the policy process (such as prior consultation). By 2006, activists helped set the agenda for an anti-mining bill that was presented by FMLN legislators, which reflected the mediated effects of La Mesa on this process.

The ban encountered resistance from corporate mining interests and was limited by the post-civil war political settlement that established neoliberal market-oriented policies and deregulation. Efforts at policy change in the mining sector moved past agenda-setting legislative approval for the ban was achieved in 2017. This can be attributed to continued pressure and shift in public opinion as well as a change in external pressures. A large international investment dispute filed by a Canadian mining firm, which had put legal pressure on anti-mining efforts for almost eight years, was resolved in October 2016. The 2017 moratorium was a political consequence of unified resistance that linked local and national concerns, an increase in political opportunities, and a reduction in the structural constraints that blocked political support for anti-mining efforts (Spalding 2018).

By contrast, gold mining has expanded in Nicaragua and resistance has had no significant political consequences. In this case, there are multiple factors that limit anti-mining impacts. Although advocates have been able to stop local projects and, much like El Salvador, mining opposition in Nicaragua is multi-sectoral. Fragmentation within the movement around its demands and goals as well as its localized orientation has made influencing national policy, let alone achieving a moratorium, difficult. Moreover, without a unified movement, it is difficult to counter Nicaragua's elite consensus that has consistently favored mining across post-revolutionary administrations, through the commodity boom, and during the Ortega government since 2007. Nicaragua's pro-mining political elite has consistently supported the unified business elite, so there has been heavy state promotion of mining since the 1990s. Institutional openings for movement influence in the policy process are especially limited in Nicaragua, which has restricted access for citizen claims in general. The narrow institutional and political access that existed during the post-revolutionary era has become even more slender with the country's democratic backsliding since the 2010s. Without the institutional space necessary for mediation to occur, it is evident that the lack of political opportunity explains why mobilization has resulted in no substantive political outcomes (Spalding 2019). Institutional changes that increase citizen access and participation can generate increased political opportunities and more possibilities for policy change, but progressive outcomes are not guaranteed. This is illustrated in the following discussion of prior consultation in two Andean countries.

Prior consultation and policy change in Bolivia and Ecuador

Prior consultation is an international norm that has spread through Latin America since 1989, when the ILO included it in Convention 169. Although prior consultation was designed as a mechanism to protect indigenous rights and territories against infrastructure projects and extractive activities that impact their lands and resources therein, it has been conceptualized more broadly as an institution for citizen participation in policy formulation and implementation. Thus, it is highly relevant for environmental policy in the region. Latin American countries have adhered to Convention 169 more than those of any other region, but there is considerable variation in how it has been implemented and included in the policy process. Comparing this institution in the context of extractive activities in Bolivia and Ecuador illustrates the consequences of prior consultation for aspects of environmental governance.

In Bolivia and Ecuador, coordinated indigenous movements and their resistance to extractive activities influenced the adoption of prior consultation. In Bolivia, the state ratified Convention 169 in 1991 in response to intense mobilization for indigenous political and economic inclusion. Furthermore, indigenous movements in Bolivia were at the center of larger political consequences of this institution. Thus, prior consultation was adopted early and was further amended and extended through social mobilization during the gas war in 2003 and strengthened with the political incorporation of indigenous groups after the election of the political left and Evo Morales in 2006. In Ecuador, the state ratified the Convention in 1998 as a result of the political power of the indigenous confederation (CONAIE) and the Pachakutik party, but prior consultation remains a weak institution (Falleti and Riofrancos 2018).

Resistance to extractivism helps explain the adoption of prior consultation in both countries as well as variation in its institutionalization (Falleti and Riofrancos 2018). In Bolivia, prior consultation has played a strong role in the regulation of natural gas extraction, which was a central demand of indigenous movements. In response to the prolonged episode of contention known as the gas war, Bolivia's Hydrocarbons Law 3058 was passed in 2005 that institutionalized prior consultation. Prior consultation was included in the law as a tool for indigenous communities. However, political mediation was necessary for prior consultation to have substantive effects. Only after the election of Evo Morales did the government pass an enabling law to make the hydrocarbons law operational (Humphreys Bebbington 2012).

Prior consultation has contributed to environmental governance, but the crucial cases of Bolivia and Ecuador demonstrate that the impact of this institutional innovation on the channeling of community and social movement demands is mediated by political and economic factors. In Bolivia, this has become an institutional constraint that limits political opportunities for social mobilization. While prior consultation has become a mechanism for indigenous communities in Bolivia to negotiate with companies around extractive projects, it has also allowed the state to intensify extractive activities without the threat of protest. Prior consultation has transformed the relationship between indigenous movements and the state, replacing the contentious politics that characterized this relationship in Bolivia during the 1990s and early 2000s (Falleti and Riofrancos 2018). This mechanism has mainly been implemented in Bolivia's natural gas sector, and the Morales government has tended to disregard social and environmental safeguards in favor of extractivism. As a result, implementation of prior consultation itself has been the target of contentious politics and is central to mobilization against projects that impact indigenous lands in Bolivia such as oil extraction and the TIPNIS highway (Humphreys Bebbington 2012).

Prior consultation has not been institutionalized in Ecuador to the same extent as Bolivia because the relative strength of indigenous movements vis-à-vis the state is weaker and more contentious. Policy consequences of the implementation of prior consultation, which would give indigenous communities a voice in environmental governance, were blocked as Rafael Correa's government adopted a technocratic approach to participation in Ecuador (Falleti and Riofrancos 2018). Moreover, political and economic constraints have limited the policy impacts of mobilization around environmental issues. Local communities rely heavily on oil extraction where it is already established, which makes unified support for resistance movements more difficult. This context shifts movement demands away from environmental concerns, such as limiting the expansion of extraction, to calls for job creation, which implies further extractive activities and environmental degradation (Pellegrini and Arsel 2018). In Bolivia and Ecuador, prior consultation was a political consequence of mobilization against extractivism, but its role in shaping further environmental policy outcomes, transforming the relationship between social movements and the state, depends on broader political and economic contexts.

Policy implementation and rollback in Brazil

Historically, state absence and high criminality in the Amazon provided openings for unregulated extractive activities and fostered high rates of deforestation for lumber and agricultural expansion. Beginning in the mid-1970s, in a well-documented case of resistance, rubber tappers working in Brazil's northern Acre region organized to challenge land speculators and ranchers who were responsible for heavy deforestation (Hochstetler and Keck 2007). To preserve the forest and their livelihoods, they built networks with local and national political allies in Brazil and fostered international pressure. They did not directly change laws or put a stop to extractive activities, but their actions had clear political consequences for environmental governance by influencing conservation policy. Rubber tappers put extractive reserves, a new form of conservation unit, on the political agenda, which was widely implemented after political opportunities increased in part related to attention to the issue after the murder of iconic leader Chico Mendes in 1988. Moreover, these efforts inserted allies of rainforest protection in state structures as activists entered Brazil's Ministry of Environment and mainstream politics,[5] which helped create the conditions for the expanded implementation of rainforest protections (Hochstetler and Keck 2007).

In the context of Brazil's redemocratization and new 1988 constitution, expanded space for social movements and political pluralism and the increased salience of environmental issues produced relatively favorable conditions for advancing environmental governance. This produced further policies in the Amazon region during the 1990s and 2000s. However, there have been serious rollbacks of hard-won protections. For example, economic pressures limited existing political opportunities in Brazil during the tenure of Worker's Party governments (2003–2016), which relaxed environmental licensing and prior consultation practices to push through extractive development projects. The agribusiness sector lobby has become a dominating force in Brazilian politics and has eroded the strength of existing environmental protections. Brazil also slowed its process of recognizing indigenous lands. An executive decree signed by Dilma Rousseff in 2012 reduced the size of conservation areas, and legislation limited the rights of traditional and indigenous groups in favor of mineral exploration (Humphreys Bebbington et al. 2018). Social and environmental safeguards have been viewed as obstacles to continued development, so environmental and related commitments are being undermined by new laws and norms that scale back licensing procedures and consultation processes, making the conditions for extractive activities more favorable (Silva et al. 2018).

Across the Amazon region, political opportunities for movements to influence forest protection have become heavily constrained by economic factors. Since the end of the global commodity boom, efforts to maintain competitiveness in global trade have pushed right- as well as left-wing governments to relax regulations. This has resulted in an increase in contentious politics as well as attacks on organizations, activists, and local communities (Humphreys Bebbington et al. 2018). Rollback of protections in the Amazon region has occurred in Brazil, Bolivia, Ecuador, and Peru, and protests continue to have political consequences, but impacts have mostly been indirect in the form of transforming public opinion or placing an issue on the policy agenda. Examples include Bolivia's TIPNES highway protest, Peru's Camisea natural gas project, and Brazil's Tapajos hydroelectric complex. As political opportunities are unfavorable and economic pressures mount, movements are excluded from the policy process and resistance to extractive activities is consigned to mobilization in the streets with little impact on policy (Humphreys Bebbington et al. 2018).

The period with more safeguards for forest communities and more political opportunity structures has given way to a period with weakened environmental protections that facilitate large-scale private investment in protected areas. In the previous period, movements had a more mediated influence on environmental policy, but in the context of current rollbacks, their effect is mostly indirect. Contentious action can avoid or mitigate rollback depending on shifts in the balance of power in economic, political, social, and ideational realms. Although there are still periodic opportunities for politically mediated outcomes, optimism about progressive policy change should be tempered. Sudden or incremental rollback of environmental governance is common, and generally impacts of mobilization on policy and related institutional change are not linear (Silva et al. 2018, 31).

Concluding remarks

The cases introduced in the chapter highlight how high political salience garners crucial attention for environmental policy, but the economic weight of extractive sectors can generate large obstacles that oppose progressive policy and institutional change. Depending on the importance of relevant natural resources and the balance of power among the state, related economic interests, and groups resisting extractive activities, there will be variation in political opportunities and structural impediments for mobilization to have policy and institutional consequences. Although environmental governance in Latin America includes adopting international norms, and governments have established rigorous policies to protect natural resources, this is insufficient to ensure that institutional arrangements and laws are implemented and enforced. Social mobilization can have consequences for environmental governance at local and national levels, but this requires close attention to broader political and economic factors.

In the context of democratization and increased political pluralism in Latin American societies, the tension between environmental protections and economic development plays out in dynamic national contexts. Furthermore, tensions at the international level between pressures to extract natural resources and to adhere to global environmental norms shape this interaction. This is a broad area of inquiry that entails consideration of all stages of the policy process and multi-level analyses. Large-scale development projects require political support, so mediation is crucial for social mobilization to impact environmental policy. Other important avenues for research not considered in this chapter include how the increased role of courts in implementation of environmental policies has altered the nature of such mediation. A case study approach is useful for further analysis that can account for how decision-makers and institutional actors respond to mobilization, how they are persuaded, and how they make calculations. This can help illuminate the significance of recent institutional changes such as consultation practices, which Latin America has championed with varying results.

Notes

1 Elaborated by the authors.
2 For a discussion of the concept of extractivism in Latin America see de Castro et al. (2016) or Silva et al. (2018).
3 This is a common path for environmental issues to gain political salience.
4 As of 2020, Chile, Mexico, and Colombia are the only Latin American members of the OECD.
5 Brazil was the first country in Latin America to form a Green Party (1986) according to the Federation of Green Parties of the Americas.

References

Amenta, Edwin, Neal Caren, Elizabeth Chiarello, and Yang Su. 2010. "The Political Consequences of Social Movements." *Annual Review of Sociology* 36: 287–307.

Amenta, Edwin and M.P. Young. 1999. "Making an Impact: The Conceptual and Methodological Implications of the Collective Benefits Criterion." In *How Movements Matter: Theoretical and Comparative Studies on the Consequences of Social Movements*, edited by Marco Giugni, Douglas McAdam, and Charles Tilly, 22–41. Minneapolis, MN: University of Minnesota Press.

de Castro, F., B. Hogenboom, and M. Baud 2016. *Environmental Governance in Latin America.* New York: Palgrave Macmillan.

Delamaza, Gonzalo, Antoine Maillet, and Christian Martinez Neira. 2017. "Socio-Territorial Conflicts in Chile: Configuration and Politicization (2005–2014)." *European Review of Latin American and Caribbean Studies.* 104: 23–46.

della Porta, Donatella and Mario Diani (eds). 1999. *Social Movements: An Introduction.* Oxford: Blackwell Publishers Ltd.

Falleti, Tulia and Thea Riofrancos. 2018. "Endogenous Participation: Strengthening Prior Consultation in Extractive Economies." *World Politics* 70(1): 86–121.

Giugni, Marco. 2004. *Social Protest and Policy Change: Ecology, Antinuclear, and Peace Movements in Comparative Perspective.* Lanham, MD; Oxford: Rowman & Littlefield.

Hochstetler, Kathryn and Margaret Keck. 2007. *Greening Brazil.* Durham, NC: Duke University Press.

Humphreys Bebbington, Denise. 2012. "Consultation, Compensation and Conflict: Natural Gas Extraction in Weenhayek Territory, Bolivia." *Journal of Developing Societies* 28(2): 133–158.

Humphreys Bebbington, Ricardo Verdum, Cesar Gamboa, and Anthony J. Bebbington. 2018. "The Infrastructure-Extractives-Resource Governance Complex in the Pan-Amazon: Roll Backs and Contestations." *European Review of Latin American and Caribbean Studies* 106: 183–208.

McAdam, Doug and Hilary Schaffer Boudet. 2012. *Putting Social Movements in Their Place: Explaining Opposition to Energy Projects in the United States, 2000-2005.* New York: Cambridge University Press.

Pellegrini, Lorenzo, and Murat Arsel. 2018. "Oil and Conflict in the Ecuadorian Amazon: An Exploration of Motives and Objectives." *European Review of Latin American and Caribbean Studies* 106: 209–218.

Silva, Eduardo. 2016. "Patagonia without Dams! Lessons from a David vs. Goliath Campaign." *Extractive Industries and Society* 3: 947–57.

———. 2017. "Pushing the Envelope? Mega-Projects, Contentious Action, and Change." *Research Group MEGA Working Paper* No. 1: 1–18.

————. 2018. "Mega-Projects, Contentious Politics, and Institutional and Policy Change: Chile, 1994–2017." *European Review of Latin American and Caribbean Studies.* 106: 133–156.

Silva, Eduardo, Maria Akchurin, and Anthony Bebbington. 2018. "Policy Effects of Resistance Against Megaprojects in Latin America." *European Review of Latin American and Caribbean Studies* 106: 25–46.

Spalding, Rose J. 2018. "From the Streets to the Chamber: Social Movements and the Mining Ban in El Salvador." *European Review of Latin American and Caribbean Studies* 106: 47–74.

————. 2019. "Mining in Nicaragua: The Political Dynamics of Extractivism." Paper prepared for delivery at the 2019 Congress of the Latin American Studies Association, Boston USA, May 24–27.

Part VI

Toward oppression-free futures

32 Feminist thought and environmental defense in Latin America

Diana Ojeda

Introduction

Despite the salient contributions from various feminist perspectives to Latin American critical thought, scholarly work on gender and feminism tends to be sidelined. While this is common to different fields of knowledge, it is particularly problematic when it comes to environmental issues, given not only current environmental crises' disproportionate *effects* along gender lines (as it has become evident in the case of climate change or the Covid-19 pandemic), but also their *origins and logics*, rooted in a masculinist order that provides fertile ground for capitalist accumulation. The last two decades have brought a remarkable upsurge of feminist thought and action in Latin America that have positioned Latin American feminisms as a political force in environmental defense. Various kinds of feminism—e.g. communitarian, popular, Indigenous, Black, ecological—have become key points of articulation of critiques and alternatives to environmental devastation and dispossession in the region and beyond. I argue their contributions should be incorporated into environmental studies and other related fields.

Women's issues are fundamental when understanding environmental issues and vice versa. Given that women have been socially designated as caregivers, their roles in agriculture, water management, and conservation need to be taken seriously, as several studies have shown (e.g. Carney 1992; Rocheleau 1995; Sultana 2006). While it has been demonstrated that women bear the worst of environmental inequalities and crises, they also deal most directly with the mounting pressures of sustaining life in such precarious scenarios (Arriagada Oyarzún and Zambra Álvarez 2019; Seager 1993). To cite a few figures, in Latin America women account for just 18% of landed property owners (FAO 2018; see also Deere, Alvarado and Twyman 2012); they contribute 73% of the unpaid work at home (Piras 2020); and, by the end of 2020, comprised 118 million if the estimated 200 million living in poverty (Blandón Ramírez 2021). At the same time, while women often are the first ones to report on and mobilize against extractivism and its consequences on the environment and communities, their right to participate in the decisions affecting their territories is systematically denied (*Fondo de Acción Urgente—América Latina* 2015).

Nevertheless, while attending women's issues, feminist approaches prevailing in the region go further, highlighting interlocked forms of oppression that, in order to be dismantled, cannot leave intact capitalism, racism, colonialism, imperialism, ableism, and homo/lesbo/transphobia. In doing so, they have also challenged the cooptation of gender discourse, breaking with the boom of feminist NGO-ization and depoliticization of the 1990s and pushing forward a radical environmental agenda.[1] As I examine in this chapter, denouncing gendered forms of discrimination, inequality, and violence has been at the center of new and strengthened convergences between feminisms and ecologisms in Latin America.

DOI: 10.4324/9780429344428-38

Drawing from a long tradition of critical thought in relation to issues such as capitalist exploitation, development, and colonialism, the articulation of feminisms and ecologisms in the region has been forged by popular and communitarian feminisms in a critical context that can be characterized by three processes.

First, the expansion and intensification of extractivism in the region and its devastating social and environmental effects have brought to the fore a strong critique of the ways in which capitalism and other forms of domination, such as patriarchy, have long been deadly bedfellows (Mies 1999). In Latin America, both neoliberal and progressive governments have pushed forward an extractivist agenda based on capitalist demands for natural resources (McKay et al. 2021). As it has been widely documented, this has been done at the expense of the environment, local communities, and workers where hydrocarbon, mining, agribusiness, and other "greener" extraction projects are located. Destruction and dispossession produced by extractivism impacts populations differently along gender lines, as various authors have noted (Bermúdez Rico 2012; Bolados García and Sánchez Cuevas 2017; Erpel Jara 2018; *Fondo de Acción Urgente—América Latina* 2016; Ulloa 2016). In fact, extractivism heavily relies on gendered inequalities, even as it reproduces them (León Araya 2017; Ojeda 2021; Silva Santisteban 2017). This has resulted in the crucial relevance of women in the development of diverse strategies of environmental defense across the region, including peasant, Indigenous, and Black women, as exemplified by women like Berta Cáceres (Honduras), Francia Márquez (Colombia), Macarena Valdés (Chile), and Máxima Acuña (Perú)—to name a few, as well as multiple women's movements and community-based organizations. While some of them do not openly recognize themselves as feminists, perhaps refusing to be aligned with a white and liberal version of the feminist movement, they rely on the principles of popular and communitarian feminisms, as well as ecofeminism, to contest extractivist projects and their effects on their territories, bodies, and lives. At the same time, there has been a growing presence of women and, more importantly, of an explicitly feminist and environmentalist agenda in formal political spaces across the region, as in the remarkable case of the recently formed Constitutional Assembly in Chile.

Second, the feminist movement across Latin America/Abya Yala has gained important terrain in struggles around a variety of issues such as feminicide, sexual and reproductive justice, and police violence. That is, women and gender dissidents have risen up against diverse forms of oppression and suffering. This moment has generated rich opportunities for understanding gender issues as central to capitalist exploitation and the destruction it creates at different scales. An important point within this issue has been the ongoing escalation of violence against environmental defenders—in particular, women. Such violence includes death threats and assassinations, as well as gender-based forms of violence that include sexual violence (Esguerra et al. 2019; Navas et al. 2022).

Lastly, environmental crises associated with climate change, large-scale deforestation, pollution, and the COVID-19 pandemic have laid bare a profound crisis of care. As basic care infrastructures—e.g. food, health, housing, and education—have been ransacked under the decades-long neoliberal age, pressures to sustain life have increased dramatically, along with the working hours of those primarily in charge of caring for others (Pérez Orozco 2011). The profound crisis of reproduction we are facing today reveals the ways in which the capitalist logic of accumulation is incompatible with the sustenance of life and the multiple social and ecological networks it depends on.

This chapter examines the contributions of Latin American feminist thought to environmental defense. In order to do so, I bring in elements from feminist perspectives that have taken seriously social reproduction and the everyday. Most of this work bridges divides between intellectual fields and comes from spaces of political action that sustain dialogue with academic circles in various countries in Latin America and the Caribbean. I focus on academic work that relates three salient, interconnected themes: The critique of hegemonic productions of nature; care of the body-territory *(cuerpo-territorio)*; and the defense of the commons. Rather than a comprehensive account of the contributions of feminist thought to different environmental struggles in the region, this chapter represents an invitation to engage with relevant work coming from Latin America on the ways in which gender issues are fundamental to better understanding current environmental crises and finding ways out.

Critiques of hegemonic productions of nature

> ... *lo femenino/cuerpo/sentimiento/naturaleza está subordinado a lo masculino/mente/razón/ cultura, con las nuevas formas de opresión de las mujeres y el actual deterioro de los ecosistemas como consecuencias.*[2]

(LasCanta 2017)

The connections between feminist and environmentalist agendas became evident in the 1970s and 1980s. In *Le Féminisme ou la Mort* (1974), Françoise d'Eaubonne coins the term and signals to what became one of the most important concerns of the field—how the patriarchal subordination of women stands in close relation to the domination of nature and its destruction. In dialogue with critiques of the environmental and social consequences of development projects sponsored by international institutions such as the World Bank and the International Monetary Fund, ecofeminism gained strength in various places in the global South. This is the case of Vandana Shiva's (1988) salient work and its analysis of colonial impositions brought by Western paradigms of progress, technology, and development.

Since then, ecofeminist thought became an important voice in denouncing the conjoined forms of oppression of women and nature with studies by Val Plumwood (1993) and Karen Warren (1997), which analyzed, respectively, the "master model" and the "logics of domination" over feminized subjects.

While consistently signaling the ways in which capitalist productions of nature rely on anthropocentrism and, more specifically, on androcentrism, ecofeminist work has offered a powerful critique of dualistic conceptualizations of society and nature. Such dualism, Plumwood reminds us, implies both separation and inferiority of a feminized Other—e.g. women, children, racialized groups, animals or the body—providing an efficient foundation for their subordination. The "master model," as she names it, thus positions of the feminine as inferior in relation to a masculinist model of rationality and domination. This is made evident, for example, in hegemonic notions of conservation and how they currently draw from colonial constructions of Latin American and Caribbean nature as 'paradise,' recycling old tropes of a feminized entity that needs both to be protected and controlled (Nouzeilles 2003; Sheller 2003; Ulloa 2004).

Despite ecofeminism's key contributions to environmental thought and its relevance to thinking the different dimensions of environmental crises, it has been heavily criticized for leveraging essentialist notions of women as biologically closer to nature or as innately capable of caring for others. In Latin America, the confluences between ecofeminist and Indigenous struggles have often been caricaturized around the figure of the Pachamama, equating nature, mother, and women (c.f. Zaragocín and Ruales 2020). Nonetheless, authors such as Gaard (2011) have demonstrated the variety and richness of different ecofeminist trajectories and their overall relevance for environmental thought (see also *Capitalism Nature Socialism* 2006). Of particular relevance is Ariel Salleh's work (1997), its clarification of ecofeminism's political strategy, and her insistence that ecofeminism can be understood as a political synthesis of four revolutionary struggles: ecologist, feminist, socialist, and post-colonial.

Ecofeminist perspectives have also been in dialogue with a long tradition of critical thought in Latin America/Abya Yala, supported by social movements and tradition, such as liberation theology (Arriagada Oyarzún and Zambra Álvarez 2019) and agroecology (Trevilla Espinal 2018). As such, it has posed important questions in relation to colonial domination and its consequences in terms of environmental destruction and the continuation of violence over Indigenous peoples and women (Rivera Cusicanqui 2010). More recently, ecofeminist thought has gained force in the region, connecting anti-extractivism with a critique of depoliticized views of climate change that disregard differentiated responsibilities in the current environmental crisis and disproportionate effects on women. Insisting on the intertwined subordination of nature and feminized subjects, authors like the Venezuelan collective *La Danta Las Canta* insist on the role of masculinist domination in what they consider 'the Phallocene', a combined "feminicide, ethnocide, ecocide, and geocide" (Las Canta 2017).

Such critiques urge the recognition of the multiple relations of interdependence that make human existence possible and argue for an ethics of care from a vision that decenters humans and expands the meaning of the political (de la Cadena 2015). As we will see in the next section, a Latin American critique of patriarchal domination's role in normative productions of nature has also allowed for a poignant revision of the ways in which capitalism's expansion has been championed on the domination of feminized bodies and territories.

The defense of the body-territory

> … *la enunciación cuerpo-territorio es una epistemología latinoamericana y caribeña hecha por y desde mujeres de pueblos originarios que viven comunidad; es decir, la articulación cuerpo-territorio pone en el centro lo comunitario como forma de vida.*[3]

(Cruz, 2016, 9)

Ecofeminist accounts of capitalism's effects on land, water, plants, animals, and humans have entered into conversation with different feminist marxist analyses, as in the case of ecosocialist ecofeminism (see Brownhill and Turner 2020). Of outmost importance is their common concern around the inseparability between the spheres of production and social reproduction. Social reproduction includes all that is needed in order to sustain life and maintain society—that is, material resources, social relations, and everyday practices. Care provision is central to it, involving activities as essential as cleaning, cooking, tending to the sick, and childrearing. In dialogue with the women's liberation movement, reflections on the reliance of the capitalist mode of production on the exploitation of nature and women have gained traction since the 1970s (Giménez 1975). Such an approach turned political attention to the centrality of care activities in sustaining life, and to the effects of capitalist exploitation on the body, as a key terrain of capitalist exploitation.

As part of such analyses, communitarian, Indigenous, and Black feminisms of Latin America/Abya Yala have been crucial to the conceptualization of the harm that capitalism, racism, colonialism, and patriarchy have jointly inflicted on bodies and territories. For example, in Guatemala and Bolivia, communitarian feminists have contributed to theorizing and denouncing the synchronized operations of extractivist and state violence (Cabnal 2010; Guzmán 2019). Similarly, in Colombia, Afro-descendant environmental defender Francia Márquez has regarded this as the imposition of a "politics of death" (Márquez 2019).

This perspective has insisted on the inseparability of body, land, and territory and how this articulation reflects the need to recover the body as a central place of struggles for territorial defense, as the Indigenous intellectual Lorena Cabnal (2015, 2018) points out. Such recovery (or recuperation) of the territory-land and the territory-body—both from colonial violence coming from outside communities and from patriarchal violence coming from within them—has put on the table an important discussion regarding the articulation of the colonial invasion based on patriarchal domination and patriarchal formations within Indigenous communities. As part of this articulation, nature "as a system historically built upon the sexed body of women" becomes a key site of struggle against patriarchal oppression, exploitation, violence and discrimination (Cabnal 2010, 122, my translation).

Taken as a central axis of militant and intellectual action, the body-territory suggests a conceptualization of bodies as historical and living territories, as the *Colectivo Miradas Críticas del Territorio desde el Feminismo* (Cruz Hernández 2016) asserts. In turn, territories are social bodies integrated into the web of life and, in that sense, demand a recognition of interdependence and an ethics of co-responsibility.

This analytical perspective has invited a potent conceptualization of gender as a key structure of domination that sustains violence and capital, including extractivism. Within such arrangements, different forms of violence coalesce, including the environmental destruction of 'sacrifice zones' and the genocide of environmental leaders, which represent the localized and embodied materialization of uneven geographies of capitalist accumulation, enabled by the designation of poisonable territories and disposable bodies. These bodies-territories are deeply interwoven by material and symbolic ecologies that are essential for life and that need to be protected from old and new waves of enclosure and control grabbing, as the next section further details.

The production of the commons in common

Lo común se produce, se hace entre muchos, a través de la generación y constante reproducción de una multiplicidad de tramas asociativas y relaciones sociales de colaboración....[4]

(Gutiérrez, Navarro and Linsalata 2016)

Intensified forms of dispossession occurring in the region during the last two decades have translated into important contributions from feminist Latin American thought in relation to strategies of survival, struggle, and resistance, from within territories of increasing environmental pressures; old and new forms of enclosure; and the privatization of spaces and ecologies essential to sustain life. The already mentioned attention to care and social reproduction has proven to be a rich space of intellectual and political debate in terms of the potential for collectivizing those practices that make life possible (Pérez Orozco 2011). As various authors suggest, the conversation surrounding social reproduction has enabled powerful connections to strategies of environmental defense (Vega et al. 2018, 16), in which care practices emerge as practices for producing life in common. The common and the commons are thus understood beyond property, public goods, or collective tenure to think about the production of the common—i.e. the everyday practices of maintenance and production of the communitarian or the production of commons in common (Aguilar et al. 2017; Gutiérrez and López 2019; Gutiérrez 2017; Villamayor-Tomas and García-López 2021).

A significant part of this debate has been nourished by a continuous conversation with autonomist Marxist feminist thinker Silvia Federici (Navarro and Gutiérrez 2018). Silvia Federici's work (2004) has been fundamental in the understanding of how capitalism's oppression by capitalism has entailed the expropriation of women's bodies, labor, and spaces. Through her writing and engagement with women's and labor collectives on issues of capitalist exploitation in various countries in Latin America/Abya Yala, Federici has been a source of inspiration for many of these reflections that—relying on feminist political economy—point to the central role of social reproduction in making capitalist accumulation possible. She points to how historical processes of privatization of land and the expansion of monetary relations largely affected women's autonomy and forms of social organization for food production. The direct connection between the expropriation of women's bodies and common goods thus becomes a very powerful meeting point for diverse currents of feminist thought in the region and an important motivation for searching for communitarian alternatives to capitalist exploitation (Flórez and Olarte 2022).

In her conceptualization of a "communitarian-popular horizon" of political action, Raquel Gutiérrez Aguilar (2017) re-signifies the communitarian as process, thus giving centrality to social reproduction as a key political site within current struggles against capitalism, as Verónica Gago (2018) suggests. Far from romanticizing the communitarian or care work, this approach signals the need to understand power relations that both shape and are shaped by these practices, while also recognizing the political potential of care work in the struggle to place life at the center.

Conclusions

In this chapter I have highlighted the important contributions of Latin American feminist thought in relation to environmental defense. Such contributions examine the connections between capitalist exploitation and patriarchal domination, as well as their effects on territories and bodies. While historically—and geographically—specific, the articulation between feminisms and ecologisms pose a very important contribution to better understanding current environmental crises and mapping out present and future pathways. This is the case of different initiatives which, inspired by communitarian, popular, Indigenous, and Black feminisms, have carved spaces for defending life against capitalist siege, drawing from a vast variety of forms of collective work and exchange, as well as diverse strategies of commoning and re-commoning. Some of these can be evidenced by cooperatives, community gardens, food banks, reproductive justice networks, and mutual aid societies all across the region.

Important trends in Latin American feminist thought have included a focus on social reproduction and struggles for sustaining life against capitalism, in its alliance with patriarchal domination, racism, and colonialism. This insistence on "putting life at the center" constitutes a powerful way of building alliances and spaces of dialogue, enabling critique and struggle across places. As communitarian feminisms suggest, environmentalist and feminist struggles correspond to lived localized and embodied realities that, at the same time, advance the necessity to articulate struggles across distance and difference—or what Zaragocín (2018) calls hemispheric dialogues.

In that sense, Latin American feminist thought has continued to provide important theoretical elements for a more nuanced understanding of environmental issues and their articulation with power relations, where gender plays a central role. In doing so, feminist contributions have insisted on the existence of tightly-woven networks and relations of interdependence, that is, the ecologies that connect us and make us accountable to other beings. As such, feminists have contributed to carving out spaces for building a dignified life, thus mapping out alternatives—other relations, other spaces, other realities, other worlds.

Notes

1 A good part of academic literature, multilateral institutions, NGOs and other organizations tend to reduce gender to women. This not only erases the power dynamics that locate women, and other subjects such as children, youth and gender and sexuality dissidents in subordinate positions, but also risks their reinforcement by presenting women's access to market mechanisms (read as "participation" and "empowerment") as the silver bullet to resolving inequalities that are rather structural. In so doing, such reductionist frameworks end up serving neoliberal principles (Álvarez, 2009).

2 "…the feminine/body/feeling/nature is subordinated to the masculine/mind/reason/culture, resulting in new forms of oppression of women and the current deterioration of ecosystems".

3 "…the body-territory enunciation is a Latin American and Caribbean epistemology made for and from women of native peoples who live in community; that is, the body-territory articulation puts at the center the communitarian as a way of life".

4 "The common is produced, it is done among many, through the generation and constant repr du ti n of a multiplicity of associative networks and collaborative social relationships…".

References

Aguilar, Raquel G., Lucia Linsalata, and Trujillo, Mina Lorena Navarro. 2017. "Producing the common and reproducing life: Keys towards rethinking the Political." In *Social Sciences for an Other Politics: Women Theorizing Without Parachutes*, edited by Ana Cecilia Dinerstein, 79–92. Springer International Publishing. https://doi.org/10.1007/978-3-319-47776-3_6.

Álvarez, Sonia. 2009. "Beyond NGO-ization?: Reflections from Latin America." *Development* 52: 175–184.

Arriagada Oyarzún, Evelyn, and Antonia Zambra Álvarez. 2019. "Apuntes iniciales para la construcción de una Ecología Política Feminista de y desde Latinoamérica." *Polis. Revista Latinoamericana* 54: 12–26.

Bermúdez Rico, Rosa Emilia. 2012. "Impactos de los grandes proyectos mineros en Colombia sobre la vida de las mujeres." In *Minería, territorio y conflicto en Colombia*, edited by Catalina Toro Pérez, Julio Fierro Morales, Sergio Coronado Delgado, and Tatiana Roa Avendaño. Bogotá: Universidad Nacional de Colombia.

Blandón Ramírez, Daniela. 2021. "Cepal: mujeres de América Latina perdieron una década de avances en materia laboral." *France24*, February 2, 2021. https://www.france24.com/es/programas/economía/20210211-economia-cepal-desempleo-mujeres-america-latina.

Bolados García, Paola, and Alejandra Sánchez Cuevas. 2017. "Una ecología política feminista en construcción: El caso de las "mujeres de zonas de sacrificio en resistencia", Región de Valparaíso, Chile." *Psicoperspectivas. Individuo y Sociedad* 16 2: 33–42. https://doi.org/10.5027/psicoperspectivas-Vol16-Issue2-fulltext-977.

Brownhill, Leigh and Terisa Turner. 2020. "Ecofeminist ways, ecosocialist means: Life in the post-capitalist future." *Capitalism Nature Socialism* 31(1): 1–14. DOI: 10.1080/10455752.2019.1710362.

Cabnal, Lorena. 2010. "Feminismos diversos: el feminismo comunitario." *ACSUR-Las Segovias* 1–35.

Cabnal, Lorena. 2018. "Acercamiento a la construcción de la propuesta de pensamiento epistémico de las mujeres indígenas feministas comunitarias de Abya Yala." In *Momento de paro tiempo de rebelión. Miradas feministas para reinterpretar la lucha*, edited by Minervas Ediciones, 116–134.

Capitalism Nature Socialism (2006). Symposium: Ecofeminist Dialogues, 17(4).

Carney, Judith. 1992. "Peasant women and economic transformation in the Gambia." *Development and Change* 23: 67–90.

Cruz Hernández, Delmy Tania. 2016. "Una mirada muy otra a los territorios-cuerpos femeninos." *Solar* 12 (1): 35–46.

de la Cadena, Marisol (2015). *Earth Beings. Ecologies of Practice Across Andean Worlds*. Durham: Duke University Press.

Deere, Carmen Diana, Gina E. Alvarado, and Jennifer Twyman. 2012. "Gender inequality in asset ownership in Latin America: Female owners vs household heads." *Development and Change* 43 (2): 505–530.

Erpel Jara, Ángela. 2018. *Mujeres en defensa de territorios. Reflexiones feministas frente al extractivismo*. Fundación Heinrich Böll.

Esguerra, Camila, Diana Ojeda, Tatiana Sánchez, and Astrid Ulloa. 2019. "Introducción." In *Dossier: Violencias contra líderes y lideresas defensores del territorio y el ambiente en América Latina*. LASA FORUM 50:4.

Federici, Silvia. 2004. *Caliban and the Witch*. https://books.google.com/books?hl=en&lr=&id=4-PvMvdVqp0C&oi=fnd&pg=PA7&dq=silvia+federici&ots=0X8yDE55mF&sig=bG3ZYdvXUfRb_opTvoWX7QDvAoA.

Fondo de Acción Urgente - América Latina. 2015. *Mujeres defendiendo el territorio. Experiencias de participación en América Latina.*

Fondo de Acción Urgente - América Latina. 2016. *Extractivismo en América Latina. Impacto en la vida de las mujeres y propuestas de defensa del territorio.*

Gaard, Greta. 2011. "Ecofeminism Revisited: Rejecting Essentialism and Re-Placing Species in a Material Feminist Environmentalism." *Feminist Formations* 23 (2): 26–53.

Gago, Verónica. 2018. "Neo-comunidad: circuitos clandestinos, explotación y resistencias." In *Cuidado, comunidad y común: experiencias cooperativas en el sostenimiento de la vida*, edited by Cristina Vega Solís, Raquel Martínez Buján, and Myriam Parede Chauca: 75–91. Madrid: Traficantes de Sueños.

Giménez, Martha. 1975. "Marxism and Feminism." *Frontiers: A Journal of Women Studies* 11: 61–80.

Gutiérrez Aguilar, Raquel, Mina Lorena Navarro, and Lucia Linsalata. 2016. "Repensar lo político, Pensar lo común. Claves para la discusión." In *Modernidades Alternativas y nuevo sentido común: ¿hacia una modernidad no capitalista?*, edited by Márgara Milám, Lucia Linsalata and Daniel Inclán, 377–417. Ciudad de México: FCPyS-UNAM.

Gutiérrez, Raquel, and Claudia López. 2019. "Producir lo común para sostener la vida. Notas para entender el despliegue de un horizonte comunitario-popular que impugna, subvierte y desborda el capitalismo depredador." In *¿Cómo se Sostiene la Vida en América Latina? Feminismos y re-existencias en tiempos de oscuridad*, edited by Karin Gabbert, and Miriam Lang. Quito: Ediciones Abya-Yala/Fundación Rosa Luxemburg.

Gutiérrez, Raquel. 2017. *Horizontes comunitario-populares. Producción de lo común más allá de las políticas estado-céntricas.* Madrid: Traficantes de Sueños.

Guzmán, Adriana. 2019. *Descolonizar la memoria. Descolonizar los feminismos.* La Paz: Tarpuna Muya.

Las Canta, LaDanta. 2017. "El Faloceno. Redefinir el Antropoceno desde una mirada ecofeminista." *Ecología Política* 53: 26–33.

León Araya, Andrés. 2017. "Domesticando el despojo: palma africana, acaparamiento de tierras y género en el Bajo Aguán, Honduras." *Revista Colombiana de Antropología* 53 (1): 151–185.

Márquez, Francia. (2019). "Francia Márquez: 'estamos cansados de la política de la muerte'" [We're tired of the politics of death]. *El Espectador*, YouTube video, 2:38. May 14, 2019. https://www.youtube.com/watch?v=VjIJGkYkOQk.

Mies, Maria. 1999. *Patriarchy and Accumulation on a World Scale: Women in the International Division of Labour.* London: Zed Books.

Navarro, Mina Lorena, and Raquel Gutiérrez. 2018. "Diálogos entre el feminismo y la ecología desde una perspectiva centrada en la reproducción de la vida. Entrevista a Silvia Federici." *Revista Ecología Política. Cuadernos de Debate Internacional* 54: 119–122.

Nouzeilles, Gabriela. 2003. *La naturaleza en disputa: retorica del cuerpo y el paisaje en America Latina.* Buenos Aires: Paidós.

Ojeda, Diana. 2021. "Social reproduction, dispossession and the gendered workings of agrarian extractivism in Colombia." In *Agrarian Extractivism in Latin America*, edited by Ben M. McKay, Alberto Alonso-Fradejas, and Arturo Ezquerro-Cañete. New York: Routledge.

Pérez Orozco, Amaia. 2011. "Crisis multidimensional y sostenibilidad de la vida." *Investigaciones Feministas* 2: 29–53.

Piras, Claudia. 2020. "Las mujeres en América Latina y el Caribe enfrentan mayores riesgos ante el Coronavirus." *Interamerican Development Bank*, March 23, 2020. https://blogs.iadb.org/igualdad/es/mujeres-enfrentan-mayores-riesgos-ante-coronavirus.

Plumwood, Val. 1993. *Feminism and the Mastery of Nature.* New York: Routledge.

Rivera Cusicanqui, Silvia. 2010. *Violencias (re) encubiertas en Bolivia.* La Paz: Editorial Piedra Rota.

Rocheleau, Dianne. 1995. Gender and Biodiversity: A Feminist Political Ecology Perspective. *IDS Bulletin* 26 (1): 9–16. https://doi.org/10.1111/j.1759-5436.1995.mp26001002.x.

Salleh, Ariel. 1997. *Ecofeminism as Politics: Nature, Marx and the Postmodern.* London: Zed Books.

Seager, Joni. 1993. *Earth Follies: Feminism, Politics, and the Environment.* Earthscan.

Sheller, Mimi. 2003. *Consuming the Caribbean: From Arawaks to Zombies.* New York: Routledge.

Shiva, Vandana. 1988. *Staying Alive: Women, Ecology and Development.* London: Zed Books.

Silva Santisteban, Rocío. 2017. *Mujeres y conflictos ecoterritoriales. Impactos, Estrategias, Resistencias* (First edition).

Sultana, Farhana. 2006. "Gendered waters, poisoned wells: political ecology of the arsenic crisis in Bangladesh." In *Fluid Bonds: Views on Gender and Water*, edited by Kuntala Lahiri-Dutt, 362–386. Kolkota: Stree Publishers.

Trevilla Espinal, Diana Lilia. 2018. "Ecofeminismos y agroecología en diálogo para la defensa de la vida." *Radio Temblor. Por Colectivo Voces Ecológicas*, December 9, 2018. http://www.radiotemblor.org/?p=12479.

Ulloa, Astrid, 2004. *La construcción del nativo ecológico. Complejidades, paradojas y dilemas de la relación entre los movimientos indígenas y el ambientalismo en Colombia*. Bogotá: Instituto Colombiano de Antropología e Historia.

Ulloa, Astrid. 2016. "Feminismos territoriales en América Latina: defensas de la vida frente a los extractivismos." *Revista Nómadas* 45: 123–139.

Vega, Cristina, Raquel Martínez Buján, and Myriam Paredes (eds). 2018. *Cuidado, comunidad y común. Experiencias cooperativas en el sostenimiento de la vida*. Madrid: Traficantes de Sueños.

Villamayor-Tomas, Sergio, and Gustavo A. García-López. 2021. "Commons Movements: Old and New Trends in Rural and Urban Contexts." *Annual Review of Environment and Resources* 46, 511–543.

Warren, Karen. 1997. *Ecofeminism: Women, Culture, Nature*. Bloomington, IN: Indiana University Press.

Zaragocín, Sofía. 2018. "Espacios acuáticos desde una descolonialidad hemisférica feminista." *Mulier Sapiens. La Mujer Resistencia: Apropiación Del Agua, Territorios En Conflicto y Atentados Contra La Vida* 10: 6–19.

Zaragocín, Sofía and Gabriela Ruales. 2020. "De-género y territorios ¿Tiene género la Tierra?" In *Cuerpos, territorios y feminismo. Compilación Latinoamericana de teorías, metodologías y prácticas políticas*, edited by Delany Tania Cruz and Manuel Bayón, 303–314. Ciudad de México: Bajo Tierra Ediciones.

33 Decolonizing time through communalizing spatial practices

María Carolina Olarte-Olarte and María Juliana Flórez Flórez

Introduction. From commodities to the common

This chapter approaches the effort sustained by Latin-American social movements to resist the commodification of natural wealth by extractive industries. We discuss how the resistance to processes of dispossession of the region's wealth have increasingly mobilized around the commons.

The chapter opens by outlining three axial premises for understanding Latin-American struggles. It then identifies a set of spatializing practices grounding four horizons for those struggles: (1) resisting the subalternization of the region; (2) defending the relational character of the territory; (3) repoliticizing the reproduction of life; and (4) recentering use value. We then identify a decolonizing temporal approach for each horizon, which destabilizes the temporality of progress underlying developmentalist approaches to nature. Finally, we close by highlighting four ways to expand post-developmentalist agendas.

Axial premises

Three axial premises provide the basis from which to approach the possibilities of decolonizing time through communalizing spatial practices.

A first premise is the region's renewed interest in the communal. Following Julieta Paredes' work (2010), we consider that there is an interest, not only in recording, but also in theorizing the multiple initiatives of struggles intended to rethink and mobilize the communal. This interest is expressed in a vast polysemy, ranging from the communal, to the common, commoning, the commons, communalization, community, communitarian, and more. This chapter identifies some contributions of movements to broadening these debates.

The second premise is to consider social movements, not only as reflexive actors (as in resource mobilization theory or political process theories) or as critics of modernity (see, for example, the work of Giddens, Touraine, Melucci), but also as *producers of critical theory* (Flórez 2014; Flórez and Olarte-Olarte 2023). According to this decolonizing premise, we will cite intellectuals from both the academy and movements. We highlight their contributions to the study of the commons in terms of spatializing practices that involve complex knowledge, as well as sustained and reflexive modes of action. These practices open up horizons of temporality different from those offered by developmentalist modernity.

Finally, we highlight *the strengthening of struggles linked to the territorial* vis-à-vis the impact of extraction in the region during the last two decades and its associated socio-environmental conflicts. We draw upon what Svampa (2012) calls the eco-territorial turn of the region's struggles, or what Raúl Zibechi and Michael Hart (2013) understand as an essential commitment of

DOI: 10.4324/9780429344428-39

the region's movements with the territorial. Hence, we formulate the explanation of spatializing practices in terms of the defense of territories as something indivisible from livelihoods.

Resisting the subalternization of the region

Latin America is often referred to as a "global pantry" feeding the world or a "great container" of economic resources whose wealth is often confined to an inventory of commodities. Against this reduction of the region to a subaltern space—ready to be exploited and devastated (cf. Alimonda 2015)—various movements posit alternative representations of their territories embedded in *the communal.* Two spatializing practices are particularly telling in this regard.

The first is to *suspend the equivalence between nature and goods or resources through land-bound concepts.* Amongst the most widely circulated of these concepts are that of the *Madre Selva* (Amazonas), the *Ñuke Mapu* of the Mapuche people (Chile and Argentina), or the *Pachamama* of Aymaras and Quechuas (Andes). According to these conceptions, *wealth* implies a set of elements and their relationships indispensable for sustaining life. By emphasizing use value (over exchange value) and concrete work (over abstract work), they resist the meaning of territorial wealth, and of the planet as a whole, in terms of exploitable goods. These land-bound notions also amplify the complexity of the most well-known theories of the commons—such as the pioneering works of Elinor Ostrom (1990)—in terms of shared resources and goods, whose property and management are collectivized. The semiotic-material force of these notions expands from this contribution to engage with the communal as a possibility of planetary survival.

The second practice is to *promote territorial representations of the region alternative to the subaltern ones.* The communal here has entailed disputing the often-unquestionable character of expert knowledge and exposing its role in modelling nature in terms of goods and resources. For example, the *Ríos Vivos* movement in Antioquia (Colombia), an organization dealing with the damage stemming from the Hidroituango Dam project and the violence associated with the armed conflict, have mapped the likely sites of the burials (*entierros*) of people who were disappeared and murdered in the area where the mega project was implemented. Based on the local knowledge of the burials threatened by the construction of the dam, the movement radically asserted the limitations of the contrasting forms of georeferencing provided by experts from public institutions and corporations (Audiencia-JEP 2019). Not only did the social mapping mobilize an inseparable relationship between their livelihoods and that particular territory, but it disputed the conversion of the river and its waters into an economic resource.

These two practices resisting subalternization have also decolonized the *stratified temporality of development.* For example, while the projection of profit dominating the public justification of the hydroelectric relies upon a notion of time as well as of progress, movements identify *connections between development and different forms of violence* (Ríos Vivos n.d.). Such connections expose *the continuity of violence,* and, in doing so, as Rios Vivos has posited (Audiencia-JEP 2019), highlight that plundering by infrastructure development cannot be separated from the history of the country's armed conflict. Further, these strategies reveal that such continuity is also inscribed, in many cases, in the long-term temporality of the colonial. For example, Alimonda's genealogy (2015) of mining in Peru highlights the continuity between colonial domination and the activities of mining companies resisted by local organizations (such as Conacami or Aidesep).

Defending the relational character of territories

A fragmented representation of territory is at the root of the extractivist development followed by the region's governments, whether progressive, leftist, or neo-liberal. A central spatial assumption enables extractive intervention; namely, the depiction of territory as a container of resources, an object whose elements are easily sectioned off (Olarte-Olarte 2019). Such representation permeates the modes of relating to, and experiencing, land use and its elements. Hydrocarbon extraction projects through hydraulic fracturing in several countries of the region are a telling instance. In contrast to this fragmented assumption, several movements are defending the relational character of their territories by articulating *processes of communalization*. For this horizon of struggle, one spatializing practice has been central: *revealing interdependences among diverse entities*.

A relational understanding of territory means that "things and beings are their relations; they do not exist prior to them," and, therefore, "living beings of all kinds constitute each other's conditions for existence" (Escobar 2015, 131). Importantly, an experience of nature based upon the interdependence between humans, humans and non-humans, organic and inorganic entities, and amongst these entities themselves is at the center of this relational understanding. Here, territory does not denote an inert and fragmentable entity, but a living and related whole whose defense is inseparable from the struggles to remain in it. As Brazilian Yanomami leader Davi Kopenawa explains (Villaverde 2017), "the environment is not independent of us; we are within it, just as it is within us; we create it and it creates us."

Regarding the central place of other-than-human entities within this relational approach, the debates in the region (including, among others, the work of Marisol de la Cadena, Gladys Tzul Tzul, Silvia Rivera Cusicanqui, and Arturo Escobar) converge with critical studies of science and technology (among others, the work of Donna Haraway, Michel Serres, and Bruno Latour). Latin-American debates have, however, extended the scope of such heterogeneity beyond that addressed by this latter group of thinkers. Silvia Rivera Cusicanqui, for instance, highlights the labor conducted by non-humans, including forces of nature. Thus, food production in Andean communities involves the creation of a community of "humans and non-humans [...] between humans and products of human labor, between humans and products of the labor of other species" (Salazar 2015, 145). In turn, Marisol de La Cadena (2015) has contrasted the state's limited understanding of land as a supplier of agricultural products with that of the Andean people in their struggle to maintain the *ayllu*: the place where *runakuna* (human beings) and *tirakuna* (earth-beings, other-than-human-beings) emerge in interdependence. In this far-reaching interconnection, earth-beings do not represent the territory, but inhabit it as presences. Kristina Lyons' (2016) work in Putumayo, Colombia, highlights the work undertaken by the soil to enable the reproduction of the communities' lives and of the soil itself. Lyons warns of the clash between the rhythms of the slow work of the soil and the urgency of the timeframes these communities are subject to if they want to take political action. In all three approaches, the relational character of the territory can lead not only to a criticism of the definition of the commons as resources (as we saw earlier) but also of the supposition that they exist as separate and autonomous entities.

Multiple examples account for interdependences among different entities. The peasants and Mayans against the genetically-modified soy agroindustry in the region of Los Chenes, Mexico, have denounced the delusion of private property borders when accounting for the extension of socio-environmental damages. The effects of genetically modified organisms seriously compromise—beyond the limits of property—the air, their model of self-subsistence, and their main productive activity, beekeeping. Another example is expressed by the philosophy of the *Jotï* or *Hoti* of the Venezuelan Amazon with their *jkyo jkwainï* (principle of life), according to which we should respect, take care of, and love all the *biotic* and *abiotic* components that surround us and thus increase our awareness of the interdependence we are all subject to (Zent, 2014). Principles of this kind inspire indigenous struggles in this region (*Hoti, Yabarana, E'ñepá* and *Piaroa*) against the Arco Minero del Orinoco, a state-led gold-mining megaproject, and the multiple and intense forms of civil and military violence, as denounced by, among others, the *Coordinadora de Organizaciones Indígenas de la Amazonía Venezolana* (cf. Observatorio de Ecología Política de Venezuela, 2018).

In defending the relational character of the territory, movements are also decolonizing the alleged *temporality of equilibrium* at the basis of development. Such alleged equilibrium assumes a balance growth and compensation for environmental and social damages. In contrast, the region's movements outline a *temporality of co-dependency networks*, which situates damage and historicizes how elements existing in networks of relations become resources *vis-à-vis* the geopolitics of extraction and their insertion, measurement, and representation in production and value chains. Further, this alternative temporality warns against global visions of the commons which overlook the local distribution of the consequences of both their exploitation and their conservation.

Repoliticizing the reproduction of life

Productive work, such as wage-based labor, has been central to the Marxist analysis of the processes of value creation and accumulation through the expanded reproduction of capital. Latin-American feminists (socialist-Marxists, anarchists, and communitarians), along with other feminists in different latitudes, have questioned the centrality of such a take on productive labor by *repoliticizing the reproduction of life*. This entails understanding reproduction as an indispensable form of labor for *the production of the common*, which constitutes the core of this horizon of struggle.

This repositioning of reproductive work involves a decolonizing engagement with the feminist and Marxist traditions these authors came from. Regarding feminism, the authors find common ground with the Wages for Housework Campaign and the Bielefeld feminists in interpellating the centrality of productive work. Thus, both elaborated a link between dispossession, female reproductive and care work, and the common. Both also assumed a broad notion of the source of value or wealth that exceeds the monetary (cf. Vega 2019). Regarding Wages for Housework, in line with Silvia Federici, Latin American readings maintain that the feminist struggle aims not only to include women in the sphere of productive work, but also to recognize that reproductive work, with its dual material and symbolic dimension, is also a source of value or wealth and, therefore, a terrain of exploitation *and* resistance (Federici 2004). In turn, their work shares the Bielefeld feminists' concern with the "subordination of unproductive work" of both women and nature (Mies and Bennholdt-Thomsen 2001 cited in Vega 2019).

We highlight three specific contributions of Latin-American authors to this debate on reproductive work. First, the incorporation of the work undertaken by non-humans (as we saw earlier). Second, the recognition of not only the home but also the community—and the interactions between them—as scenarios of reproductive work. Finally, the region's feminists place landbound material demands at the center of their political struggles and therefore of the analysis of reproductive work. While the claim that the material is political has been gaining centrality in current distributive disputes throughout the world, it has always been at the heart of struggles in the region. This approach enhances the rupture with colonial perspectives that have deployed physical proximity to land—and the dependence of livelihoods on it—as a criterion by which to view peasant movements in the South as if they were in a prior stage of struggle.

As for Marxism, various feminist movements in the region have repoliticized the reproduction of life. They argue that "the radical contradiction between life and its ever-renewing possibilities of reproduction, and capital and its violent logics of separation, exploitation and, dispossession" (Gutiérrez, Linsalata and Navarro 2017, 88–89) have also sustained the current economic system. It is not the first time that the contradiction between capitalism and reproduction has been highlighted (cf. Harvey 2014).[1] Yet, these authors point to that contradiction by counteracting the colonial gesture of certain Marxists, for whom the territorial struggle of the movements in the region is confined in time and space, and, therefore, still has to contend with capitalism.[2] Temporally, as their demands are considered pre-modern or archaic. Spatially, because they are located in areas categorized as "remote." In contrast, communitarian feminists show the synchronicity of the Indigenous people's struggles as contemporaries of their own modernity (Rivera Cusicanqui 2010), the temporal validity of reproductive work thanks to complex intergenerational links (Gutiérrez, Linsalata and Navarro 2017), and the reproductive dynamics in everyday spaces that guarantee these communities the possibility of dignified resistance.

From this twofold decolonizing approach to feminism and Marxism, these Latin-American authors are broadening the discussion of the commons by showing how various movements are *re-repoliticizing the reproduction of life*. As explained by Raquel Gutiérrez and others, organizations and communities guarantee the material and symbolic conditions of their own reproduction (Gutiérrez and López 2019). This is a form of reproductive labor whereby the common is produced by communities themselves, and which authors refer to as the *production of the common*: material and symbolic processes by which social ties and relations create concrete wealth (Gutiérrez, Linsalata & Navarro 2017). Following this analysis, three spatializing practices have been key to repoliticizing the reproduction of life.

One is to *defend communitarian autonomy vis-à-vis the state and capitalism*. A telling instance is the defense of autonomy carried out by the network of community radio stations belonging to the *Asociación de Cabildos Indígenas del Norte del Cauca*, in Colombia. Community radios here do not just fulfil a cultural or media function; they also enact a form of reproductive labor by means of which indigenous communities produce the common (create and maintain emotional ties, social relations, and local knowledge) that guarantees their autonomy in facing the violent Colombian military-industrial complex (army-sugarcane industry) in their region.

A second practice to repoliticize the reproduction of life is *to redistribute communitarian labor* to undo the naturalization of overburdening women with this kind of work. As sharply argued by Gladys Tzul-Tzul (2016), *kax kol*, or the communal work of the Chuimekená, in Guatemala, is undertaken mainly by women; it is they who, in these communities, organize the collective workforce and thus sustain the reproduction of life. To recognize the feminization of such overburdened labor can further contribute to repositioning communitarian work as a matter pertaining to everyone willing to be part of the struggle. Here it is critical to recognize that, as Gutiérrez, Linsalata, and Navarro (2017) point out, the production of the common is not an automatic process; it requires the daily and extraordinary commitment of the collective, rendering this reproductive work an eminently political practice.

Finally, *defending the possibility of political organization* is the third practice to repoliticize the reproduction of life. It entails a demand to understand how the permanent struggle of movements facing direct threats to their lives and integrity becomes part of the reproductive work of survival. Following Gutiérrez, Linsalata, and Navarro (2017), if commodification (reducing commons to goods and natural resources) separates humans from non-humans, the undermining of organizations is the equivalent process by which humans are separated from each other. Thus, the protests against the widespread assassination of leaders, for example in Mexico, Brazil, or Colombia, are not only important as condemnations of the death of particular individuals; they also entail a radical rejection of the destruction of the very possibility of political organization as a guarantee for the search for alternative futures.

The kind of practices employed for re-politicizing the reproduction of life so far approached, also decolonize the *temporality of innovative entrepreneurship*. As illustrated by discourses of sustainable developmental, innovation is often posited both as a way out of the planetary environmental crisis and as a profitable means by which to incorporate environmental sustainability.

In contrast, many Latin-American movements claim a *temporality of long-standing resistances*, which connects to the community's intergenerational work to reproduce life. This work, far from being archaic, emerges today as indispensable in dealing with the irreversibility of environmental depredation.

Recentering use value

Accumulation by use has emerged as a sophisticated form of capital accumulation through a kind of right to "absolute" use which is increasingly threatening the commons. It consists of a transformation of place to such an extent that the geographical location of the exploited site ends up being the only characteristic by which such a site can be identified following the extraction process. The result is a sort of geographical ghost, as the transformation of the material and socio-environmental relations subjected to extraction turn the respective area into a completely different place. Importantly, the prominence of rights of use is at the very least indifferent to the attributes of possession and exclusion, generally ascribable to a property title over a plot of land. Such indifference supersedes the traditional approaches to property rights to the extent that it does not emerge from the limitations to ownership, but from the prevailing autonomy of the rights of use to protect particular uses of, and activities performed upon, a given resource (Olarte-Olarte 2019). Here, the irreversible transformation of soils, landscapes, and ecosystems can derive in complex forms of what Ojeda (2016) refers as "dispossession in situ."

In recentering the use value of what is produced (over its exchange value), many movements in the region are resisting accumulation by use through *processes of collectivization of knowledge*. These processes are land-bound, transgenerationally transmitted, and go beyond private property claims. Here, we highlight three spatializing practices.

The first is to *support and disseminate alternative forms of exchange to those based on individual profit*, such as reciprocity, equity, solidarity, or mutuality—non-capitalocentric principles, according to Gibson-Graham (1996, 2006). For example, the food autonomy practices of the *Red Agroecológica de la Sabana de Bogotá* incorporate intergenerational knowledge that guarantee seed reproduction as a common. As such, it has sustained use value-based seed exchange practices to recover varieties that are threatened or that were annihilated by the expansion of monoculture.

The second practice is *to preserve forms of labor alternative to capitalist ones*. The use value activated by these alternative forms of work displaces the centrality of capitalist salaried labor—along with its exchange value and abstract wealth—aimed at large-scale production and economic growth; that is, the maximum use of space in the shortest possible time. Such is the case of transgenerationally-transmitted forms of work; for instance, the *mingas* of the Andean Indigenous world, the *tonga* of the Afro-diasporic organizations, and the peasant *mano vuelta*, among others. Payment for these forms of work takes a different form to salary, and its use value and concrete wealth have greater relevance.

Recognizing simultaneous and diverse rhythms of work is the last practice. It focuses on concrete and heterogeneous wealth production (according to its qualities) rather than abstract and homogenized profits (according to the labor time socially necessary to produce them). As a result, the knowledge of different rhythms of work undertaken by humans and non-humans, such as soil and water, emerge as central for the self-creative capacity of the communities; Escobar (2018) calls this *autonomous design*. For example, the collective work of recuperating degraded soils undertaken by the Nasa Indigenous people (in Colombia) is based on forms of cultivation that acknowledge the diverse rhythms of the soil, in opposition to the exhaustive timeframe of sugarcane monoculture. Here, the centrality of non-humans' rhythms for the political possibilities of communal organization becomes explicit.

By recentering the use value of what is produced, Latin-American movements are simultaneously decolonizing the *temporal confinement of territories* entailed by the accumulation by use. For example, while fracking reduces territorial disputes to questions of shortage—under the guidance of "good" management practices regarding the exploitation of non-renewable sources or resources (gas or oil)—various struggles mobilize an awareness of the *temporality of the irreversible*. That is, the irreversibility of socio-environmental consequences, the loss of organic and inorganic networks of relations, and the knowledges associated with their survival. This is achieved via an awareness of soil depletion that changes the spatial-temporal scale of productivity, measured by maximum extraction in the shortest time. This consideration provides an important material standard for the establishment of reparation measures: instead of focusing on the exhaustion of resources, it targets the relationships that guarantee their existence.

The potential of Latin-American movements to expand post-developmentalist agendas

This chapter identified four spatializing practices through which Latin-American movements are communalizing the conditions that sustain life. In doing so, the practices show how the movements resist the subalternization of the region; defend the relational character of territories; repoliticize the reproduction of life in those territories; and, finally, recenter the use value of what is produced therein.

These spatializing practices demand an engagement with the meaning and scope of "territory" for social struggles. This, in turn, contributes to a better understanding of the complexity of the commons. First, by approaching the territory as an integrated whole, thus undermining the commodification processes that separates humans from non-humans. Then, by abandoning an understanding of wealth as a set of fragmentable resources, to approach it instead through the human and non-human relationships that make their existence possible. Third, by acknowledging the arduous labor of reproducing life undertaken by the communities themselves. Finally, by reflecting upon the transgenerational collectivization of knowledge sustaining their struggles.

We further argued that, in deploying these spatializing practices around the communal, the region's movements are simultaneously tracing horizons to decolonize the temporalities of developmentalist progress. In particular, Latin-American movements are challenging: the stratified temporality that isolates successive and overlapping processes of violence; the alleged equilibrium between economic growth and environmental sustainability; the obsession with innovation and entrepreneurship and its corresponding compulsion for sustainability branding; and, finally, the temporal confinement derived from enclosure through accumulation by use.

To close, we offer concrete alternatives for a post-developmentalist agenda, following the path paved by Latin-American movements, including: First, reconnecting different forms of violence, including colonial violence, to their continuity in development projects. Second, showing the temporalities underlying co-dependency networks between humans and non-humans. Third, revealing the temporality of long-standing resistances during which communitarian processes have sustained the interdependencies between humans. Finally, warning against the irreversible character of spatial damage so as to open up minimal but potent possibilities for alternative futures.

Notes

1 For Harvey, a contradiction of contemporary capitalism revolves around social reproduction, understood as the progressive privatization and commodification of household tasks driven, in part, by the growing intensity of capital in domestic technologies.

2 This is clear from Harvey (2014) when he states that, although he agrees with Braudel that in the late medieval period the material reproduction of ordinary people had little to do with capital or even the market, such a formulation has no relevance for our time: "*except* in the increasingly scarce and remote areas of the world (for example, indigenous societies or very remote peasant populations) where capital does not yet exercise its dominant influence" (190–191).

References

Alimonda, Hector. 2015. "Ecología política latinoamericana y pensamiento crítico: Vanguardias arraigadas." *Desenvolvimento e Meio Ambiente* 35: 161–68.

Audiencia-JEP. 2019. Audiencia para indagar por los restos de los desaparecidos en zona de influencia de Hidroituango. Justicia Especial para la Paz October 2019.

Escobar, Arturo. 2015. "Commons in the pluriverse." In *Patterns of commoning*, edited by David Bollier and Silke Helfrich, 348–360. Amherst: Commons Strategies Group and Off the Common Press.

———. 2018. *Designs for the pluriverse: Radical interdependence, autonomy, and the making of worlds.* Durham and London: Duke University Press.

Federici, Silvia. 2004. *Calibán y la bruja. Mujeres, cuerpo y acumulación originaria.* Madrid: Editorial Traficantes de Sueños.

Flórez, Juliana. 2014. *Lecturas emergentes. El giro decolonial en los movimientos sociales.* Bogotá: PUJ Editorial.

Flórez-Flórez, Juliana and María Carolina Olarte-Olarte (2023). 'Decolonizing approaches to Latin American social movements.' In *The Oxford Handbook of Latin American Social Movements*, edited by Federico M. Rossi. Oxford: Oxford University Press. https://global.oup.com/academic/product/the-oxford-handbook-of-latin-american-social-movements-9780190870362?cc=co&lang=en&.

Gibson-Graham, J.K. 1996. *The end of capitalism (As We Knew It): A feminist critique of political economy.* Minneapolis-London: University of Minnesota Press.

———. 2006. *A poscapitalist politics.* USA: University of Minnesota Press.

Gutiérrez, Raquel, Lucia Linsalata, and Mina Navarro. 2017. "Producing the common and reproducing life: Keys Towards rethinking the political". In *Social Sciences for an Other Politics: Women Theorizing Without Parachutes*, edited by Ana Dinerstein, 79–92. Cham: Palgrave Macmillan.

Gutiérrez, Raquel y Claudia López. 2019. "Producir lo común para sostener la vida. Notas para entender el despliegue de un horizonte comunitario-popular que impugna, subvierte y desborda el capitalismo depredador". In: *¿Cómo se Sostiene la Vida en América Latina? Feminismos y re-existencias en tiempos de oscuridad*, edited by Karin Gabbert and Miriam Lang, 387–417. Quito: Fundación Rosa Luxemburg and Ediciones Abya-Yala.

Harvey, David. 2014. *Diecisiete contradicciones y el fin del capitalismo.* Quito: IAEN.

de La Cadena, Marisol. 2015. *Earth Beings. Ecologies of Practice Across Andean Worlds.* Durham and London: Duke University Press.

Lyons, Kristina. 2016. "Decomposition as life politics: Soils, *Selva* and small farmers under the Gun of the U.S.-Colombian War on Drugs." *Cultural Anthropology* 31 1: 57–81.

Mies, Maria and Veronika Bennholdt-Thomsen. 2001. "Defending, reclaiming and reinventing the commons." *Canadian Journal of Development Studies* 22 4: 997–1023, cited in Vega, Cristina. 2019. "Reproducción social y cuidados en la reinvención de lo común. Aportes conceptuales y analíticos desde los feminismos." *Revista de Estudios Sociales* 70: 49–63.

Observatorio de Ecología Política de Venezuela "Una mirada estructural del megaproyecto Arco Minero del Orinoco." Report, June 28, 2018.

Ojeda, Diana. 2016. "Los paisajes del despojo: propuestas para un análisis desde las reconfiguraciones socioespaciales." *Revista Colombiana de Antropología* 52 2: 19–43.

Olarte-Olarte, María Carolina. 2019. "From territorial peace to territorial Pacification: Anti-Riot Police Powers and Socio-Environmental Dissent in the Implementation of Colombia's Peace Agreement." *Revista de Estudios Sociales* no. 67: 26–39. https://revistas.uniandes.edu.co/doi/full/10.7440/res67.2019.03.

Ostrom, Elinor. 1990. *Governing the Commons. The Evolution of Institutions for Collective Action.* Cambridge: Cambridge University Press.

Paredes, Julieta. 2010. *Hilando Fino. Desde el feminismo comunitario*. La Paz: Comunidad de Mujeres Creando Comunidad.

Ríos vivos (web), n.d. 'Colombia: Victimas de megaproyectos y la naturaleza como víctima', available at https://riosvivoscolombia.org/wp-content/uploads/2019/05/comunicado-colombia-victimas-de-megaproyectos-y-la-naturaleza-como-victima.pdf.

Rivera Cusicanqui, Silvia. 2010. *Ch'ixinakax utxiwa. Una reflexión sobre prácticas y discursos descolonizadores*. Buenos Aires: Tinta Limón.

Salazar, Huáscar. 2015. "Entrevista a Silvia Rivera Cusicanqui. Sobre la comunidad de afinidad y otras reflexiones para hacernos y pensarnos en un mundo otro." *El Apantle. Revista de estudios comunitarios* 1: 141–168.

Svampa, Maristella. 2012. "Extractivismo neodesarrollista y movimientos sociales ¿un giro ecoterritorial hacia nuevas alternativas? Más allá del desarrollo." In *Más allá del desarrollo. Grupo permanente de trabajo sobre alternativas al desarrollo*, compiled by Miriam Lang and Dunia Mokrani, 185–218. Quito: Abya Yala y Fundación Rosa Luxemburgo.

Tzul-Tzul, Gladys. 2016. *Sistemas de gobierno comunal indígena. Mujeres y tramas de parentesco en Chuimeq'ena'*. Guatemala: SCEE, Tz'i'kin, CIPJ y Maya'Wuj Editorial.

Vega, Cristina. 2019. "Reproducción social y cuidados en la reinvención de lo común. Aportes conceptuales y analíticos desde los feminismos." *Revista de Estudios Sociales* 70: 49–63.

Villaverde, Noemí. 2017. *Una antropóloga en la luna. Las historias más sorprendentes de la especie humana*. Madrid: Anaya Multimedia.

Zent, Eglee. 2014. "Ecogonía III. Jkyo jkwainï: La filosofía de la vida de los jotï del Amazonas venezolano." *Etnoecológica* 10, no. 3: 122–150.

Zibechi, Raul and Hardt, Michael. 2013. *Preservar y compartir: bienes comunes y movimientos sociales*. Buenos Aires: Mardulce.

34 Environmental thought in movement

Territory, ecologisms, and liberation in Latin America

Melissa Moreano Venegas, Diana Carolina Murillo Martín,
Nadia Romero Salgado, Karolien van Teijlingen,
Iñigo Arrazola Aranzabal, Manuel Bayón Jiménez, Angus Lyall,
and Diana Vela-Almeida

Introduction

In this chapter we provide a brief survey of Latin American ecological thought, highlighting theoretical contributions and emancipatory practices among Indigenous, ecologist, and feminist movements. We speak of Latin American ecological thought, distancing ourselves from hegemonic *environmentalism*, which tends to fixate on quixotic, technological, and market fixes to our planetary crisis (Latorre-Tomás 2009). To some extent, each of the traditions described in this chapter combines the struggle against ecological destruction with the struggle against social oppression, in its particular colonial formations in contemporary Latin America (in particular, we emphasize experiences in Ecuador). In fact, each of these critical traditions locates the roots of ecological crisis in European conquest, insofar as colonialism shaped modern Eurocentric subjectivities and material practices; the origins and global expansion of capitalism; and the international division of labor, coupled with racialized, gendered, and class hierarchies of domination (Dussel 1996; Quijano 2014). Colonial forms of spatial segregation (Fanon 2007; Mariátegui 2007), the organization of national economies for exporting natural resources (Furtado 2001), and steep racial and cultural hierarchies (Casanova 1963; Stavenhagen 1968) have persisted long after the formal independence of Latin American nations. These legacies are clearly imprinted in Latin American societies and landscapes today, and give form to Latin American resistance movements and traditions of critical thought, such as liberation theology, dependency theory, popular pedagogy, and, more recently, decolonial theory.

Ecological thought, in turn, has developed in parallel with these currents (Moreano, Molina, and Bryant 2017). In this chapter, we discuss liberation theology, which has theorized the oppression of both nature and the poor since the 1960s. Secondly, we characterize the conceptual and political shift that took place in the 1980s, as peasant demands for "land" as a means of production transformed into demands for "territory" among Indigenous and Black movements. We then characterize the diversity of popular ecologist movements that have articulated their own pursuit of social and ecological justice. Finally, we describe the thought and praxis of Latin American eco-feminism, which signals connections between the oppression of women and nature by patriarchy, capitalism, and racism. In the face of the region's history of colonial subordination, social movements from among popular, peasant, feminist, Indigenous, and Afrodescendant groups have developed important traditions of critical thought that are fundamental for political action in the region and beyond. Thus, we highlight the role of Latin American social movements as producers of theories and practices that challenge ecological destruction on a planetary scale, along with the Eurocentric world order that propels it.

DOI: 10.4324/9780429344428-40

Eco-theology of liberation

Liberation theology is a tradition that critiques capitalism through a revised or renewed vision of religious faith and the mission of the Catholic Church (Mainwaring 1986). It questions the notion that the Church is removed from social reality, insofar as it has been complicit with exploitation and the reproduction of capital. In turn, liberation theology adheres to dialectical materialism as a method for understanding reality and transforming it—particularly within Marxist branches of the tradition—and promotes popular education as a method of raising social and political consciousness (Mena et al. 2007).

Contributions of liberation theology to Latin American ecological thought revolve around the further observation that both the poor and the Earth are oppressed by capitalism (Boff 1996). Eco-theology proposes to build a relationship with the Earth in an emancipatory praxis, whereby the poor are not subjects of charity but rather revolutionary subjects that make manifest the Kingdom of God on Earth (Dussel 1994). Eco-theology calls for the restoration of the sacredness of the Earth and the bonds of belonging within a community of life (Boff 1985).

In the case of Brazil, for example, upheaval related to the assassinations of trade unionists and community leaders in the second half of the 20th century, as well as forced displacements and slave-like labor conditions, led to the formation of the Indigenous Missionary Council (CIMI) and the Pastoral Land Commission (CPT) in the 1970s (Stédile and Mançano 1999). The CPT worked to strengthen communities and defend the sacredness of nature; it supported peasant productive autonomy and livelihoods that were harmonious with nature; and it incorporated the celebration of popular religiosity in their work. The Landless Rural Workers' Movement (MST), created from the CPT in 1984, continues this work, as it defends agroecology as an alternative to agribusiness.

In the 1970s and 1980s, the Movement of Rubber Tappers, or *Seringueiros*, created by Chico Mendes, began promoting the conservation of the Amazon and the practice of extracting rubber without killing trees, as opposed to the destructive practices of large companies and large landowners. These sustainable rubber tappers chained themselves to trees, linking the defense of their own livelihoods with the defense of nature. Shortly after the assassination of Mendes in 1988, the first extractivist reserves were created, in which peasant and Indigenous collective property were conceived as territorial conservation units (Porto-Gonçalves 2016). The *Seringueiros* thus argued that extractivism is not negative unless and until it becomes incorporated into capital accumulation (Galafassi and Riffo 2018). In 2009, the movement founded the National Council of Extractivist Populations.

In Ecuador, liberation theology is associated with Monsignor Leonidas Proaño and the work of the Church of the Poor. It was also key to the formation of some of the strongest Indigenous organizations in the country, such as the Confederation of Peoples of the Kichwa Nationality of Ecuador (ECUARUNARI) in 1972 and the Confederation of Indigenous Nationalities of Ecuador (CONAIE) in 1986. The convergence of liberation theology and ecologisms arose from support for the defense of the rights of Indigenous and peasant communities to "vital territory" (Proaño 1986, 80). This convergence would have the effect of drawing liberation theology into resistance to oil and industrial mineral extraction and large-scale agriculture. In turn, liberation theology influenced the formation of several grassroots ecological organizations that, similar to the *Seringueiros*, have promoted the *sustainable use* of ecosystems, such as the mangrove forests (Gerber and Veuthey 2010). In the area of Intag, located in the northern Andes, liberation theology has played an important role in a peasant movement critical of both social marginalization and large-scale mining (Kuecker 2007). Finally, liberation theology has also inspired spaces in which urban activists have been trained—for example, in the most emblematic ecologist organization in Ecuador, *Ecological Action*. Each of these examples reflects a commitment to combating social domination and ecological destruction, as a unified, moral project.

The territorial turn, autonomies, and nature

As we mentioned in the introduction, colonialism in Latin America established ethnic and racial classifications, which sustained exploitation, plunder, and the unequal distribution of power. Indigenous and Black peoples and movements in the region, historically subordinated within this structure, have confronted it through discourses and practices of open rebellion, such as that of Tupac Amaru in Peru (Mires 1988) or the Haitian revolution (James 2001); the formation of Indigenous and fugitive Black communities throughout the Americas (Bledsoe 2017); and everyday forms of resistance and cultural production (Muratorio 1991).

In the mid-20th century, this resistance took a particular form as peasant movements fought for land redistribution. Subsequent land reforms in the 1960s and 1970s were limited (Kay 2001) and, in the 1980s and 1990s, neoliberalism facilitated the reconcentration of land among modernizing elites; favored transnational capital; and weakened unions, provoking the beginning of another cycle of resistance. In light of the 500th anniversary of the European conquest, explicitly *ethnic* regional and national movements began to (re)consolidate. Thus, the Central of Indigenous Peoples of Beni (CPIB) in Bolivia, the Coordinator of Indigenous Organizations of the Brazilian Amazon (COIAB), the National Indigenous Organization of Colombia (ONIC), the Confederation of Indigenous Nationalities of Ecuador (CONAIE), and the National Indigenous Congress (CNI) in Mexico, among many others, re-focused their demands on cultural respect and an expanded and decolonized conception of land as *territory* (Porto-Gonçalves 2009).

This notion of territory is widely conceptualized as a material space for social reproduction and as a social space for autonomous political and cultural creation (Moreano and Vela-Almeida 2019; Machado-Aráoz 2012; Escobar 2014). This conceptual turn has aimed to decolonize or recover territory from its entanglement with the nation-state (Porto-Gonçalves 2006). The demand for formal recognition of ancestral Indigenous territories in the hemisphere is a struggle against colonial oppression, ethnic discrimination, and the commodification of land and nature. Emblematic Black and ethnic movements, such as the Process of Black Communities (PCN) in Colombia, the CONAIE in Ecuador, the Zapatistas in Mexico, the CIDOB in Bolivia, and the Mapuche people in Chile, have driven this "territorial turn" (Offen 2003) in their relationship with the state. As a result, states have been forced to recognize certain territorial autonomy rights. In Ecuador and Bolivia, ethnic movements have succeeded in forcing the state to recognize the plurinational character of the nation in their constitutions.

Indigenous movements frequently claim their stewardship over territory in terms of human responsibilities toward non-human beings. For example, in the 1990 "Agreement on the territorial rights of the Quichua, Shiwiar and Achuar peoples in Ecuador," the Indigenous movement claimed to "speak in the name of all the lives of the forest," the beings of the waters, fertility, cultivation, harvest, and medicine—that is, the "gods that maintain life" on the land where their ancestors lived (Larco and Espinosa 2012, 208). Thus, the reference to non-Western conceptions of nature in the struggle for territory constitutes a common strategy in the region (Escobar 2014).

We conclude by highlighting the example of the Zapatistas. The particular importance for ecological thought of the Zapatista struggle for autonomy in Chiapas lies in its attempts to transform its own socio-political and territorial relations through an anti-hierarchical praxis that includes nature. While state sovereignty is a form of hierarchical rule—that is, the domination of humans over other humans and nature, the Zapatistas explicitly seek to not reproduce sovereignty ("power over"), but rather cultivate decision-making from below ("power to") (Holloway 2002). Through their revolutionary schools, institutional dispersion, and rotation of representatives, they have constituted a process of political participation and an ethic called "commanding by obeying." In this context, the Zapatistas have theorized and defended territory, not only as a resource but also as a space to disarticulate Western hierarchies, between humans and with respect to nature.

Popular ecologisms

Popular ecologies understand ecological problems as interconnected with capitalism, inequality, and social injustice. They have emerged from struggles against extractive projects and the implementation of neoliberal policies in the region since the 1980s, articulating impoverished sectors, non-governmental organizations (NGOs), and intellectuals from the global North and South. These types of struggles have been conceptualized as "ecologism of the poor" or "popular ecologism" (Martínez-Alier 2009). Popular ecologists defend and promote *buen vivir* or "good living" as an alternative to development; the defense of the "Pachamama"; and the rights of nature or Mother Earth, stimulating debates around the world over the meaning of development, well-being, and the relationship between humans and nature.

Pachamama, as well as Mapu (a Mapuche concept) and *kawsak sacha* (living forest, for the Sarayaku people in Ecuador; see Gualinga 2019), refer to the interconnection between human beings and non-human entities. Pachamama and Mapu denote "a whole that goes beyond visible nature, that goes beyond the planets, that contains life, the relationships established between living beings, their energies, their needs and their desires" (*Feminismo comunitario* 2010), thus questioning modern dualistic, capitalist, and anthropocentric imaginaries that would separate nature and society.

Demands regarding the rights of nature suggest that nature is a *subject*. Since the formal recognition of these rights in the 2008 constitution of Ecuador, social movements and activists around the world have debated their meaning, scope, and limits (Cullinan 2019), and they have been used to challenge extractive projects in Ecuador (Moreano 2017) and in the legal defense of the Whanganui and Te Urewera rivers in New Zealand; the Atrato, Magdalena, Cauca, Amazon, Coello, Combeima, and Cocora tributaries in Colombia; and all rivers in Bangladesh. Furthermore, rights of nature have been recognized by Mexico City, the Brazilian municipality of Paudalho, and various states and municipalities of the United States (CELDF 2019).

Finally, "good living" (*sumak kawsay*) or "living well" (*suma qamaña*) has been defined as a way of living in harmony with the human and non-human environment (Hidalgo-Capitán and Cubillo-Guevara 2017). It is a concept that is not born exclusively from indigenous thought, although this has been essential in its construction (Chuji, Rengifo, and Gudynas 2019). Hidalgo-Capitán and Cubillo-Guevara (2014, 2017) identify three currents of *buen vivir* thinking: the indigenist, which emphasizes the self-determination of peoples and an Andean cosmovision; the statist current that focuses on social equity; and the ecologist current, in which it is conceived as an alternative to capitalist development.

These concepts have emerged and circulated in Latin America through exchanges between marginalized populations, NGOs, and academics. These articulations have also shaped shared political praxis, for example, through the sharing of strategies to negotiate with and halt extractivist industries or to draw attention to socio-environmental conflicts (Bebbington 2007). Articulations have also generated South–South ecologist networks, which have been essential in promoting moratoriums on the expansion of extractive projects. Oilwatch, for example, is a network of "resistance to oil activities in tropical countries" that has promoted leaving oil underground. Since the 1990s, this proposal has spread from Ecuador through Nigeria to Costa Rica to Europe and the United States (Martínez 2000). In Ecuador, it materialized in 2007 in the Yasuní–ITT Initiative to leave oil under the Yasuní National Park (Acosta and Martínez 2010), although the government suspended the initiative in 2013. Other organizations that have promoted moratoriums are the Latin American Network for the Fight against Dams and in Defense of Rivers, their Communities, and Water (REDLAR); the Observatory of Mining Conflicts in Latin America (OCMAL); the Latin American Alliance against Fracking; the Latin American Network of Women Defenders of Social and Environmental Rights; and the Network for a Transgenic-Free Latin America (RALLT).

Latin American eco-feminisms

Ecofeminism in Latin America originated as a concept and praxis in the 1980s. Initially, distinct strands of eco-feminism were influenced by ecofeminist theology which called for an intimate relationship with God in order to recover our relationships with non-human nature (Gebara 2000). Subsequently, women's movements against extractivism and in defense of 'body-territories' have become influential (LaDanta LasCanta 2017). They are characterized by their opposition to the domination of women and nature, with their origins in a patriarchal system, and their attention to the body, care for life, and territory (Svampa 2015). Here, we highlight ecofeminisms that have become iconic in the region.

Amazonian women's movements have emerged in recent years with an anti-extractive message, drawing attention to how the destruction of bioregions is linked to violence against their bodies. In Ecuador, Amazonian women have been organizing since 2013, demanding the halt of mining and oil concessions and respect for the self-determination of Indigenous peoples. In Brazil, thousands of Indigenous women from 130 peoples gathered in 2019, denouncing extractivist policies with the slogan "Territory: our body, our spirit." In Peru, in 2017, the Tribunal of Justice and Defense of the Rights of Pan-Amazonian and Andean Women took place, putting forth demands regarding climate change, violence against women due to the destruction of the Amazon, and gender inequality in climate change mitigation and adaptation programs.

The anti-patriarchal community feminism of Bolivia and the Iximulew-Guatemala Network of Women Healers represent a process of decolonization of white-bourgeois feminism. These community feminist networks fight for the depatriarchalization of social relations from the position of Indigenous plurality and communal life, reconceiving their relationship with nature as one of interdependence and recognizing themselves as part of a unitary "body-territory." Communitarian feminism questions the folkloric and reductive vision of Indigenous peoples, which assumes Indigenous women as defenders of nature without understanding the structures of domination to which they are subject (Guzmán 2015). In turn, for the Network of Women Healers, the depatriarchalizing and decolonization process extends to the emotional and spiritual recovery of women and the defense of body-territories in their diverse forms (Cabnal 2017).

The Central American ecologist struggle was made widely visible by Berta Cáceres and the Civic Council of Popular Indigenous Organizations (COPINH). This Indigenous leader fought for the recovery of Lenca territory, opposing a hydroelectric dam that violated territorial autonomy and access to water. This struggle reflected a particular notion of body-territory. According to Berta, "in our cosmovisions we are beings emerged from the earth, water and corn" and in the rivers reside the spirits of girls, which is why women are their main guardians (Curiel 2019). In turn, her struggle was explicitly directed toward patriarchal power structures.

She advocated for the end of femicides, founded a shelter for women, and strengthened female leadership by establishing the COPINH Women's Assembly (García Rojo 2018). After her assassination in 2016, Berta became a symbol of women's struggles for nature and against the violence faced by female defenders of nature (Curiel 2019). Her image now appears in protests and marches across the continent (see Méndez 2018; Lakhani 2020).

Zapatista women have also greatly impacted ecofeminist thought in Latin America since the publication of the "Revolutionary Laws of Women" in 1994 and 1996, in which they advocate for the rights of Indigenous women in the face of the patriarchal state (Padierna Jiménez 2013). Their struggle explicitly links the recognition of Indigenous women's rights to territorial and cultural autonomy and to ecologist demands (Comandanta Esther 2001; Marcos 2011). They have forced the restructuring of the Zapatista movement, as they have become commanders, spokeswomen, and members of the "Good Government Councils." Since 2018, they have organized the "International Meeting of Women in Struggle" to share experiences of struggle against the multiple forms of violence they face.

Conclusions

Latin America has rich critical-theoretical traditions that have nurtured ecological thinking. These critical traditions have been shaped by the legacies of European colonization; the realities of a capitalist world system geared toward depleting natural resources; and widespread cruelty towards women and other feminized and racialized bodies. Here, we have described Latin American social movements as loci of knowledge production and political action regarding alternative forms of social existence as well as coexistence with the non-human world. Their production of novel concepts such as "body-territory" in the midst of their struggles reflects the fact that Latin American ecological thought is produced in and by social organizations and articulations in movement. Collective action-thought challenges common academic notions of a division between social movements as objects of study and as producers of knowledge and theory.

The ideas collected in this chapter have inspired, and will continue to inspire. action elsewhere in the world. Proposals such as Pachamama, *buen vivir*, and body-territory articulate political imaginaries and contribute to global critical thinking that questions capitalist development and divisions between nature and society. They offer new horizons of possibility for coexistence.

Finally, we offer two brief reflections for moving forward. First, it is perhaps more important than ever to reinvigorate liberation theology in Latin American ecological thought in order to counterbalance the contemporary dominance of conservative branches of Church in society and politics. Second, we believe that theories of decoloniality must distance themselves from essentialized notions of social movements and, instead, study reality as it is in order to transform it. Here, Latin American ecofeminisms have much to say, as they have challenged the tendencies of certain ecological traditions to essentialize their relationship with nature and generated new ways of doing politics within movements and communities.

To learn more about key movements and individuals in Latin American ecologist thinking see the following sources:

Acción Ecológica, Ecuador:
https://www.accionecologica.org

Bertha Cáceres and el Consejo Cívico de Organizaciones Indígenas Populares (COPINH), Honduras:
https://copinh.org/

Central de Pueblos Indígenas del Beni (CPIB), Bolivia:
www.amazonia.bo

Chico Mendes and el Conselho Nacional das Populações Extrativistas, Brasil:
http://www.memorialchicomendes.org/quem-somos/

Comissão Pastoral da Terra (CPT), Brasil:
https://www.cptnacional.org.br

Congreso Nacional Indígena (CNI) en México:
www.congresonacionalindigena.org

Confederación de Nacionalidades Indígenas del Ecuador (CONAIE):
https://conaie.org

Confederación de Pueblos Indígenas de Bolivia (CIDOB):
www.cidob.org

Conselho Indigenista Missionário (CIMI), Brasil:
https://cimi.org.br

Coordenação das Organizações Indígenas da Amazônia Brasileira (COIAB), Brasil:
www.coiab.org.br

Feminismo Comunitario Antipatriarcal, Bolivia:
Facebook: @feminismo.comunitario.Antipatriarcal

Movimento dos Trabalhadores Rurais Sem Terra (MST), Brasil:
https://mst.org.br

Mujeres Zapatistas, México:
http://enlacezapatista.ezln.org.mx/2019/12/31/palabras-de-las-mujeres-zapatistas-en-la-clausura-del-segundo-encuentro-internacional-de-mujeres-que-luchan/

Organización Nacional Indígena de Colombia (ONIC):
www.onic.org.co

Proceso de Comunidades Negras (PCN), Colombia:
www.renacientes.net

TZ'KAT Red de Sanadoras Ancestrales del Feminismo Comunitario Territorial desde Iximulew, Guatemala:
https://pbi-guatemala.org/es/qui%C3%A9n-acompa%C3%B1amos/tzkat-red-de-sanadoras-ancestrales-del-feminismo-comunitario-territorial-desde

Regional networks:

Oilwatch - Network of resistance to oil activities in tropical countries
https://www.oilwatch.org/

Observatorio de Conflictos Mineros de América Latina (OCMAL):
https://www.ocmal.org/

Red Latinoamericana contra represas (REDLAR):
http://www.redlar.org

Red Por una América Latina Libre de Transgénicos (RALLT):
http://www.rallt.org/

References

Acosta, Alberto, and Esperanza Martínez. 2010. *ITT-Yasuní: Entre El Petróleo y La Vida*. 1st ed. Quito: Abya Yala and Universidad Politécnica Salesiana.

Bebbington, Anthony. 2007. *Minería, Movimientos Sociales y Respuestas Campesinas: Una Ecología Política de Transformaciones Territoriales*. Vol. 2. Lima: Instituto de Estudios peruanos.

Bledsoe, Adam. 2017. "Marronage as a past and present geography in the Americas." *Southeastern Geographer* 57(1): 30–50.

Boff, Leonardo. 1985. *Jesucristo El Liberador. Ensayo de cristología crítica para nuestro tiempo*. Santander: Sal Terrae.

———. 1996. *Ecología: grito de la Tierra, grito de los pobres*. Madrid: Trotta.

Cabnal, Lorena. 2017. "Tzk'at, Red de Sanadoras Ancestrales del Feminismo Comunitario desde Iximulew-Guatemala". *Ecología Política*, 98–102.

Casanova, Pablo González. 1963. "Sociedad plural, colonialismo interno y desarrollo." *América Latina* 6(3): 15–32.

CELDF. 2019. "Rights of Nature Timeline | CELDF | Protecting Nature and Communities." Advancing Legal Rights of Nature: Timeline. 2019. https://celdf.org/advancing-community-rights/rights-of-nature/rights-nature-timeline/.

Chuji, Mónica, Grimaldo Rengifo, and Eduardo Gudynas. 2019. 'Buen Vivir'. In *Pluriverse: A Post-Development Dictionary*, edited by Ashish Kothari, Ariel Salleh, Arturo Escobar, Federico Demaria, and Alberto Acosta, 111–14. New Delhi: Tulika Books and Authorsupfront.

Comandanta Esther. 2001. "Discurso de la Comandanta Esther en la tribuna del Congreso de la Unión." https://enlacezapatista.ezln.org.mx/2001/03/28/discurso-de-la-comandanta-esther-en-la-tribuna-del-congreso-de-la-union/.

Cullinan, Cormac. 2019. "Nature Rights." In *Pluriverse: A Post-Development Dictionary*, edited by Ashish Kothari, Ariel Salleh, Arturo Escobar, Federico Demaria, and Alberto Acosta, 243–46. New Delhi: Tulika Books and Authorsupfront.

Curiel, Ochy. 2019. "Berta Cáceres y El Feminismo Decolonial." *LASA Forum* 50 (4): 64–69.

Dussel, Enrique. 1994. "Teología de la liberación y marxismo." In *Mysterium liberationis: conceptos fundamentales de la teología de la liberación*, edited by Ignacio Ellacuria and Jon Sobrino, Vol. 1, 115–144. Madrid: Trotta.

———. 1996. *Filosofía de la liberación*. Bogotá: Nueva América.

Escobar, Arturo. 2014. *Territorios de diferencia: lugar, movimientos, vida, redes*. Popayán: Editorial Universidad del Cauca.

Fanon, Frantz. 2007. *The Wretched of the Earth*. Translated by Richard Philcox. New York: Grove/Atlantic, Inc.

Feminismo comunitario. 2010. "Pronunciamiento del Feminismo Comunitario latinoamericano en la Conferencia de los pueblos sobre Cambio Climático." Biodiversidad en América Latina. May 6, 2010. http://www.biodiversidadla.org/Documentos/Pronunciamiento_del_Feminismo_Comunitario_latinoamericano_en_la_Conferencia_de_los_pueblos_sobre_Cambio_Climatico.

Furtado, Celso. 2001. *La economía latinoamericana: formación histórica y problemas contemporáneos*. México, DF: Siglo XXI editores.

Galafassi, Guido Pascual, and Lorena Natalia Riffo. 2018. "Una Lectura Crítica Sobre El Concepto de 'Extractivismo' En El Marco de Los Procesos de Acumulación." *Trama, Revista de Ciencias Sociales y Humanidades*. 7 (2): 108–17. https://doi.org/10.18845/tramarcsh.v7i2.3939.

García Rojo, Laura. 2018. *El papel de liderazgo de las mujeres indígenas defensoras de derechos humanos en los conflictos ecoterritoriales relacionados con el agua en Honduras. Estudio a través del Caso de Berta Cáceres*. Valladolid: Universidad de Valladolid. http://uvadoc.uva.es/bitstream/handle/10324/34101/TFM_F_2018_67.pdf?sequence=1&isAllowed=y.

Gebara, Ivone. 2000. *Intuiciones ecofeministas. Ensayo para repensar el conocimiento y la religión*. Madrid: Editorial Trotta.

Gerber, Julien-François, and Sandra Veuthey. 2010. "Plantations, resistance and the greening of the Agrarian Question in Coastal Ecuador: The greening of the Agrarian Question in Coastal Ecuador." *Journal of Agrarian Change* 10 (4): 455–81. https://doi.org/10.1111/j.1471-0366.2010.00265.x.

Gualinga, Patricia. 2019. "Kawsak Sacha." In *Pluriverse: A Post-development Dictionary*, edited by Ashish Kothari, Ariel Salleh, Arturo Escobar, Federico Demaria, and Alberto Acosta, 223–26. New Delhi: Tulika Books and Authorsupfront.

Guzmán, Adriana. 2015. "Feminismo Comunitario-Bolivia. Un Feminismo Útil Para La Lucha de Los Pueblos." *Revista Con La A*, March 2015. https://conlaa.com/feminismo-comunitario-bolivia-feminismo-util-para-la-lucha-de-los pueblos/?output=pdf.

Hidalgo-Capitán, Antonio Luis, and Ana Patricia Cubillo-Guevara. 2014. "Seis Debates Abiertos Sobre El Sumak Kawsay." *Íconos: Revista de Ciencias Sociales* 48: 25–40.

———. 2017. "Deconstruction and Genealogy of Latin American Good Living (Buen Vivir). The (Triune) Good Living and Its Diverse Intellectual Wellsprings." *International Development Policy* 9 (9): 23–50.

Holloway, John. 2002. "*¿Es La Lucha Zapatista Una Lucha Anticapitalista? (Pregunta Dirigida al Autor Por Los Editores de La Publicación)*." *Revista Rebeldía*, 2002.

James, Cyril Lionel Robert 2001. *The Black Jacobins: Toussaint L'Ouverture and the San Domingo Revolution*. London and New York: Penguin Books.

Kay, Cristobal. 2001. Estructura agraria, conflicto y violencia en la sociedad rural de América Latina. *Revista mexicana de sociología* 63 (4): 159–95.

Kuecker, Glen David. 2007. "Fighting for the forests: Grassroots resistance to mining in Northern Ecuador." *Latin American Perspectives* 34 (2): 94–107. https://doi.org/10.1177/0094582X06299081.

Lakhani, Nina. 2020. *Who Killed Berta Caceres?: Dams, Death Squads, and an Indigenous Defender's Battle for the Planet*. London and New York: Verso Books.

Larco, Carolina, and León Espinosa, eds. 2012. *El Pensamiento Político de Los Movimientos Sociales*. Pensamiento Político Ecuatoriano. Quito: Ministerio de Coordinación de la Política y Gobiernos Autónomos Descentralizados.

LasCanta, LaDanta. 2017. "De la teología al antiextractivismo: ecofeminismos en Abya Yala." *Ecología Política* 54: 35–41.

Latorre-Tomás, Sara. 2009. *El ecologismo popular en el Ecuador: pasado y presente*. Quito: Instituto de Estudios Ecuatorianos, Facultad Latinoamericana de Ciencias Sociales. https://www.researchgate.net/profile/Sara-Latorre/publication/265670619_El_ecologismo_popular_en_el_Ecuador_pasado_y_presente/links/55158b730cf2b5d6a0ea0321/El-ecologismo-popular-en-el-Ecuador-pasado-y-presente.pdf.

Machado-Aráoz, Horacio. 2012. "Los dolores de Nuestra América y la condición neocolonial. Extractivismo y biopolítica de la expropiación." *OSAL* 13 (32): 51–66.

Mainwaring, Scott. 1986. *The Catholic Church and Politics in Brazil, 1916-1985*. Stanford, CA: Stanford University Press.

Marcos, Sylvia. 2011. *Mujeres, Indígenas, Rebeldes, Zapatistas*. 1st ed. México, DF: Eón.

Mariátegui, José Carlos. 2007. *7 Ensayos de Interpretación de La Realidad Peruana*. 3rd ed. Colección Clásica 69. Caracas: República Bolivariana de Venezuela, Fundación Biblioteca Ayacucho.

Martínez, Esperanza. 2000. "Moratoria a la actividad del petróleo." In *El Ecuador post petrolero*, edited by Alberto Acosta, 1st ed. Quito, Ecuador: Acción Ecológica.

Martínez-Alier, Joan. 2009. *El ecologismo de los pobres: conflictos ambientales y lenguajes de valoración*. 3rd ed. Antrazyt 207. Barcelona: Icaria.

Mena, Marcelo, Soledad Chalco, Maritza Idrobo, and Jacqueline Artieda. 2007. *Aportes al debate sobre el socialismo del siglo XXI: el pensamiento de Monseñor Leonidas Proaño y Fernando Velasco*. Quito: Centro de Investigaciones CIUDAD.

Méndez, María José. 2018. "'The River Told Me': Rethinking Intersectionality from the World of Berta Cáceres'. *Capitalism Nature Socialism* 29 (1): 7–24.

Mires, Fernando. 1988. *La Rebelión Permanente: Las Revoluciones Sociales En América Latina*. 1a ed. Historia. Delegación Coyoacán, México, DF: Siglo Veintiuno Editores.

Moreano, Melissa. 2017. "The Political Ecology of Ecuadorian Environmentalism: Buen Vivir, Nature and Territory." PhD Thesis, London: King's College London.

Moreano, Melissa, Francisco Molina y Raymond Bryant 2017. "Hacia una ecología política global: Aportes desde el Sur". In *Ecología política latinoamericana. Pensamiento crítico, diferencia latinoamericana y rearticulación epistémica*, edited by Héctor Alimonda, Catalina Toro Pérez y Facundo Martin, Vol. 1,

197–212. Buenos Aires: Grupo de Trabajo en ecología política de CLACSO y Universidad Autónoma Metropolitana de Buenos Aires.

Moreano, Melissa y Diana Vela-Almeida. 2019. "El *lugar* de la ecología política dentro de la geografía latinoamericana: el caso de CLAG." *The Journal of Latin American Geography* 19 (1): 74–83. https://doi.org/10.1353/lag.2020.0009.

Muratorio, Blanca. 1991. *The Life and Times of Grandfather Alonso, Culture and History in the Upper Amazon.* New Jersey: Rutgers University Press.

Offen, Karl H. 2003. "The territorial turn: Making black territories in Pacific Colombia." *Journal of Latin American Geography* 2 (1): 43–73.

Padierna Jiménez, María del Pilar. 2013. "Mujeres Zapatistas: la inclusión de las demandas de género." *Argumentos* 26 (73): 133–42.

Porto-Gonçalves, Carlos Walter. 2006. *A reinvencao dos territórios: a experiencia latino-americana e caribenha. Los desafíos de las emancipaciones en un contexto militarizado.* Buenos Aires: CLACSO.

———. 2009. "Del desarrollo a la autonomía: la reinvención de los territorios." *Revista América Latina en Movimiento* 445: 10–13.

———. 2016. *O Difícil Espelho: a originalidade teórico-política do movimento dos seringueiros e a "confluência perversa" no campo ambiental no Acre en Despojos y resistencias en América Latina, Abya Yala.* Porto-Gonçalves, Carlos Walter y Hocsman, Luis Daniel. Buenos Aires: Estudios Sociológicos Editora.

Proaño, Leonidas. 1986. *Plan Nacional de la Pastoral Indígena en Cristianos Hoy: Testimonios de liberación.* Quito: Fraternidad.

Quijano, Aníbal. 2014. *Cuestiones y Horizontes: De La Dependencia Histórico-Estructural a La Colonialidad/Descolonialidad Del Poder: Antología Esencial.* 1st ed. Colección Antologías. Buenos Aires: CLACSO.

Stavenhagen, Rodolfo. (1968). "Clases, colonialismo y aculturación. Ensayo sobre un sistema de relaciones interétnicas en mesoamérica." In *Ensayos sobre las clases sociales en México*, 109–71. Mexico: Editorial Nuestro Tiempo.

Svampa, Maristella. 2015. "Feminismos del Sur y ecofeminismos." *Nueva Sociedad* 256: 127–31.

35 Agroecology and food sovereignty in the Caribbean

Insights from Cuba, Puerto Rico, and Sint Maarten

Georges F. Félix

Introduction

At a global scale, family farms produce close to 80% of the food on our plates but hold control of barely 20% of agricultural lands (Graeub et al. 2016). A growing global population demands more and more nutritious foods. To meet this demand farmers increasingly rely on external inputs, such as genetically-modified seed varieties, herbicides, fertilizers, and pesticides. This has led to environmental degradation and has undermined human and ecosystem wellbeing. Many farmers across the world—and particularly in Latin America and the Caribbean (LAC)—implement environment-friendly techniques to produce healthy foods without posing risks either to the environment or to human health. Farmer struggles and existing alternatives to extractive food production schemes have highlighted the need for political recognition (McCune et al. 2017), for the re-appropriation of local resources (Timmermann and Félix 2015), and for collective or communal governance of the environment as represented by the many pockets of resistance across the Americas (Arnold, Díaz, and Algoed 2020). The farming principles and practices of "agroecology" may be a vehicle to work toward food sovereignty, technological self-sufficiency, and social justice.

Availability, access, stability, and utilization of food products in food chains define "food security." Yet the concept of "food sovereignty" goes beyond this to ensure that local farming communities can determine and control their food systems. Despite their rich history, the struggles for agroecology and food sovereignty led by Caribbean farmers and movements on island territories have seldom been documented. Therefore, the objective of this chapter is to contextualize Caribbean food system characteristics from an agroecology and food sovereignty perspective, through the compilation of existing literature, open-access databases, and public information to highlight historical trends, similarities, and contrasts between three island territories: Cuba, Puerto Rico, and Sint Maarten.

Caribbean food systems in context

In the insular Caribbean, Cuba, Puerto Rico, and Sint Maarten showcase contrasting food systems despite their climatic similarities. However, it can be noted that each Caribbean island has a very particular history and trajectory. In the case of Cuba and Puerto Rico, their history from 1493 through 1898 was relatively similar, as both were Spanish colonies until 1898 when the United States gained control over the two islands in the aftermath of the Spanish–American War. After that date, Cuba quickly became independent (in 1902) while Puerto Rico remained a US colony until present-day (even though its official status changed to "Commonwealth" in 1952, loosely granting some autonomy to this territory). The case of Sint Maarten is quite unique within this geographical context. The northern part of the island is a French municipality, fully

DOI: 10.4324/9780429344428-41

incorporated in the parliamentary processes of metropolitan France, while the southern part was a Dutch colony, nowadays an autonomous nation within the Dutch Kingdom.

Historical disadvantages

Food provision and agricultural practices are central to the understanding of environmental and political struggles, agrarian reforms of the 1960s, and the consequent expansion of global agro-industrial projects in LAC (Altieri and Toledo 2011). Indeed, exogenous factors have contributed to poverty and land degradation throughout the region (Chappell et al. 2013), especially because colonialism is a common denominator amongst LAC countries. Since 1492, foreign settlers from Spain, France, England, The Netherlands, Portugal, and, more recently, the United States have thrived by extracting resources from LAC, regardless of the already-existing local political, economic and organizational features of native societies (Timmermann 2019).

Overall, the combined services and industry sectors contribute most to the economic development and employment opportunities of all three countries. Food self-sufficiency is indeed low, given the share that agriculture has in each country's economy: Cuba's farming represents a 5% share, Puerto Rico's farming represents a 0.8%, and Sint Maarten's farming sector barely accounts for 0.1% of economic activities (UNdata 2022). Agricultural lands in Cuba have increased since the 1960s and have reached a relatively stable plateau at 60% of the land, with 30% classified as forest cover in 2020 (Figure 35.1). Inversely, Puerto Rico experienced major declines in the agricultural land share, dropping from 70% in 1960 to a low 20% in 2020 (Figure 35.1). Because of agricultural abandonment, forest cover in Puerto Rico rapidly increased from less than 10% at the beginning of the century to a surprising 60% relative land share at present, making it a perfect case study for the desirable, yet unplanned, ecological transition theory (Parés-Ramos, Gould, and Aide 2008).

Sunny side up?

The Green Revolution introduced techniques which heavily rely on 'recipes' and external, often toxic inputs. Across the continent, the Green Revolution, promoted by the actions of multinationals, has displaced rural communities from their territories, simultaneously triggering

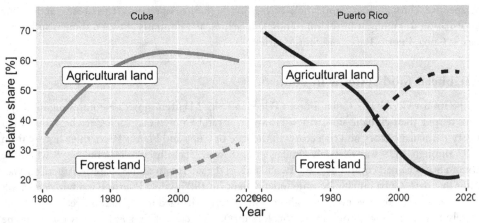

Figure 35.1 Share in land area, Cuba and Puerto Rico, 1960–2019. www.faostat.org.

adverse health effects for humans and ecosystems through the extensive use of agro-chemicals, the push for intensive mechanization, and the over-simplification of farming system processes, amongst other characteristics (Timmermann 2015). Unsustainable practices such as extensive monocultures, the concentration of land ownership, and commodity production drove the need for increased labor and technology, which further complexified the genetic and cultural pool of Caribbean nations. Institutionalized disadvantages based on the neoliberal paradigm persist, exemplified by the racism, social injustice, and unbalanced power relations within and between LAC countries (Giraldo and McCune 2019).

As a consequence, farmers have become dependent on resources and factors that are beyond their control (Table 35.1). It is precisely in Latin America and the Caribbean where strong political views on the role of agroecology have emerged (Mier y Terán Giménez Cacho et al. 2018). The historical injustices linked to the *latifundios* or landlord estates, alongside a deep respect of Indigenous peoples' practices and livelihoods, have opened ways for the development of a unique *"Pensamiento Agroecológico Latinoamericano"* (Rosset et al. 2020). In the last 40 years, agroecological thinking has evolved as a counter-proposal, by placing farming families at the

Table 35.1 Drivers to scaling agroecology in Cuba, Puerto Rico and Sint Maarten, following indicators proposed by Mier y Terán Giménez Cacho et al. (2018)

Drivers	Cuba	Puerto Rico	St. Maarten
Crises	"Período especial" US embargo Hurricanes COVID-19	Colonization Industrialization GMOs Hurricanes COVID-19	Colonization Tourism Hurricanes COVID-19
Social organization	ANAP ACTAF Indio Hatuey University	Farmer organizations Farmer markets University	Individual projects
Simple agroecological practices	Bio-intensive farming Urban agriculture Permaculture	Bio-intensive farming Urban agriculture Shaded coffee	Bio-intensive farming Urban agriculture
Teaching-learning processes	Campesino-a-campesino University curricula Technology transfer Research-for-development	Workshops University curricula Volunteering Work brigades	Workshops University curricula Volunteering
Discourse	Food sovereignty	Food security	Food enterprise
External allies (examples of)	CLOC/La Via Campesina ALBA Movimientos SOCLA CUSAN	CLOC/La Via Campesina ALBA Movimientos SOCLA CUSAN University	UNDP University
Favorable markets	Yes, state-developed and institutionalized (i.e. food for schools, local communities, and food exports).	No, although there are grassroots and/or private initiatives, including farmer cooperative markets for organic and agroecological products, commercial eco-friendly hubs and health food stores.	No, only few producer-organized selling points for selected consumers.
Political opportunities	Community self-organization	March Against Monsanto International exchanges Public policy hearings	Outreach and education Scientific events Public outreach

centerpiece of territorial struggles and recognizing important temporal and spatial variations in the experiences of Latin American and Caribbean smallholders (Chappell et al. 2013; Timmermann 2019).

Famer-to-farmer agroecology movement in Cuba

Between the 1960s and the early 2020s, the production of at least 43 different crop items have been recorded for Cuba (Figure 35.2), including tropical fruits, root crops, and vegetables. While production of papaya, beans, cabbages, plantains, tomatoes, coconuts, cocoyam, watermelons, potatoes, and cassava have increased, other crops, such as coffee, groundnuts, avocados, tobacco, and sugarcane, have dramatically decreased. Nonetheless, the case of food production in Cuba can be analyzed in two periods: before and after 1989. Before 1989, Cuban agriculture was characterized by high-external input agriculture featuring large-scale monocultures of export commodities, such as sugar cane and tobacco. Capital, agro-chemicals, and machinery were provided via the socialist trading bloc with the USSR. Cuban agriculture was doomed to failure due to high dependency on external inputs. By the 1980s, yields had already been declining, soil quality was severely degraded, and farmers struggled to control pests and diseases (Altieri, Funes-Monzote, and Petersen 2012). Increasing fertilizer use and import to sustain agricultural production dropped suddenly after the fall of the Soviet bloc in 1989 (Figure 35.3), forcing Cuban farmers and technicians to re-design their farming methods and practices.

The collapse of the Soviet Union and the simultaneous (and still-in-place) instauration of the US trade embargo, also known as the "*período especial*" (the special period), can be regarded as a moment of social and material austerity in Cuba. Industrial sugarcane production was no longer possible without external inputs and has decreased by two-thirds since the 1990s (Figure 35.4). One important factor contributing to the increase of agroecological practices in Cuba was indeed the slowdown of chemical fertilizer importation from eastern Europe. Crop productivity and crop yields (excluding sugar cane) critically dropped in the last 20 years of the 20th century. After 1989, the advent of an authentic smallholder-driven "agroecological revolution" catalyzed an exponential increase of local food production through a horizontal farmer-to-farmer knowledge transfer methodology (Rosset et al. 2011). Cubans excelled in systematizing the farmer-to-farmer dynamics of knowledge sharing, even though this methodology actually originated among Indigenous peasants in Guatemala (Holt-Giménez 2002). Its rapid successes and visible impacts were facilitated by a well-structured strategy and the institutionalization of agroecology within the ANAP, the National Smallholder Farmers' Association (Fernandez et al. 2018; Funes-Aguilar and Vázquez-Moreno 2016).

In 2019, the country produced nearly 7 million tons of agricultural produce (excluding sugar cane and tobacco), an equivalent of 0.6 tons per capita (FAOstat 2022). This has been possible, not only through the efforts of ANAP, but also through intensive research investments in development strategies (Giraldo and McCune 2019). The organic and agroecological farmers responsible for an increase in healthy and local food items includes the over 400 members of ANAP, who are organized in different cooperative schemes, as well as the massive network of urban and permaculture gardeners across the country. As a side note, urban farms (*organopónicos*) in La Havana make accessible more than 80% of the food consumed by urban dwellers, the great majority of which is produced using organic and agroecological methods by law (Funes-Aguilar and Vázquez-Moreno 2016).

Additionally, there are several "lighthouse" projects that are worth mentioning, both for their impeccable implementation of agroecological practices and approaches and, for their outreach beyond Cuba, which has been outstanding and inspiring for farmers across latitudes (Funes-Aguilar and Vázquez-Moreno 2016). These projects include *Finca Marta* in the

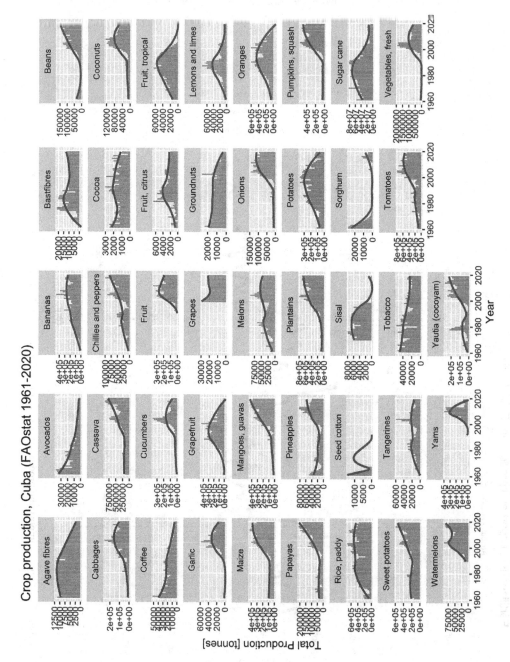

Figure 35.2 Crop production in Cuba, 1960–2019. www.faostat.org.

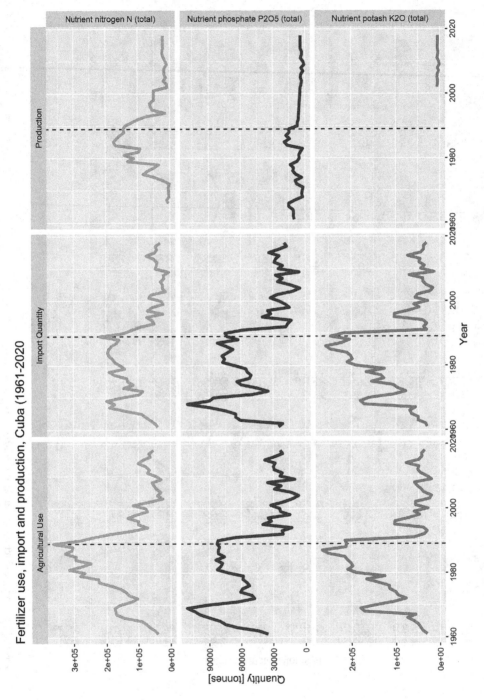

Figure 35.3 Fertilizer use, import and production per nutrient base in Cuba, 1960–2018. www.faostat.org.

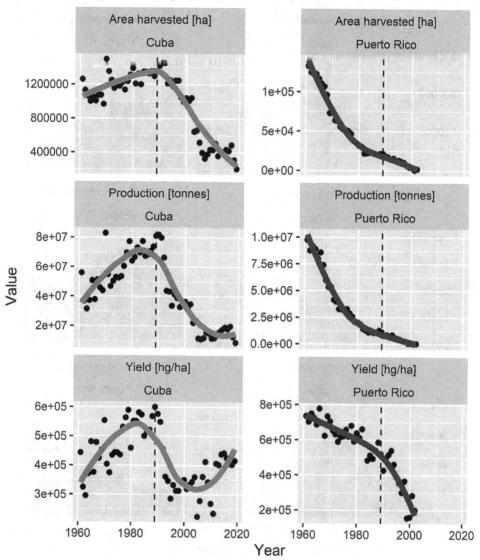

Figure 35.4 Agricultural production (area, production and yield) in Cuba and Puerto Rico, 1960–2019 (Left: only sugarcane; Right: sugarcane excluded;). www.faostat.org.

province of Artemisa led by Dr. Fernando Funes-Monzote; *Finca del Medio* in the province of Sancti Spiritu, a family farm fostered by the Casimiro's; and *Organóponicos Alamar*, an urban farm in La Havana that produces crops and animal products on a formerly abandoned 10-ha. piece of land now run and owned by more than 150 cooperative workers. The role of the Cuban agroecology movement has been indeed central in showcasing the potential of agroecology to work toward food sovereignty under tropical conditions. Moreover, the coordination of regional gatherings through Cuba's involvement in the international farmer federation CLOC-La Via Campesina has been fundamental to sharing agroecological knowledge, both regionally and internationally.

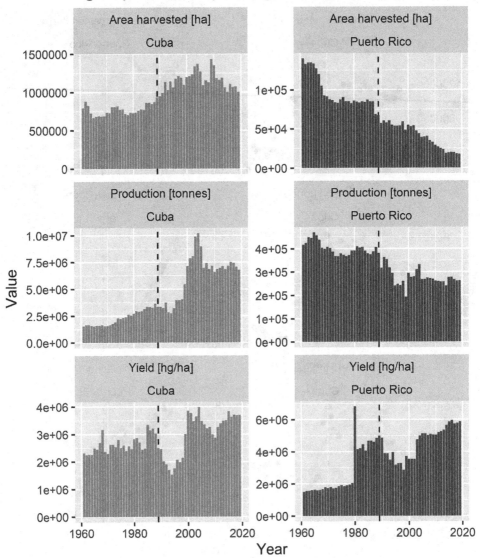

Figure 35.4 (Continued)

Supplementary efforts to support and increase local food production are needed in order to reduce food imports from more than 50% of the food consumed in Cuba (Altieri and Funes-Monzote 2012). Nonetheless, Cuba is one of the few countries in Latin America (alongside Brazil, Bolivia and Ecuador) where food sovereignty has been promoted through specific laws and national government-led projects. Indeed, the 2020 Food Sovereignty and Nutritional Education Plan (*Plan de Soberanía Alimentaria y Educación Nutricional de Cuba*) sets the bases for accelerated development of local food systems through a territorial approach. The Ministry of Agriculture collaborated with the UN Food and Agriculture Organization (FAO) and other organisms to establish the aforementioned national plan as an official law. While the fruits of this project still need to be assessed, the intention and the breadth of it are encouraging. Food Sovereignty Tours organized by advocacy think-tanks such as Food First and CUSAN (Cuba-USA Network) have contributed to knowledge systematization, sharing, and dissemination (Funes-Aguilar and Vázquez-Moreno 2016; Fernandez et al. 2018).

Re-peasantization dynamics in Puerto Rico

Puerto Rico (Borikén, its native name) is an archipelago that includes the main island of Puerto Rico (the smallest of the Greater Antilles) and a number of smaller islands within its jurisdiction, the largest of which are Mona (uninhabited), Vieques, and Culebra. Puerto Rican diets were self-sufficient in the 19th century with regards to diverse crops for local diets such as rice, tubers, and maize, such that 70% of croplands were dedicated to food production (Hill 1898). By the first half of the 20th century (nearly 50 years after the US invasion), Puerto Rico was importing 70% of its food, while the United States Merchant Marine Act of 1920 ensured that all goods imported into Puerto Rico by sea should be aboard American ships, consequently increasing food costs and decreasing quality (Carro-Figueroa 2002). At the same time, "Operation Bootstrap" moved the Puerto Rican economy away from agriculture, resulting in a linear decrease in the number of hectares destined to agricultural activities as well as an important reduction in total food production (Figure 35.4).

Between 2012 and 2018, farming area decreased by 16%, the number of registered farmers reduced by 37%, and the average farm size increased by 33% (USDA 2020). Fewer farmers, less agricultural land and bigger farms represent worrisome trends for a country that depends on food imports (more than 85% on food consumed) to sustain its population needs (Comas 2009). The traditional 'rice and beans' that Puerto Ricans enjoy daily are nowhere to be found in local agriculture in recent decades. Crop production and diets have indeed considerably changed despite having historical registries for more than 37 food items (Figure 35.5).

Commercial production of native crops such as avocados, beans, cassava, maize, pigeon peas, pineapple, tobacco, sweet potato, and yams have almost disappeared since the 1960s. Nonetheless, the commercial cultivation of products like tomatoes, papaya, plantain, bananas, and mangoes has increased. These crops are grown principally under industrial paradigms in the south-eastern region of the island, where the use of toxic inputs is widespread, particularly the use of herbicides, causing major health issues in the municipalities where urban land use and agriculture are most intensive on the island (Figure 35.6). In the south-eastern region of Puerto Rico, experimental and commercial production of genetically modified organisms (GMOs) further complicates the environmental and social panorama of local communities (García-López 2014). Government incentives tend to favor large holders (>50 acres, referring to approximately 2,000 registered farms between 2012 and 2018) and fail to recognize and support the role that smallholder farmers play in Puerto Rico's food system (<50 acres, referring to 11,000 farms in 2012 and to 6,000 farms in 2018).

The agroecology and food sovereignty movements in Puerto Rico consider that these and other issues need to be addressed in order to attain sustainable and resilient food systems. Many initiatives work toward the goal of the re-peasantization of Puerto Rico's economy, from individual farm projects to organic markets and alternative education strategies, as well as seed-saving projects and NGO-led programs. One of the most important groups for agroecology advocacy and outreach is the 30-year-old grassroots organization *Boricuá*, composed of more than 100 members, including farmers, consumers, scholars, and activists (Organización Boricuá de Agricultura Ecológica 2022). This organization has been part of the international farmer federation CLOC-La Via Campesina since 2012 and has been key in making visible the struggles of smallholder farmers in Puerto Rico. Puerto Rico's participation in the worldwide

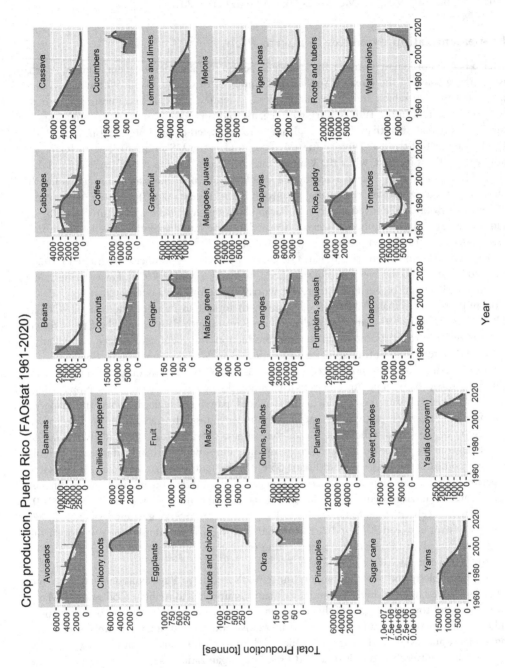

Figure 35.5 Crop production in Puerto Rico, 1960–2020. www.faostat.org.

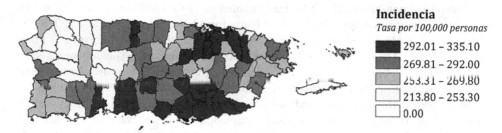

Incidencia
Tasa por 100,000 personas

■ 292.01 – 335.10
■ 269.81 – 292.00
□ 253.31 – 269.80
□ 213.80 – 253.30
□ 0.00

Figure 35.6 Cancer prevalence between 2005 and 2010 per 100,000 persons in Puerto Rico. http://www.estadisticas.gobierno.pr/iepr/LinkClick.aspx?fileticket=XNd4xX_dMjg%3D&tabid=186.

"March Against Monsanto" has also contributed to the education of the general public on the harmful effects to health and the environment of GMOs and toxic agricultural pesticides (Ruiz Marrero 2013). *Boricuá*'s collaboration, alongside researchers and practitioners, with the Latin American Scientific Society for Agroecology (SOCLA) has facilitated agroecology as an important topic of discussion in scientific debates through the Annual (now bi-annual) Agroecology Symposium, which has been celebrated at the University of Puerto Rico at Utuado since 2007 (12th edition programmed for March 2022).

Nowadays, there are several organic markets across the island, where farmers and consumers meet (Fonseca 2014). The most important and the first one to have opened nearly 20 years ago is the *Mercado de la Placita Roosevelt*, run by the Organic Cooperative *Madre Tierra*. This market, celebrated twice a month in the capital city, has been an important factor in the organization of various sectors around issues linked to sustainable food systems and environmental protection. Innovative farm-to-table strategies such as the direct selling of fresh produce and community-supported agriculture schemes have been flourishing in many different locations across the island. Some of these initiatives are led by the organizations *Proyecto Agroecológico Campesino* and *Finca El Timón* in Lares; *Armonía en la Montaña* in Aibonito; *La Colmena Cimarrona* in Vieques; and *Finca Agroecológica El Josco Bravo* in Toa Alta, amongst others.

In the aftermath of Hurricanes Irma and María in 2017, the importance of agroecology and farming in general became clear (Rodríguez-Cruz, Moore, and Niles 2021). Not only were agroecological projects more resistant to the environmental hazard as a consequence of high diversification, but these also recovered faster than conventional/industrial farming systems (Félix and Holt-Giménez 2017; Álvarez-Febles and Félix 2020). Additionally, the agroecological farmer networks demonstrated a great capacity to quickly respond and overcome damages through well-structured methodologies of farm work brigades and international solidarity (McCune et al. 2019).

In response to the COVID-19 pandemic that began in March 2020, initiatives such as *Apoyo Mutuo Campesino* in Utuado also contributed to channeling healthy food products to marginal communities during the island's lockdown period (Tittonell et al. 2021). In general, direct sales are a major strength of agroecological farmers.

In terms of alternative education around agroecology, three projects are worth highlighting. In 2016, the Department of Agricultural Technology at the University of Puerto Rico at Utuado (UPRU) initiated a four-year BA program entitled "Sustainable Agriculture" that focuses on agroecology and agroforestry. This program is the first of its kind in Puerto Rico, as it counterbalances the historical (conventional) Agronomy BSc curriculum held at UPR Mayagüez, in which most agronomists study. In 2021, the third cohort graduated from UPRU, with promising profiles of future farmers focused on agroecological production and product transformation.

Another education project that has already trained more than 600 new farmers in the last seven years is the agroecology school *El Josco Bravo*. Initially started with one site in Toa Alta, then expanded to Mayagüez, Gurabo, and, more recently, Ponce, Utuado, Orocovis, Toa Baja, Cayey, and Ciales. The *"Productores y Promotores"* program features three components: theoretical, practical, and volunteering. This structure is very attractive and informative for beginners, as well as for experienced producers who want to deepen their knowledge of agroecology.

Another project that integrates food and education, *Desde Mi Huerto*, has championed the production and commercialization of locally-adapted, organic seed varieties and has been instrumental in the dissemination of eco-friendly urban farming techniques throughout the territory. Their seeds and products are available through their website, a first of its kind on the island, called the "Puerto Rican organic seed catalog" (Desde Mi Huerto 2022). These and other innovative agroecological projects are documented in short films, for example *Agroecología en Puerto Rico* by JuanMa Pagán and Mariolga Reyes (Producciones Buruquena 2011). Whereas agroecology may be accused of being utopian, one example to the contrary demonstrates its possibilities. The many "living farm labs" that exist provide inspiration for the conceptualization and materialization of agroecological food systems in Puerto Rico and beyond.

Reclaiming healthy diets in Sint Maarten/Saint Martin

Information about Sint Maarten and, particularly its food production is scarce (see Table 35.1). The island of Sint Maarten/Saint Martin (Soualiga, its indigenous name) is indeed a territory in dispute at many levels: politics, languages, economies, and food systems. In the northern part (French side), the supermarkets and the food system in general are dominated by products imported from Metropolitan France, a vestige of long-settled colonial rule. However, on the southern part (Dutch side), food items are imported from the United States, The Netherlands, and elsewhere. In this context, tourism dominates the economy and few people are involved in local food production. The active farmers that can be encountered in this territory fight a daily struggle to cultivate their land, acquire seeds and plant material, and sell their produce in local markets. Nonetheless, agroecology and organic farming in a place like Sint Maarten makes plenty of sense, since it fosters more autonomy, makes healthier diets more accessible, and supports local economic and social development. An interesting observation is that Rastafarian culture is common to the few farm projects on the island that base their production systems on agroecology principles and solidarity economies. Thus, no pesticides or synthetic inputs are mobilized in these projects, as they cultivate health and self-sufficiency.

Examples of agroecology-based commercial farming initiatives include those of Farmer Shola, Touzah Jah Bash, Roland M. Joe, and Denicio Wyatte, amongst others. Farmer Shola is a Nigerian woman who has lived in St. Maarten for several decades and grows a diversity of food crops that she sells on-site. Touzah Jah Bash is a Rastafari elder based on the French side who also grows a diversity of food crops, such as leafy greens, mangoes, root crops, and medicinal plants that are sold on-site. Farmer Roland M. Joe, also known as Bushman, manages a steep piece of land nearby Salty Pond, where he not only grows a diversity of food items on terraces without any kind of external inputs, but also sells in his restaurant and café ("Ital Shack") located on the road at the entrance of the property. All of the people behind these initiatives are profoundly committed to providing healthy and nutritious diets to their communities: "I-tal is Vital!"

Since there is no government office for food and agriculture affairs, farmer Denicio Wyatte has declared himself the "Minister of Agriculture," both as a joke and as a serious statement to underline the importance of reclaiming organic food and farming on the island (SXM Agriculture 2021). Wyatte's initial project *Spaceless Gardens* started out as a landscaping company that quickly became the farming project "ECO St. Maarten Agriculture Research and Development Foundation" ("ECO" refers to "Educating, Communicating and Outreaching"), which sells fresh food on-farm and prepares Sunday lunches with produce from the farm. Through this foundation, international funding and local collaborations have been channeled to support hurricane-resistant farming infrastructure and civil society education activities.

Partnerships with the University of St. Martin (USM), the Netherlands Red Cross, United Nations Development Programme (UNDP), and others have catalyzed two major agroecology training programs on the island, which include: a formal university curriculum in agroecology at USM and an informal on-farm agroecology course in the St. Peters district. The first one is designed to reinforce theoretical and practical approaches to food production with university students. It is facilitated by mixed-profile instructors Tanisha Guy (biologist), Georges Félix (agronomist), and Denicio Wyatte (farmer). This innovative agroecology course has been taught in 2020 (face-to-face) and 2021 (online) as part of an optional module within the BA in Education degree. The second program, also known as "United Farmers of SXM," was completely coordinated and carried out by farmer Denicio Wyatte for community members with the objective of motivating "new" farmers and developing skills for seed saving, organic crop production, sustainable marketing strategies, healthy cooking and hurricane-resilient building.

In terms of research, "Islanders at the Helm" is a recently funded Dutch Government (NWO) program that uses community engagement and multidisciplinary research to help co-create sustainable solutions for climate change challenges. Within the five-year period of funding, four PhD candidates and four Postdocs in the fields of architecture, anthropology, musicology, paleoethnobotany, archaeology, policy, and governance will be hosted at universities throughout Sint Maarten, Aruba, and Curaçao. Albeit few, these initiatives slowly build their way to proving that self-sufficiency and food sovereignty are important. They are geared toward developing alternative livelihoods beyond the tourism industry in the Caribbean, which can be beneficial to farmers and society at large.

Going forward

Agroecology-based farm design principles

Agroecology can be defined as "the analysis of ecological principles applied to the study and design of sustainable agricultural production systems" (Francis et al. 2003). This ecological approach to farming provides many opportunities for the livelihoods of farming families to thrive. It requires transdisciplinary scientific principles and farmer innovations and practices, as well as participatory-action analyses regarding profound social and political issues of food systems (Wezel et al. 2009).

The three Caribbean examples in this chapter showcase very different points in the development of agroecological food systems. While the food sovereignty movement in Cuba is largely based on agroecology and fully supported by the government, in Puerto Rico, agroecology represents a form of resistance against colonial heritage and simultaneously constitutes a proposal for autonomy. Similarly, in Sint Maarten the lack of government support for food and farming has shaped the terrains for a few "rebels" who base their livelihoods on agroecological food production rather than on tourism and services.

Agroecological research and/or development projects have called attention to the rural and Indigenous reality of environmental problems, by highlighting the potential of rural actors as agents of transformation for a different way of relating to the environment (van der Ploeg 2014). Traditional ecological knowledge has endured from ancestral societies to modern times, rescued through oral traditions as well as from historical written archives, and it permeates amongst rural families in the Caribbean. Archaeological evidence indicates that the initial inhabiting of the Caribbean islands, registered between 5,000 and 8,000 years ago, was significantly influenced by plant domestication events (Pagán-Jiménez et al. 2015). Indigenous societies in the Caribbean had developed complex religious and spiritual systems, had built adapted housing and maritime transport strategies, and were farming a great variety of crops (Pagán-Jiménez et al. 2015; Rodríguez-Ramos et al. 2013).

Ancestral agricultural practices to cope with drought and other extreme atmospheric events are still present amongst modern Caribbean and Mesoamerican farming communities (Moreno-Calles et al. 2016; Rodríguez-Ramos et al. 2013). Cultural and agricultural diversity in the Caribbean results from its biophysical isolation, with island geographies, including coastal regions, high-mountain plateaus and even drylands, mainly under sub-tropical conditions. While specific farming practices are often site-specific, the principles of agroecology remain universal across latitudes (Box 35.1). These principles have recently been endorsed and expanded by the UN Food and Agriculture Organization (FAO) to include social and political elements that can guide the transition toward more sustainable and resilient food systems, including the optimization of intergenerational benefits such as good nutrition and food sovereignty, rather than solely annual profits, as well as the securing of intergenerational knowledge transfers (Barrios et al. 2020). Thus, agroecology scaling requires a re-appropriation of local resources through environment-friendly farming techniques, and simultaneously the re-establishment of socially-just relations between peers that foster financially-healthy markets.

Social and political values of food sovereignty

Food sovereignty is a term that was coined by *La Via Campesina* in 1996 as a response to the invisibilization of smallholder farmers' contribution to global food production (Claeys 2015). In this sense, global farmer movements have emphasized the need to protect local production

---

Box 35.1 Principles of agroecology (Altieri and Nicholls 2008)

(i)   Diversity: features the combination of species and genetic varieties in space and time through practices such as crop rotations, crop associations and agroforestry systems.

(ii) Recycling: looks out for nutrient flows and circular cycles of energy and matter by improving capture, retention and reduction of wastes through the re-utilization of "wastes" and the implementation of complex annual-perennial plant interactions.

(iii) Self-regulation: of pests and insects is a function catalyzed by habitat provision in complex perennial structures through practices such as agroforestry, atmospheric N-fixation and landscape-scale approaches to promote biological interactions.

(iv) Synergies: between system components appeals to the idea that securing favorable soil conditions can improve plant health via permanent soil cover, soil erosion control measures and crop associations in order to enhance ecosystem multifunctionality.

(v)   Efficiency: looks into achieving reasonable farm productivity through minimization of internal losses, reduction of external inputs and increase and diversify outputs.

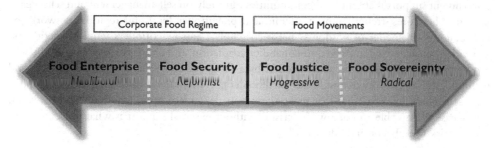

Figure 35.7 Continuum of food system discourses within corporate food regime and food movements. Based on Holt-Giménez (2010).

systems from corporate enterprises by declaring "the right of each nation to maintain and develop its own capacity to produce its basic foods respecting cultural and productive diversity" (Nyéléni 2007). *La Via Campesina*, a global farmer movements' federation, embraces agroecology as the sole vehicle to effectively achieve food sovereignty in its most radical form. Social movements and grassroots community organizations have indeed, influenced the relationship between people, food, and environments throughout the Americas and elsewhere. There are, however, a variety of discourses around the food system (Figure 35.7) that recognizes either the importance of corporate food regimes or the role of food movements (Holt-Giménez 2010).

The colonial push for certain products (e.g., sugarcane, coffee, tobacco) and the subsequent introduction of industrial farming practices (e.g., monocultures, herbicides, synthetic fertilizers, pesticides) have not only proven to be ill-adapted to the Caribbean's needs but have also had negative consequences on biodiversity, soil quality, and human health. The vulnerability of the agriculture sectors to reduced water availability, droughts, and extreme weather events such as hurricanes, is evident. These current concerns may become more acute in the future as the climate changes and temperatures rise.

On the one hand, neoliberal and reformist views acknowledge the importance of corporations in the food system, viewing consumers as mere recipients of what corporations can distribute, and ultimately, can make profits from. Whereas the term 'food security' is widespread amongst policy-makers, its four pillars of food availability, access to food, use and utilization, and stability, are limited to fulfilling population calorie intake but fail to address nutritional needs or the sources of food (Holt-Giménez 2010). Year-round access to specific imported foodstuffs considers consumer demands but does not respect the seasonality of local production systems nor the benefits that localized food systems can offer to consumers (Timmermann and Félix 2016).

On the other hand, the political discourse around 'food sovereignty' is transformative and radical by design. Progressive and radical views place the producer and the consumer as essential actors that can shape the means of food production and make local and healthy food more accessible to populations. Farmer organizations and consumer groups have found ways of (re) introducing nutrition-sensitive approaches to farming and diets (Ruel, Quisumbing, and Balagamwala 2018), and have underlined the need for a "dialogue of wisdoms" between actors of the food system (Martínez-Torres and Rosset 2014).

The quest for food self-sufficiency in the Caribbean islands has been hindered by several historical crises. Going forward with agroecology and food sovereignty in the Caribbean territories of Cuba, Puerto Rico, and Sint Maarten requires further focus on how agroecology-based farming systems can serve as vehicles for climate justice, especially if based on design-driven climate change resilience, disaster risk reduction, and the important role of animal farming systems. Indeed, food sovereignty is a tool for social justice, particularly where government support is absent, and communities must rely on self-management and self-organization. Therefore, strengthening academic and practitioner local and regional networks by encouraging "dialogues of wisdoms" between scholars and practitioners would benefit food movements in achieving improved sustainability, adaptation, and transformative capacity. Finally, food sovereignty needs to be accompanied by a longer-term vision alongside energetic and technological sovereignty to ensure that smallholder farmers in the Caribbean are fully equipped with self-reliant infrastructure that make them non-dependent upon external resources when crises arise. Being able to cope with "crises" without external support is what prevents situations of risk from becoming "disasters."

References

Altieri, Miguel A., and Fernando R. Funes-Monzote. 2012. "The Paradox of Cuban Agriculture." February 2012. https://climateandcapitalism.com/2012/02/04/the-paradox-of-cuban-agriculture/.

Altieri, Miguel A., Fernando R. Funes-Monzote, and Paulo Petersen. 2012. "Agroecologically Efficient Agricultural Systems for Smallholder Farmers: Contributions to Food Sovereignty." *Agronomy for Sustainable Development* 32 (1): 1–13. https://doi.org/10.1007/s13593-011-0065-6.

Altieri, Miguel A., and Victor Manuel Toledo. 2011. "The Agroecological Revolution in Latin America: Rescuing Nature, Ensuring Food Sovereignty and Empowering Peasants." *Journal of Peasant Studies* 38 (3): 587–612. https://doi.org/10.1080/03066150.2011.582947.

Álvarez-Febles, Nelson, and Georges F. Félix. 2020. "Hurricane María, Agroecology, and Climate Change Resiliency." In *Climate Justice and Community Renewal*, edited by Brian Tokar and Tamra Gilbertson, 131–46. London: Routledge. https://doi.org/10.4324/9780429277146-9.

Arnold, Pierre, Jerónimo Díaz, and Line Algoed. 2020. "Collective Land Tenure in Latin America and the Caribbean, Past and Present." In *On Common Ground: International Perspectives on Community Land Trust*, edited by John Emmeus Davis, Line Algoed, and María E. Hernández-Torrales, 167–88. Terra Nostra Press.

Barrios, Edmundo, Barbara Gemmill-Herren, Abram Bicksler, Emma Siliprandi, Ronnie Brathwaite, Soren Moller, Caterina Batello, and Pablo Tittonell. 2020. "The 10 Elements of Agroecology: Enabling Transitions towards Sustainable Agriculture and Food Systems through Visual Narratives." *Ecosystems and People* 16 (1): 230–47. https://doi.org/10.1080/26395916.2020.1808705.

Carro-Figueroa, Vivian. 2002. "Agricultural Decline and Food Import Dependency in Puerto Rico: A Historical Perspective on the Outcomes of Postwar Farm and Food Policies." *Caribbean Studies* 30 (2): 77–107. https://doi.org/10.2307/25613372.

Chappell, M. Jahi, Hannah Wittman, Christopher M. Bacon, Bruce G. Ferguson, Luis García Barrios, Raúl García Barrios, Daniel Jaffee, et al. 2013. "Food Sovereignty: An Alternative Paradigm for Poverty Reduction and Biodiversity Conservation in Latin America." *F1000Research* 2. https://doi.org/10.12688/f1000research.2-235.v1.

Claeys, Priscilla. 2015. "Food Sovereignty and the Recognition of New Rights for Peasants at the UN: A Critical Overview of La Via Campesina's Rights Claims over the Last 20 Years." *Globalizations* 12 (4): 452–65. https://doi.org/10.1080/14747731.2014.957929.

Comas, Myrna. 2009. "Vulnerabilidad de Las Cadenas de Suministros, El Cambio Climático y El Desarrollo de Estrategias de Adaptación: El Caso de Las Cadenas de Suministros de Alimento de Puerto Rico." Universidad de Puerto Rico en Mayagüez. https://myrnacomas.com/tesis/.

Desde Mi Huerto. 2022. "Tienda de Semillas." 2022. https://en.desdemihuerto.com/.

FAOstat. 2022. "Food and Agriculture Country Data." 2022. https://www.fao.org/faostat/en/#data.

Félix, Georges F., and Eric Holt-Giménez. 2017. "*Hurricane María: An Agroecological Turning Point for Puerto Rico?*" Food First Backgrounder. Vol. 23. Oakland, California.

Fernandez, Margarita, Justine Williams, Galia Figueroa, Garrett Graddy-Lovelace, Mario MacHado, Luis Vazquez, Nilda Perez, Leidy Casimiro, Graciela Romero, and Fernando Funes-Aguilar. 2018. "New Opportunities, New Challenges: Harnessing Cuba's Advances in Agroecology and Sustainable Agriculture in the Context of Changing Relations with the United States." *Elementa* 6. https://doi.org/10.1525/elementa.337.

Fonseca, Verónica. 2014. "¿Dónde Están Los Mercados Agrícolas En Puerto Rico?" *Diálogo UPR*, 2014. https://dialogo.upr.edu/donde-estan-los-mercados-agricolas-en-puerto-rico/.

Francis, C., G. Lieblein, S. Gliessman, T. A. Breland, N. Creamer, R. Harwood, L. Salomonsson, et al. 2003. "Agroecology: The Ecology of Food Systems." *Journal of Sustainable Agriculture* 22 (3): 99–118. https://doi.org/10.1300/J064v22n03_10.

Funes-Aguilar, Fernando, and Luis J. Vázquez Moreno. 2016. *Apuntes de La Agroecología En Cuba*. La Habana, Cuba: Editora Estación Experimental de Pastos y Forrajes Indio Hatuey.

García-López, Gustavo A. 2014. "Decimos No a Los Transgénicos." *80 Grados*, 1–8.

Giraldo, Omar Felipe, and Nils McCune. 2019. "Can the State Take Agroecology to Scale? Public Policy Experiences in Agroecological Territorialization from Latin America." *Agroecology and Sustainable Food Systems* 43 (7–8): 785–809. https://doi.org/10.1080/21683565.2019.1585402.

Graeub, Benjamin E., M. Jahi Chappell, Hannah Wittman, Samuel Ledermann, Rachel Bezner Kerr, and Barbara Gemmill-Herren. 2016. "The State of Family Farms in the World." *World Development* 87: 1–15. https://doi.org/10.1016/j.worlddev.2015.05.012.

Hill, Robert Thomas. 1898. *Cuba and Porto Rico, with the Other Islands of the West Indies*. New York: The Century Co. https://archive.org/details/cubaportoricowit00hill.

Holt-Giménez, Eric. 2002. "Measuring Farmers' Agroecological Resistance after Hurricane Mitch in Nicaragua: A Case Study in Participatory, Sustainable Land Management Impact Monitoring." *Agriculture, Ecosystems and Environment* 93 (1–3): 87–105. https://doi.org/10.1016/S0167-8809(02)00006-3.

———. 2010. "Food Security , Food Justice , or Food Sovereignty?" *Backgrounder* 16 (4): 1–4.

Martínez-Torres, María Elena, and Peter M Rosset. 2014. "Diálogo de Saberes in La Vía Campesina : Food Sovereignty and Agroecology." *The Journal of Peasant Studies* 0 (0): 1–19. https://doi.org/10.1080/03066150.2013.872632.

McCune, Nils, Ivette Perfecto, Katia Avilés-Vázquez, Jesús Vázquez-Negrón, and John Vandermeer. 2019. "Peasant Balances and Agroecological Scaling in Puerto Rican Coffee Farming." *Agroecology and Sustainable Food Systems* 43 (7–8): 810–26. https://doi.org/10.1080/21683565.2019.1608348.

McCune, Nils, Peter M. Rosset, Tania Cruz Salazar, Antonio Saldívar Moreno, and Helda Morales. 2017. "Mediated Territoriality: Rural Workers and the Efforts to Scale out Agroecology in Nicaragua." *Journal of Peasant Studies* 44 (2): 354–76. https://doi.org/10.1080/03066150.2016.1233868.

Mier y Terán Giménez Cacho, Mateo, Omar Felipe Giraldo, Miriam Aldasoro, Helda Morales, Bruce G. Ferguson, Peter Rosset, Ashlesha Khadse, and Carmen Campos. 2018. "Bringing Agroecology to Scale: Key Drivers and Emblematic Cases." *Agroecology and Sustainable Food Systems* 42 (6): 637–65. https://doi.org/10.1080/21683565.2018.1443313.

Moreno-Calles, Ana Isabel, Alejandro Casas, Alexis Daniela Rivero-Romero, Yessica Angélica Romero-Bautista, Selene Rangel-Landa, Roberto Alexander Fisher-Ortíz, Fernando Alvarado-Ramos, Mariana Vallejo-Ramos, and Dídac Santos-Fita. 2016. "Ethnoagroforestry: Integration of Biocultural Diversity for Food Sovereignty in Mexico." *Journal of Ethnobiology and Ethnomedicine* 12 (1): 1–21. https://doi.org/10.1186/s13002-016-0127-6.

Nyéléni. 2007. "*Forum for Food Sovereignty*." Sélingué, Mali.

Organización Boricuá de Agricultura Ecológica. 2022. "Quiénes Somos." 2022. http://organizacionboricua.org/.

Pagán-Jiménez, Jaime R., Reniel Rodríguez-Ramos, Basil A. Reid, Martijn van den Bel, and Corinne L. Hofman. 2015. "Early Dispersals of Maize and Other Food Plants into the Southern Caribbean and Northeastern South America." *Quaternary Science Reviews* 123: 231–46. https://doi.org/10.1016/j.quascirev.2015.07.005.

Parés-Ramos, Isabel K., William A. Gould, and T. Mitchell Aide. 2008. "Agricultural Abandonment, Suburban Growth, and Forest Expansion in Puerto Rico between 1991 and 2000." *Ecology and Society* 13 (2): 20. https://doi.org/10.5751/ES-02479-130201.

van der Ploeg, Jan Douwe 2014. "Peasant-Driven Agricultural Growth and Food Sovereignty." *Journal of Peasant Studies* 41 (6): 999–1030. https://doi.org/10.1080/03066150.2013.876997.

Producciones Buruquena. 2011. "Agroecología En Puerto Rico." 2011. www.agroecologiapr.org.

Rodríguez-Cruz, Luis Alexis, Maya Moore, and Meredith T. Niles. 2021. "Puerto Rican Farmers' Obstacles Toward Recovery and Adaptation Strategies After Hurricane Maria: A Mixed-Methods

Approach to Understanding Adaptive Capacity." *Frontiers in Sustainable Food Systems* 5 (July): 1–16. https://doi.org/10.3389/fsufs.2021.662918.

Rodríguez-Ramos, Reniel, Jaime Pagán-Jiménez, Jorge Santiago-Blay, Joseph B. Lambert, and Patrick R. Craig. 2013. "Some Indigenous Uses of Plants in Pre-Columbian Puerto Rico." *Life: The Excitement of Biology* 1 (1): 83–90. https://doi.org/10.9784/leb1(1)rodriguez.09.

Rosset, Peter M., Lia Pinheiro Barbosa, Valentín Val, and Nils McCune. 2020. "Pensamiento Latinoamericano Agroecológico: The Emergence of a Critical Latin American Agroecology?" *Agroecology and Sustainable Food Systems* 00 (00): 42–64. https://doi.org/10.1080/21683565.2020.1789908.

Rosset, Peter M., Braulio Machín Sosa, Adilén María Roque Jaime, and Dana Rocío Ávila Lozano. 2011. "The Campesino-to-Campesino Agroecology Movement of ANAP in Cuba: Social Process Methodology in the Construction of Sustainable Peasant Agriculture and Food Sovereignty." *Journal of Peasant Studies* 38 (1): 161–91. https://doi.org/10.1080/03066150.2010.538584.

Ruel, Marie T., Agnes R. Quisumbing, and Mysbah Balagamwala. 2018. "Nutrition-Sensitive Agriculture: What Have We Learned so Far?" *Global Food Security* 17 (September 2017): 128–53. https://doi.org/10.1016/j.gfs.2018.01.002.

Ruiz Marrero, Carmelo. 2013. "La Agroecología Como Proyecto Nacional," 2013. https://www.80grados.net/la-agroecologia-como-proyecto-nacional/.

SXM Agriculture. 2021. "SXM Agriculture." 2021.

Timmermann, Cristian. 2015. "Pesticides and the Patent Bargain." *Journal of Agricultural and Environmental Ethics* 28 (1): 1–19. https://doi.org/10.1007/s10806-014-9515-x.

———. 2019. "A Latin American Perspective to Agricultural Ethics," 203–17. https://doi.org/10.1007/978-3-030-17963-2_11.

Timmermann, Cristian, and Georges F. Félix. 2015. "Agroecology as a Vehicle for Contributive Justice." *Agriculture and Human Values* 32 (3): 523–38. https://doi.org/10.1007/s10460-014-9581-8.

———. 2016. "Food Sovereignty and the Global South." In *Encyclopedia of Food and Agricultural Ethics*, 1–6. Dordrecht: Springer Netherlands. https://doi.org/10.1007/978-94-007-6167-4_524-1.

Tittonell, P., M. Fernandez, V. E. El Mujtar, P. V. Preiss, S. Sarapura, L. Laborda, M. A. Mendonça, et al. 2021. "Emerging Responses to the COVID-19 Crisis from Family Farming and the Agroecology Movement in Latin America—A Rediscovery of Food, Farmers and Collective Action." *Agricultural Systems* 190 (March). https://doi.org/10.1016/j.agsy.2021.103098.

UNdata. 2022. "Country Statistics." 2022. http://data.un.org/.

USDA. 2020. "Puerto Rico Agriculture: Results from the 2018 Census of Agriculture." *San Juan*. https://www.nass.usda.gov/Publications/Highlights/2020/census_puertorico.pdf.

Wezel, Alexander, S. Bellon, T. Doré, C. Francis, D. Vallod, and C. David. 2009. "Agroecology as a Science, a Movement and a Practice. A Review." *Agronomy for Sustainable Development* 29 (4): 503–15. https://doi.org/10.1051/agro/2009004.

36 Re-existence struggles and socio-ecological alternatives for reproduction of dignified and sustainable life in territories affected by the extractivist offensive in Latin America

Mina Lorena Navarro, Sandra Rátiva Gaona, and Talita Furtado Montezuma

Introduction

The starting point of this text is the concern for the impacts that the new *extractivist offensive* is leaving in its wake in all Latin American countries on human and non-human populations in their territories, attacking the basic conditions of their subsistence and thus deepening the epochal and socio-ecological crisis that we are facing at the planetary level.

In the context of the proliferation of socio-ecological conflicts, we are interested in analyzing the processes of organization and transformation underway, that is, the *know-how* that multiple and diverse rural and urban community networks have recreated to defend and affirm their modes of existence. To this end, we trace the collective political strategies that have been built at the community and organizational level, in light of socio-ecological conflicts, mainly in Colombia, Brazil, and Mexico, countries in which an alarming extractivist violence is expressed, as documented in recent human rights reports.

Own elaboration; with the technical support of Kevin Hernández. (1) the struggle of the P'urhépecha people of Cherán; (2) the resistance of the Agrupación un Salto de Vida in Jalisco; (3) the struggle of the Organización Popular Francisco Villa de Izquierda Independiente in Mexico City; and (4) the Juntas Logramos Más Campaign. In Colombia, (5) the territorial defense of the Asociación de Cabildos Indígenas del Norte del Cauca (ACIN); (6) the Comité Ambiental del Tolima; (7) in the Sararé region, the Empresa Comunitaria de Acueducto, Alcantarillado y Aseo de Saravena (ECAAAS-ESP). In Brazil, (8) the experience of the communities of Serra do Brigadeiro in Minas Gerais; (9) the communities of the central sertão do Ceará against uranium mining; (10) the communities of Altamira/PA affected by Belo Monte.

Map showing important centres in Mexico, Columbia and Brazil.

Extractivist offensive and violent appropriation of nature

The *extractivist offensive* that today besieges territories and livelihoods (Seoane 2012, 123; Composto and Navarro Trujillo 2014, 48) is part of the civilizing strategy of capital where the labor and energy of *human and non-human natures is violently appropriated*. The historical root of this violence dates back to the times of the plundering of Abya Yala and in the last two decades it has intensified in all Latin American countries, with the so-called "commodity consensus" (Svampa 2013), further deepening the colonial, dependent, and subordinated position of the continent in the world system.

In this context, Latin American governments, including the so-called Fourth Transformation led by Andrés Manuel López Obrador in Mexico, the right-wing Iván Duque Márquez in

DOI: 10.4324/9780429344428-42

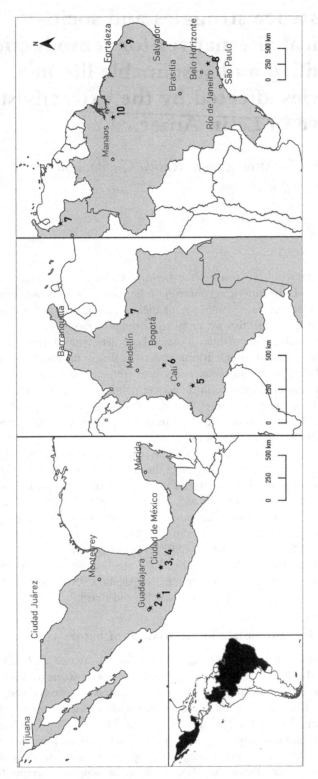

Figure 36.1 Locator of the struggles and processes mentioned in Mexico, Colombia and Brazil.

Colombia, and the ultra-right-wing Jair Bolsonaro in Brazil, seek to expand the extractivist frontier and thus guarantee the material and energy required by capitalist dynamics, regardless of the socio-ecological costs and impacts that result from this processes of dispossession.

Worse still, in the face of falling commodity prices and the end of the cycle of some progressive governments, between 2013 and 2015, pressures to expand the extractivist frontier were further exacerbated and, with it, violence against populations and territories in dispute (Svampa and Teran 2019, 183). In turn, those territories not fully connected to the circuits of the logic of value and distinguished by an abundant biocultural wealth, face the terror associated with the expansion of illegal economies, including drug trafficking; both by state anti-drug security policies (Paley 2020) and also by the control of trafficking routes or production of inputs in rural territories.

In the face of this violence, the proliferation of socio-ecological conflicts is notorious and, with it, the activism of thousands of indigenous, peasant, Black, or quilombola communities, as well as organizations in urban centers, who fight to defend and sustain their ways of life and livelihoods. In the Atlas of Environmental Justice ("EJOLT") we can see that Brazil occupies second place worldwide with 159 registered environmental conflicts, Colombia in fifth place with 129, and Mexico in sixth place with 122 (EJOLT 2020).

Survival re-existence struggles

I am a woman of African descent. I grew up in an ancestral territory that dates back to 1636. From a young age we are taught the value of the land. We know that the territories on which we build our community and recreate our culture are not a gift, as it cost our elders many years of work and suffering in the mines and slave estates. The upbringing in my community is based on values such as solidarity, respect and honesty. We are taught that dignity has no price, that to resist is not to endure. We are taught to love and value the territory as a living space, to fight for it, even at the risk of our own lives.

(Remarks by Francia Márquez on receiving the Goldman Awards, 2018)

The perspective we propose to think about the transformations and alternatives that are set in motion in the contexts of socio-ecological conflict is that, in the process of opposition to extractivism, critical know-how is articulated against what threatens the territory, hand in hand with a desire that affirms what is collectively known to be necessary to guarantee the reproduction of life.

Under this perspective, struggles in defense of life, as Carlos Porto Gonçalves and Enrique Leff put it, are not only a resistance or reaction to the invader, but also a form of re-existence because they incorporate new horizons of meaning born of the circumstances and challenges of the present. "They resist because they exist; therefore, they r-exist" (Porto-Gonçalves 2016, 8).[1] In this sense and recovering the words of Francia Márquez, the defense of life is not centered on a horizon of *resistance as endurance*, but on the perseverance of a desire to sustain an existence with dignity.

Threats associated with dispossession are usually experienced as the arrival of a time that endangers life and its reproduction (Composto and Navarro Trujillo 2014; Navarro Trujillo 2015; Linsalata 2016). This instant of danger activates a collective sense of emergency and a willingness to strengthen ties and generate alliances to make common defense in the face of shared needs and problems. It is there that renewed forms of participation and collective co-responsibility are being established to defend and guarantee the usufruct of shared material wealth in the midst of the processes of appropriation, separation, and mediation imposed by extractivist violence. This effort to reconstitute political capacities to guarantee the reproduction of life collectively is expressed in a range of actions: strengthening the right of self-determination in the midst of territorial threats; deepening connections to the land through productive and agroecological projects that strengthen material autonomy; managing community and collective water systems in rural and urban contexts in the face of policies of extraction and privatization of this good; safeguarding and regenerating biodiversity through the implementation of internal regulations for the protection of the territory; recovering and reaffirming ancestry and spirituality; or naming the experiences of macho violence within the living spaces that women inhabit and proposing other terms of relationship at the community and organizational level.

The reconstitution of a politics centered on the care of life

Carolina Márquez, in her research on *the struggle of the P'urhépecha people* of Cherán in Mexico, analyzes community reconstitution from the process of *revaluation of life* that was set off by the organized community defense in 2011, to stop the illegal logging of forests, extortion, kidnappings, rapes, and murders of the population at the hands of organized crime. For this researcher, the revaluation of life is expressed in the strengthening of the Communal Government and their own security model, called Ronda Comunitaria, both under a system of uses and customs; the reforestation actions through the Forest Ranger Team and the promotion of a communal nursery for pine production that have sought to alleviate environmental damage. The work that a community collective, mainly made up of women, does to document oral history through audiovisual recording, reflecting, and researching from the community itself to collectively produce knowledge, also expresses this revaluation of life (Márquez 2016).

An urban experience, which contrasts with the conditions in which the Cherán autonomy exercise is taking place, are the self-convened councils of residents of the municipalities of El Salto and Juanacatlán. These municipalities on the banks of the Santiago River in the Guadalajara metropolitan area are identified as one of the Environmental Emergency Regions of Mexico. In particular, *Un Salto de Vida*, an organization of inhabitants of different ages, has been meeting since the mid-1990s to take action against the health effects and environmental damage caused by the contamination of the Santiago River, just a few meters from their homes (Navarro 2020).

In the face of the socio-ecological disaster they are facing, links have been forged between the inhabitants, called ejidos, and affected communities to promote various strategies to make the conflict visible and generate alternatives for life by conserving the ecological refuges that remain in the watershed and demanding the definitive closure of the Los Laureles landfill, which receives approximately 3,500 tons of garbage daily. In addition, a vegetable garden and nursery for the agroecological production of vegetables for consumption by the community and endemic plants for reforestation, have been implemented. These efforts seek to promote a process of community reappropriation of the riverside ecosystem that is capable of restoring living conditions and stopping the infamous devastation of their territory.

In both experiences we can see that the relationship of care with the territory is not given beforehand, but depends on a practical activity that we call *production of the commons*, which is established between a set of people capable of intertwining their doings and establishing cooperative links to address shared problems and needs (Gutiérrez Aguilar, Navarro, and Linsalata 2017). Both in the communities of Cherán and in those of the Ribera del Río Santiago this is expressed in the persistent dispute to conserve and guarantee the spheres that guarantee life, but also to try to regain control over the means of existence that have been expropriated and contaminated by the extractivist offensive.

The production of the commons as a strategy for the defense of life occurs in a context of dispute and profound asymmetry with respect to the economic, ideological, and coercive resources of governments and companies. Within these frameworks of inequality and hierarchy, collective struggles have been learning to creatively combine multiple strategies in the fields of legal defense (injunctions, consultations, lawsuits, bills, territorial planning), along with; dialogue with governments and companies through negotiation; increasing the visibility of the conflict; articulating alliances, self-organization; the mobilization of art and culture; and direct action.

In the light of the lessons learned by various organizations, it can be seen that the processes in defense of life are generating popular mandates to force those who govern to obey (Castro 2019, 13). Such mandates are sustained through the articulation of a collective subject in struggle that is linked to other actors at different scales of action, with the capacity to establish vetoes and produce their own decisions.

A paradigmatic experience in this regard is the resistance against the La Colosa mining project, which has managed to mobilize broad sectors of the department of Tolima, in central Colombia. It is a process that began in the rural community of Piedras, which, in 2013, refused the entry of mining machinery to its territory, demanding the implementation of the constitutional process of popular consultation (Velandia Perilla 2015). The movement subsequently grew throughout the region through the actions of the *Environmental Committee of Tolima*, which organized "marches, cultural takeovers, door-to-door informative campaigns, visits to schools, academic forums, conferences, seminars, assemblies, citizen talks, etc." (García Parra 2012, 462).

The reconstitution of political capacities in contexts of siege has led to the recomposition of interdependent relationships with rivers, mountains, land, seeds, and, in general, with all coexisting beings and abiotic environments that are part of the territories. In this regard, *Un Salto de Vida* in Mexico points out that it is not possible to rebuild the social fabric if at the same time the territory of life and the community relationship with it is not regenerated (Navarro 2020). In Tolima, on the other hand, not only have social relations been reactivated to organize the defense of water and the struggle against mining, but debates have also been generated on land use planning and climate change as a social problem.

Emergence of an ecological and interdependent sensibility

In the contexts of eco-community recomposition, the struggles in defense of life are articulating an ecological agenda which, in dialogue with Martínez Alier (2004), we would say does not emerge from an ideological process in the hands of professionals but from a pressing need of those affected to put a limit on the expansion of extractivist frontiers. Water for life, not for death! In defense of life and against the projects of death! Without gold we live, without water we die! This territory is not for sale, it is recovered and defended! Women, water, and energy are not merchandise! These are among the slogans that condense the nuclei of good sense that are produced from the struggle to signify the terms of the dispute being faced (Tischler and Navarro 2011, 69).

The struggles in defense of life are forcefully questioning the techno-scientific paradigms with their premises of control, domination, and separation from the fabric of life, as well as the anthropocentric fantasy, which has placed the human species as the measure and center of the universe, denying its interdependence and hierarchizing its relationship with the rest of the companion species (Haraway 2016).

Perhaps for this reason, a practice of cultivating the spiritual in these struggles is growing with increasing strength, which is expressed in the tasks of safeguarding the territories or in the rituals at the beginning of a meeting or political gathering, to gather strength, and honor and protect life against the sieges and affectations of extractivism on communities and ecosystems.

The spiritual dimension of the defense of life stands out in the experience of *the communities of Serra do Brigadeiro*, in Minas Gerais, Brazil, whose resistance to the mining company, in defense of water, strengthened their affection for the place. The resistance has been articulated with the vindication of the peasant way of life and a dimension of ecological spirituality that is expressed in the Franciscan walks.[2] This spirituality has also been expressed by the Church, the *Movement for Popular Sovereignty in Mining (MAM)*, and the local communities. One of the results of this social mobilization process was the approval, in 2018, of an unprecedented municipal law that declares more than 10,000 ha. as a protected Water Heritage site in the city of Muriaé, in Minas Gerais. It is estimated that the law will have the capacity to protect approximately 2,000 water sources in Belisario, a district of Muriaé. In addition, it was also possible to include in the Muriaé Plan the creation of a sustainable environmental macro-zone covering important areas of the region where mining will be restricted, creating an important reference for communities to declare a mining-free territory.

In Colombia, in the midst of the barbarity of war, the forms of spiritual and ancestral authority of the Nasa indigenous people in the department of Cauca have been strengthened. In particular, we refer to the *Indigenous Guard*, composed primarily of male members of the various communities that inhabit the region and mandated by the assemblies and ancestral authorities organized by the *Association of Indigenous Councils of Northern Cauca (ACIN)*.

The Indigenous Guard has the task of keeping the peace in *the territory* by traveling, observing, and intervening in any situation that affects daily life, such as the presence of the insurgency, paramilitary groups associated with drug trafficking and illegal gold mining, new armed groups that are appearing in the region, and the private security of the sugarcane agro-industry. The ancestral knowledge that sustains their journey through the mountains and valleys means that in many other regions this path is being replicated, as is the case of the peasant guard, or the Black communities that have initiated their *Maroon guard* process (Rojas and Useche 2019*).* These recent examples correspond to processes that have been happening for years with the peasant patrols of the Peruvian Andes, or the Regional Coordinator of Community Authorities—Community Police of Guerrero, Mexico (Fini 2018), which perform similar functions and have the popular mandate to safeguard the territory through collective bodies responsible for producing and guaranteeing community security and justice.

In all these experiences we note that there is a broadening of the recognition of the terms of relationship with other species and of what we call the condition of interdependence inscribed in the heart of all life (Navarro and Linsalata 2021). In this field, the leading and historical role of women stands out, which we will examine in more detail below.

The strength of women: the emotional and affective work of re-existence

In recent times, the confluence between women who are part of some process of defense of life (whether it be territorial, indigenous, community, or anti-extractivist) and Abya Yala feminists is notorious. These dialogues are materialized in a series of meetings and diverse modes of agreement between women (Gutiérrez Aguilar, Sosa, and Reyes 2018) that have been positioning the need to think about anti-patriarchal and anti-colonial content in the struggle against extractivisms.

Astrid Ulloa, for example, documents how mega-mining privileges male presence in the labor space, accentuates income inequalities, and increases rates of sexual violence (Ulloa 2016, 128, 131). At the same time, gender inequalities are deepened or transformed, which does not mean that they are inaugurated, but rather that processes of dismantling of institutions and social fabrics that previously protected women are experienced, hand in hand with the intensification of hierarchical relations that tend to masculinize the ancestral positions of men (Segato 2012, 110).

In these contexts, women have been positioning the body as a space of resistance to expropriation and territorial dispossession, and the territory as a living and extended body to which they themselves are integrated. From community and territorial feminisms, indigenous women speak of the *body-territory* to recognize one's own body, in connection with the fabric of life and in interdependence with the territory that the body inhabits (Cabnal 2018; Colectivo Miradas Críticas 2017). Undoubtedly, through expressing and naming violences, these experiences weave into the struggle for re-existence of what modern fragmentation separates: the meanings between life, body, territory, affect, and political action.

The cultivation of an emotional and affective dimension in the struggle in defense of life enables a form of knowledge of what bodies experience and their relationship with the environment in which they are situated. To approach this question, Alice Poma and Tomasso Gravante (2015) talk about a kind of work that collectives carry out daily to manage their own emotions. We add that this work is generally carried out by women, and is aimed at managing the pain, grievance, hopelessness, helplessness, repression, and the relationship with death that extractivist projects impose.

In addition, in recent years, various collective formats have been generated for women to meet and discuss what they are facing and to focus on a practice of collective care that strengthens them. In Mexico, a group of women defenders of their territories from all over the country have launched the collaborative campaign "Juntas Logramos Más" (Together We Achieve More) in which they have been building an extended narrative community from which to create, share, and weave stories about women and their struggles against dispossession, in order to name and make visible their contributions to the defense of territory.

Another experience is that of spaces for political healing and integral feminist protection, as is the case of the work promoted by the Mesoamerican Women Human Rights Defenders Initiative and the Oaxaca Consortium for collective self-care for women defenders and activists who are members of national networks and organizations in El Salvador, Guatemala, Honduras, Nicaragua, and Mexico. Through these experiences it has become clear that the impacts of the violence of the capitalist-patriarchal and colonial complex cannot be addressed individually and much less can the community networks, of which the defenders are a part, be ignored.

Undoubtedly, the struggles in defense of life are sustained by the multiple and intense battles that women are waging, both in ordinary and extraordinary times of existence, to guarantee the reproduction of life, which is known to be fragile and at the same time interdependent. We call attention to the renewed forms that these actions are acquiring, such as taking the floor in an assembly, giving an interview to the media, assuming organizational tasks or intervening in an unprecedented way in spaces they had not traditionally inhabited, putting their bodies in front of the imminent threat of the machines of the extractivist projects, and preventing the arrival of corporate or state personnel or other strangers to the community.

The images of the women of Cherán in 2011, who, tired of the violence of organized crime and loggers, organized the uprising of their communities; or the Black women of Suárez in Colombia, taking the Pan-American Highway to demand the government in office stop the legal and illegal mining activity in the Ovejas River, where they practice their artisanal mining vocation of barequeo; or the first march of indigenous women in Brazil, which brought together more than 2,500 women to denounce machismo as a colonial legacy and the authoritarianism of the current Brazilian government.

The prefiguration of anti-capitalist horizons

As we have shown, the struggles in defense of life in the face of extractivist projects in Latin America have been socially and ecologically prefiguring other terms for the management of interdependent relations. The prefigurative is understood as, in dialogue with Hernán Ouviña, those practices that, in the present moment, 'anticipate' the germs of the future society, without waiting to take the power of the State to transform it: "it is necessary to build popular spaces and organizations within society, based on a new universe of symbolic and material significance antagonistic to the capitalist one" (Ouviña 2007, 180).

From this perspective, the anti-capitalist, anti-patriarchal and anti-colonial content does not appear as the culmination of a complete process in a given and guaranteed struggle in advance, but as part of a daily and contradictory becoming of prefigured relations of another type.

Suma Kawsay, Vida Digna, Buen Vivir, Vida Sabrosa, Suma Qamaña, and *Yeknemillis* are all terms that have sought to name from diverse realities and worldviews a different sense of existence in opposition to the colonial and capitalist idea of development, economic growth and progress (Gudynas 2011; Svampa 2013). From these horizons of meaning, agroecology projects, local production and consumption cooperatives, community water management, community ecotourism, popular education and communication, habitat construction and cooperative housing are germinating, effecting the logics of valorization and the mechanisms of circulation of socially produced wealth.

Even in urban contexts as hostile as Mexico City, we find processes of habitat production, such as the one promoted by the Francisco Villa Organization of the Independent Left, which has built a community framework in the eight housing cooperatives that provide housing for more than 1,100 families in the most impoverished areas. In their territories they advance the collective management of their security, the production of food on domestic scales, the reuse of wastewater for latrines and irrigation, the use of renewable energies, and resistance to the speculative extractivism of urbanizing capitals (Navarro 2016; Cruz Meléndez 2018).

Throughout the continent, agroecology is materialized as the production of food without pesticides, in the care and reproduction of native seeds, in productive patios, in the rotational management of crops and agroforestry systems, in the integral management of territories, in the ethical care of livestock species, and in other expressions that contribute to the socio-ecological transformation of production networks in rural areas throughout Latin America. In this sense, agroecology is a strategy of anti-capitalist struggle in hundreds of production cooperatives, as is the case of the experiences articulated in the Latin American Coordination of Rural Organizations CLOC-Via Campesina.[3]

Communities in the *central sertão do Ceará*, Brazil, who are resisting an exploration project of the country's largest uranium deposit, report that they maintain productive yards, seed houses, composting projects; agricultural, livestock, hunting and fishing consortiums, in addition to using social technologies that coexist with the semi-arid ecosystem (Montezuma 2015).

In Brazil, quilombola fisherwomen have been active in the Cumbe community mangrove, in conflict with shrimp farming and continuing their history of activism in defense of the "natural commons: mangroves, rivers, gamboas, periodic lagoons, dunes, carnaubais, beaches and traditional territory" (Nascimento and Lima 2017). Women coco babaçu breakers in the face of land conflict in the community of Centrinho do Acrísio, in Maranhão, identify the vitality and opportunity of common forms of land use and the resignification of private property.

In the region of Sararé, Arauca, a border area between Colombia and Venezuela strongly affected by oil extraction, criminalization by the state, and the armed conflict; we can observe a network of solidarity economy and self-management practices involving cooperatives of transporters, food producers, communication processes, and even financial services, which are also cooperative (Miller 2017). All this is part of a strategy of resistance to war and extractivism, with explicitly anti-capitalist orientations, embodied in the Plan de Vida or Plan de Equilibrio Regional created by the communities of the region (Moncayo Santacruz 2017). In the same area, peasant, union, community, youth, and women's organizations have managed to develop a community water management scheme for a town of 70 thousand inhabitants through the *Empresa Comunitaria de Acueducto, Alcantarillado y Aseo de Saravena (ECAAAS-ESP)* (Rátiva-Gaona 2020; Gutiérrez Aguilar and Rátiva Gaona 2021).

The ECAAAS-ESP has combined assembly forms with institutional mechanisms for the management of common goods such as water sources and the infrastructure of public services. Due to its size and level of institutionalization, the ECAAAS-ESP has managed to consolidate processes of mobilization and dispute against the State, and processes of redistribution of socially produced surpluses, both for its workers and for the communities and organizations that make up the assembly and the municipality (Rátiva-Gaona 2020).

All these community and popular networks that organize themselves against the extractivist offensive in their territories constitute the vital force of the prefiguration of social relations towards anti- and post-capitalist horizons. Many of the experiences we have pointed out here have moved from the negation of extractive projects toward the construction of less patriarchal, more ecological, more loving, less hierarchical and more beautiful forms of reproduction of life.

Conclusions

In this text we set out to trace and analyze the processes of organization and transformation underway, through the *know-how* that multiple and diverse rural and urban community networks have recreated to defend their livelihoods and affirm their modes of existence. The outlines of this know-how were composed from the recognition of a series of common dimensions that diverse community networks organizing in defense of life are producing for the transformation and generation of socio-ecological alternatives. Situating these efforts as struggles of *re-existence* called us to trace the double movement of a praxis of resistance that at the same time affirms a way of life that is denied by the sieges of the extractivist offensive.

In all these re-existence struggles we note the reconstitution of a political capacity that is capable of putting into play a series of strategies and mandates in order to put a limit to the extractivist offensive. The emergence and realization of a politics centered on the care of life has led activists to problematize and rethink the terms in which interdependent relationships are organized among community members, including coexisting beings and abiotic environments that are part of the territories they are defending. Hence, in the voices of the defenders of the territories it is possible to recognize an understanding of what it means to put life at the center of the dispute in order to organize life with dignity and sustainability.

In all these experiences, it is worth highlighting the multiplicity of interventions that women cultivate on a daily basis to sustain the defense of life not only at the level of productive and reproductive activities, but also in political-organizational tasks. The historical work of women in the defense of life has been gaining recognition, which, among other things, is due to the encounters and diverse modes of agreement between women in struggle and feminists who have been denouncing the multiple forms of violence committed against their bodies and expressing the need to think of anti-patriarchal and anti-colonial content in the struggle against extractivism.

Certainly, in the midst of the profound epochal crisis we are facing, we know that the possibility of continuing to exist as a species is at stake. The community networks that are affected and besieged by the transformations of the extractivist offensive are mobilizing all their political, spiritual, physical and emotional capacities to re-exist, reconstructing the terms of interdependence and the conditions that make dignified life possible in their ecosystems.

Notes

1 In the authors' words, "these populations do not only resist against dispossession and de-territorialization: they redefine their forms of existence through emancipation movements, by reinventing their identities, their ways of thinking, their modes of production and their livelihoods" (Porto-Goncalves & Leff, 2015: 73).
2 Las Caminatas Franciscanas, tienen lugar en julio, durante seis días en los que los caminantes pasan por varios manantiales y comunidades de la Serra do Brigadeiro. Al pasar por las comunidades, son recibidos con presentaciones culturales locales, como las Quadrilhas, y son recibidos en las casas de los vecinos para el descansar, comer, ser parte de las actividades de celebración religiosa. Información disponible en https://www.cedefes.org.br/eventos/caminhada-franciscana-na-serra-do-brigadeiro/, consultado: 07.09.2020.
3 Para conocer más de la CLOC, consultar https://cloc-viacampesina.net/que-es-la-cloc-via-campesina/.

References

Cabnal, Lorena. 2018. "Acercamiento a la construcción de la propuesta de pensamiento epistémico de las mujeres indígenas feministas comunitarias de Abya Yala." In *Momento de paro. Tiempo de rebelión Miradas feministas para reinventar la lucha*, compiled by Minervas, Colectivo de mujeres, 116–134. Uruguay: Minervas Ediciones, Una Editorial Propia.

Castro, Diego. 2019. "Self-determination and political composition in Uruguay.Una mirada a contrapelo de dos luchas pasadas que produjeron mandatos." Sociology PhD Thesis, Institute of Social Sciences and Humanities BUAP, Puebla.

Colectivo Miradas Críticas del Territorio desde el Feminismo. 2017. *Mapeando el cuerpo-territorio. Methodological guide for women defending their territories.* Ecuador: Colectivo Miradas Críticas del Territorio desde el Feminismo- Red Latinoamericana de Defensoras de Derechos Sociales y Ambientales- Instituto de Estudios Ecologistas del Tercer Mundo- CLACSO.

Composto, Claudia and Mina Lorena Navarro. 2014. *Territorios en disputa. Capitalist dispossession, struggles in defense of the natural commons and emancipatory alternatives for Latin America.* Mexico City: Bajo Tierra Ediciones, 33–74 pp.

Cruz Meléndez, Claudia. 2018. "Struggle for housing, building a new world. Las experiencias del Movimiento de Pobladores en lucha y la Organización Popular Francisco Villa de Izquierda Independiente." *Thesis to opt for the degree of Master in Political and Social, Studies* CDMX: Universidad Nacional Autónoma de México. https://repositorio.unam.mx/contenidos/luchar-por-la-vivienda-construir-un-mundo-nuevo-las-experiencias-del-movimiento-de-pobladores-en-lucha-y-la-organizaci-3504165?c=ywgJBW&d=false&q=★:★&i=2&v=1&t=search_0&as=1.

Environmental Justice Organizations, Liabilities and Trade- EJOLT. 2020. *Environmental Justice Atlas*, Available at: http://ejatlas.org/ Accessed 20.09.2020.

Fini, Daniele. 2018. "What can the communitarian? Method for the analysis of community struggles and their emancipatory potentialities: the case of the CRAC-PC in Guerrero." Sociology PhD Thesis, Institute of Social Sciences and Humanities BUAP, Puebla.

García Parra, Renzo Alexander. 2012. "La Colosa, the first cyanide leaching open-pit mining project in central Colombia. An alternative reading from the social and environmental movement." In *Mining, territory and conflict in Colombia*. Bogotá, D.C., Colombia: Universidad Nacional de Colombia, Facultad de Derecho, Ciencias Políticas y Sociales: Instituto Unidad de Investigaciones Jurídico-Sociales Gerardo Molina.

Gudynas, Eduardo. 2011. "Debates on development and its alternatives in Latin America: A brief heterodox guide". In *Más allá del desarrollo*, edited by Miriam Lang and Dunia Mokrani, Grupo Permanente de trabajo y alternativas al desarrollo. Quito, Ecuador: Rosa Luxemburg Foundation/ Abya Yala.

Gutiérrez, Raquel, Mina Lorena Navarro, and Lucia Linsalata. 2017. "Repensar lo político, pensar lo común. Keys for discussion." In *Modernidades alternativas*, edited by Daniel Inclán Lucia Linsalata, Márgara Millán, 377–417, Mexico City: UNAM-Ediciones del Lirio.

Gutiérrez, Raquel, Noel Sosa, and Itandewei Reyes. 2018. "The *among women* as negation of the forms of interdependence imposed by capitalist and colonial patriarchy. Reflections on violence and patriarchal mediation." *Heterotopías* No 17 UNC, Córdaba, Argentina. https://revistas.unc.edu.ar/index.php/heterotopias/article/view/20007.

Gutiérrez Aguilar, Raquel and Sandra Rátiva Gaona. 2021. "La producción de lo común contra las separaciones capitalistas: hilos de una perspectiva crítica comunitaria en construcción." In *La Lucha por los comunes y las alternativas al desarrollo frente al extractivismo: miradas desde las ecología(s) política(s) latinoamericanas*, edited by Denise Roca-Servat and Jenni Perdomo Sanchéz, 41–65. Buenos Aires, Argentina: CLACSO.

Haraway, Donna. 2016. *El Manifiesto de las especies de compañía*. Buenos Aires: Sans Solei ediciones.

Leff, Enrique, and Carlos Walter Porto-Gonçalves. 2015. "Political Ecology in Latin America: the Social Re-Appropriation of Nature, the Reinvention of Territories and the Construction of an Environmental Rationality." *Revista Desenvolv. Meio Ambiente* 35: 65–88.

Linsalata, Lucia. 2016. *Lo popular-comunitario en México: desafíos, tensiones y posibilidades*, Mexico: Instituto de Ciencias Sociales y Humanidades "Alfonso Vélez Pliego".

Márquez, Carolina. 2016. "Revaloración de la vida: la comunidad p'urhépecha de Cherán, Michoacán ante la violencia, 2008-2016." *Tesis de Maestría en Acción Pública y Desarrollo Social*, México: El Colegio de la Frontera Norte, A.C.

Martínez Alier, Joan. 2004. *El ecologismo de los pobres. Conflictos ambientales y lenguajes de valoración*. Barcelona: Icaria.

Miller, Amy. 2017. *Tomorrow's Power*. Documentary.

Moncayo Santacruz, Juan Eduardo. 2017. *El territorio como poder y potencia: relatos del piedemonte araucano*. Bogotá, DC: Pontificia Universidad Javeriana, Bogotá.

Montezuma, Talita De Fátima. 2015. "Licenciar e Silenciar: análise do conflito ambiental nas audiências públicas do Projeto Santa Quitéria, CE." Dissertação (Mestrado em Direito). Universidade Federal do Ceará. Fortaleza: UFC.

Moore, Jason W. 2020. *El capitalismo en la trama de la vida. Ecología y acumulación de capital*. Madrid: Traficantes de sueños.

Nascimento, João L and Iván C. Lima. 2017. *Na Pesca e na luta: mulheres quilombolas prescadoras do mangue do cumbre contra as injusticas ambientais*. Seminário Internacional Fazendo Gênero, Anais Eletrônicos, Florianópolis. Brasil.

Navarro, Mina Lorena. 2015. *Luchas por lo común: antagonismo social contra el despojo capitalista de los bienes naturales en México*. Puebla : Mexico, D.F: Benemérita Universidad Autónoma de Puebla, Instituto de Ciencias Sociales y Humanidades "Alfonso Vélez Pliego"; Bajo Tierra Ediciones.

Navarro, Mina Lorena. 2016. *Making common against fragmentation in the city: experiences of urban autonomy*. Puebla: Benemérita Universidad Autónoma de Puebla, Instituto de Ciencias Sociales y Humanidades "Alfonso Vélez Pliego".

Navarro, Mina Lorena. 2020. "Violencia biocida sobre los cuerpos-territorios en resistencia en la Cuenca Alta del río Santiago." *Diálogos Ambientales*, Semarnat, Mexico,

Navarro, Mina Lorena and Luica Linsalata. 2021. "Capitaloceno, luchas por lo común y disputas por otros términos de interdependencia en el tejido de la vida. Reflexiones desde América Latina." *Revista de Relaciones Internacionales*, Madrid.

Ouviña, Hernán. 2007. "Hacia una política prefigurativa. Algunos recorridos e hipótesis en torno a la construcción de poder popular". In *Reflexiones sobre el poder popular*. Argentina: Editorial El Colectivo.

Paley, D. M. 2020. *Neoliberal warfare. Disappearance and search in northern Mexico*. Parole. https://libertadbajopalabraz.files.wordpress.com/2020/07/paley-guerra-neoliberal.pdf.

Poma, Alice and Tomasso Gravante. 2015. "Emotions as an arena of political struggle. Incorporating the emotional dimension to the study of protest and social movements." *Revista Especializada en Estudios sobre la Sociedad Civil*, No. 4, Mexico: UAEM.

Porto-Gonçalves. 2016. "Struggle for the Earth. Metabolic rupture and social reappropriation of nature." *Polis. Revista Latinoamericana* 45.

Rátiva-Gaona, Sandra. 2020. "The power of water. Community water management and popular struggle against capitalist separations: the case of the Empresa Comunitaria de Acueducto, Alcantarillado y Aseo del municipio de Saravena, Colombia, 2019." *Revista Trabalho Neccesario* 1836: 399–403. https://doi.org/10.22409/tn.v18i36.38678.

Rojas, Axel, and Vanessa Useche. 2019. *Indigenous, Afro-descendant and peasant guards in the department of Cauca. Political history and territorial defense strategies.* Semillero Taller de Etnografía. Grupo de Estudios Lingüísticos, Pedagógicos y Socioculturales-GELPS-. Popayán: University of Cauca.

Segato, Rita. 2012. *Gênero e colonialidade*: em busca de chaves de leitura e de um vocabulário estratégico descolonial. *E-cadernos CES* 18: 106–131.

Seoane, José. 2012. "Neoliberalismo y ofensiva extractivista: actualidad de la acumulación por despojo, desafíos de Nuestra América." *Theomai* 26, Buenos Aires, November 2012, p. 123.

Svampa, Maristella. 2013. "'Commodity Consensus' and languages of valuation in Latin America." *Revista Nueva Sociedad*, 244: 30–46.

Svampa, Maristella, and Emiliano Teran. 2019. "On the frontiers of epochal change. Scenarios of a new phase of extractivism in Latin America." In *How is life sustained in Latin America? Feminisms and re-existences in times of darkness*, 169–217. Quito, Ecuador: Ediciones Abya Yala: Rosa Luxemburg Foundation.

Tischler, Sergio and Mina Lorena Navarro. 2011. "Memoria y antagonismo en las luchas socio-ambientales." *Revista Desacatos* 37: 67–80.

Ulloa, Astrid. 2016. "Territorial feminisms in Latin America: defenses of life against extractivisms." *Nómadas* 45: 123–139.

Velandia Perilla, Rubén. 2015. *Consulta popular en Piedras, Tolima: un mecánismo alterno para restringir la actividad minera.* Cali, Colombia: Facultad de Ciencias Económicas y Sociales. Universidad del Valle.

37 The dimensions of life

Environment, subject, and the wild thought

Ailton Krenak and Felipe Milanez

Epistemology and the savage life

Unlike the moral understanding of civilized and savage, I have observed the savage as life. The expression of life is wild. Life does not seek our species; life runs through our species. That's why a tree once was a stone and a river once was a cloud. It is so wonderful. You look at a cloud and you see a river. That is an experience of evolution. But not in the sense that it was thought of in the 20th century, as something that happens outside of us, an evolution that is a prisoner of these two grids: culture and nature. For a 17th-century naturalist, wild or savage was the opposite of civilized. Wilderness was that natural truth of life that they sought. Where did such a profusion of life come from? By turning to exist in the mountains, in Everest, in the Himalayas, in the African deserts, in the Andes. They were looking for the source of life, searching for where life is. It is very interesting because it is a childish human experience to search for the origin of life, when in fact it is life that speaks through us. The naturalists were hunting for life. But we are already life (Krenak 2021). We don't need to look for it anywhere. I think this is so wonderful because it gives us confidence, a firm trust that life is greater than any observation we can produce, including that of science.

For a long time, I ran away from school. For a long time, I wanted to get away from schools, any Western school, because I knew that we already had the best school. I resisted government schools as much as I could. The Krenak nation understands that we must preserve our children from negative contact with so-called "education" because to educate is to domesticate. We want to preserve our savage soul. We still want to be able to react and to be able to move within changes and events. If we accept to settle into an education system that trains us for a certain type of survival, when the sky falls on our heads, we won't know what to do. And this has already happened a few times in history, as Davi Kopenawa Yanomami has said (Kopenawa and Albert 2010). In our history, the sky has fallen on our heads a few times, and our ancestors had to create another world anew. This sky falling on our heads, we can interpret as change. In our very old narratives, our first humans, their world ended, and they had to reinvent other worlds—a second, third edition of the world (Krenak 2019).

Maybe what this brutal colonization that has hit the planet in the last millennium has done to the world means exactly that: The sky has fallen on our heads. It has confused our convictions. But if we continue, those of us with memory, clinging to that memory, we are able to search again for our place on Earth, and we are also able to reconstitute the life forms that we love and value. We love ourselves and want to continue living in these places.

DOI: 10.4324/9780429344428-43

Many of our territories have been disfigured. But we still know who we are. That is my vision. Of all the promises that the whites have made, the only one they have kept is that they would take our land. And that seems like a common expression for all nation-states. Of all the promises they have made to us, the one they insist on really keeping is to take our territories. This makes our battle all over the world a constant claim for territory, either because we cannot hunt or because we cannot have water. Any restriction means that we don't have autonomy; we don't have full access to nature's resources; and we are being enclosed. And that is what the original peoples all over the world are rising up against. The oldest colonial expression is the assault on our homes. The rest is a consequence, including the epistemologies of the North, which are only the deepening of the assault. And we have to resist.

The gift of life

We live in times of disillusionment with the human management of life on earth that has left people without direction. Life runs through everything; it is in everything and also in us. Life passes through us; it crosses us. The evolution that Darwin saw is not limited to the 19th century. It happens to us on a daily basis, just like a butterfly in the backyard. Life is always being recreated. Life never ceases to exist; it just moves from place to place. That is the example I have in mind, that of the caterpillar and the butterfly.

There is a narrative of celebration of life among the Desana and Tukano, two Amerindian Amazonian peoples, that says that these bodies in which human beings are configured today were, in another time, bodies of waters, of amphibians, of fish, and that this transformation from life in water to life on land has occurred within the history of the "canoe of transformation." This would be the event that encapsulates the understanding—the cosmovision of these peoples about the transformation of life that arrived on the continent from the water in the form of fish—of a variety of fish; a multitude of species—and that were later configured as human bodies, the human populations that today inhabit the forest.

But beyond the observation of the metamorphosis that happens in the beings that surround us—in the plants, in the animals—there is our idea of the body (Krenak and Coccia 2020). The body is what we identify as our anatomy, which has been and can be, again, other bodies. This life that goes through us also goes through other bodies.

The idea of life—of the potency of life relating to everything that exists in Gaia—this wonderful organism of the Earth is permeated by this sense of life as a creation of the world that did not happen in a remote event, but happens in every instant. This metamorphosis is the translation of this constant creation of life. It is close to what Eduardo Viveiros de Castro (1996) calls "Amerindian perspectivism." According to this idea, form and content may be changing all the time. The content is the life that runs through us, while the forms can be many, various, and can be anything from plants to organisms that we consider to be more active than a plant, like a cat or a dog that we see jumping all the time and we understand that it is alive by interacting with it.

The life of a plant or a tree, such as a kapok or Sumaúma that can live 200 or 300 years, standing on the edge of a river, where generations of people can pass by and suggest that there is a presence there. It turns out that in the giant Sumaúma tree live many other beings, and many other beings pass by there. Some Amazonian cultures will say that the Sumaúma is inhabited by an infinite number of beings, which is another idea of metamorphosis whereby the form may be constant for some time, but the content is continuously renewed. This is the idea of the re-creation of life.

What I mean to say, in a very precise way, is that we are nature. Contrary to what they teach in schools or what I read in Western literature that we have a body separate from nature. We—our bodies—are nature.

Us and the plants

Plants, as forms through which the content of life passes through, have a special relationship with human subjectivity. I'll give an example. The body painting that the Amazon peoples do with the genipap plant, a beautiful graphic art. In this sense, I want to reflect on the use of plants in the practices of care in Indigenous medicine—in traditional Indigenous medicine—based on the ideas expressed above that content traverses matter, life crosses our bodies, and we are nature.

We are used to admitting that if we take a plant and prepare a bath, a tea, or extract the active ingredient from that plant, it is possible to produce a medicine. But what I am proposing to think about is the various uses that Indigenous medicine doctors make of the association between humans and plants in healing practices. In these practices, they do not just use teas or active components of the plants for phytotherapies. They go beyond such physical and chemical practices. I am proposing the idea that spiritual masters can access the healing power of the plant in practices that affect human subjectivity. In different relationships with plants—or, more precisely, with the "masters" or "owners" in emotional relationships with plants—people can summon the extra-physical qualities of plants to function in care practices that go beyond the therapeutic.

These care practices with plants that I refer to—without applying any active component of plants other than their symbolic expressions—include, for example, bathing with the tincture of *jenipap* fruit, as part of an initiation rite that will open new perceptions for the healer. From this bath, the healer will be able to count on the help of this plant—in this case, the *jenipap*—when they are working on the health of the person being treated.

A painting on the body—an imprint on the body with black marks from the *jenipap*—will summon the healing nature of this plant to re-establish balance and activate self-healing capacities of the patient. The healing relationship between the plant and the patient unfolds in a different dimension than that of a biochemical relationship, acting on the subjectivity of the patient on an emotional plane.

Plants can also act as aids in healing through dreams with these plants—or, to be more precise, their manifestation in dreams, where they take the place of masters in the treatment of various disturbances or various states (i.e. mental, physical), in which the recipient of this healing will receive this care, establishing a communion or a connection with the nature of this plant.

The plant is an active subject, and the patient relates to it and develops an affective relationship with it. That is why the plants usually choose the healer, the agent that will work with them. A knowledgeable person can be visited by a plant in a dream, but he can also be walking along a trail in the forest and be surprised by the presence of a plant which is not endemic to that place—a plant which should not be there by that stream or on the top of that hill, but that appears nonetheless. And the knowledgeable person will recognize that plant because they will be contacted and feel the plant's communication. Of course, a person that is open to this kind of contact is already on the path and already has knowledge of this relationship. It is not very common for a person who is not open to this contact to be surprised by a plant talking to him. But plants have this power of communication.

The different relationships that the plants establish with humans do not occur in physical forms, but rather through the 'subtle body' (*corpo sutil*). This aspect that is imprinted on the body through painting—like a tattoo or drawing, like the *jenipap* on the arm of a Kayapó. This means that the *jenipan* is on the arm, but it is also in another place or another dimension of the self. It is on another place of which the one who is painted may not be aware, which is the subtle body.

Some other nonwestern medicines—different from biomedicine, like ayurvedic, acupuncture, and what is called 'Chinese medicine'—also have a sensitivity to this subtle body and other vital centers like chakras that balance the circulation of energy between the subtle and physical body.

The subtle body is like a sensitive layer of this physical body that is very active when we are children, but with the development of the intellect and these other nervous reactions of the body, it is weakened from its attachment to the physical body. One might think that a sensible body is organically related to the head, torso, and limbs or to what is sensory. But the subtle body, in the way we are referring to it, is the capacity that a person maintains throughout his or her physical development to activate the senses. For example, when someone is frightened or has an accident, it is very common for the person to go through something that psychoanalysis calls trauma—an emptiness or the detachment of this sensitive body, a loss of balance and well-being. The person's physical, mental, and subtle alignment, which we could call psychological or spiritual, we instead refer to as subjectivity.

A primary relationship to plants is through subjectivity. This relationship can be established through different uses, like picking the plant, activating the plant, and leaving its essence in the air and the environment. As we go around the garden and pick up the citronella leaves and rub them, creating a positive aura and building a beneficial relationship with our physical body. It gives a good sensation that we are breathing in, breathing out, feeling, thus amplifying our frequency waves from the physical body to the subtle body. Of course, for someone who has long experienced pressure from the surrounding reality, which forces them to think all the time and pay attention to time (e.g. if it's daytime or nighttime), the field of subjectivity becomes very obstructed, which diminishes the flow between the physical body and the subtle body.

This relationship of plants to us is prior to the direct use of the active ingredient of the plant, whether through tea or any other application of a plant. It is about broadening our understanding that *all* plants are "power plants". There is no list of plants that are or are not power plants—all plants are power plants.

In the Northeast region of Brazil, many people have a tradition with the *jurema* plant. In the Amazon, we will find peoples that use other plants, like *ayahuasca*, to stay in two well-demarcated and known fields. These relations are informed by the cultural matrix of each people, who carry forth their heritage, and it is in these communities that we find a vast repertoire of uses of healing plants. These uses range from the use of cotton to restore the physical body affected by some accident or illness to the use of some plants to invoke the power of other plants. The term 'blessing' may have ended up being common to many cultures because it is the easiest way to communicate what happens—because it is a subtle, indirect application of the plant.

For example, when you bless someone with a branch of rue (*Ruta graveolens*), which is very sacred in many cultures in Brazil, you do not have to tell the patient to eat the rue or to drink the rue tea. But when you bless with rue, you are using or, better yet, invoking the power of that plant and its healing power to communicate with the subtle body of that person. I think that everybody knows that this is effective across a variety of cultures.

A variety of Indigenous cultures refer to the owners of these plants as other beings. They are beings that control this domain and they open the domain or access to knowledge of the domain to selected humans. They are people that these plants admit as if they were their affiliates, appearing in their path, in their yard, or in their dreams. Thus, there are different ways of contact that the plant, by its own initiative, makes with us, the humans. It is different from you moving to get a plant in a botanical garden, in a nursery, or in a herbarium. It will leave the place where it is and will make contact with you. It would be as if she had assumed a place as master and subject of that caring action, working together with the human person. In that sense, the plant is not working on the person, but rather is calling the human person to work. And this is very wonderful because it opens up a perspective that these other beings have their own initiative and have their own fields of action and mastery.

When perception is opened to this field of subjectivity, the human person is opened to all these possibilities of interaction, communion, and contact. In the case of plants, this occurs even despite the long history of using plants to produce medicines and the drug industry, which is based, for the most part, on the use of the active ingredient of plants and from animal and mineral sources. Much medicine comes from plants. The medicines that people consume in their daily lives come from plants.

The colonization of nature has been driven by the search for minerals from rocks and mountains, among other sources. This relationship of coloniality with the mineral kingdom also occurs with respect to the plant kingdom, the vegetable kingdom, and the animal kingdom. These are distinctions that are made in Western sciences.

But in the interaction of plants with the subtle field—with the subjectivity of humans—what is affected are other senses. They are, literally, guided by plants. By the same token, people who have worked or still work with plants, who are alive and still work with the plants as an active mobilizer of these subjectivities, also activate the subjectivity of the plants, just as they affect these people. Therefore, they relate to and affect each other.

This entails expanding the field of subjectivity relations between beings beyond this colonizing anthropocentric use—for example, the colonization of corn or tobacco by modern industry. Tobacco has such a representative history of this agency of plants that it is a history of the Mbya Guaraní. When families of this people moved through vast regions, they felt welcomed by places. The first sign they had that they could make a *Tekohá* (sacred place of life) was the instruction that *Ñanderu* (a guarani entity) gave them that they should open a clearing in the forest for a temporary camp. In the place that they felled, they would stay there in the *tapiri* and, after a few days of camping there, they would walk through the *coivara* (traditional farm) in a place where it was already cleared. They would walk around looking carefully where the fallen trunks had sprouted a tobacco plant, the *pety*. The word *petyngua refer to* the furnace where tobacco is burned, whereas *pety* is the tobacco. For us, in the Krenak language, it is *kumam*, the smoke. The little plant appears spontaneously in the corner of the forest, in the place where a clearing was opened. And if it appears, you can build an *opy*, you can build a house of religion, you can start to plant, and you can build a house inside the area. This is a balanced settlement model, obeying the Ñande *Rekó* principle of 'the path'. The settlement is born from the land. It is not implanted in the land; rather, it springs from the land. If we were to speak in the language of the subtle field, we could say that the land demarcates a place for us and not that we demarcate the land. It is the land that chooses who it wants. This is another kind of communion or contact.

We highlight tobacco, but there are other examples. Among the Yawanawa in Acre, my dear friend Nixiwaka says that when they settled in the village down in the Gregório river after living for years in the Seringal Kaxinawa village, the Nova Esperança, they did not visit the old village for a long time. And one time he went walking up in the Seringal Kaxinawa village and was surprised to see a little plant catching his attention. He got close and saw that it was a *muká*, which is a kind of shamanic potato and did not exist anywhere else.

This plant or *muká* chose where there was an old dwelling and appeared there. Thus, this plant activated for him and his relatives a field of subjectivity that was hidden or that was sleeping, and it provided a true rebirth in the culture of the Yawanawa and in the formation of new doctors and new shamans, the people who started to receive dreams, songs, music, and painting. It was a true rebirth of something they already had because their ancestors already had this knowledge. But they were not applying it. The Yawanawa reapplied this whole repertoire of knowledge that the plants had stored for a long time to deliver to them in the sacred village of Seringal Kaxinawa.

Conclusion

We are nature, we live in territories that have been disfigured, but we still know who we are. We love ourselves and want to continue living in these places. Territories of life, as life runs through everything and happens daily. I have briefly spoken about different forms of knowing and different modes of being. And I gave an example of practices of care, based on the interaction of vegetables and humans, in which the subjectivity of humans is affected by plants, guided by plants. This entails expanding the field of subjectivity relations between beings beyond a colonizing anthropocentric use, also known as domestication. For example, the relationship between the sacred *muká* and the Yawanawá nation—which consists of hosting knowledge and experience and providing the rebirth of their culture—shows a broader dimension of what can happen when humans establish a different communion with life. As I have said, a person that is open to this kind of contact already has knowledge of this relationship.

References

Kopenawa, Davi and Bruce Albert. 2010. *The Falling Sky Words of a Yanomami Shaman*. Translated by Nicholas Elliott and Alison Dundy. Cambridge, MA: Harvard University Press.

Krenak, Ailton. 2019. *Ideas to Postpone the End of the World*. Trad. Anthony Doyle. Toronto: House of Anansi Press.

Krenak, Ailton. 2021. *A Vida é Selvagem [Life is Wild]*. Rio de Janeiro: Cadernos Selvagem, Dantes Editora.

Krenak, Ailton, and Emanuele Coccia. 2020. "Conversa Selvagem [*Wild Conversation*]." YouTube video, September 16, 2020. https://www.youtube.com/watch?v=0LvAauH3tfw.

Viveiros de Castro, Eduardo. 1996. Os pronomes cosmológicos e o perspectivismo ameríndio. *Mana* 2 (2) https://doi.org/10.1590/S0104-93131996000200005.

38 Environmental justice movements as movements for life and decolonization

Experiences from Puerto Rico

Katia R. Avilés-Vázquez, Gustavo García-López,
Carol E. Ramos Gerena, Evelyn Moreno Ortiz,
Elga Vanessa Uriarte-Centeno, Roberto Thomas Ramírez,
Jesús J. Vázquez-Negrón, Marissa Reyes Díaz,
José Santos Valderrama, and Angélica M. Reyes Díaz

Introduction: Confronting the death politics of colonial-capitalism

Colonialism has inflicted mass suffering and death upon many Indigenous and racialized peoples throughout Latin America and across the globe. While nation-state colonialism ended across most of the Latin American region by the second half of the 20th century, neocolonial relations or coloniality, enmeshed with global capitalism, have continued to shape socio-ecological dynamics in the region. This system is founded on the subhumanization of colonized populations and the imposition of a necrocapitalism or the accumulation of wealth through permanent war and conquest (Maldonado Torres 2016, 12). Colonized territories are "sacrificed" for extraction, treated as disposable, and are killed or left to die (Bolados, Chap. 18 this volume). The accumulation of wealth is thus the accumulation of extinction, leading to "mass extermination" (McBrien, 2019). Colonialism is, in sum, a politics of death, that occurs as both slow and spectacular forms of war and violence. The hundreds of environmental defenders killed every year and the countless others silenced by threats exemplify the most recent politics of death (Navas et al., Chapter 25 this volume).

Scholars and movements have recently coined terms such as "environmental colonialism" (Atiles Osoria 2014; O'Connell and Silva, Chap. 7 this volume) and "eco-imperialism" (Clark and Foster 2009) to describe the environmental implications of this system of death. These concepts are used to identify how fossil fuel corporations have become 'the new master' of territories of formerly enslaved Black and brown peoples (Bullard 1993; de Onís 2018a); how powerful countries and corporations from the North seek control over oil and other energy resources across the global South (Foster 2008); the conversion of Indigenous territories into uranium extraction sites and/or dumps for nuclear waste (LaDuke and Churchill 1985); the global trade of waste dumped from richer countries into former colonies of the global South (Pellow 2007; Demaria 2016); and the promotion of intensive extraction (extractivism)[1] of resources like minerals, timber and agricultural products, leading to enclosures, dispossession, and uneven development (Svampa 2015, Grandía, Chap. 6 this volume; Leguizamón, Chap. 9 this volume; Navarro, Chap. 36 this volume). Eco-imperialism and environmental colonialism are also used to describe how the international climate regime and its mitigation and adaptation programs promote the continuation of obscene inequalities in consumption and greenhouse gas emissions (Sultana 2022); and the management of climate-related disasters as opportunities for the accumulation of more power and profits, also referred to as "disaster capitalism" and "disaster colonialism" (Klein 2016, McCune et al. 2018, Rivera 2022).

DOI: 10.4324/9780429344428-44

Puerto Rico, occupied by Spain in 1493, and now by the United States since 1898 has never been formally independent. The island nation presents an extreme case of a long-standing coloniality, intersected with the more formal institutions of colonialism. The Spanish and United States imperialism over Puerto Rico has been characterized as a space/regime where all forms of life are intentionally exploited and expended through toxic accumulation (Lloréns and Stanchich 2019). New forms of the politics of death and war are entangled with the institution-alization of a permanent "colonial state of exception" and state-corporate crimes (Atiles Osoria 2020), through which populations—and their environments—are excluded from even the most basic rights and protections ascribed under liberal democracies in the name of 'development,' debt repayment, or disaster recovery.

The Puerto Rican experience discussed in this Chapter is interconnected—both conceptu-ally and materially—to other contexts across Latin America and the Caribbean. The Caribbean, in particular, has always been used as a "pivotal space" in the extractive geographies of colonial capitalism – from plantations and industrialization, to tourism, and banking. In the 20th cen-tury, the region had extensive investment in the mining and refining of bauxite; ports, military bases, and weapons-testing ranges; and it was the world's largest exporter of refined petroleum products, mostly for the United States (Sheller 2020). This has been coupled in recent decades with the expansion of mass visitor-centered tourism, air travel, and high-tech and service industries (ibid.). Somewhat ironically, these Caribbean islands are simultaneously sold as remote, isolated, depopulated paradises; and less important, 'poor' places to be used and then 'let drown' (Klein 2016) as sea level rises.

Latin America and the Caribbean are also home to a rich history of movements[2] that con-front the extractivism and death of colonial capitalism—be it manifested in mining, agro-indus-try, forest plantations, or urban-industrial expansion—and create alternatives to this system, grounded on communal practices for the reproduction of the material and symbolic conditions of life (Gutierrez-Aguilar 2018, Composto and Navarro 2014, Navarro, Rátiva and Furtado, Chap. 36 this volume; Roca-Servat and Perdomo-Sanchez 2020). These movements, com-posed of marginalized communities (including Black, Indigenous, Latino, peasants and women), have coined the term "system of death" to categorize colonial capitalism. They understand the environment not as something abstract for conservation, but as their home, a matter of human rights, social justice (Merlinsky 2017), and life itself. In this approach, territory, human bodies, and other living beings are intrinsically interconnected. Therefore, the violences—or the nur-turing—of one impacts the other (Zaragocin and Caretta 2021). Given the centrality of life in these struggles, Zibechi (2016) referred to them as *"comunidades en pie de vida"* ("communities standing in/for life"). 'Life' in these movements is interconnected with the defense of the "commons": those social and natural elements which communities care for collectively to ensure the wellbeing of their bodies-territories (Roca-Servat and Perdomo-Sanchez 2020). As such, EJ movements in the region frequently involve the resistance against the privatization of nature, and the development of communal 'alternatives' based on a praxis of *autogestión* (self-management) and territorial autonomy, to collectively restore, share, and sustain these commons (Villamayor-Tomás and García-López 2021; also Lopez Flores Chap. 23, this volume; Mora Chap. 22, this volume; Ojeda Chap. 32, this volume; Olarte and Flores Chap. 33, this volume). Finally, given the interconnections between colonialism, capitalism, extractivism, and environmental injustices, many EJ movements are also struggles for freedom, sovereignty, and decolonization (Connell 2020, Pulido and de Lara 2018, Sze 2020).

In Puerto Rico, the long history of colonialism has been countered by a diversity of environ-ment-related resistance movements with varying degrees of success. These include resistances to mining, industrial complexes, waste incineration, landfill facilities, agro-toxics, genetical-ly-modified seeds, and (sub)urban expansion. In this Chapter, we provide key examples of these

EJ movements and reflect on their resonance with Latin American debates on decolonial environmental justice. We consider EJ movements in Puerto Rico as organized efforts to defend life against the deadly forces of environmental colonialism and its interrelated oppressions across class, race and gender lines. We further understand these movements as enacting a liberatory form of EJ through a decolonial praxis: one that restructures social-environmental dynamics at the local scale and fosters sovereignty over the means of the reproduction of life. As elsewhere in Latin America, these movements have roots in longer histories of working-class, anti-racist, feminist, and anti-imperialist struggles, generating multiple paths towards territorial sovereignty. In what follows, we discuss three historicized periods to highlight diverse EJ movements in Puerto Rico and their lessons: (1) the resistance to toxic military-industrial projects; (2) the defense of public lands and beaches against urbanization enclosures; and (3) environmental sovereignty and just transition initiatives.

Resisting toxic military-industrial projects and fossil fuel colonialism

Environmental justice movements in Puerto Rico first emerged in the 1960s–1970s, in opposition to the intensive industrialization, militarization, and urbanization plans of "Operation Bootstrap" (*Manos a la Obra*, in Spanish), carried out from the 1940s onward by the United States to 'modernize' and 'develop' Puerto Rico. The project vilified traditional local agriculture as a symbol of ignorance and backwardness, and it uprooted thousands from rural communities to create cheap labor in the cities, while corporations were given large subsidies (McCune et al. 2018). It also sought to turn vast regions of Puerto Rico into extractive zones with military bases; petrochemical, pharmaceutical, and electronics industries; mining pits; and road and water pipeline infrastructures to supply these industries (Massol González 2019, 31–32; Figure 38.1).

A key project which became a central focus for EJ movement opposition was a proposed open-pit copper mine in 14,000 hectares of the main island's central mountainous range (*Cordillera Central*). Different groups, from progressives close to the colonialist government to the socialist pro-independence movement (Movimiento Pro Independencia), and Misión Industrial (the first EJ organization in Puerto Rico), fought to stop this mining project for two decades. Activists charged that this plan would amount to a genocide and would produce "Puerto Rico without Puerto Ricans" (Figure 38.2)—similar to how the military bases and nuclear testings made Guam and the Marshall Islands practically uninhabitable. Misión Industrial linked the EJ discourse not only with class and racial issues, but also with what the organization termed "environmental colonialism" or the use of colonial territories as zones of extraction and pollution (Concepción 1995; Atiles-Osoria 2014). In the 1970s, socialist leader Juan Antonio Corretjer declared that the independence movement had to oppose the mines because they would make Puerto Rico unlivable (Anazagasty 2015). Casa Pueblo, a community organization from the central mountain town of Adjuntas, formed in 1980 to continue this struggle. Their slogan was "Yes to Life, No to the Mines" and one of their concerns was the environmental and health impacts of mining on water and agriculture. Environmental justice was thus defined as a struggle to sustain socioecological territories—what Casa Pueblo has called the "geographic homeland,"—coupled with decolonization and sovereignty.

In the same period (70-80s), many other EJ movements emerged. Fisherfolk and other residents from the eastern island-municipality of Vieques organized against the US Navy military training range; which had occupied two-thirds of this island and carried out live bombings that have polluted the land, the sea, and its people since the 1940s. In the late 1990s, the Vieques struggle became a nationwide movement with broad international support, which successfully kicked the Navy out of the island (McCaffrey 2008). Other movements, led by Misión

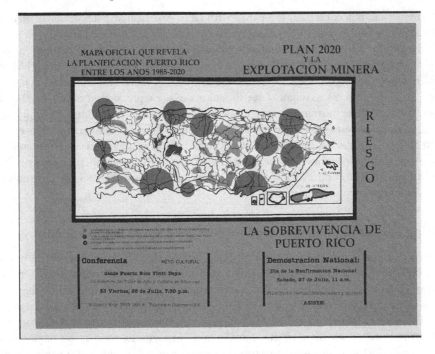

Figure 38.1 Poster for a conference and protest about Plan 2020, organized by the New Movement in
    Solidarity with Puerto Rican Independence in San Francisco, CA. The poster is based on the
    cover of a report from Casa Pueblo/ Taller de Arte y Cultura (c. 1985). The map, dated July
    1974, showed 11 proposed industrial parks (circles in the image) to be developed from 1985
    until 2020.

Source: https://ourvisionary.prcc-chgo.org/plan2020-commentary-alexismassol/. Original document available at the
Colección Puertorriqueña, University of Puerto Rico- Rio Piedras: https://issuu.com/coleccionpuertorriquena/docs/
plan_2020-_taller_arte_y_cultura_adjuntas.

Industrial, confronted the petrochemical industries of the (Berman Santana 1996) and other
megaprojects, such as a nuclear power plant, a superport, and a Monsanto factory (Concepción
1995). These movements were characterized by their strong education, organization and collaboration with the communities directly impacted by these projects. While the mines and
some of the other projects were defeated, EJ movements continue to confront the toxic-
extractive industrial 'developments' planned by Operation Bootstrap. The connections with
anti-colonialism persist in these EJ movements, even if not always front-and-center (Atiles-Osoria
2014; de Onís 2018b).

Over the last decade, the continued activism against "energy colonialism" (de Onís 2018a)
have made visible the operations of environmental colonialism, and generated new alternative
imaginaries of energy justice and sovereignty. For example, during governor Luis Fortuño's
administration (2008–2012), Casa Pueblo and other EJ local organizations successfully stopped
a proposed gas pipeline (*Gasoducto del norte*) that would have crossed the main island from south
to north (Massol Deyá 2018).

Meanwhile, the Anti-Incinerator Coalition has been fighting for a decade and, until now,
have prevented the construction of an enormous waste incinerator in the northern coastal town
of Arecibo, where communities already faced high levels of air pollution. Despite strong local
opposition, both projects were approved 'fast-track' by the government, under a declaration of
a "state of emergency" that reduced mandatory environmental analysis and public participation

Figure 38.2 Poster of a solidarity campaign against the mines in the US. The text on the right hand emphasizes that 63% of the country was owned by the US and concludes that, by 2020, it will be a "Puerto Rico without Puerto Ricans" (Terry Forman, Fireworks Graphics Collective, 1983).

Source: Oakland Museum of Contemporary Art, republished with permission.

(Torres Asencio 2017). Both the waste incinerator and pipeline were marred by corruption scandals and dubbed by activists as projects of death. These projects reminded us of what EJ activists from Naples (Campania, Italy) have called "biocide" (de Rosa, 2018). The mobilizations against the incinerator and pipeline have similarities with the long histories of struggles against 'biocide' since the 1960s in the southern region of Puerto Rico: against petrochemicals and pharmaceuticals and their legacy of highly-contaminated wastelands, to the murderous pollution by AES coal plant and its toxic ashes (Lloréns and Stanchich 2019).

Similar projects have been incentivized and facilitated by the government through the imposition of full imperial control since 2016. The 2016 federal PROMESA Act created a US government-appointed 'Fiscal Management' Supervisory Board (locally known as *la Junta*) to directly control Puerto Rico's budget and implement austerity policies to pay a highly corrupt public debt of more than $74 billion USD (Martínez Mercado 2017). After hurricanes Irma

Figure 38.3 Protest in front of the US White House with a banner reading "No to the Gas Tube, Yes to Life, Forests, Water, and People." The protest was in alliance with EJ activists against the Keystone XL pipeline in the US.

Photo by Casa Pueblo (Sept. 3, 2011). Source: https://miatabey.com/2012/01/21/obama-descarta-el-oleoducto-de-keystone-xl/.

Figure 38.4 Protest by island and diaspora activists in front of the AES shareholders meeting in Virginia, US, calling the company a murderer for polluting the air and water with heavy metals, leading to a cancer epidemic and other health problems. Protesters also made connections between the struggle for liberation and against US imperialism in Puerto Rico, as well as in other island geographies, such as the Philippines.

Photo by Frente Independentista Boricua (April 18, 2019). Source: https://elfrentepr.org/initiatives.

Figure 38.5 Protest by the Campamento contra las Cenizas (Encampment Against Coal Ashes) in Peñuelas.

Photo by Ramón Toñito Zayas - GFR Media. (November 25, 2016). Source: *El Nuevo Día* newspaper. https://www.elnuevodia.com/noticias/locales/fotogalerias/cientos-protestan-en-penuelas-contra-el-deposito-de-cenizas-de-carbon/.

and María hit in September 2017, la Junta and the Puerto Rico government proposed the "fast-track" approval of construction, energy and other infrastructure projects, together with the privatization of lands and the public energy company, and budget cuts for environmental protection, as "critical" changes for the country's recovery, ushering a new phase of "disaster (colonial) capitalism." At the same time, it has reinvigorated local environmental activism, highlighting the interconnected life-threatening toxicity of polluting corporations and the colonial institutions fostering them. The Resistance Against Coal Burning (*Resistencia Contra la Quema de Carbón*), the Anti-Incineration Coalition, and other groups have organized protests against la Junta. The anti-Junta group *Jornada Se Acabaron las Promesas* (Promises are Over Journey), for instance, brought coal ashes from the AES plant to the Capitol in Puerto Rico and displayed a banner stating: "The only ashes we want are from the Junta" (Figure 38.6).

The pipeline and incinerator were ultimately cancelled, and in 2020 a law prohibiting coal ash deposits was approved, but the struggles against environmental colonialism continue. After María, then-governor Ricardo Rosselló moved ahead to privatize the local public electric utility company, the Electric Power Authority (PREPA). This plan was continued by current governor Pedro Pierluisi, a former lawyer of la Junta. The privatization of PREPA has had immediate disastrous results for residents, including frequent death-threatening blackouts and rising electricity prices (de Onís and Lloréns 2021). PREPA presented a draft plan (Integrated Resources Plan - PIRA in Spanish), which proposed four new marine and terrestrial gas ports, gasification of all existing power plants, and the construction of a new 200-Megawatts gas plant—perpetuating our dependence on fossil fuels for another 60 years. Meanwhile, the fight over coal ashes continues, as the government resists closing the AES plant and the company. Alongside corporate interests, the local government has approved the installation of dozens of industrial-scale solar-powered farms on prime agricultural lands, threatening Puerto Rico's already fragile food security and sovereignty, a basic element for life.

Figure 38.6 Protest of *Jornada Se Acabaron las Promesas* (The Promises are Over Campaign) outside the Capitolio in Puerto Rico (August 17, 2017). Photo by Víctor Birriel - Inter News Service.

Source: *Metro* newspaper. https://www.metro.pr/pr/noticias/2017/08/17/tiran-cenizas-escalinatas-del-capitolio.html.

Defending public lands and beaches against urbanization enclosures

In the 1990s, urbanization primarily for tourist-residential projects intensified the privatization and destruction of Puerto Rico's coastline and the displacement of communities, thus becoming a key source of EJ conflicts (Valdés Pizzini 2006). In response, activists spoke of a "second Vieques" with localized struggles across the country against such construction projects (Altieri 2004). These movements reclaimed the *"playas pa'l pueblo"* ("beaches for the people") slogan, coined in the late 1960s and early 1970s by the former Pro-Independence Movement (MPI) and Puerto Rican Socialist Party (PSP) to denounce the privatization of the country's coasts at the hands of foreign hotel chains and the US Navy. The movements visibilized the connection between colonial relations and the injustices of the privatization of these commons: the beach in Puerto Rico is legally a common property owned by 'the people'.

A leading example was Amigxs del M.A.R. (*Movimiento Ambiental Revolucionario*/ Friends of the Sea), which in 2005 set up a civil disobedience camp in the tourist area of Isla Verde to prevent the privatization of a portion of the beach for the expansion of the Courtyard Marriott hotel (Figure 38.7). This led to the creation of the *Playas Pa'l Pueblo* Coalition, which gathered nearly a dozen organizations and hundreds of individuals that occupied the beach for 14 years. Through this collective re-appropriation, the community regenerated the urban coastal forest by reintroducing endemic and native coastal beach plants, and restored its dunes, making the area more resilient to hurricanes, swells and floodings. In this fight, EJ was combined with the protection of common goods for public benefit and wellbeing through a democratic, collective self-management of the territory. Moreover, it was linked to the anti-colonial fight. The beaches—with their free access for citizens' leisure and for livelihood activities—are an essential cultural-geographic component of what defines us as a people, connects us to the greater Caribbean, and distinguishes us from the empire or continent. The struggle came to an end in 2019 with a partial victory for the Coalition. All private contracts were cancelled and the land was restored as a public beach, guaranteeing the prohibition of construction in the area.

Figure 38.7 Mural in the Isla Verde camp reading "Beaches are of the people, they are not to be sold".
Source: Amigxs del Mar (Friends of the Sea) Facebook page.

The Coalition's was the first of several camps established in various beaches across Puerto Rico. Most recently, in 2021, a camp was formed against an apartment complex in the western town of Rincón, as powerful individuals, including a cousin of governor Pierluisi, tried to build a pool on the beach.

The struggles against urbanization and the dispossession of public lands and common resources intensified in recent years—particularly in the context of the debt/housing/post-María disasters, special laws for attracting vulture investors and creating "opportunity zones" for these vultures, the avalanche of federal reconstruction funds (CDBG-DR), and a law that "fast-tracks" construction permit approvals. In the summer of 2019 and after historic protests that forced then-governor Ricardo Roselló (2017–2019) to resign in the midst of corruption scandals, the Puerto Rico Planning Board published a nationwide land-use zoning map and a new Joint Regulation that eliminated protections to hundreds of thousands of acres of existing agricultural and conservation lands. The map and regulation would legalize gentrification and the intensification of tourist-residential projects being opposed by EJ movements since the 1990s, with many new such projects being advanced by disaster capitalists. Both the map and regulation resembled Bootstrap's policies (Cotto 2017), except with much more emphasis on urbanization, showing both the continuity and the adaptation of colonial extractivism that generates systemic environmental injustices.

In response, a coalition of more than 70 environmental, agroecological, sovereignty, science, communitarian, and artist collectives joined to oppose the process. The movement denounced the corruption networks trying to displace local communities and steal and destroy resources. It also emphasized our right to stay in place and the need to protect our greatest source of life: land (Figure 38.8). Multiple protests and media discussions led to a legislative inquiry and the creation of a governor-appointed commission to review the process (although the commission never released its findings publicly). Additionally, on two occasions—in 2019 and 2021—the

Figure 38.8 Flyer for one of the protests for "Our Right to Stay" and against the Zoning Map and the "destruction of Puerto Rico".

Source: JuntesPR Twitter page. https://twitter.com/JuntesPR/status/1170341059072790528/photo/1.

courts declared null and void the Joint Regulation and consequently the Map. This was a partial victory for the movement. While the public lands under threat were protected, the business-as-usual corrupt permitting processes continue under the previous regulations.

Environmental sovereignty and just transition initiatives as decolonial life alternatives to the disasters of colonial-capitalism

Events over the past four years—from the imposition by the US of the colonial control board to the aftermath of Hurricane María—have "opened eyes" to the disasters generated by the politics of death of colonial-capitalism and also to the power of self-assembly and autogestión, as part of a politics of life (Rodríguez Soto, 2020; Villarubia-Mendoza and Vélez-Vélez. 2019).

EJ movements show a long history of such dynamics between protest and the development of grassroots autogestión projects, combining environmental protection, research and education, and solidarity economies (Torres-Abreu et al. 2023). In other words, environmental justice in Puerto Rico has moved between a politics of surviving/existing and a politics of thriving/re-existing. Given that the colonial matrix is based on creating a system of toxic-extraction, which make us dependent precisely on what kills us, autogestión initiatives have been understood by various Puerto Rican EJ groups as ways of re/producing the geographic homeland to sustain it as a viable place to live and for generating concrete alternatives ("realizable utopias") that can dismantle relations of bondage.

Casa Pueblo was one of the pioneers in these utopias. In confronting the mines and later the gas pipeline, the main slogans of "the people have decided" and "yes to forests/water/people" prefigured the idea of a sovereign people's power—also present in the 2021 slogan of "Ricky we fired you." Casa Pueblo coupled these enunciations with concrete proposals for a "People's Forest" (Massol González 2019, 101–102), a forest with and for the people in the area of the mine and a project of "communitarian autogestión", based on the principles of democratic self-government, self-sufficiency, social economies, and solidarity. Casa Pueblo viewed this model as one for autonomous spaces (with "their own voice" and initiatives) to break the bonds of political dependency, manipulation, and imposition, while transforming the country from below to create their/our "realizable utopia" (Massol González 2019, 129–141, 196). With this perspective, they developed regional coffee production and processing, a community store, ecotourism, and forest-based education, music, cinema, and a radio station, as well as renewable energy. This praxis has challenged hegemonic common senses about community-economy-ecology, "living/performing a form of life that is not at the expense but rather in support of other (human and non-human) lives, a common(s) sense of life" (García López et al. 2017).

In the march of more than 20,000 people organized by Casa Pueblo against the gas pipeline, on the symbolic May Day of 2011, an activist clearly expressed the centrality of popular sovereignty in this life-and-death struggle:

If Fortuño, his government or whoever decides to impose the construction of this nefarious project, leaving (us with) no alternative to choose between death or life, I will choose life… in my sovereign right to self-defense of the waters, the forests, the karst, and the security of our people…. (cited in Delgado Esquilin 2011)

After the victory against the gas pipeline, Casa Pueblo expanded its focus on renewable energy and, in 2017, before Hurricane María, launched the "50% with Sun" campaign, which proposed to move 50% of residences in Puerto Rico to solar energy. María made this proposal more tangible and urgent in Adjuntas and across the island. With this goal in mind, Casa Pueblo developed more than 150 solar "oasis" in restaurants, small colmados (markets), and houses across all the barrios (neighbors) of Adjuntas, particularly among people with special health conditions requiring electricity (many energy-dependent patients died after María). In April, 2019, they organized the "March of the Sun," which mobilized thousands of people under the slogans of "energy insurrection; resurrection of the planet" and "for an energy future that is ours" (Figure 38.9), creating our own destiny from below. As Arturo Massol, executive director of Casa Pueblo, stated in the press conference, the march was meant to "affirm our destiny of a future of energy self-sufficiency" in the face of the government's plans to perpetuate what he has called our "energy slavery" to imported fossil fuels. In the march, Casa Pueblo presented and validated with the crowd – in a sort of spontaneous assembly—their initiative to make Adjuntas the first fully-solar town of Puerto Rico.

The Jobos Bay Eco-Development Initiative (IDEBAJO) has also been organizing autogestión projects for a long time, as part of their struggles against toxic industries in their region. In 2014, as a direct response to their successful opposition to a proposed offshore gas port, IDEBAJO began developing a community solar project (Coquí Solar) before Hurricane María, as a way to not only provide energy but also regenerate communal ties and provide employment to local youth (de Onís 2018a). After the hurricane, they continued the solar project and added projects in solidarity housing reconstruction and urban food gardens. Roberto Thomas (2019), one of the leaders of IDEBAJO, explains that these autogestión projects, by generating new sources of livelihoods, seek to confront the dual tragedy of people being dependent on working in what kills them and not being mobilized to fight with a clear vision of an alternative.

Figure 38.9 March of the Sun. Photo by Ben Moon.

Source: Honnold Foundation. https://www.honnoldfoundation.org/adjuntas-a-solar-community.

Yet, autogestión is not only a localized effort; it is also part of broader struggles for multi-scalar transformations. For example, IDEBAJO, El Puente-ELAC, Sierra Club, PREPA workers' union (UTIER), and CambioPR launched a coalition called *Queremos Sol* (We Want Sun), which elaborated a clear plan for a national renewable energy transition, centered on the principles of participation, transparency, public governance, and auditing of PREPA's debt. This campaign has been crucial in fighting against the many facets of energy colonialism in recent years, and it has demonstrated the importance, as well as the contemporary challenges, of building multi-sectoral coalitions between worker unions, community EJ organizations, and other civic organizations.

The local agroecology movement has also been crucial in self-management praxis as an alternative to disaster colonialism: from struggles against transgenic crops and protecting farmland from urbanization, to building food sovereignty from below. The work that agroecological activists have been carrying out in PR since 1989 has focused on a long-term vision that rejects the globalized and corporate agro-industrial model. Agroecology has been a way to resist assimilation, reclaim culture and lands, and to relearn how to live with our resources rather than without them, as the city has been built. Agroecology has become an organizing tool and source of knowledge to bridge increasing disparity in access to land and subsidies, especially for young farmers. Grassroots efforts have generated legislation such as the Bosque Modelo (Model Forest) Law in 2014, the Land Use Plan in 2015, the approval of a Sustainable Agriculture bachelor's degree in the University of Puerto Rico-Utuado, stopping United States-backed island-wide fumigation, and against genetically-modified crop testing and its deadly pesticide Roundup, struggles that have parallels in other parts of the region (see Castro and Mempel Chap. 13, this volume).

Organización Boricuá de Agricultura Ecológica de Puerto Rico (Boricuá Agro-Ecological Organization) was the first agroecology organization in the archipelago to integrate mutual support through a "Campesino to Campesino" framework, through which experienced and beginner farmers learn from each other by doing hands-on work in the farms. Boricuá was

Figure 38.10 The 2019 annual March Against Monsanto, organized by Nada Santo Sobre Monsanto (Nothing Saintly About Monsanto), with the message "This is what Monsanto does with our island… MUERTO RICO" (DEAD RICO).

Source: Nada Santo sobre Monsanto Facebook page. https://www.facebook.com/nadasantopr/photos/a.58012012868 9361/2608207575880596/?type=3&theater.

founded in 1989 in the mountains of Orocovis, PR. Members at the time highlighted the need to: (1) protect and share local "jíbaro" (peasant) knowledge; (2) use an intergenerational approach with emphasis on the youth; and (3) form a base that is mostly composed of "jíbaros", farmworkers and food sovereignty activists from all over the archipelago. Boricuá has based their organizing strategy doing Agroecological and Solidarity Brigades, a method founded by the organization that centers on four pedagogical principles: (1) technical work (farming), (2) political education and dialogue (workshops), (3) cultural grounding (history), and (4) shared reflection (collectivity). Brigades allow participants to learn agroecological practices for various landscapes while deepening relationships and doing meaningful work collectively.

Founding members of Boricuá and the brigades had gained experience from social movements such as Amigos de la Montaña y la Tierra Alta, simultaneously in the movements against

Figure 38.11 From the disasters of capitalism to disaster capitalism: Resistances and alternatives. Forum organized by a self-assembly of university professors (PAReS) with the support of The Intercept and CJA. Art by Agitarte.

Source: JunteGente. http://juntegente.org/2019/02/16/video-del-los-desastres-del-capitalismo-al-capitalismo-del-desastre-resistencias-y-alternativas/.

mining in the Cordillera Central, and organizing communities and fighting for the independence of PR. After a long and continuous tradition of organizing through brigades, Boricuá has formed a network of agroecological farms, working groups and educational initiatives that practice agroecology to achieve food sovereignty and social justice. Throughout the years, the brigades have contributed to other movements, organizations and communities in PR, particularly after María.

As in Casa Pueblo, IDEBAJO, and others, the agroecological network's long history of self-organization facilitated a rapid response after María. Immediately after the hurricane, Boricuá began to organize solidarity brigades across farms to support rebuilding infrastructure and crops, also integrating political training and other convivial activities (McCune et al. 2018). This organizing was part of a grassroots-led "just recovery" strategy with the support of "Our Power Puerto Rico" campaign, led by the US-based Climate Justice Alliance (CJA), in collaboration with the Black Dirt Farm Collective and La Vía Campesina International (CJA 2019). The initiative sought to confront disaster capitalists seeking to profit from the devastation after María, calling for a "Puerto Rico recovery designed by Puerto Ricans" (Yeampierre and Klein 2017). It also demanded repealing local austerity policies and the colonial PROMESA and Jones acts, and approving a Just Recovery US Aid Package. As Jesús Vázquez from Boricuá explained, the objective of this organizing is not a 'recovery' or reconstruction of the system

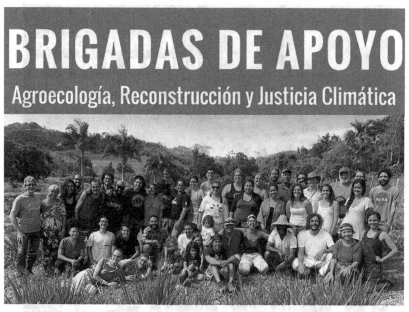

Figure 38.12 Solidarity brigades organized by Boricuá with CJA.

Source: Boricuá Facebook page. https://www.facebook.com/organizacionboricua/photos/a.207033052654156/1781011008589678/?type=3.

that caused the disaster in the first place, but to have "systemic change, starting with our own communities and territories… For us, this work is resistance, as well as the solution we are seeking at the same time" (in CJA 2019).

The vision of EJ that emerges from these movements, more expanded and visibilized after María, is one strongly based in fighting colonialism and constructing diverse sovereignties through autogestión (García López 2020; Thomas 2019). In a context of contested or 'unsettled' sovereignty of the nation state (Bonilla 2020), energy and food are seen as the basic elements of a territorial, community or 'people's' sovereignty. Breaking free from the toxic fossil-agro-industrial cartel through self-organized initiatives, creates material freedom for basic needs of life, and demonstrates the possibility of a sovereign people. Agroecological activists use slogans such as "agroecology or death" and "our machete will liberate us," transposing and transforming the slogans ("independence or death") used by independence movements of the past in Puerto Rico. It is simultaneously an affirmation of the struggle for liberation, and between life and death, between two systems that cannot coexist. Rather than the previous sense of giving one's life for the "patria", a utopia of national political independence, this new affirmation invites us to make a different, decolonized life, in order to survive and thrive. This is not only a material struggle, but also a decolonization of discourses and imaginaries,

Figure 38.13 Poster for "One single struggle" protest in the Department of Natural Resources (DRNA in Spanish) against Puerto Rico's "environmental disaster". The crosses make again reference to the death of Puerto Rico as well as the movements' demands for clean energy, protection of forests and beaches. Art by: Unknown.

Source: Amigxs del Mar Facebook page. https://www.facebook.com/amigxsdelmarpr/photos/a.2263670537193007/3308071806086203/.

a breaking of what activists call the "colonial blackmail": the idea that we are too small, resource-poor and inept to govern and develop ourselves (Berman Santana, 1996). Longtime Puerto Rican EJ activist Juan Rosario, reflecting on the challenge of confronting this blackmail, adds the "sovereignty of the spirit"—being able to take charge of one's own governance and future. Thus the emphasis to "speak with our own voice," a liberated voice that is not bound by hegemonic political-economic interests, and to demonstrate, with actions, what is possible to achieve, as empowered, liberated peoples: grow our own food, produce our own energy, make our own decisions, collectively, for the common good.

The impetus for EJ has continued to expand in recent years, connected to the globalization of climate activism. In 2019, there was a strong public participation in the global climate strike week, including a sit-in against the AES coal plant, the 7th annual El Puente-ELAC "Climate Walk"; including a forum and gathering of organizations and citizens to discuss actions linking community struggles for just recovery with the climate emergency and ongoing debates about a Puerto Rican grassroots-led version of the Green New Deal (Garcia-Lopez and Cintrón-Moscoso 2019), and an ongoing campaign on just transitions by some of these organizations. Most recently, various local environmental movements have begun to strengthen their alliances. The *Queremos Sol* coalition is allying with the food sovereignty and agroecology groups to fight corporate renewable projects that threaten agricultural lands, and also with the just recovery initiatives to fight the use of the post-hurricane disaster recovery funds to rebuild the fossil electricity infrastructure and make Puerto Rico dependent on methane gas. The slogan "One struggle" ("*Una sola lucha*") has become popular in some protests, showing the interconnections between the protection of forests and beaches under threat from 'development,' just access to land and water, the demands for renewable energy and opposition to the AES coal plant, and the cancellation of the energy privatization contract to the transnational consortium LUMA Energy (Figure 38.13). As Amigxs del Mar put it in a social media post announcing one of the protests, this is "for every one of those struggles to live in this country." A Coalition of Coastal Communities, promoted by Amigxs del M.A.R., has also met several times to face the increasing construction craze destroying and privatizing beaches. These emerging networks seek to counter the persistent fragmentation of struggles that has been recurrently identified as a challenge within the Puerto Rican environmental movements.

Still, there is much to do, as the struggles against colonialism—as manifested in la Junta and the debt—have not been clearly integrated with this network. It is also important to recognize the specificity of the context and limitations of the victories, which, by not confronting directly the root of the problem (colonial-racist-patriarchal capitalism), are faced with recurrent onslaughts of new toxic-extractive projects. While coal ash deposits are prohibited by law, coal production is not, and so now the resistance groups are struggling to "operationalize" this law, while the government drags its feet. In the same way, although the zoning map was revoked, the government has continued approving construction projects under the 2010 Joint Permit Regulation, and in 2022, proposed a new Regulation that further weakens environmental and citizen participation provisions.

Conclusions: connecting EJ struggles across the Caribbean and Latin American geographies

There are clear connections between the Puerto Rican experience and the ways in which movements across Latin America and the Caribbean resist the politics of death of colonial-capitalist extractivism, through communal projects grounded in specific territories, and decolonial practices of reproducing life-in-common. Moving ahead, deeper intersecting and expanding the scope, scale, and pace of alternatives in this archipelago will be needed to transcend the exterminating system we face. Making these connections allow different struggles to see and target the systemic causes of their problems: capitalism, colonialism, racism (Sze 2020). This process involves building collective protection and healing which creates hope for island-futures of 'humanity as a global archipelago,' to paraphrase Mimi Sheller (2020). Since its beginnings, the transnationalization of EJ movements in Puerto Rico has been key to their activities. Misión Industrial linked the movements in Puerto Rico with those of latino, Black, and Indigenous communities in the United States, participating in the First National People of Color Environmental Leadership Summit in 1991, where the first principles of EJ were elaborated – linkages which remain until today (see Sze, 2020). Later, during the national movement to get the US Navy out of the island of Vieques, there were strong links forged with other island nations fighting military bases, a key element of eco-imperialism.

More recently, the agroecology organization Boricuá and other local environmental organizations have created connections to climate justice and food sovereignty movements across the globe. After María, these networks were crucial for the struggle for a just recovery across the Caribbean, including farm-to-farm solidarity brigades in other islands hit by the hurricane. Casa Pueblo, in their struggle against the gas pipeline, joined forces with the indigenous activists against the Keystone XL in protests in the United States. They have also built bridges with other communities fighting against mining and for territorial sustainability through different events, such as an encounter of women environmental defenders that they organized in collaboration with the Latin American Mining Monitoring Project (LAMMP) in Adjuntas in 2016 (Massol Deyá 2016); and their participation in the Latin American Model Forest Network, which, for instance, inspired the creation of a "forest-school" in Ecuador emulating the one Casa Pueblo has developed in Adjuntas (Periodico La Perla 2017).

The movement against the coal plant AES has also linked local EJ struggles to various geographies. AES confronted strong opposition (and a lawsuit) in the Dominican Republic, where it dumped toxic ashes from the Puerto Rico plant; this, in turn, led to the cancellation of additional planned dumping there, and the dumping of the ashes in Puerto Rico. The connection extended to Colombia, where the coal burnt by AES in Puerto Rico is mined from (and also sent as far away as Turkey—see Cardoso and Turhan, 2018). There, a movement for a just energy transition beyond coal, connecting energy with issues of sovereignty and autonomy, has been solidifying over the last decade (Roa Avendaño 2021). In recent years, activists from across these geographies have created transnational alliances. For example, Puerto Rican journalists were key in documenting and publicizing the crimes of AES in the Dominican Republic. Additionally, a field visit to Colombia's Cerrejon region by Puerto Rican researcher Hilda Lloréns, and activist Ruth Santiago (Lloréns and Santiago 2018), longtime comrade of IDEBAJO and the anti-coal and *Queremos Sol* struggles, opened a process of collaboration. In December 2018, Colombian, Indigenous Wayu, and Dominican labor and environmental activists and researchers participated in an international conference on energy justice in San Juan, together with their counterparts from Puerto Rico.

The colonial extractivist vision of nature which generates ecocidal practices based on the creation of profits and the notion of (white, cis, hetero, male) human superiority, will ironically have its most harmful effects on the populations most excluded from geopolitical decision-making. Nation-states born from that vision have left people to die as disposable objects, without any accountability, justified by class, gender, and race differences. In this sense, Caribbean communities face an enormous challenge to create new ways of organizing spatial, political, and economic interactions, while we shake off the colonial customs that brought us here and at the same time demand responsibility, reparations, and systemic change from the global North, the major contributor to the climate crisis we live today. In other words, we need to break the cycle of aspiring to solve the problem with the very tools that caused it.

The construction of a new way of being in the world without neglecting the knowledge and technique that the historical development of the West can bring, but complemented by alternative ways of knowing that provide a different ethical and methodological content, is an imperative of life. Environmental justice movements in Puerto Rico, and the transnational networks in which they are embedded, have visibilized and confronted the projects of dispossession-and-destruction planned for Puerto Rico since the 1960s, showing the links to systemic corruption, capitalist accumulation, and colonialism. Decades of experience have culminated in an empowerment process where the people were not only fending for themselves or reacting to government atrocities as part of a system reset cycle, but viewing themselves as planners and architects of a different future.

Beyond the dominant discourse of emergency and crisis, which foresees only more crises (Garriga-López 2020), they have created the emergence of other possible worlds, understanding that all of the shared justice struggles—for education, for health, for housing, for food, for clean water, the sun, for anti-debt futures, for women and Black liberation, for national sovereignty—are interconnected, because life is not fragmentary. Decolonizing, as Adriana Garriga-López (2020) puts it, "is a future-oriented disposition" (123) full of creative and destructive possibilities, building island worlds in which we thrive, where "all that we have is what we owe to one another" (Moten and Harney 2016, as cited in Garriga-López 2020, 125). These movements have shown the way to self-assembly, to enact through a praxis of alternative "projects of life," based on autogestión, which integrate social justice and ecological sustainability, and the construction of sovereignties from below. They have thus reframed the struggle for survival and the hegemonic calls for 'resilience' into a resistance, and a re-existence, against-within-and-beyond the structures of colonial disasters.

Acknowledgments

Sections of this text draw on a modified version of García López (2020). The authors thank Beatriz Bustos and Diana Ojeda for highly valuable feedback on earlier versions of this essay. Garcia-Lopez's work was supported by the Stimulus Program for Scientific Endeavors (CEEC) of the Portuguese Foundation for Science and Technology (FCT), under contract CEECIND / 04850/2017 / CP1402 / CT0010.

Notes

1 Extractivism is a mode of accumulation of capital based on the large-scale, intensive extraction of natural resources, primarily for export. See also: Riofrancos, this volume; Svampa 2015.
2 Variously referred to as environmentalism of the poor, popular or socio-environmentalism, "territorial struggles", or "anti-extractivist movements."

References

Altieri, J. R. (2004). Bajo Fuego el Sector de la Construcción en Puerto Rico. *Indymedia*, April 28.
Anazagasty, J. (2015). "Por razones prácticas y éticas". *80 grados*, 18 December. Available at: https://www.80grados.net/por-razones-practicas-y-eticas/ (accessed December 18, 2015).
Atiles-Osoria, J. M. (2014). Environmental colonialism, criminalization and resistance: Puerto Rican mobilizations for environmental justice in the 21st century. *RCCS Annual Review*, 6: 3–21.
Atiles Osoria, J. (2020). Exceptionality and colonial-state–corporate crimes in the Puerto Rican fiscal and economic crisis. *Latin American Perspectives*, 47(3): 49–63.
Berman Santana, D. (1996). *Kicking off the bootstraps: Environment, development and community power in Puerto Rico*. Tucson: University of Arizona Press.

Bonillla, Y. (2020). Postdisaster futures: Hopeful pessimism, imperial ruination, and La futura cuir. *Small Axe* 24(2): 147–162.

Bullard, R. (1993). Introduction. In R. Bullard (Ed.), *Confronting environmental racism: Voices from the grassroots* (pp. 15–40) Boston: South End Press.

Cardoso, A., and Turhan, E. (2018). Examining new geographies of coal: Dissenting energyscapes in Colombia and Turkey. *Applied Energy* 224: 398–408.

Casa Pueblo/Taller de Arte y Cultura de Adjuntas (c. 1985) *Plan 2020 y la Explotacion Minera*. Available at Coleccion Puertorriquena, Universidad de Puerto Rico. Available at: https://issuu.com/coleccionpuertorriquena/docs/plan_2020-_taller_arte_y_cultura_adjuntas.

CJA- Climate Justice Alliance (2019) *Our Power PR: Moving towards a Just Recovery*. Available at: https://climatejusticealliance.org/our-power-puerto-rico-report/ (accessed July 17, 2019).

Clark, B., and Foster, J. B. (2009). Ecological imperialism and the global metabolic rift: Unequal exchange and the guano/nitrates trade. *International Journal of Comparative Sociology* 50(3–4): 311–334.

Composto, C., and Navarro, M. L., eds. (2014). *Territorios en disputa: Despojo capitalista, luchas en defensa de los bienes comunes naturales y alternativas emancipatorias para América Latina*. México, DF: Bajo Tierra Editores.

Concepción, C. M. (1995). The origins of modern environmental activism in Puerto Rico in the 1960s. *International Journal of Urban and Regional Research* 19(1): 112–128.

Connell, R. (2020). Maroon ecology: Land, sovereignty, and environmental justice. *The Journal of Latin American and Caribbean Anthropology*, 25(2): 218–235.

Cotto, C. (2017). Un nuevo Plan 2020. *Claridad*, September 12. Available at: https://www.claridadpuertorico.com/un-nuevo-plan-2020/.

de Onís, C. M. (2018a). Fueling and delinking from energy coloniality in Puerto Rico. *Journal of Applied Communication Research* 46(5): 535–560.

de Onís, C. M. (2018b). 'Es una lucha doble': Articulating environmental nationalism in Puerto Rico. In K Hester-Williams and L Nishime (eds), *Racial Ecologies* (pp. 185–204). Seattle, WA: University of Washington Press.

de Onís, C. M., and Lloréns, H. (2021). "Fuera LUMA": Puerto Rico Confronts Neoliberal Electricity System Takeover amid Ongoing Struggles for Self-Determination. *Georgetown Journal of International Affairs*, June 21. Available at: https://gjia.georgetown.edu/2021/06/21/fuera-luma-puerto-rico-confronts-neoliberal-electricity-system-takeover-amid-ongoing-struggles-for-self-determination/.

Delgado Esquilin, G. (2011). Consideran al gasoducto Vía Verde el próximo Vieques. *Mi Puerto Rico Verde*, May 2. Available at: http://www.miprv.com/sera-el-gasoducto-via-verde-el-proximo-vieques/.

Demaria, F. (2016). Can the poor resist capital? Conflicts over 'accumulation by contamination' at the ship breaking yard of Alang (India). In N. Ghosh, P. Mukhopadhyay, A. Shah, and M. Panda (Eds.), *Nature, economy and society* (pp. 273–304). New Delhi: Springer.

De Rosa, S. P. (2018). A political geography of 'waste wars' in Campania (Italy): Competing territorialisations and socio-environmental conflicts. *Political Geography*, 67, 46–55.

Foster, J. B. (2008). Peak oil and energy imperialism. *Monthly Review*, 60(3), 12–33.

García López, G. A. (2020). Environmental Justice Movements in Puerto Rico: Life-and-death Struggles and Decolonizing Horizons. *Society and Space Magazine*, February 25. Available at: https://www.societyandspace.org/articles/environmental-justice-movements-in-puerto-rico-life-and-death-struggles-and-decolonizing-horizons.

García López, G. A., Velicu, I., and D'Alisa, G. (2017). Performing counter-hegemonic common (s) senses: Rearticulating democracy, community and forests in Puerto Rico. *Capitalism Nature Socialism* 28(3): 88–107.

Garriga-López, A. (2020). Debt, Crisis, and Resurgence in Puerto Rico. *Small Axe* 24(2): 122–132.

Gutierrez-Aguilar, R., ed. (2018). *Comunalidad, tramas comunitarias y producción de lo común. Debates contemporáneos desde América Latina*. Oaxaca: Editorial Pez en el Árbol.

Klein, N. (2016). Let Them Drown: The Violence of Othering in a Warming World. *London Review of Books* 38(11). Available at: https://www.lrb.co.uk/the-paper/v38/n11/naomi-klein/let-them-drown (accessed March 15, 2020).

LaDuke, W., and Churchill, W. (1985). Native America: The political economy of radioactive colonialism. *The Journal of Ethnic Studies* 13(3): 107.

Lloréns, H., and Santiago, R. (2018). Traveling on Coal's Death Route: From Puerto Rico's Jobos Bay to La Guajira, Colombia. *Latino Rebels*, August 14. Available at: https://www.latinorebels.com/2018/08/14/coaldeathroute/.

Lloréns, H., and Stanchich, M. (2019). Water is life, but the colony is a necropolis: Environmental terrains of struggle in Puerto Rico. *Cultural Dynamics* 31(1–2): 81–101.

Maldonado Torres, N. (2016) *Outline of ten theses on coloniality and decoloniality*. Frantz Fanon Foundation. Available at: https://fondation-frantzfanon.com/wp-content/uploads/2018/10/maldonado-torres_outline_of_ten_theses-10.23.16.pdf (accessed June 12, 2020).

Martínez Mercado, E. (2017). Controversiales propuestas recicladas en la lista de "proyectos críticos" de Rosselló. *Centro de Periodismo Investigativo*, April 10. Available at: https://periodismoinvestigativo.com/2017/04/controversiales-propuestas-recicladas-en-la-lista-de-proyectos-criticos-de-rossello/.

Massol Deyá, A. (2016). Encuentro latinoamericano de mujeres por el ambiente y el desarrollo sostenible. *80grados*, September 9. Available at: https://www.80grados.net/encuentro-latinoamericano-de-mujeres-por-el-ambiente-y-el-desarrollo-sostenible/.

Massol Deyá, A. (2018). *Amores que luchan: Relato de la victoria contra el gasoducto en tiempos de crisis energética*. San Juan: Ediciones Callejón.

Massol González, A. (2019). *Casa Pueblo Cultiva Esperanzas. Proyecto de Autogestión Comunitaria*. Adjuntas: Editorial Casa Pueblo.

McCaffrey, K. T. (2008). The struggle for environmental justice in Vieques, Puerto Rico. In Carruthers, D. (ed.), *Environmental justice in Latin America: problems, promise, and practice* (pp. 263–285).

McBrien, J. (2019). This is not the sixth extinction. It's the first extermination event. *Truthout*, September 14. Available at: https://truthout.org/articles/this-is-not-the-sixth-extinction-its-the-first-extermination-event/ (accessed September 14, 2019).

McCune, N., Perfecto, I., Vandermeer, J., and Avilés-Vázquez, K. (2018). Disaster colonialism and agroecological brigades in post-disaster Puerto Rico. In: *Emancipatory Rural Politics Initiative (ERPI) 2018 Conference*. Available at: https://www.tni.org/files/article-downloads/erpi_cp_31_mccune_et_al.pdf (accessed March 18, 2018).

Merlinsky, G. (2017). Los movimientos de justicia ambiental y la defensa de lo común en América Latina. Cinco tesis en elaboración. In Alimonda, H., Toro Pérez, C., and Martín, F. (2017). *Ecología política latinoamericana: Pensamiento crítico, diferencia latinoamericana y rearticulación epistémica*, Vol. 2, (pp. 241–264). Buenos Aires: Facundo Martín, CICCUS.

Pellow, D. N. (2007). *Resisting global toxics: Transnational movements for environmental justice*. Cambridge, MA: MIT Press.

Periodico La Perla (2017, May 12). Casa Pueblo inspira red de Bosques Escuelas en Ecuador. Available at: https://www.periodicolaperla.com/casa-pueblo-inspira-red-bosques-escuelas-ecuador/.

Pulido, L., and De Lara, J. (2018). Reimagining 'justice' in environmental justice: Radical ecologies, decolonial thought, and the Black Radical Tradition. *Environment and Planning E: Nature and Space*, 1(1–2): 76–98.

Rivera, D. Z. (2022). Disaster colonialism: A commentary on disasters beyond singular events to structural violence. *International Journal of Urban and Regional Research*, 46(1): 126–135.

Roa Avendaño, T. (2021). Soberanía y autonomía energética. Treinta años de debates alrededor de asuntos cruciales. In Roa Avendaño (Ed.), *Energías para la transición. Reflexiones y relatos* (pp. 27–64). Bogota, Colombia: Censat Agua Viva and Heinrich Böll Foundation. Available at: https://co.boell.org/sites/default/files/2021-06/Energ%C3%ADas%20para%20la%20transici%C3%B3n.pdf.

Roca-Servat, D., and Perdomo-Sanchez, J., eds. (2020). *La lucha por los comunes y las alternativas al desarrollo frente al extractivismo*. Buenos Aires: CLACSO.

Rodríguez Soto, I. (2020). Mutual aid and survival as resistance in Puerto Rico. *NACLA Report on the Americas* 52(3): 303–308.

Sheller, M. (2020). *Island futures: Caribbean survival in the Anthropocene*. Durham, NC: Duke University Press.

Sultana, F. (2022). The unbearable heaviness of climate coloniality. *Political Geography*, 102638 (online).

Svampa, M. (2015). Commodities consensus: Neoextractivism and enclosure of the commons in Latin America. *South Atlantic Quarterly* 114(1): 65–82.

Sze, J. (2020). *Environmental Justice in a Moment of Danger*. Berkeley: University of California Press.

Thomas, R. (2019). Ending colonialism. In Morales, I. (ed.), *Voices from Puerto Rico: Post-Hurricane Maria*. New York: Red Sugarcane Press, pp. 107–112.

Torres-Abreu, A., García-López, G. A., and Concepción, C. M. (2023). De la protesta a la propuesta: Articulations between environmental justice movements and community-based management of protected areas in Puerto Rico. *CENTRO Journal* 35(1): in press.

Torres Asencio, L. J. (2017). La Ley 76-2000 y nuestro estado permanente de emergencia. *80grados*, February 17. Available at: https://www.80grados.net/la-ley-76-2000-y-nuestro-estado-permanente-de-emergencia/ (accessed February 17, 2017).

Valdés Pizzini, M. (2006). Historical Contentions and Future Trends in the Coastal Zones: The Environmental Movement in Puerto Rico. In: Baver, S. L. and Lynch, B. D. (eds.), *Beyond Sand and Sun: Caribbean Environmentalisms*. Brunswick: Rutgers University Press, pp. 44–64.

Villamayor-Tomás, S., and García-López, G. A. (2021). Commons movements: Old and new trends in rural and urban contexts. *Annual Review of Environment and Resources* 46: 511–543.

Villarubia-Mendoza, J., and Vélez-Vélez, R. (2019). Puerto Rican People's Assemblies Shift from Protest to Proposal. *NACLA*, August 20. Available at: https://nacla.org/news/2019/08/22/puerto-rican-people's-assemblies-shift-protest-proposal (accessed August 20, 2019).

Yeampierre, E., and Klein, N. (2017). Imagine a Puerto Rico Recovery Designed by Puerto Ricans. *The Intercept*, October 20. Available at: https://theintercept.com/2017/10/20/puerto-rico-hurricane-debt-relief/ (accessed October 25, 2017).

Zaragocin, S., and Caretta, M. A. (2021). Cuerpo-territorio: A decolonial feminist geographical method for the study of embodiment. *Annals of the American Association of Geographers*, 111(5): 1503–1518.

Zibechi, R. (2016) Comunidades en pie de vida. *Rebelion*, June 25. Available at: https://rebelion.org/comunidades-en-pie-de-vida/ (accessed January 10, 2020).

39 Community contributions to a Just Energy Transition

Juan Pablo Soler Villamizar

An energy model to be transformed

Every year, the number of people harmed and displaced by the climate crisis is increasing due to the destruction of their territories and livelihoods caused by torrential rains, floods, or hurricanes. These situations have highlighted the obsolescence of an energy model based on burning fossil fuels. The International Energy Agency (IEA 2013) points out that the energy sector is by far the largest source of energy consumption in the world. This sector is also a large contributor to climate disorder, environmental pollution, social inequality, and energy waste. In this regard, the International Energy Agency (IEA 2013) points out that the energy sector is by far the largest source of greenhouse gas emissions, accounting for more than two-thirds of the total in 2010 and highlights that global energy consumption continues to increase, led by fossil fuels, which account for more than 80% of the world's energy consumption.

The effects of the fossil fuels model are exacerbated by the distortion of the capitalist system on energy production, that is, energy as a common good and its access as a right has been replaced by energy as a commodity and access to it has been conditioned by people's ability to pay. Under the agreements by governments and companies, a corporate architecture in the production and management of energy has been built that maximizes profits and dividends and socializes losses and social and environmental liabilities.

In the face of this situation, without paying attention to the distortion introduced by capitalism, governments and companies have positioned Energy Transition (ET) as a proposal to change the ways of producing energy and thus reduce greenhouse gas (GHG) emissions. However, the notion of ET tends to be given a variety of meanings depending on who enunciates it and, above all, with what intentions (Nuñez 2020). The majority of governmental proposals respond more to corporate ET that ends up exacerbating the climate, environmental, and social crisis, to the point that some governments have limited their vision of ET and even consider gas[1] or fracking oil as the transition fuels, and renewable energies as a new niche market.

This has been in concert with the path outlined in the United Nations Framework Conferences on Climate Change. Since the establishment of the Kyoto Protocol (UNFCCC 1998), companies have envisioned the reduction of GHG emissions and the rise of renewable energies as a business opportunity, where the Clean Development Mechanism, Emissions Trading, and carbon credits were created. Then the proposals derived from the Paris Agreement (UNFCCC 2015) for the decarbonization of the energy system show that the structural transformation of the energy system is far from its goals and objectives due to the greedy attempts to generate dividends from the crisis. This is why, for example, there are companies determined to exploit the energy potential of countries to the maximum, without even stopping to assess whether this energy is really needed or whether local communities are in agreement.

DOI: 10.4324/9780429344428-45

The architecture of the renewable energy business has gone beyond expectations.

In the case of Colombia, the National Development Plan 2018–2022 had established plans to expand the installed capacity of renewable energy to 1500 MW; however, with only one auction, contracts for 2250 MW were concluded, exceeding the governmental expectation (Presidency Colombia 2019). These actions have been so successful that companies have been proposing *private energy auctions* in order to not depend on governments.

Outlines for a "just" energy transition

There is social consensus on the need for an energy transition to face the climate crisis, but a series of requirements must be met to ensure that the changes generated actually provide a solution to the problems generated by the fossil fuel model. To this end, trade unions and social and environmental organizations have proposed incorporating the dimension of justice into the transition.

Thus, a Just Energy Transition poses as conditions to achieve a structural transformation of the fossil fuel model: intersectoral dialogue, respect for local cultures, real evaluation of alternatives, de-growth and reduction of consumption, among other things. Concepts such as a Popular Energy Transition based on socio-environmental, participatory, and cooperative justice are also linked to a Just Energy Transition (Bertinat et al. 2021).

Within this framework diverse organizations have advanced in the construction of proposals or practices for the transformation of the model energy. In other words, from a social perspective, climate change is fundamental, but it is not the only adversity that generates the new energy model. In addition to the fossil fuel model, there are other impacts that must also be addressed with the same urgency.

For example, a Just Energy Transition advocates universal access to energy goods through community self-management of energy, reduction of environmental impact by reducing consumption, decentralization, and energy efficiency which promotes practices conducive to other paradigms of society that respect the customs and traditions of all peoples, promote gender equity and the dismantling of patriarchy.

The following points highlight the main elements or guiding principles that an energy transition should contemplate in order to be just, which emanate from the implementation of a popular energy model produced by social organizations in Latin America (Soler Villamizar 2020).

Changing the energy model, not just the way energy is produced

The predominant notion of ET is to replace oil, gas, and coal with renewable energies such as solar, wind, and geothermal. For this reason, a popular energy model proposes an integral vision of all the components and characteristics of the energy system, not simply the technology used to produce or transform energy. In this way, it is proposed to stop the violation of the rights of communities, irreversible environmental impacts such as the availability of drinking water, forced displacement, unjustified electricity and fuel prices, targeted assassinations of environmental defenders, among other injustices.

Energy for what, for whom and at what cost?

These three simple questions demand complex answers that must necessarily guide decision-making in terms of redesigning public policies in the energy sector. This exercise has been put into practice by the Colombian movement, Ríos Vivos (2014), which found that the use of energy from new projects is the export or supply of the demand of mining extraction projects in the future; they also point out that energy is produced for industrial sectors at a lower cost than residential users.

Therefore, "¿Energy for what?" is linked to knowing what use the energy will be put to in order to prevent waste and avoid spending the energy potential before it is needed. For example, a dam has an average useful life of 50 years, so it makes no sense to build it 20 years earlier than it is needed.

The question "for whom?" allows for the definition of issues of scale and form of generation. That is to say, a locality can opt for one or another small-scale generation proposal without the need to connect to Colombia's National Interconnected System (SIN), which implies a percentage of losses associated with the transport of energy between the site where it is generated and where it is consumed, as well as other problems which can be avoided with decentralized projects. Here it is also necessary to review the energy policy in depth, given that the differentiated tariffs in favor of large consumers show that the energy system responds to private and commercial interests over public and social needs.

In addition to this assessment, it is necessary to quantify and qualify the real costs of energy; that is, not only the cost referring to the kilowatt hour, the gallon of fuel, or the cubic meter of gas. This exercise involves the application of ecological economics to establish the real or approximate costs of energy generation activities, including environmental mitigation variables, cultural impact, unforeseen externalities, and social impact, among others.

Energy as a human right and the common good

The wave of privatization that took place in Latin America since the 1990s marked a turning point in the provision of public utilities, especially water and energy. Water and energy **users became customers**, and access to these common goods became conditioned on the ability to pay.

In this process, loans from international banks demanded the participation of private companies in the management of public infrastructure (Urrea and Camacho 2007). This came together with the unfounded myths of liberalization, among which were: that liberalization would improve efficiency; electricity would be cheaper; it would bring environmental benefits; governments could choose (privatization would not be an imposition); it would strengthen democracy; and it would benefit the poor (Chavéz & Roa 2002).

In response, social organizations have demanded that water and energy be declared as common goods, never merchandise, as is the case of the Atarraya Nacional en Defensa del Agua y la Energía in Colombia, or the Argentinean organizations of the Observatorio Petrolero Sur (OPSUR), that have agreed that energy should be established as a human right at the constitutional level (Álvarez 2021).

Democratize energy and universalize its access

It is not possible to produce wealth with energy poverty. Democratization implies access; capacity building for construction, management and maintenance; replacing external dependence with local independence; as well as respecting local uses and customs. In the words of Bertinat (2021), democratization aims to generate spaces for active citizen participation in energy-related decision-making processes. It seeks to balance power relations in the energy sector, to ensure that access to information is free and unbiased, and to build counter-hegemony against the large energy multinationals.

Nationalization of companies in the energy sector energy sector

The liberalization of the energy sector in Latin America has meant that foreign companies and national private economic groups predominantly control the sector despite the fact that the countries have sufficient capacity to operate and maintain it and even expand it. The current architecture of the system results in capital flight, since the profits generated are invested in places far away from where the greatest impacts were generated. Therefore, we propose the nationalization of companies dedicated to energy generation, where communities and local governments where projects are built have real participation in decision making.

Justice for working men and women

The trend of precarious labor conditions through outsourcing, lower wages, and increased working hours in renewable energy projects must be reversed. According to the Brazilian Electricity Sector Workers Union, the cleaner the energy source, the more precarious the conditions of the workers.

The notion of a just transition was born in the 1980s in the United States when a chemical company was sealed due precisely to environmental impacts. Faced with this sudden situation, the company's labor union raised the importance of acting in favor of environmental preservation but warned of the need to carry out a closure while providing guarantees to workers who did not know any other trade, in other words: the bill for the closure of the factories due to operational failures should not be paid by the workers and because of their situation they should have the necessary conditions to be reintegrated into the labor market without affecting their livelihoods. This discussion was finally taken up in the text of the Paris agreement (UNFCCC 2015) by recognizing the need to carry out a fair reconversion of the labor force and the creation of decent work and quality jobs.

Stop the violation of human rights

A new energy model must start by eliminating human rights violations, including assassinations, forced displacements and disappearances, and carrying out the corresponding individual and collective reparations processes. A large number of union leaders in the energy sector, working to defend collective rights, as well as leaders of communities affected by energy projects in Latin America, have been murdered with impunity. A series of abuses have also been committed against nature that affect the living conditions of local communities, such as the destruction of livelihoods, access to water, disrespect for beliefs, and the breaking of social fabrics. Avoiding these violations starts with revealing the truth about these abuses as well as reconciling with and respecting decisions made by local communities.

Cultural transformation

The construction of another energy model must be accompanied by a cultural process that implies new behaviors or practices in the use of energy, change in production/consumption logics as part of the search for a new economic paradigm which, until recently in Latin America, has focused on the extraction of minerals, hydrocarbons and biodiversity.

The current way of life, in large cities and rural areas, is mediated by time and efficiency. The aim has been to produce faster, move around in less time, limit leisure and recreation, intensify the working day, and encourage consumerism. All these activities are associated with a greater and intensive use of energy, so that the transformation of the system is mediated by new cultural practices that are related to the care of oneself, of others, of nature, so that speed or less time is not the variable that defines relationships and quality of life.

Social or fair energy tariffs

Throughout Latin America the people demand a fair price for electrical energy in relation to socio-economic conditions. It is worth noting that one of the unfulfilled promises of the privatization of the electricity sector in the continent was the reduction of electricity tariffs.

In Uruguay, there has been a call for the establishment of a **Social Tariff** that implies paying a fair price. In Colombia, it has been proposed to establish a **free vital minimum** that provides access to energy necessary for decent living conditions.

In terms of a fair tariff, organizations debate the unsustainability of the model in which companies that consume larger amounts of energy pay lower prices than residential users, as is the case in Chile, Peru, and Guatemala (see Table 39.1). In addition, there are markets where large consumers are allowed to reach agreements directly with generators, achieving even lower prices, as is the case in Colombia and Brazil.

Conflicting discussions for the integration of proposals

Subjectivity, uses and customs, the lived experiences of groups and communities, the cosmovision, whether it be the cosmogony of the peoples or the results of academic research, set an important standard when proposing methodologies and discussions aimed at generating consensus and articulating social agendas around the energy transition.

The use of unfamiliar terms, or terms that have different meanings according to cultures or beliefs, can create distances or differences that are not necessarily dissensions but that can delay or hinder decision-making. What is proposed then is not to label these discussions as

Table 39.1 Electricity prices in Latin America in 2019

Country	Residential Rate	Commercial Tariff
	Cents/kWh	Cents/kWh
Venezuela	0,00000000724	0,0000000085
Paraguay	2,79	5,17
Mexico	4,6	17,99
Argentina	7,09	8,57
Ecuador	7,23	10,09
Brazil	8,26	14,07
Bolivia	10,54	21,58
Colombia	12,23	14,4
Costa Rica	14,27	19,92
Panama	14,35	23,41
Chile	15,01	14,16
Peru	16,56	11,79
Guatemala	17,95	14,65
El Salvador	20,77	21,41
Uruguay	23,08	16,93

Source: BNamericas, 2019.

"dissensions," which from the outset mark a distance, but to call them "conflictive debates" which, with an appropriate approach, can generate strong consensus and strengthen intersectoral organizing.

As an example, we can cite the discomfort and debate generated by the use of the expression "no to dams" at an International Meeting of People Affected by Hydroelectric Projects, where Asian communities stated that they could not oppose dams because they guaranteed their access to water. It took a long time to understand that they were referring to artisanal water collection systems similar to the Jagüeyes built by the Wayúu indigenous communities in the Colombian Guajira, and another time to understand that these traditional systems did not correspond or were not categorized with the "dams" that were being discussed at the meeting.

Another example can be found in discussions about the rejection of large-scale metal mining. Many *barequeros*, or *garimpeiros* (a person who searches for gold and extracts it) as they are known in Brazil, often feel that they are being blamed or that their activity is being rejected even when there is no comparison: on the other hand, many demonize this work even though they are unaware of the techniques they employ, which do not use mercury or cyanide to separate the gold, nor the cultural dimensions of their work.

Some of the conflict debates identified in Latin America are highlighted below (Soler Villamizar 2020):

- **Nationalization of companies**: The conflicts generated by the companies that manage the current mining-energy model cannot be solved by nationalizing or renationalizing the companies. The debate can be overcome to the extent that in addition to "nationalizing" there is a dialogue on the need to recover the logic of the public in the actions of the companies instead of a commercial logic.
- **Extractive transition**: the construction of solar panels, windmills, batteries, etc., requires minerals, so a call is made to evaluate to what extent the proposed techniques contribute to the required transition or if, on the contrary, they lead to a new face of the extractive model in which environmental and cultural liabilities are once again assumed by local communities while the climate crisis worsens.
- **Closure of mines and other projects**: Usually workers in the sector distance themselves from energy transition discussions because they fear the possibility of losing their jobs. The dialogue should allow them to understand that the transition will not happen overnight, that it is a global process determined by the economy and the crises that are already taking place and that what the social movements are really proposing is to find solutions with decent jobs within the framework of a just transition or reconversion.
- **Mining yes, but not like this**: Some groups propose that mining done right is possible. Others say that any mining, regardless of scale, is not sustainable. The whole system of mineral extraction should be studied and categories established according to the uses that are actually needed. An example can be the discussion of gold: a large part of its extraction goes to federal reserves and for jewelry, uses that generate an unjustifiable burden of environmental and social liabilities; however, another small part of that gold is destined for medical use and communication, which are defended by some sectors as necessary mining.
- **Jobs and fiscal resources**: It is argued that the mining and energy sector produces the jobs and fiscal resources that countries need, making opposition unfeasible. Rudas and Espitia (2013) found that, for the Colombian case, employment generation is marginal despite its accelerated growth and dynamism between 2000 and 2012 and that the share in total employment was just over 1% compared to 13% for industry and 18% for the agricultural sector. Regarding the generation of resources, Rudas (2017) points out that the Colombian Constitutional Court declared Article 229 of the Mining Code, in Ruling

C-1071 of 2003, which established that royalties were incompatible with other taxes, so that mining companies, for more than a decade, would have deducted the concept paid in these from their income tax. In this way, almost 50% of the value of the royalties paid would have been deducted from the companies' income tax. This is why it is necessary to review the entire structure of taxes, exemptions and tax breaks in each country so as not to analyze the amounts of royalties alone, which at first glance may seem a lot, but the exemptions should be deducted as well as the public expenditure arising from environmental stabilization and social care derived from the impacts of the mining and energy sector activities.

- **The role of technology**: Some sectors have the tendency to grant truth and knowledge to technicians or experts, ignoring the ancestral knowledge of other cultures. For this reason, it is necessary to recognize the knowledge of other cultures and ancestral peoples, and the recognition of pluri-nationalities.
- **Extractivism in progressive governments**: For some groups, the increase in extractive projects and activities in progressive governments is justifiable and necessary; for others, they do not correspond to the energy transition required to mitigate climate challenges and other crises. This can be overcome by intersectoral dialogues convened by these governments, a cost-benefit analysis of what has been done and of what has been planned, and developing a policy to address the negative impacts generated and unforeseen externalities.
- **The need for development**: Some groups have their own vision of development or of the good-this-or-that, others perceive it as an empty, imposed, and discursive conception. Development is generally assimilated to economic growth, and this to the growth of the set of goods and services that are produced, regardless of the type of goods or for what or for whom they are intended (Bertinat 2016). In many cases it is not necessary to delve into what development means, or whether the correct way is to propose endogenous development, *sumak kwasay* (a concept from the quechua language that refers to *buen vivir* or good living) or build alternatives to development, because regardless of the category used, what should prevail is that the organizations articulate their practices, such as agroecology, real participation, democracy in decision making, self-care, respect for diversity, etc.

Community energy for a just transition

In Latin America, several social organizations have advanced in the implementation of practical initiatives for the transformation of the energy model, through which they have promoted community energy management, and the creation of decent living conditions to mitigate the social, environmental, economic, and cultural impacts generated by energy projects in their territories.

These actions emerged long before the global discourses of energy transition were on the public climate agenda and, in general, have not been supported by the government. Here we highlight the work carried out by Fundación Centro para la Investigación en Sistemas Sostenibles de Producción Agropecuaria (CIPAV[2]) in Valle del Cauca, Colombia, regarding the implementation of biodigesters since the 1980s; the installation of biodigesters associated with agroecological production in the province of García Rovira by CENSAT Agua Viva[3] in Santander, Colombia in the 1990s; the adaptation of solar collectors for hot water by the Movement of People Affected by Dams in Brazil in the 2000s; and the diffusion of the practice and knowledge associated with the use of Biogas in Cuba (Chacón Guardado et al. 2017).

To make these and other more recent experiences visible, CENSAT Agua Viva and other social organizations have created a virtual space for meeting and dissemination of energy alternatives underway in Latin America. In this virtual space interested people and movements can, from seeing what has been done, be inspired and create their own local initiatives or improve existing ones. This space is hosted on the site: http://energiasparalavida.censat.org.

In the last twelve years there has been a significant qualification of community-based social organizations that have begun to develop in their small plots of land experiences of self-management of energy, making their proposals for sustainable occupation of the territories more comprehensive. Among them are the Colombian Biomass Energy Network, Red Biocol,[4] the Latin American Network of Biodigesters for Latin America and the Caribbean, Red Biolac,[5] the Colombian Movement, Ríos Vivos, the Guatemalan indigenous communities that have implemented community turbines with the support of the Madre Selva Collective, Hidrointag in Ecuador, the Colectivo de Reservas Comunitarias y Campesinas de Santander, CRCCS,[6] and the Comunidades Sembradoras de Territorios, Aguas y Autonomía, Comunidades SETAA, affected by the Hidroeléctrica Hidroituango hydroelectric dam in Antioquia, Colombia.

These last two organizations, together with CENSAT Agua Viva, have been making a joint effort during the last eight years to generate a process of promotion and dissemination of community alternatives that generate local solutions to social, environmental, and economic needs and problems as an alternative to the prevailing energy model. In this way, a process of dissemination and promotion of **community energies** has been created, which represents the possibility for social movements, communities, peasant groups and the like, to become energy generators within the framework of a sovereignty or popular energy model (Soler 2021).

Principles of community energies

Reflecting on practical work, we can say that a community energy project is one that incorporates the following principles (Soler Villamizar 2021):

- Generation of energy from sources other than fossil fuels, damming of rivers or nuclear energy; therefore, it is based on non-conventional renewable energies: solar, wind, biomass, human, and hydro sources, among others.
- It should consider an analysis of alternatives and choose the one with the least environmental impact, taking into account the availability of environmental assets, the socio-cultural context, gender aspects, etc.
- It must have the participation and approval of the communities in the impacted area, which must have undergone a process of empowerment in their vision of energy sovereignty, in order for participation to be real and effective.
- It must transform power relations, therefore, decisions must be made by the community and have an institutional framework that facilitates participation and democracy.
- The energy generated should be used for family or community self-supply, encouraging new uses and applications of energy within the organizational processes; bartering or commercialization of surpluses with neighbors; and the sale of surpluses.
- Community initiatives that have the possibility of commercializing their surpluses should direct the investment of their profits to improving the living conditions of the community, environmental preservation, etc.

Community training school

Because access to energy-related knowledge has been restricted to those with the possibility of entering academic institutions and conditioned on the investment of economic resources, time, and travel; when communities are beneficiaries of alternative energy projects, they are subordinated to the assistance of a technician from outside the community. This outside technician is unable to provide the community with the necessary knowledge which, over the course of time, implies higher economic costs in the operation and maintenance of the project. This also limits the possibility of generating democratic processes of participation in the access to energy.

Faced with these challenges, in 2014, an informal popular education process was initiated, in communities from the Colombian departments of Antioquia, Santander and Córdoba, called Escuela de Formación de Técnicos y Técnicas Comunitarias en Energías Alternativas (School for the Training of Community Technicians in Alternative Energies), in which peasants from grassroots community processes participate, with the particularity that some of the participants do not know how to read and write, did not finish high school or have not studied for a long time. These considerations have involved the development of a methodology based on learning- by-doing around four basic technologies: biodigesters, photovoltaic systems, solar dehydration and efficient wood stoves. In all cases, the implementations have sought to add value to rural production and find new or new applications for energy.

Community energy dates for a just transition

From the analysis of the implementations carried out in the theoretical-practical training process, the proposal of **Community Harvests for a Just Energy Transition** has emerged. These harvests, in turn, are a process of self-recognition of the contribution of traditional peasant practices and neighborhood communities for the mitigation of the climate crisis and for the construction of another energy model.

The community energies are grouped in with the harvests of water, energy, sun and new relationships, therefore, the hope is that, regardless of the place, any person or community begins to sow actions today, to obtain a good harvest of energies for life tomorrow.

Sun harvests are carried out by the implementation of solar dehydrators, efficient wood stoves and ovens, woody forests or wood energy crops, biodigesters, photovoltaic systems, solar collectors, agroecological crops, and fodder banks. **Water harvesting** is carried out through the installation of pico- and micro-hydroelectric power plants, rainwater collection systems, river crossings, water mills, and reforestation. **Human energy harvesting** is achieved through the use of bicycles, bicimáquinas, the organization of mingas and convites (reunions of people that gather for a common objective), or the recovery of practices such as the "brazo cambiado" (changed arm). The **harvesting of new relationships** is accomplished by promoting the commercialization or use of local products, agroecological farmers' markets, by encouraging at the individual and collective level the consumption of what is necessary, by the recovery of recipes and typical dishes of the regions, advocating that the ingredients are produced locally, by permaculture, by the diversification of crops and by the implementation of processes of self-care and self-management of health such as synergistic medicine and the use of medicinal plants.

The harvests are open to implementation, so if you do not yet have any practice, you can start with one and then continue sowing new practices, activities, or actions to increase the harvest. They are open to complementation since there are certainly many more practices and actions that can be increased in each one of them, or even recognizing existing practices since usually the actions that have been ongoing for centuries are not categorized as alternative energies, or sustainable energy self-sufficiency, or as part of a Just Energy Transition.

However, all actions or activities that are framed within the harvests have a great potential to contribute significantly to the reduction of GHGs, favor carbon sequestration, and ensure the conditions for the reproduction of the water cycle. They also enable rural and neighborhood communities to have access to sustainable energy and to improve the quality of life and environmental conditions. Specifically, these benefits derive from reducing the burning of fossil fuels in the transport of agricultural inputs, concentrates or animal feed, mobility, and/or the cooking of food and discouraging energy wastage; as well as from the creation of added value to peasant production, fair trade and increased productivity and quality of agricultural products; reducing deforestation, strengthening the social fabric, mitigating odors in livestock production, reducing health care costs, and decentralizing water and energy management.

Concluding remarks

For Latin American social movements and organizations that have been working for decades toward the transformation of the energy model, the energy transition represents only one part of the **energy sovereignty** proposal that corresponds to the construction of a **new paradigm of society** in which all forms of life and worldviews are respected.

However, it is noted that the energy transition will only achieve its objective of facing climate and environmental challenges to the extent that it involves all sectors of society in decision-making, understanding that it is not merely a technical issue. This is why the organizations propose the need to carry out a **Just Energy Transition**.

While this is being achieved, it is necessary to multiply exponentially the community energy proposals and the spaces for deliberation, meeting, and sharing of experiences among the different sectors and organizations of society. In addition, there are issues that require a deep reflection to guide decision-making for the construction of public policy related to the energy sector and related areas. For example, the transition per se does not make sense if new technologies demand equal or greater amounts of materials and energy for their manufacture. We cannot fall into the error of promoting an **Extractive Transition** oriented by the opportunity of new green businesses.

Nor should short-term or low-impact solutions be proposed in light of the time available for humans to carry out a transition that generates dignified living conditions for all and guarantees the minimum vital needs of all species of fauna and flora. On the contrary, we must advocate for long-term solutions that transform the structural causes of the climate crisis, but also of the biodiversity, food and financial crises, among others, which implies rethinking the fiscal dependence of the state on the mining and energy sector. At the same time, it implies challenges for all sectors of society in order to carry out the necessary transformations in power relations, energy uses and culture: only in this way will another energy model based on social and environmental justice emerge.

Notes

1 https://www.larepublica.co/especiales/colombia-potencia-energetica/el-gas-natural-sera-fundamental-para-la-transicion-de-la-matriz-energetica-nacional-2966365; https://www.semana.com/pais/articulo/como-van-las-tarifas-de-gas/307780/.
2 www.cipav.org.co.
3 http://www.censat.org/.
4 www.redbiocol.org.
5 http://redbiolac.org/.
6 http://www.fundaexpresion.org/Espa/colectivo.html.

References

Álvarez, Rafael. 2021. ¿Energía para qué y para quien? About the International Weekly of Energy Transition in Latin America. Available at: https://www.facebook.com/watch/?v=455651275850209.
Avendaño, Roa, Tatiana, Juan Pablo Soler Villamizar, and José Aristizábal. 2018a. "*Transición energética en Colombia: aproximaciones, debates y propuestas*." Fundación Heinrich Böll.

Bertinat, Pablo. 2016. Transición energética justa: Pensando la democratización energética. *Montevideo, Uruguay, Friedrich Ebert Stiftung. ANÁLISIS Magazine NO 1,* December 2016.

Bertinat, Pablo, Jorge Chemes, and Lyda Fernanda Forero. 2021. *Energy Transition: Contributions for collective reflection.* Transnational Institute and Taller Ecologista.

BNamericas. 2019. "Precios de la electricidad en Latinoamérica: Comparación de países." *BNamericas,* Retrieved on 12/20/2020. https://www.bnamericas.com/es/noticias/precios-de-la-electricidad-en-latinoamerica-comparacion-de-paises.

Chavéz, Daniel, and Tatiana Roa. 2002. ¡Apagón! Los Mitos de la Liberalización de la Energía Eléctrica. Proyecto de la Energía. CEUTA, CENSAT and Transnational Institute. TNI Briefing Series No 2002/5. Available at: https://www.tni.org/es/briefing/apagon.

Chacón, Guardado, José Antonio and Jorge Santamarina Guerra. 2017. *El Movimiento de Usuarios de Biogás en Cuba.* Playa, Cuba: Cubasolar.

International Energy Agency - IEA. 2013. *Redrawing the Energy-Climate Map, World Energy Outlook Special Report.* Paris: OECD/IEA.

Movimiento Ríos Vivos. 2014. ¿Energía para qué, para quién y a qué costo? Paper by Movimiento Ríos Vivos at the National Mining, Energy and Environmental Constituent Assembly. Available at: https://prensarural.org/spip/spip.php?article15606.

Nuñez, Jonatan. 2020. *Transición Justa: Debates Latinoamericanos.* Observatorio Petrolero Sur. Available at: https://opsur.org.ar/2020/11/26/transicion-justa-debates-latinoamericanos-para-el-futuro-energetico/.

Presidency Colombia. 2019. "Con nueva subasta, Gobierno Nacional superó en más del 50% la meta en energías renovables." Retrieved December 30, 2020 from: https://id.presidencia.gov.co/Paginas/prensa/2019/Con-nueva-subasta-Gobierno-Nacional-.

Avendaño, Roa, Tatiana, Juan Pablo Soler Villamizar, and José Aristizábal. 2018b. "*Transición energética en Colombia: aproximaciones, debates y propuestas.*" Fundación Heinrich Böll.

Rudas, Guillermo. 2017. ¿Son deducibles las regalías como costo en el impuesto a la renta? *Razón Pública,* November 6, 2017. Retrieved March 15, 2020 from: https://razonpublica.com/son-deducibles-las-.

Rudas, Guillermo, and Jorge Espitia. 2013. "La paradoja de la minería y el desarrollo. Análisis departamental y municipal para el caso de Colombia. En Minería en Colombia." In *Minería en Colombia: Institucionalidad y territorio, paradojas y conflictos,* 27–70. Bogotá: Contraloría General de la República.

Soler, Villamizar, Juan Pablo. 2020. *Transición energética en América Latina.* CENSAT Agua Viva. Living Rivers Movement. MAR.

Soler, Villamizar, Juan Pablo. 2021. *Energías comunitarias: oportunidades y desafíos en Colombia.* Bogotá: CENSAT Agua Viva.

United Nations Framework Convention on Climate Change. UNFCCC. 1998. Kyoto Protocol.

United Nations Framework Convention on Climate Change. UNFCCC. 2015. Paris Agreement.

UPME (Mining and Energy Planning Unit). Ministry of Mines and Energy. 2015. Integración de las energías renovables no convencionales en Colombia. ATN/FM-12825-CO AGREEMENT. Republic of Colombia.

Urrea, Danilo, and Juana Camacho. 2007. *Agua y Trasnacionales en la Costa Caribe. Laboratorio experimental del modelo privatizador en Colombia.* Bogotá: CENSAT Agua Viva.

Index

Printed in the United States
by Baker & Taylor Publisher Services